Sacred Plants for Sustenance of Mankind

The Authors

Dr. S.K. Sood is Professor and Dean Life Sciences, Himachal Pradesh University, Shimla. He obtained his M.Sc. (1974) and M.Phil. (1976) from Himachal Pradesh University and Ph.D. (1981) from Nagarjuna University, Guntur, specializing in Plant Systematics and Embryology. For his post-doctoral, he underwent specialized training in Ethnobotany. He has had a brilliant academic career (triple gold medallist) followed by over 30 years of teaching and conducting research in Botany at post-graduate as well as research levels. Dr Sood has guided about 30 M.Phil. and 15 Ph.D. students and published numerous research papers in various National and International Journals of repute. He has attended and participated in many Seminars, Workshops and Conferences. He is also associated with many educational and research bodies in India and abroad. A prolific writer, his books '*Ethnobotany of Cold Desert Tribes of Lahoul-Spiti (N.W. Himalaya)*', 2001; '*Ethnobotany of Rewalsar Himalaya*', 2004; '*Ethnic Aphrodisiac Plants*', 2005;'*Ethnic Indian Plants in Cure of Diabetes*', 2005; '*Sacred and Magico-religious Plants of India*', 2005; '*Ethnic Plants of India Used in Cancer Cure – A Compendium*', 2005; '*Orchidaceae and Mankind*', 2005;'Honey Plants of the North-West Himalaya', 2006'; Edible Roots and Underground Stems of Ethnic India', 2006; 'Fibre and Fodder Resources of India', 2008; 'Root Drug Plants of Ethnic India', 2008; 'Ethnobotanical Studies on Trees, Shrubs and Climbers of Himalaya', 2009; 'Dye, Masticatory and Fumitory Resources of India', 2009; 'Flavouring and Fragrant Resources of India', 2010, 'Ethnic Healing Herbs for Cold, Flu and Lungs Ailments', 2011; 'Indian Herbs for Perfumery and Beauty Care', 2011 were widely acclaimed. His current major interests are in the theory and practice of systematics, flora of North-West Himalaya, ethnic tribes and conservation research.

Dr. Suresh Kumar received his Ph.D. in the year 2010 from Himachal Pradesh University, Shimla. At present, he is teaching in one of the academic institutions in the State. He is co-author in large number of books in ethnobotany written by the senior author (S.K. Sood). His current major interests involve floristics, ethnobotany and management of biodiversity.

Dr. T.N. Lakhanpal is former Director, Institute of Integrated Himalayan Studies, Himachal Pradesh University, Shimla. He is a Gold Medallist from Punjab University, Chandigarh and received Ph. D. from the University of Delhi. He has published more than 150 research papers. He has co-authored "Taxonomy of Indian Myxomycetes, Rural Energy Alternatives in the Hilly Areas and the Amanitaceae of India". Dr. Lakhanpal is very well known for his research contributions in India as well as abroad.

Sacred Plants for Sustenance of Mankind

S.K. Sood
Suresh Kumar
T.N. Lakhanpal

Department of Biosciences
Himachal Pradesh University
Shimla – 171 005, H.P.

2013
Daya Publishing House®
A Division of
Astral International Pvt. Ltd.
New Delhi – 110 002

ISBN 9789351241706

Published by : **Daya Publishing House®**
A Division of
Astral International Pvt. Ltd.
ISO 9001:2008 Certified Company
4760-61/23, Ansari Road, Darya Ganj
New Delhi-110 002
Ph. 011-43549197, 23278134
E-mail: info@astralint.com
Website: www.astralint.com

Laser Typesetting : **Classic Computer Services**
Delhi - 110 035

Printed at : **Salasar Imaging Systems**
Delhi - 110 035

PRINTED IN INDIA

Abbreviations

AD	*anno domini*
Anon.	Anonymous
Aq.	Aqueous
B.C.	Before Christ
C. Ar.	Central Area
cm	Centimeter
E.	East/Eastern
EBH	Ethnobotanical Herbarium
Eng.	English
et al.	(*et alia)* and other authors
etc.	(*et cetera*) and so on
Extr.(s)	Extract (s)
Fig. (s)	Figure (s)
g	Gram
H.	Hindi
H.P.	Himachal Pradesh
i.e.	(L. *id est*) that is
m	metre
ml	millilitre
N.	North/Northern
N.W.	North West
Off.	Offerings
Pl. (s)	Plate (s)
S.	South/Southern
S.E.	South-East
Sans.	Sanskrit
Sp.	Species
Spp.	Species
Sq m	Square Metre
Sub sp.	Subspecies
Syn.	Synonym
Tem.	Temple
Tot.	Total
U.P.	Uttar Pradesh
Var.	Variety
viz.	*Vide licet* (namely)
Vol. (s)	Volume (s)
W.	West/Western

Preface

Hinduism, born and nurtured in the lap of mother nature, has been greatly influenced by plants, creatures, planets and five elements since time immemorial. All our religious scriptures (Vedas, Upanishads, Arthasashtra, etc.) are replete with information signifying the mythological, religious and cultural importance of plant species in human life. Our ancestors not only worshipped the trees and plants, but they also venerate mountains, water and animals. Evidently, none of the religious ceremonies of Hindus is complete without the use of plants/plant parts as it is believed to sanctify the surroundings for devotional ambience and provided a link with the ethereal, with eternity, with the universe, and ultimately helped to create a sense of reverence for nature and natural resources. Undeniably, plants that are used extensively as temple offerings and associated with religio-cultural beliefs are commonly grown in the courtyards of temples or in the patches of forest land attached to the holy shrines. Most of the plants in the temple-yards are also protected for their ecological, economical, medicinal and social importance as they are considered representatives of the ecology, species and genetic diversity. Thus, interlinking religion with nature promote the conservation of biodiversity up to a large extent. Unfortunately, the temple-yards despite their enormous diversity have not been yet recognized as rich centres of plant genetic resources. So far no work has been done in this regard to record enormous diversity of plant genetic resources conserved in the temple-yards of Himachal Himalaya. The present book is timely and is the most authoritative reference works of high and lasting value providing first-hand information of over 305 species (edible-77 spp; fodder-32 spp; material culture-60 spp; medicinal-96 spp; ornamental-232 spp; sacred-114 spp) belonging to 223 genera under 95 families, systematic enumeration enriched with such details as synonyms, family, vernaculars, English, Hindi and Sanskrit Names, distribution, habitat ecology, reproductive cycle, material examined and botanical account. In addition to the information pertaining to the name of temple, main deity, other deities, total area of temple-courtyard, location (including place, approach, altitude), management, historical background, temple architect, offerings to God, offerings to people ('prasadam'), fairs and festivals, observations related to courtyard plants (protected as well as for landscaping), predominant plants of the courtyard, maintenance, ethnobotanical informations regarding sacred uses and associated myth, medicinal, edible, ornamental, material uses, status of courtyard plants in the surrounding areas, etc. has been incorporated. Besides, appendices providing useful information on the families, local names and glossary of words, about 522 photographs in 101 plates, 21 figures and 53 tables have also been incorporated to enhance the utility of the text. As a source of reference it is indispensable and provides valuable information of great utility in a single volume. No library should be without it.

The authors are grateful to Prof. A.D.N Bajpayee, Vice-Chancellor, Himachal Pradesh University, Shimla for the support and incessant encouragement. Well deserved appreciation is due to Prof. D.C. Kalia, Prof. D.C. Gautam and Prof. V.K. Mattu for their co-operation and well wishes; to Dr Sushil Bassi, Mr Munish, Ms Dhriti, Ms Meenakshi, Ms Seema, Ms Suman and Ms Romita for their ungrudging help in critically reading the various drafts of this manuscript, and to Mr Bal Krishan for his help in various ways. Thanks are also due to authors of books, papers and figures listed in the treatise. Finally, we would like to thank the members of our families for moral support and unflinching patience during the course of the work.

Suggestions for further improvement of the treatise will always be welcomed and entertained whole heartedly.

S.K. Sood

Suresh Kumar

T.N. Lakhanpal

Contents

Introduction

India-Richness of Floristic, Ethnic and Agro-Diversity

There is no gainsaying the fact that India (Lat. 8°–04' N to 37°–06' N; Long. 68°–07' E to 97°–25' E) is a veritable emporium of floristic, ethnic variability and agricultural diversity. Its botanical canopy comprises more than 15,000 species out of the total 4,22,000 flowering plants reported from the World (Govaerts, 2001); and the flora having the various constituent elements like south-east Asian and Malayan (35 per cent), Temperate (8 per cent), Steppe (1 per cent), African (2 per cent), Mediterranean Iranian (5 per cent), adventive weeds and naturalized aliens (18 per cent), exhibits close affinity to that of Indo-Malayan and Indo-Chinese region (Nayar, 1977). In India, the floristic diversity is spread in natural habitats in different vegetation or forest types, and interestingly, it is in such areas that native communities dominate, use native plant wealth and practice traditional agriculture, in keeping with their own lifestyles, customs, rituals and beliefs (Arora, 1997); northeastern part is the richest in floristic diversity. Of the estimated floristic wealth, 7,500 are medicinal, 3,900 are edible, 700 are culturally important, 525 are used for fibre, 400 for fodder, 300 for pesticides and insecticides, 300 for gum, resin and dye and 100 provide incense and perfumes (Anon., 1994). The diversity in plant genetic resources numbers around 166 species (Zeven and de Wet, 1982) whereas about 320 species of wild relatives of crop plants distributed in eight phyto-geographical zones (Arora and Nayar, 1984) comprise the diversity of agriculturally important wild relatives of crop plants. According to Nayar *et al.* (1989) about 20 per cent of Indian flowering plants are useful to man directly. Unfortunately, the wealth of plant genetic resources is alarmingly depleting due to habitat destruction and extension of agriculture to forest areas and gradual diversion of forest land for non-forestry uses (Paroda and Arora, 1991). It has been reported that an average of 1 in 10 of the species of vascular plants on this earth are endangered due to unplanned human activities (Lucas and Synge, 1978). India with its 53 million tribal people belonging to 550 native or tribal communities under 227 ethnic groups covering 5000 villages, is perhaps the largest country possessing a good treasure of accumulated indigenous/tribal knowledge (Anon., 1994; Pushpangadan, 1994). They constitute about 7.7 per cent of India's population (Sinha, 1996), mostly located in central and peninsular region and in the northeast, with sporadic pockets in the northwestern plains, the Himalayas in the north and in the Andaman and Nicobar Islands in the extreme south. As many as 106 different languages and 227 subsidiary dialects are spoken by tribals in India (Arora, 1995, 1996). Over 10,000 wild plant species have been reported to be used by Indian tribals for meeting their varied requirements. Living in harmony with nature is inherent in the culture and way of life of the tribal societies. However, due to the advent of modern civilization in tribal pockets, the life and culture of Indian tribals are also acculturated and changing to urban pattern.

Indigenous Societies, Traditional Culture, Biological Diversity and Sustainable Development

The term indigenous people/tribals/ethnical societies/rural communities referred to people who have close relationship with the land where they are resident (or regularly migrate across)–including the waters, forests, sky, hills and mountains associated with those lands (Rao, 1990; Alcorn, 1996). Indigenous societies all over the world in different geographical regions have discovered multiple uses of natural resources around them in the form of traditional knowledge (Jain, 2004). The word tradition is defined as beliefs and customs handed down, usually orally, from one generation to another. This knowledge is not only "technical" in nature, but also incorporates integrative insights, wisdom, ideas, perceptions and innovative capabilities which pertain to ecological, biological, geographical and other physical phenomena (Mashelkar, 2001). Traditional knowledge survives usually among the indigenous and local communities. Their knowledge of plants is important not just for the tribal people themselves, but for the wider world (Maheshwari, 1987). Moreover, these people are not to be pitied for their primitive existence, but honoured and conserved for their richness of human existence in harmony with nature. In fact, these are God's children left to guide us to a better destiny than await us if we continue to follow the present path of exploitation (Chandra, 1990). Agenda 21 of the Rio Earth Summit (1992) stated that indigenous people have a vital role in environmental management and development because of their knowledge and traditional practices. Janaki Ammal (1956) lamented that the tribal traditions are fast disappearing due to urbanization, rapid industrialization and changes in sustenance economy.

It is a fact little realized that much of what remains of our rich biological heritage has been preserved through the ages because of a wealth of conservation-oriented cultural and religious traditions (Gadgil and Vartak, 1981). The cultural diversity of ethnic groups modifies the component and distributive patterns of biological resources through human practices in the process of production and cultural beliefs, and this always had a positive impact on biological diversity in the environment (Shengji,

1991). The relationship between natural resources and the people has been forged within moral, cultural, political-economic, and ecological boundaries. Respect for these boundaries by different communities and social groups was the result of historically accepted formal and informal rules and norms (Gupta and Ura, 1990). Both indigenous culture and biodiversity are inseparable. Also, the principle of biodiversity conservation and the conservation of cultural diversity should be considered as concomitant process and as integral factors in the conservation of biological diversity today (Shengji, 1991). All over the world, the indigenous people have protected the biodiversity (forest and wildlife) with which they have a symbiotic relationship. Studies on cultural and biological diversity are highly relevant for natural conservationists (Mc Neely, 1995; Dasmann, 1995). The 15th session of the General Assembly of IUCN held in Christenchurch, New-Zealand, in October 1991, recognized the importance of the cultural heritage of mankind and the role of traditional cultures in conservation of nature (Mc Neely and Pit, 1985). The recommendations highlighted the need for an active involvement of the indigenous people in planning and implementation of conservation programmes. However, the practice related to applying traditional culture to serve present-day conservation and management systems is a growing field, and needs to be widened (Aizhong *et al.,* 1999).

Due to depletion of natural resources at a fast rate, biodiversity conservation has been a major topic of concern to the scientists all over the world. The main aim of conservation is to ensure perpetual propagation of these resources and may also include sustainable utilization, as is found possible. Presently, global concern for sustainable development and the conservation of biological diversity is dominated by the strategies and styles suitable for today's critically degraded environment, as no single method of conservation is optimal for all situations. Since pre-historic times many societies of indigenous people have formed and established their own traditional conservation methods extending to the protection of plants, animals and ecosystems, on the basis of ethics, religion and cultural beliefs of the society, thus ensuring sustainable utilization of our vast resources (Shengji, 1991). Need of the hour is to tap and utilize this aspect of traditional experience on biodiversity and ecosystem management for managing our natural resources effectively.

The Hindus–True Conservationists

Plants have a special role in religious and social ceremonies of every rural society (Manilal, 1989), and offered for worshipping several deities by the people of our country right from the Vedic period. One can find thousands of references to the uses of plants by man, in the time-honoured Indian epics and other ancient Sanskrit literature. Almost all the religious communities and tribal societies consider some plant(s) as holy in origin and essential/important in religious functions (Sensarma, 1995). Hindus believe that the plants used in various rituals are endowed with peculiar medicinal, culinary and other properties that account for their reverence in Hindu ceremonies (Banerjee, 1979). There is no part of the world, in which, this greatest gift of God to mankind has not been regarded with special reverence. Closely allied with plant-worship is the sacred and superstitious reverence, which from time immemorial, has been paid by various communities to certain trees and plants. Many flowers, fruits, or whole plants and vegetations have been used by the Indians in their various religious rituals and worshipping as incense, votive offerings to deities in temples, pujas ('Ganesh', 'Mangal-gauri', 'Navaratri', 'Ghatmahalakshmi', 'Rishi', etc.), ceremonial sacrifices, warding-off evil spirits; and for providing fertility to barren women, boon of a male child, good luck, prosperity and protection to the worshippers. It is customary to use leaves of *Piper betle, Mangifera indica* and *Cynodon dactylon* in all religious ceremonies in India (Singh, 2002).

A noticeable aspect in India is that the courtyards of temples or shrines, hermitages (centres of spirituality and learning) and temple groves maintain significant level of biodiversity of tree species (unconventional fruit/nut crops or serving other functions); medicinal (to meet the demand no longer met by the shrinking natural ecosystems), edible, fibre, fodder, ornamental plants, etc; hosts for beneficial insects/birds that support integrated pest management in addition to cultivars and species of cultivated crops. Usually, these areas act as reservoirs of biological diversity and contribute directly towards its conservation and sustainable utilization. The rural inhabitants of H.P. (a State with 55,673 sq km area; over 205 temples; 16,997 inhabited villages of different sizes under 12 districts, Bilaspur, Chamba, Hamirpur, Kangra, Kinnaur, Kullu, Lahul and Spiti, Mandi, Shimla, Sirmaur, Solan, Una; soothing landscapes and snow-clad mountains; 37,591 sq km forest area; high endemism; abode of 'Rishies' and 'Munies'; and tribals inhabiting the entire districts of Kinnaur and Lahul and Spiti, the Pangi and Bharmour tehsils of Chamba district) are the inheritors and developers of rich traditions of plants and ecosystem management. These indigenous people maintain courtyards of temples or shrines at convenient places to raise plants that are frequently used or are considered valuable aesthetically and economically and also require protection on religious grounds. Unfortunately with the development of society and economy, traditional culture is declining largely due to influences of mainstream culture in this part of Himalaya falling in the Western Himalayan biotic province within the Sino-Himalayan subzone (Rodgers, 1985), and needs to be documented for devising any comprehensive conservation strategy for the region in particular.

Floristic Studies in India

Floristics in India–A Historical Perspective

The earliest credit for scientific beginning of plant exploration work in India goes to Garcia da Orta, a great pharmacist from Portugal who recorded this information in his book '*Colloquious des Simples*' published in 1565. But, it was Heinrich Van Rheede–the Governor of Dutch possession in Malabar in 1667, who made the first real attempt to explore systematically the plant wealth of India. His monumental work '*Hortus Indicus Malabaricus*', published in 12 volumes from Amsterdam, is still regarded as a valuable contribution to floristics in India. This book attracted the attention of numerous workers to the richness of flora of India. Thereafter, account of Indian plants in the form of lists was included in the works entitled '*Flora Indica*' (Burman, 1768) and '*Flora Malabarica*' (Burman, 1769), followed by other valuable contributions, *viz.*, '*Plants of Coast of Coromandel* (1795-1819)', '*Hortus Bengalensis*' (1814) and '*Flora Indica*' (1832) by William Roxburgh; '*Prodromus Florae Peninsulae Indiae Orientalis*' (Wight and Walker-Arnolt, 1834); '*Icones Plantarum Indiae Orientalis*' (Wight, 1838-1853); '*Illustrations of Botany and Other Branches of the Natural History of the Himalayan Mountains and Flora of Cashmere*' (Royle, 1833-1840); '*A Catalogue of the Plants Growing in Bombay and its Vicinity*' (Graham,1839); '*Plantae Rarios Quas in Indae Orientali Collegit Victor Jacquemont*' (Cambessedes, 1844); '*Plants of Canara and Neighbouring Areas of Maharashtra and Malabar*' (Hohenacker, 1849); '*Notulae und Plantas Asiaticas*' (Griffith, 1854); '*Flora Indica*' (Hooker and Thomson, 1855); '*Catalogue of the Indigenous Flowering Plants of the Bombay Presidency*' (Dalzell, 1858); '*The Bombay Flora*' (Dalzell and Gibson, 1861); '*Handbook of Indian Flora*' (Drury, 1864); '*Report on the Vegetation of the Andaman Islands*' (Kurz, 1870); '*Forest Flora of North-West and Central India*' (Brandis, 1874); and '*A Sketch of the Vegetation of the Nicobar Islands*' (Kurz, 1876). Publication of Hooker's '*Flora of British India*' in seven volumes during 1872-1897 formed basis of all the later provincial or local Indian floras published till date. Since then, many comprehensive floristic studies on Indian flora had been those of Brandis (1874, 1906), Gamble (1898), Coventry (1901), Cooke (1901-1908), Kanjilal (1901, 1911, 1933), Collett (1902), Prain (1903), Duthie (1903-1929), Rama Rao (1914), Fyson (1915, 1932), Bamber (1916), Osmaston (1927), Blatter and Hallberg (1918), Parker (1918), Haines (1921-1925), Parkinson (1923), Kanjilal and Gupta (1928), Kanjilal *et al.* (1934-1940), Santapau (1953, 1958, 1962), Bor (1960), Maheshwari (1963, 1966), Champion and Seth (1968), Ramaswamy and Razi (1973), Saldanha and Nicolson (1976), Singh and Kachroo (1976), Srivastava (1976), Babu (1977), Kachroo *et al.* (1977), Oommachan (1977), Singh and Tiagi (1979), Rao and Razi (1974), Deb (1981, 1983), Manilal and Sivarajan (1982), Rao and Verma (1982), Dhar and Kachroo (1983), Chowdhery and Wadhwa (1984), Saldanha (1984), Haridasan and Rao (1985), Rao (1985-1986), Joseph (1986, 1987), Manilal (1988), Henry *et al.* (1989), Roy *et al.* (1992), Aswal and Mehrotra (1994), Saxena and Brahmam (1994-1996), Deorani and Naithani (1995), Chowdhery (1998), Gaur (1999), Bedi (2000), Hynniewta *et al.* (2000), Kumar (2001), Sharma (2002) and Pullaiah (2004). Besides, an exhaustive treatise, '*Dictionary of Angiosperm Families and their Genera*' (Sambamurty, 2002) provided detailed information on the floristic diversity of the country represented by about 13,600 genera under 510 families and arranged according to the different accepted systems of the classification.

Status of Floristic Diversity in Himachal Pradesh

For the state of H.P., the first real attempt to carry out floristic investigations was made by William Spencer in 1817. However, the real credit to initiate systematic floristics in H.P. goes to William Moorcroft who collected plants from the areas of Kangra, Kullu, Lahaul and Spiti in 1819. Thereafter, Countess of Dalhousie had been credited for collecting about 600 plant species from Shimla regions, falling under the erstwhile undivided Punjab or the territories of H.P. before its reorganization in 1971. Some other plant explorers of the nineteenth century who made collections from various localities of H.P. include V. Jacquemont, J.F. Royle, R. Inglis, M.P. Edgeworth, G.T. Vigne, R.S. Simpson, W.H. Parish, W. Grant, H. Collett, J.S. Gamble, D. Brandis, J.F. Duthie and G.A. Gammie (*see* Moorcroft and Trebeck, 1841). During the 20th century, notable collectors from the state were, R.N. Parker, C.E. Parkinson, W. Koelz, R.S. Troup, N.L. Bor, M.B. Raizada, M.A. Rau, U.C. Bhattacharya, M. Sharma and B.S. Aswal. Besides, the other notable contributions which added to the wealth of literature on floristics pertain to compilations, like '*The Himalayan District of North West Provinces of India*' (Atkinson, 1882), '*Flora of British India*' (Hooker, 1872-1897), '*Flora Simlensis*' (Collett, 1902), '*Flora of Bashahr Himalaya*' (Nair, 1977), '*Flora of Himachal Pradesh*', Vols I-III (Chowdhery and Wadhawa, 1984), '*Concise Flowers of the Himalaya*' (Polunin and Stainton, 1987), '*Flowers of the Himalaya–A Supplement*' (Stainton, 1988), '*Flora of Lahaul-Spiti-A Cold Desert in North West Himalaya*' (Aswal and Mehrotra, 1994), '*Flora of Kullu District*' (Dhaliwal and Sharma, 1999), '*Flora of Great Himalayan National Park, Himachal Pradesh*' (Singh and Rawat, 2000), '*Flora of Chamba District, Himachal Pradesh*' (Singh and Sharma, 2006) as well as a large number of research contributions more particularly from the different parts of H.P., namely, *Chamba* (Gupta, 1961,

1964, 1971), *H.P.* (Sarin and Chopra, 1985; Chauhan, 1983, 1989a, b, c, 1990, 1999, 2003; Singh and Aswal, 1992; Sharma and Sood, 1997; Singh, 1999; Badola, 2001), *Kangra Valley* (Ahluwalia, 1952; Uniyal and Chauhan, 1971, 1973; Kapur, 1985), *Kullu* (Rastogi, 1960; Uniyal and Chauhan, 1972; Dobriyal *et al.*, 1997), *Lahul-Spiti* (Rau, 1961; Sarin, 1967; Uniyal *et al.*, 1973; Arora *et al.*, 1980; Chauhan, 1989a, b, c; Chaurasia *et al.*, 2001), *Mandi* (Singh, 1993), and *Una* (Chauhan, 1974). Moreover, Chauhan (1995) described the economically useful plant resources of H.P., and Sood *et al.* (2001), Kala (2003) and Sood and Thakur (2004) provided a comprehensive ethnobotanical account of cold desert tribes of Lahaul-Spiti (129 ethnobotanically important plants belonging to 92 genera and 44 families), Trans-Himalayan area (220 medicinal plants) and Rewalsar Himalaya (262 plant spp belonging to 82 families and 220 genera), respectively. Also, attempts made by Ved and Tandon (1998) and Ved *et al.* (2003) listed various categories of threat status to about 60 plant species in North-West Himalaya.

Overview of Floristics in N.W. Himalaya

Hardwick (1798; cited in Burkill, 1965) seems to be the first botanist who collected plants from the Himalayan region. It was during the first half of the nineteenth century, Royle (1833-1840) based upon his extensive collections in Bushar, Kinnaur and Cashmere Himalaya, gave an illustrated account of species collected during these expeditions in '*Illustrations of Botany and Other Branches of the Natural History of the Himalayan Mountains and Flora of Cashmere*'. In the mid nineteenth century, Hooker and Thomson (1855) collected about 2000 plant species from North-Western Himalaya. Duthie (1882) documented plants of Western Himalaya from Kashmir to Kumaon region. Collett's (1902) '*Flora Simlensis*' described 236 species of flowering plants from Shimla and adjoining areas. In the year 1927, Osmaston published forest flora of Kumaon, whereas Kanjilal and Gupta (1928) compiled '*Forest Flora of the Chakrata, Dehra Dun and Saharanpur Forest Divisions*'. Pennel (1943) carried out in depth studies on the Scrophulariaceae of the W. Himalayas. Later, Singh and Kachroo (1976) documented '*Forest Flora of Srinagar and Plants of Neighbourhood*'. In 1977, Kachroo *et al.* compiled Flora of Ladakh and C.R. Babu documented and described herbaceous flora of Dehradun. The work on Flora of Mussoorie has been compiled by Raizada and Saxena (1978). '*Flowers of Himalaya*' by Polunin and Stainton (1984, 1987) with 689 colour photographs and 316 line-drawings and description of over 1,500 plant species has been considered one of the most useful books for studying the floristic diversity of Nepal and parts of India falling in the Western Himalaya. Work on critical taxonomic treatment of 401 weed species represented by 223 native and 178 introduced species has been published by Kaul (1986). Present status of the range in diversity of plants occurring in hill districts of Garhwal division has been given by Gupta (1989). Notably, Maithani *et al.* (1991) provided important details for the identification of some important Himalayan shrubs. Useful information pertaining to 985 species belonging to 353 genera under 79 families of gymnosperms and angiosperms has been published by Aswal and Mehrotra (1994) in their book entitled '*Flora of Lahaul Spiti (A Cold Desert in North-West Himalaya)*'. An extensive account of the forest flora of Pir Pinjal has been presented by Singh and Kachroo (1994). A concise field guide to the flowering plants of the Himalayas has been contributed by Dang (1998). Gaur (1999) compiled flora of District Garhwal, North-West Himalaya with ethnobotanical notes. A comprehensive floristic account of 930 plant species belonging to 504 genera under 126 families of Kullu Valley has been dealt by Dhaliwal and Sharma (1999). Singh and Rawat (2000) studied flora of Great Himalayan National Park. Important publication '*Flora of Cold Deserts of Western Himalaya, Vol. 1 (Monocotyledons)*' by Murti (2001) provided detailed floristic information of 347 species belonging to 103 monocotyledonous genera under 16 families. A comprehensive account of flora of Tons valley enlisting over 761 species from the region has been compiled by Rana *et al.* (2003). Likewise, Kaur and Sharma (2004) exhaustively enumerated floristic account of 911 plant species belonging to 544 genera under 139 families of higher plants in their book on '*Flora of Sirmaur (Himachal Pradesh)*' and Sood *et al.* (2006) gave a beautiful account of 290 species under 221 genera of the honey plants of the north-west Himalaya. Another important publication on trans-Himalaya (Chaurasia *et al.*, 2007) provided in depth study on the ethnobotany and plants of the region.

Status of Work on Ornamental, Religio-cultural Practices and Conservational Aspects

Considerable literature on ornamental and horticultural aspects have been published earlier. Among them mention may be made of '*Flowering Trees and Shrubs in India*' (Cowen, 1950); '*Flowering Trees of the World (for Tropics and Warm Climates)*' (Menninger, 1962); '*Flowering Shrubs*' (Pal and Krishnamurthi, 1967); '*A Field Guide to the Flowers of Europe*' (Polunin, 1969); '*Flowering Vines of the World*' (Menninger, 1970); '*Beautiful Climbers of India*' (Pal, 1972); '*Ornamental Horticulture in India*' (Randhawa, 1973); '*A Dictionary of the Flowering Plants in India*' (Santapau and Henry, 1973); '*The Illustrated Encyclopaedia of Flowering Plants*' (Krisa, 1982); '*Flowering Trees*' (Randhawa, 1983); '*Field Guide to the Common Trees of India*' (Bole and Vaghani, 1986); '*Shrubs for Tropical and Subtropical Gardens*' (Oakman, 1990); '*Tropical Garden Plants In Colour*' (Bose and Chowdhury, 1991); '*The American Horticultural Society A-Z Encyclopedia of Garden Plants*' (Brickell *et al.*, 1997); '*Encyclopaedia of Garden Plants*' (Booth, 1999); '*Some Beautiful Indian Climbers and Shrubs*' (Bor and Raizada, 1999); '*Trees of India (Medicinal, Commercial, Religious and Ornamental): A Colour Atlas*' (Warman, 1999); '*Illustrated Encyclopedia: Garden Plants*' (Berry and Bradley, 2000); '*Tropical Flowering Plants*' (Llamas, 2003); '*The Complete Encyclopedia of Trees and Shrubs: Descriptions, Cultivation Requirements, Pruning, Planting*' (Wasson, 2003); '*Garden Plants in Colour*' (Bose *et al.*, 2004); '*Advances in Ornamental Horticulture, 6 Vols.*' (Bhattacharjee, 2006); '*Ornamental Plants and Flowers of Tropical Mexico*' (Trapp, 2006); and '*Flowers and Gardens in India-A Manual For Beginners*' (Temple Wright, 2007). All these works dealt elaborately about the garden plants, shrubs, flowering trees and ornamental horticulture.

Dymock (1889) gave an account of use of turmeric in Hindu ceremonials. As early as in 1897, Brandis first reported the widespread presense of sacred groves in his book '*Forestry in Indian: Origins and Early Development*'. However, documentation of the floristic wealth of sacred groves has been done by Gadgil and Vartak (1973), Vartak *et al.* (1986), Kulkarni *et al.* (1993), Upadhye *et al.* (1997) and Kumbhojkar and Vartak (1988). Even today the practice of tree worship is being continued and reported from different parts of India (Mitra, 1928; Sastry, 1957; Chandervaker, 1965; Pande, 1964; Sahay, 1964; Bhatnagar and Lall, 1965; Gupta, 1965; Nair, 1965; Nayak, 1965; Sen Gupta, 1965, 1980; Banerjee, 1974; Sinha, 1979; Amrita, 1999; Hegde and Patil, 1999; Warman, 1999; Sane and Ghate, 2006). The work of Moldenke (1954) brings together the economic plants of the Bible whereas association of plants like *Azadirachta indica* with 'Gudhipadava', *Ficus benghalensis* with 'Vatpurnima', *Bauhinia racemosa* and / or *Prosopis cineraria* with 'Dasahara' and *Ocimum sanctum* with 'Tulsi-vivaha' have been recorded by Agharkar (1961). Plants in religious cultural beliefs of the tribals of Bastar (M.P.), Tai Khamtis and 'Koch' community of Assam, and Mithila have also been done (Jain, 1963, 1971; Banerjee, 1974; Gogoi and Borthakur, 1991; Jha *et al.*, 1994; Saikia and Nath, 2003). Pal (1970) gave an account of the use of plants especially in 'Durga-Puja'. Kapoor *et al.* (1971) described medicinal plant wealth around religious shrines of India. Gadgil and Vartak (1975) pioneered systematic studies on the sacred groves in Maharashtra; and recorded four important regions for extant sacred groves in India, namely Khasi and Jantai Hills of north-east, the Western Ghats, Aravali Hills of Rajasthan and the Sarguja, Chanda and Bastar areas in Central India (Gadgil and Vartak, 1976). Also, they reported on the beliefs and folklores related to sacred groves (Vartak and Gadgil, 1981). Since then literature pertaining to conceptual, historical, socio-cultural, conservation, productivity, management and functional attributes of the sacred groves have been mushrooming (Khiewtem and Ramakrishnan, 1989; Rao, 1990; Bachmann, 1992; Nipunage *et al.*, 1992; Marglin and Mishra, 1993; Menon and Sasidharan, 1994; Alcorn, 1996; Borrini-Feyerabend, 1996; Godbole, 1996; Induchoodan, 1996; Uniyal *et al.*, 1996; Vartak, 1996; Hajra, 1997; Nair, *et al.*, 1997; Singh, 1997; Upadhye *et al.*, 1997; Godbole *et al.*, 1998; Ramakrishnan, 1998; Ramakrishnan *et al.*, 1998; Saxena *et al.*, 1998; Ramanujam, *et al.*, 2002; Sinha, 2003; Chakraborty and Pal, 2004). Notably, Sithole (1976) and Gupta (1996a, b) described plants represented in ancient Indian sculptures and temple arts. Likewise, a number of publications reported on the plants known to be traditionally worshipped in India and used in religion, religious rites, beliefs and mythology (Rai Chaudhuri and Pal, 1981; Bhatia *et al.* 1984; Belapure *et al.*, 1987; Farooqui, 1989; Pandey, 1989; Sensarma, 1989, 1994, 1995; Gupta, 1991; Bennet *et al.*, 1992; Dagar, 1995; Amirthalingam, 1998; Pramod *et al.*, 2003). Conservational practices of tribals of Purulia and their relevance in eco-development pragrammes have been highlighted by Pal and Mudgal (1985) whereas Rana *et al.* (1996) presented the strategies for sustainable utilization of plant resources by the tribals of Tons valley, Western Himalaya. Hajra (1987) identified plants listed in Sanskrit literature for magico-religious beliefs. Jain (1989, 1992) described the significant role played by clans in conservation of nature. In 1989, Majupuria and Joshi published a book describing religious and useful plants of India and Nepal. Pandey (1989) gave an account of a number of sacred plants useful for mankind. Chandrakanth *et al.* (1990)

described temple forests in India. Kulkarni *et al.* (1993) studied floristics of a sacred forest patch from Sundargad district of Orissa State. Mao (1993) and Singh and Singh (1996) reported on the superstition in botanical folklore about 'Mao' Naga tribe and 'Meitei' culture in Manipur. Mitra (1994) discussed the cultural, ethnobotanical, medicinal environmental and economic significance of 'Asvattha' (*Ficus religiosa*)– the most worshipped tree in India. Moreover, Shah (1994) reported on the cultural and religious aspects of 'Jatamansi' (*Nardostachys jatamansi*) in the Kumaon Himalaya. Studies of Arora (1995) and Maheshwari (1995) discussed the role of tribal communities in crop domestication and crop improvement, vis-a-vis conservation of variability of native land races and prevalent species diversity; and the role of ethnobotany in development and conservation of resources, respectively. Dube (1998) discussed in detail the sacred plants associated with many festivals such as: *Ficus religiosa* 'vat' in 'Vat Savitri Vrat', *Cynodon dactylon* 'durva' in 'durvaashtami', *Ocimum sanctum* 'tulsi' in a number of rituals. Apart from helping in the conservation of biodiversity, watershed protection and moderation of the local climate, sacred groves also form the binding force of the people of an area (Unnikrishnan, 1995). Role of Brahmkamal [*Saussurea obvallata* (DC.) Edgew.] in the life and culture of Garhwalis and strategies for its conservation and sustainable utilization has been pointed out by Saklani and Rao (1996). Plants used in sacrificial rites have been documented by Dixit (1997). A typical religious practice of tree marriage in rural Orissa adds to the conservation of such species (Mohanty, 1997). According to Nayar (1997), the sacred groves represent the cost-effective community-based models for the conservation of the culture as well as forests and its biota. Work on ethno-medico-botany of some sacred plants have been carried out by Upadhye *et al.* (1997), Bhatt *et al.* (2001), Joshi *et al.* (2001) and Dhiman (2003). For 'Patra-puja' and in many traditional forms of worship in India, Ghate (1998) identified the use of forty-five plant species. Borthakur *et al.* (1998) regarded the house gardens of Assam as a repository and testing site for uncommon species and varieties of plants. Considering the groves as relics of past vegetation, Ramakrishnan (1998) advocated that these should be declared as 'World Heritage Sites' under UNESCO. Singh *et al.* (1998) worked out eco-cultural analysis of sacred species and ecosystems in Chhakinal watershed in Kullu area of north-western Himalaya. Kulkarni and Kumbhojkar (2000) reviewed plant conservation practices in Maharashtra during 17th century. Ramsankar (2001) documented 19 plant species associated with 'Soharia' festival celebrated each year to worship Goddess 'Kali'. Merat (2002) investigated on the plants of religious importance of Buldana district of Maharashtra State. Nonhare *et al.* (2002) regarded the trees growing in and around 'Deogudis' in each village of 'Bastar' (Chattisgarh) as abode of Gods and Goddesses. The religious and scientific significance of 'Govardhan puja' and 'Annakut' have been worked out by Singh and Chauhan (2002). Based upon their work, Maya *et al.* (2003) opined that the temple tanks form ideal habitats for the conservation and sustainable utilization of medicinal flora. Kumar *et al.* (2007) gave an account of the ethnobotanical aspects of religious practices in Kumaon especially with reference to 'Havan'. Jha and Goel (2004) dealt with some plant species utilized in folklore life of Mithila region (North Bihar) for preparing rosaries and non-flower garlands. Kumar and Yadav (2004) discussed significance of sacred plants in 'shraddh' ritual (Pindadan) in Gaya, Bihar whereas Singh and Chauhan (2004) gave descriptions of 23 species of the leaves used for worshipping different Gods in Brij Mandal. Recently, Sane and Ghate (2006) described sacred conservation practices at species level through tree worship in ancient India. Significantly, in order to turn vacant temple lands green, the Travancore Devaswom Board (TDB), which administers and controls over 1,200 temples in Kerala, launched a project to plant 28 species of trees and medicinal plants under a new scheme 'Nakshatraphalam' (Nair, 2006).

In spite of such an exhaustive floristic, horticultural, ethnic and religio-cultural conservation practices of the tribals and nature loving rural-based communities of H.P., no attempt has been made so far to document information on this aspect from the temple-courtyards of the State. Recognizing the paucity of information, the present investigation was undertaken.

The Study Area

Etymology, Location and Extent

The word Himachal Pradesh literally means 'Land of snowy mountains', which sprawls in the northwestern lap of the Himalaya between 30°22' to 33°12' north latitude and 75°47' to 79°04' east longitude. To the east, it forms India's border with Tibet, to the north lies State of Jammu and Kashmir, Uttaranchal in the southeast, Haryana in the south and Punjab in the west (Figure 1). The entire territory of Himachal Pradesh is mountainous with altitude varying from 350 to 7,000m above the mean sea level. Topographically, Himachal's territory from south to north can be divided into three zones–The Shivaliks or outer Himalaya, inner Himalaya or mid-mountains and Alpine Zone or the greater Himalaya. The most prominent landmark of the State is the perennial white snowline on various peaks. The Dhauladhar overlooks the Kangra valley in supreme grandeur, while the Pir Panjal, the great Himalayan and the Zanskar ranges stand guard over Chamba, Lahul-Spiti, Kullu and Kinnaur. Of the twelve districts delineated for the State, area wise, Hamirpur with 1,118 sq km (2.01 per cent) is the smallest and Lahul-Spiti covers the largest area of 13,835 sq km (24.85 per cent). The entire population of the State is estimated to be around 6,077,900 in number according to 2001 census.

Historical Perspectives

Historical evidence indicates that after the Aryan invasion and the Khasa incursion, Himachal was dotted with tribal kingdoms or republics, chief among were the Audambaras (descendants of Vishwamitra, and in existence in the 5th century B.C.), the Trigarta (Susharma, the founder of Trigarta was an ally of the 'Kaurvas' and attacked Virata–the King of the Matasya with whom Pandavas had sought

Figure 1: Map of Himachal Pradesh Showing Location of the Various Temples (For details of location codes *see* Table 1)

refuge), the Kulut [the territory mentioned to be located in upper Beas valley (now Kullu valley) as per holy scriptures.

Mahabharata, Ramayana, and Vishnu Purana] and the Kulindas (hilly areas between Beas and Yamuna, which were probably analogous to Shimla and Sirmaur areas). The present day Kanets are probably the descendants of the Kulindas. From the geneological records and from other evidence, it appears that most of the hill states were already formed before the twelfth century. Some of them as Brahmapura (Brahmaur), Trigarta (Kangra), Kuluta (Kulu), Spiti and Srughna (probably Sirmaur area) were, however, in existence for quite some centuries and also mentioned by Hiuen Tsang. The evidence of Kols having settled in Himalaya is also found from the sculptures on the rocks of Chandeshwar, Sameshwar and Chattis of Kumaon, and Kangra district. In Himachal, the present-day Kolis, Halis, Doms and Dhagis (scheduled castes) are probably their descendants.

After independence, the Chief Commissioner's Province of Himachal Pradesh came into being on April 15, 1948 with the integration of 31 hill States and received additional regions added to it in 1966; and it acquired full fledged statehood on January 25, 1971 as the eighteenth State of Indian Union.

Climate and Biogeography

Due to variation in elevation and aspect, there is great diversification in the climatic conditions of H.P. It varies from hot and subhumid tropical (450-900m) in the southern low tracts, warm and temperate (900-1,800m), cool and temperate (1,900-2,400m), and cold alpine and glacial (2,400-4,800m) in the northern and eastern mountain ranges. The climate in Lahul-Spiti and Kinnaur is of the semi-arid highland type.

The year in H.P. is divided into three seasons-cold (October-February), hot (March-June) and rainy (July-September). By October the skies are clear and mornings and evenings are bracingly cold. Humidity is low. Snowfalls usually occur in the higher hills in December and January and occasionally it may be experienced earlier or later. At elevations of about 3,000m, the average snowfall is about three metres and lasts from December-March. Above 4,500m there is perpetual snow. The severity of winter is generally over by the end of February. Temperature begins to rise gradually. Spring, from mid-February to March-April, is shortlived but extremely delightful. After this, the weather becomes hot and dusty in the areas adjoining the plains, but remains pleasant and comfortable in the places located on higher altitudes. The monsoon arrives generally towards the end of June or early July. Rainfall tapers off by the end of September which ushers in autumn. The average rainfall in the state is 152 cms (60 inches). The highest rainfall is in Kangra district at Dharamshala and lowest in Lahul-Spiti at place Spiti.

Himachal has rich flora and fauna. Legally defined forest areas constitute about 66.52 per cent of the area of H.P. The southern part of the State which is at lower elevation has both tropical and subtropical dry broadleaf forests and tropical and subtropical moist broadleaf forests with vegetation dominated by sal, sisham, chir pine, etc. The temperate region grows oaks, deodar, blue pine, fir and spruce. Alders, birches, rhododendrons and moist alpine scrubs form the regional vegetation in the uppermost region.

The soil of Himachal varies according to aspect, altitude and climate. On the whole, the soil is thin. Depth occurs in the valley or on hill slopes.

Communities and Dialects

About 95.4 per cent of the population of H.P. consists of Hindus, Muslims 2.0 per cent, Sikhs 1.2 per cent and Buddhists 1.2 per cent . Amongst Hindus, the main communities are Brahmins, Rajputs, Kannets, Rathis and Kolis. Mahajans and Soods, the business communities, are sprinkled all over. Lahulis of Lahul and Spiti region are mainly Buddhists. The other communities are Julahas, Chamars, Lohars, Tarkhans, Kumhars, etc. The population of H.P. consists of a medley of tribes, each with its own language and dialect. The tribal population of the State comprise of the Gaddis (traditional shepherds who migrate from the alpine pastures to the lower regions during the winters), Kinnars (inhabitants of the Kinnaur region and traditionally practiced polyandry and polygamy), Gujjars (nomads who rear buffalo herds), Pangawals, Lahulis and Spitians. Himachal also has a sizeable population of Tibetans. The people are simple, straightforward, hard-working and honest. Hindi is the State language but people mostly converse among themselves in Pahari. The major spoken languages include Hindi, Punjabi, Mahasui, Kulluyi, Lahuli, Kinnauri, Chambyali, Sirmauri, Bilaspuri, Pahari, Dogri, Kangri.

Occupation and Livelihood

Agriculture being the main occupation of the people of H.P. has an important role in the economy of the State. It provides direct employment to about 71 per cent of the main working population. Rice, wheat and maize are important cereal crops of the State. Groundnut, soyabean and sunflower in *kharif* and rapeseed/mustard and toria are important oilseed crops in the rabi season. Urd, bean, moong, rajmash in *kharif* season and gram in rabi are the important pulse crops of the State. Under the diversification approach, major emphasis is being laid on the production of off-season vegetables, quality vegetable seeds, potato and ginger besides soybean, oilseeds and pulses.

Faith and Worship

The State of H.P. is popularly known as 'Devabhoomi' or the land of Gods. The people of Himachal are deeply religious, God-fearing and worship different deities with all reverence. Early inhabitants of Himachal professed some form of Shaivism. Shiva is the most popular God and is called by different names, such as Mahasu, Mani Mahesh, Baijnath and others. The Vishnu or Krishna cult is confined mostly to the towns. Buddhism has taken the shape of Lamaism here and is mostly practiced in the districts of Lahul-Spiti and Kinnaur. The important Gurudwaras situated at Paonta Sahib and Rewalsar are connected with the tenth Guru, Gobind Singh. Besides, the people also worship the Godlings or the village deities, the Deotas, the Rishies, the Munies, the Siddhas, the Pandavas, the Hill Tops, the trees, the Joginis or Wood

Fairies, the Kali, the Shakti, the Nagas and even a host of devils and deities of the aborigines.

The temples and the other religious places are built in the traditional (Shikhara type, Pagoda-like, Dome-shaped or Pentroof type) style. Most of these have fascinating legends attached to them. There are innumerable temples of Shikhara type spread all over the State but the ones that deserve mention are Baijnath temple, Bajreshwari Devi temple, Masrur temple of Kangra, Bharmaur temples, Lakshmi Narayan temple of Chamba, Trilokinath temple of Lahul, Basheshwar Mahadev temple of Bajaura (Kullu), Hatkoti temple of Shimla district and Trilokinath temple of Mandi. Pentroof type of temples include Shakti temple of Chamba, Naga temple of Khajiar (Chamba), Chandika Devi temple of Kinnaur, Bhekhali Devi temple of Kullu, etc. Dome-shaped temples include Naina Devi temple of Bilaspur. Pagoda-like temples are common in Kinnaur and Kullu districts.

Fairs and Festivals

The fairs and festivals in H.P. are associated with many folklores and give valuable insight into the rural lives, customs and beliefs. Every district in H.P. has its own cycle of annual fairs related to the history and sociological background of that area. All the seasonal and religious festivals and fairs are celebrated with great fervour and enthusiasm. On the day of festival, people generally cook special food, wear neat and clean clothes and worship at the temple. The tribal festivals in H.P. are quite unique in their celebrations. The different fairs and festivals of the region include the Halda festival, Diwali, Dussehra, Fulaich, Holi, Bishu, Rakhar Puniya, Gugga Navmi, Losar, Sajo, Lohadi, Faguli, and Shivratri.

Enumeration

For collecting first hand information on the plants in the courtyards of temples of H.P., intensive and extensive surveys (Table I) were conducted in all the twelve districts over a period of five years (2002-2007).

Table 1: Temples of Himachal Pradesh

*Temple**	*Location in Map*	*Concerned Deity*	*Place*
Bilaspur			
Baba Balak Nath Temple	B1	Baba Balak Nath	Shah Talai
Baba Nahar Singh Temple	B2	Baba Nahar Singh	Bilaspur
Garuna Jhadi Temple	B3	Baba Balak Nath	Shah Talai
Hari Devi Temple	B4	Hari Devi	Hari Devi
Lakshmi Narayan Temple	B5	Lakshmi Narayan	Bilaspur
Markanday Temple	B6	Markanday	Maakadi
Naina Devi Temple	B7	Naina Devi	Kharota
Naina Devi Temple	B8	Naina Devi	Naina Devi
Naina Devi Temple	B9	Naina Devi	Nihar
Nav Durga Temple	B10	Nav Durga	Kharota
Shiva Temple (Shiv Baudi)	B11	Shiva	Bachhretu
Chamba			
Baba Lakhdata Pir Temple	C1	Lakhdata Pir	Dundiara
Bansi Gopal Temple	C2	Lord Krishna	Chamba
Bhalai Mata Temple	C3	Bhadrakali	Bhalai

Contd...

Table 1–*Contd...*

*Temple**	*Location in Map*	*Concerned Deity*	*Place*
Bharmani Mata Temple	C4	Bharmani Mata	Malkouta (Bharmour)
Champawati Temple	C5	Champawati	Chamba
Chamunda Devi	C6	Chamunda Devi	Chamba
Chaurasi Temple	C7	Ganesha, Shiva, Narsingh, Dharamraj, Lakhna Mata, Shitla, Chamunda, Hanuman	Bharmour
Hari Rai Temple	C8	Lord Vishnu	Chamba
Khajji Nag Temple	C9	Khajji Nag	Khajiar
Lakshmi Narayan Temple	C10	Lakshmi Narayan, Shiva	Chamba
Lakshmi Narayan Temple	C11	Lakshmi Narayan	Dalhousie
Maha Kali Temple	C12	Kali	Chamba
Narsingh Temple	C13	Narsingh	Chamba
Raghunath Temple	C14	Raghunath	Dalhousie
Shitla Mata Temple	C15	Shitla Mata	Chamba
Shiv-Shakti Temple (Chhatrari Temple)	C16	Goddess Shiv-Shakti	Chhatrari
Sita-Ram Temple	C17	Sita-Ram	Chamba
Suhi Mata Temple	C18	Suhi Mata	Chamba
Hamirpur			
Awah Devi Temple	H1	Awah Devi	Awah Devi
Baba Balak Nath Temple	H2	Baba Balak Nath	Deoth Sidh
Baba Sidh Chaino Temple	H3	Baba Sidh Chaino	Smaila
Gauri Shanker Temple	H4	Shiva-Parvati	Tira (Sujanpur)
Murli Manohar Temple	H5	Radha Krishana	Sujanpur Tira
Narbadeshwar Temple	H6	Shiva	Sujanpur Tira
Santoshi Mata Temple	H7	Santoshi Mata	Ladraur
Satyanarayan Temple	H8	Satyanarayan	Hamirpur
Shani Dev Temple	H9	Shani Dev	Sarlin (Lamblu)
Shirdi Sai Baba Temple	H10	Shirdi Sai Baba	Kadriyana
Tauni Devi Temple	H11	Tauni Devi	Tauni Devi
Trilokinath Temple	H12	Trilokinath (Lord Shiva)	Barah Mahino Ka Paron
Vaishno Mata Temple	H13	Vaishno Mata	Dhangota
Kangra			
Ashapuri Temple	KG1	Goddess Ashapuri	Ashapuri
Baba Balak Rupi Temple	KG2	Baba Balak Rupi	Balak Rupi

Contd...

Table 1–*Contd...*

*Temple**	*Location in Map*	*Concerned Deity*	*Place*
Baba Kalinath Temple	KG3	Baba Kalinath	Baroh
Bagulamukhi Temple	KG4	Bagulamukhi	Vankhandi
Bajreshwari Devi Temple	KG5	Bajreshwari Devi	Kangra
Bhagsu Nath Temple	KG6	Shiva, Nag Devta	Bhagsu
Budhist Temple (Thek Chen Choiling)	KG7	Gautam Buddha	Dharamshala
Budh Temple	KG8	Gautam Buddha	Tilokpur
Chamunda Nandikeshwar Temple	KG9	Chamunda Devi	Jadrangal
Jwalamukhi Temple	KG10	Jwalaji	Jwalamukhi
Kalinath Kaleshwar Mahadev Temple	KG11	Kaleshwar Mahadev (Shiva)	Kaleshwar
Lakshmi Narayan Temple	KG12	Lakshmi Narayan	Alampur
Lakshmi Narayan Temple	KG13	Lakshmi Narayan	Chanor
Machhial Mahadev Temple	KG14	Shiva	Andreta
Mahakal Temple	KG15	Shiva	Mahakal
Mangla Mata Temple	KG16	Mangla Mata	Tilokpur
Mukut Nath Temple	KG17	Mukut Nath (Shiva)	Sansaal
Nag Devta Temple	KG18	Nag Devta	Chelian
Nagni Mata Temple	KG19	Nagni Mata	Bhati
Nagni Mata Temple	KG20	Nagni Mata	Nagni
Nag Temple	KG21	Nag Devta	Kariyada
Radha Krishna Temple	KG22	Radha Krishna	Baroh
Radha Krishna Temple	KG23	Radha Krishna	Dehra (Gopipur)
Rock-Cut Temple	KG24	Ram-Lakshman-Sita	Masroor
Shibbo Than Temple	KG25	Baba Shibbo Nath	Bharmar
Shiva Temple	KG26	Shivlingam	Baijnath
Shiva Temple	KG27	Shiva	Amrella
Shri Brijraj Swami Temple	KG28	Shri Krishna	Nurpur
Shri Raghunath Teda Temple	KG29	Raghunath	Chakvan Kalidhar
Sidh Chaino Temple	KG30	Baba Sidh Chaino	Dangra
Trilokinath Temple	KG31	Trilokinath (Lord Shiva)	Tilokpur
Vindhyavasini Mata Temple	KG32	Vindhyavasini Mata	Bandla
Kinnaur			
Badri Narayan Temple	KN1	Badri Narayan	Batseri
Baring Nag Temple	KN2	Baring Nag	Sangla
Boirus Temple	KN3	Bhairav	Kothi
Chandika Devi Temple	KN4	Chandika	Kothi

Contd...

Table 1–*Contd...*

*Temple**	*Location in Map*	*Concerned Deity*	*Place*
Kamakhaya Devi Temple (Kamru Kila)	KN5	Kamakhaya Devi, Thanapati	Kamru
Mahabodi temple	KN6	Gautam Buddha	Behringi (Kothi)
Maheshwar Devta Temple	KN7	Maheshwar Devta	Thanang
Nag Devta Temple	KN8	Nag Devta	Bari
Shamsheer Devta Temple	KN9	Shamsheer Devta	Rakchham
Usha Devi Temple	KN10	Usha Devi	Nichar
Kullu			
Ambika Mata Temple	KL1	Ambika (Durga)	Nirmand
Bhekhali Mata Temple	KL2	Bhuvneshwari Mata	Bekhal
Bijli Mahadev Temple	KL3	Bijli Mahadev (Lord Shiva)	Mathan
Chandi Temple	KL4	Durga	Nirmand
Dakshini Mahadev Temple	KL5	Shiv-lingam	Nirmand
Dhankeshwar Mahadev Temple	KL6	Shiva	Bial
Gurudwara Sahib Manikaran	KL7	Guru Nanak Dev	Manikaran
Hadimba Devi Temple	KL8	Hadimba	Manali (at Dungri)
Himalayan Buddhist Monastery	KL9	Gautam Budh	Manali
Maharishi Vashisht Temple	KL10	Vashisht	Manali
Manu Temple	KL11	Manu	Manali
Murlidhar Temple	KL12	Radha-Krishna	Manikaran
Naina Devi Temple	KL13	Naina Devi	Manikaran
Parshuram Temple	KL14	Parshuram	Nirmand
Raghunath Temple	KL15	Sita-Ram	Kullu
Raghunath Temple	KL16	Raghunath	Manikaran
Rama Temple	KL17	Lord Rama	Manikaran
Shitla Mata Temple	KL18	Shitla Mata	Kullu
Shiva Temple	KL19	Shiva	Manikaran
Shiva Temple	KL20	Shiva	Nirmand
Siyali Mahadev Temple	KL21	Shiva	Manali
Vaishno Devi Temple	KL22	Vaishno Devi	Vashing
Vishveshvar Mahadev Temple	KL23	Shivlingam	Bajaura
Lahul-Spiti			
Kee Gompa	LS1	Gautam Buddha	Kee
Labrang Monastery	LS2	Gautam Buddha, Padam Sambhav, Kali	Toche
Losar Gompa	LS3	Jabru Devta	Losar
Mrikula Devi Temple	LS4	Kali	Udaipur

Contd...

Table 1–*Contd...*

*Temple**	*Location in Map*	*Concerned Deity*	*Place*
Othang Gompa	LS5	Gautam Buddha	Othang
Raja Ghepan Temple	LS6	Raja Ghepan	Shashun
Sakya Monastery	LS7	Gautam Buddha	Kaza
Shashur Monastry	LS8	Padam Sambhav, Deva Gyatso	Tingche
Trilokinath Temple	LS9	Trilokinath, Gautam Buddha	Trilokinath
Yordang Gompa	LS10	Padam Sambhav, Gautam Buddha	Biling
Mandi			
Bhootnath Temple	M1	Shiva	Mandi
Ekadash Rudra Temple	M2	Shiva	Mandi
Gurudwara Shri Guru Gobind Singh	M3	Guru Gobind Singh	Rewalsar
Hanogi Mata Temple	M4	Durga	Hanogi
Kamaksha Devi Temple	M5	Kamaksha Devi	Kao
Kamru Nag Temple	M6	Kamru Nag	Kamrah
Mahamaya Durga Temple	M7	Durga	Pangna
Mahunag Temple	M8	Mahunag	Bakhaari
Mamaleshwar Mahadeh Temple	M9	Shiva	Mamel
Naina Devi Temple	M10	Naina Devi	Sarkidhar
Nawahi Devi Temple	M11	Nawahi Devi	Nawahi Devi
Ningmapa Budh Monastary	M12	Padam Sambhav	Rewalsar
Panchvaktra Mahadev Temple	M13	Shiva	Mandi
Shikari Devi Temple	M14	Shikari Devi	Shikari Devi
Shri Lomash Rishi Temple	M15	Lomash Rishi	Rewalsar
Sidh Baba Kamlahiya Temple	M16	Baba Kamlahiya	Kamlah
Tarna Devi Temple	M17	Kali	Mandi
Trilokinath Temple	M18	Shiva	Mandi
Shimla			
Bhimakali Temple	SH1	Bhimakali	Sarahan
Bhuteshwar Mahadev Temple	SH2	Shiva	Rampur
Bijjat Maharaj Temple	SH3	Bijjat Maharaj	Saranh
Dingu Durga Temple	SH4	Durga	Sanjauli (Shimla)
Durga Temple	SH5	Durga	Jatog Cant (Shimla)
Duttatrey Temple	SH6	Duttatrey, Atri, Anusuya Mata	Dutt Nagar
Hatkeshwari Durga Mata Temple	SH7	Durga	Hatkoti
Jakhoo Hanuman Temple	SH8	Hanuman	Shimla

Contd...

Table 1–*Contd...*

*Temple**	*Location in Map*	*Concerned Deity*	*Place*
Jathiya Devi Temple	SH9	Jathiya Devi	Dhonkhri
Kalibari Temple	SH10	Kali	Kalibari (Shimla)
Kali Mata Temple	SH11	Kali	Padhai
Kamna Devi Temple	SH12	Durga	Kamna Devi (Shimla)
Lankadvir Temple	SH13	Lankadvir, Brijeshwari	Chopal
Mahasu Devta Temple	SH14	Mahasu Devta	Gajedi
Mahish Mardini Mahalakshmi Temple	SH15	Maha Lakshmi	Rampur
Narsingh Temple	SH16	Narsingh	Shingla
Shikhdu Devta Temple	SH17	Shikhdu Devta	Rohru
Shri Ayodhya Nath Temple	SH18	Ram-Sita-Lakshman	Rampur
Shri Sankatmochan Hanuman Temple	SH19	Ram-Sita-Hanuman	Shimla
Shri Sat Narayan Temple	SH20	Vishnu	Rampur
Surya Temple	SH21	Surya Devta	Nirth
Tara Devi Temple	SH22	Tara Devi	Patyud
Sirmaur			
Balasundari Temple	SR1	Balasundari	Trilokpur
Brahmchari Ashram	SR2	Gaytri Mata	Renuka ji
Gurudwara Shri Paonta Sahib	SR3	Guru Gobind Singh	Paonta Sahib
Jagan Nath Temple	SR4	Jagan Nath	Nahan
Kalisthan Temple	SR5	Kali	Nahan
Katasan Devi Temple	SR6	Katasan Devi	Bada Van
Narsingh Temple	SR7	Narsingh	Nahan
Nirwan Ashram	SR8	Guru Sunder Muni Nirwan	Renuka ji
Pujari Ghat Rama Temple	SR9	Ram, Sita, Lakshman	Paonta Sahib
Radha Krishan Temple	SR10	Radha Krishan	Naina Tikker
Renuka Mata Temple	SR11	Renuka Mata	Renuka ji
Rishi Markanda Dham	SR12	Markanda Rishi	Jogivan
Sahibji Maharaj Gurudwara	SR13	Sahibji Maharaj	Shambhuwala
Sanyas Ashram	SR14	Shiva	Renuka ji
Shirgul Devta Temple	SR15	Shirgul Devta	Churdhar
Shiva Temple	SR16	Shiva	Rukhdi
Shiva Temple	SR17	Shiva	Trilokpur
Vishwakarma Temple	SR18	Vishwakarma	Paonta Sahib
Yamuna Temple	SR19	Radha Krishan	Paonta Sahib

Contd...

Table 1–*Contd...*

*Temple**	*Location in Map*	*Concerned Deity*	*Place*
Solan			
Baduwada Temple	SL1	Sehdev	Dadlaghat
Bhadrakali Temple	SL2	Bhadrakali	Jakholi
Dev Dhar Temple	SL3	Dev Dhar Devta	Arki
Lutru Mahadev Temple	SL4	Shiva	Arki
Mutru Mahadev Temple	SL5	Shiva	Arki
Sai Baba Temple	SL6	Shirdi Sai Baba	Garkhal
Shakani Ganga Temple	SL7	Goddess Shakani	Arki
Shiva Temple	SL8	Shiva	Dadlaghat
Shiva Temple	SL9	Shiva	Jakholi
Shiva Temple	SL10	Shiva	Jatoli
Shri Sidh Baba Balak Nath Temple	SL11	Baba Balak Nath	Garkhal
Shulini Mata Temple	SL12	Shulini Devi	Solan
Sidheshwari Temple	SL13	Sidheshwari Devi	Solan
Una			
Baba Achutanand Thakurdwara	U1	Swami Achutanand	Amlehad
Baba Barbhag Singh Gurudwara	U2	Baba Barbhag Singh	Maidi
Bagula Mukhi Temple	U3	Bagula Mukhi	Kaloh
Bhadrakali Temple	U4	Bhadrakali	Bhadrakali
Dera Baba Rudranand	U5	Baba Rudranand, Lakshmi Narayan	Nari
Dron Shiv Temple (Shiv Badi)	U6	Shiva	Ambota
Gugga Jahar Pir Temple	U7	Gugga Pir	Chak Sarai
Lakshmi Narayan Temple	U8	Lakshmi Narayan	Chowki Maniar
Lohara Temple	U9	Nag Devta, Bajrangbali, Baba Balak Nath, Jatheri Mata	Lohara
Mata Chintpurni Temple	U10	Chintpurni	Chhaproh
Rama Temple	U11	Ram-Sita-Lakshman	Andora
Sadashiv Temple	U12	Shivlingam	Bahi
Santoshi Mata Temple	U13	Santoshi Mata	Khad
Saptdevi Sidh Peeth Ashram	U14	Vaishno Devi, Jwalaji, Naina Devi, Chamunda, Bajreshwari, Kali, Chintpurni	Kinnu
Shitla Mata Temple	U15	Shitla Mata	Baret
Shri Hanuman Temple	U16	Hanuman	Thatthal

* Arranged District-wise.

Field tours to these temples were planned in such a way as to collect plant species either in flowering or fruiting stage. For future reference and study, the information pertaining to flowering and fruiting season of courtyard plants had been enumerated in Table 2.

Table 2: Flowering and Fruiting Season of the Temple-Courtyard Plants of H.P.

Name of the Plant	*Flowering and Fruiting Season*
Acacia catechu (Linn.f.) Willd.	May-July
Acacia dealbata Link.	Winter, early spring
Acacia nilotica (L.) Willd. *ex* Delile subsp. *indica* (Benth.) Brenan	July-September
Acalypha wilkesiana M. Arg.	Throughout the year
Acer oblongum Wall.	February-April
Achras zapota Linn.	December-January
Aegle marmelos (Linn.) Correa	April-July
Aesculus indica Colebr.	May-September
Albizia lebbeck (L.) Benth.	April-August
Albizia stipulata (Roxb.) Boiv.	September-October
Alnus nitida (Spach) Endl.	September-October
Aloe vera (Linn.) Burm.f.	March-April
Alpinia nutans Rosc.	May-August
Alstonia scholaris (L.) R. Br.	December-March
Althaea rosea (L.) Cav.	August-September
Amaryllis reginae Linn.	Spring or early Summer
Antirrhinum majus Linn.	June-October
Araucaria columnaris (Forster) Hooker	September-October
Aristolochia elegans Mast.	Blooms throughout the Summer
Artemisia moorcroftiana Wall. *ex* DC.	August-October
Artocarpus heterophyllus Lam.	Flowers during the winter (December-March) and fruits during the rainy season (July-August)
Artocarpus lakoocha Roxb.	February-September
Asclepias curassavica Linn.	Flowers throughout the year
Asclepias physocarpa Schlecht.	April-July
Asparagus adscendens Roxb.	March-June
Asparagus densiflorus (Kunth) Jessop	May-June
Asparagus racemosus Willd.	April-August
Bauhinia vahlii Wight and Arn.	April-June
Bauhinia variegata Linn.	March-August
Beloperone guttata Brand.	Flowers almost throughout the year

Contd...

Table 2–*Contd...*

Name of the Plant	*Flowering and Fruiting Season*
Berberis lycium Royle	April-June
Bergenia ciliata (Haw.) Sternb. forma *ligulata* Yeo	March-July
Betula utilis D. Don	June-September
Bischoffia javanica Blume	March-April
Boerhaavia diffusa Linn.	Major part of the year
Bombax ceiba Linn	February-May
Bougainvillea glabra Choisy	Flowering in late summer and early autumn
Bougainvillea spectabilis Willd	Almost throughout the year
Brassica oleracea var. *capitata* Linn.	Summer
Broussonetia papyrifera (L.) L'Herit. *ex* Vent.	April-May
Bryophyllum calycinum Salisb.	May-June
Buxus sempervirens Linn.	April-November
Calendula officinalis Linn.	February-Early December
Callistemon lanceolatus DC.	May-September
Callistephus chinensis (L.) Nees	August-October
Calotropis procera R. Br.	February-May
Cannabis sativa Linn.	June-September
Canna indica Linn.	Usually in the rainy season (June-September)
Capsicum annuum Linn.	August-September
Carica papaya Linn.	Greater part of the year
Carissa carandas Linn.	April-June
Carissa opaca Stapf. *ex* Haines	May-June
Carissa spinarum Linn.	November-February
Caryota urens Linn.	May-August
Cassia fistula Linn.	March-July
Cassia glauca Lam.	April-August
Casuarina equisetifolia Linn.	February-May
Catharanthus roseus (Linn.) G. Don	During the rainy season
Cedrela toona Roxb. *ex* Rottl.	May-June
Cedrus deodara (Roxb.) Loudon	June-July
Celosia argentea var. *cristata* Haines	May-October
Celtis australis Linn.	April-October
Centaurea cyanus Linn.	June-October
Centella asiatica (Linn.) Urban	May-November
Cestrum diurnum Linn.	Flowers appear in Summer and continue till the end of the rainy season

Contd...

Table 2–*Contd...*

Name of the Plant	*Flowering and Fruiting Season*
Cestrum nocturnum Linn.	July-October
Chamaerops humilis Linn.	February-March
Chlorophytum capense Kuntze	July-September
Chrysanthemum indicum Linn.	April-October
Chrysanthemum leucanthemum Linn.	May-October
Cinnamomum camphora (L.) Nees and Ebesm.	April-August
Citrus aurantifolia (Christin) Sw.	March-June
Citrus jambhiri Lush.	April-June
Citrus limon (Linn.) Burm. f.	April-June
Citrus maxima (Burm. f.) Merrill	November-March
Citrus reticulata Blanco	August-September
Clematis grata Wall.	July-September
Clerodendrum inerme (Linn.) Gaertn.	March-June
Clerodendrum philippinum Schauer	August-October
Clerodendrum splendens G. Don.	Flowers during winter months
Clerodendrum thomsonae Balf.	June-September
Coix lachryma-jobi Linn.	Autumn
Coleus blumei Benth.	September-October
Colocasia esculenta (Linn.) Schott.	June-October
Convolvulus arvensis Linn.	December-March
Cordia dichotoma Forst.	April-July
Cosmos bipinnatus Cav.	July-October
Cosmos sulphureus Cav.	March-November
Cupressus sempervirens Linn.	April-May
Cupressus torulosa D. Don	February-April
Cycas revoluta Thunb.	February-May
Cynodon dactylon (Linn.) Pers.	August-November
Cyperus flabelliformis Rottb.	July-August
Dahlia rosea Cav.	July-September
Dalbergia sissoo Roxb.	May-August
Datura fastuosa Linn.	June-September
Datura innoxia Mill.	June-August
Datura stramonium Linn.	June-October
Daucus carota Linn.	Summer
Debregeasia hypoleuca Wedd.	May-August
Deeringia amaranthoides (Lamk.) Merr.	May-November

Contd...

Table 2–*Contd...*

Name of the Plant	*Flowering and Fruiting Season*
Delphinium ajacis Linn.	May-September
Dendrophthoe falcata (Linn. f.) Ettingsh.	November-March
Dianthus barbatus Linn.	May-August
Diospyros lotus Linn.	April-August
Diospyros montana Roxb. var. *cordifolia* (Roxb.) Hiern.	March-May
Dodonea viscosa (Linn.) Jacq.	October-March
Dolichos lablab Linn.	November-December
Duranta repens Linn.	May-August
Ehetia acuminata R. Br.	May-July
Elaeocarpus ganitrus Roxb.	April-June
Emblica officinalis Gaertn.	April-September
Erigeron alpinus Linn.	May-October
Eriobotrya japonica (Thunb.) Lindl.	July-September
Eucalyptus tereticornis Sm.	March-April
Euphorbia milii Des Moul.	March-April and throughout the Summer
Euphorbia pulcherrima Willd. *ex* Klotzsch.	November-January
Euphorbia royleana Boiss.	June-July
Ficus benghalensis Linn.	February-June
Ficus carica Linn.	March-August
Ficus elastica Roxb.	April-July
Ficus glomerata Roxb.	April-July
Ficus palmata Forsk.	April-September
Ficus religiosa Linn.	March-October
Ficus roxburghii Wall. *ex* Miq.	April-August
Ficus rumphii Blume	May-June
Ficus virens Ait.	March-August
Fraxinus xanthoxyloides (Wall. *ex* G. Don) DC.	July-September
Gaillardia pulchella Foug.	July-October
Gomphrena globosa Linn.	Flowers during Rainy and cold seasons in India
Grevillea robusta A. Cunn. *ex* R. Br.	March-July
Grewia oppositifolia Roxb.	April-June
Hamelia patens Jacq.	August-September
Hedychium coronarium Koenig.	May-July
Helianthus annuus Linn.	July-September
Hemerocallis fulva (Linn.) Linn.	May-June
Hibiscus mutabilis Linn.	September-December

Contd...

Table 2–*Contd...*

Name of the Plant	*Flowering and Fruiting Season*
Hibiscus rosa-sinensis Linn.	February-October
Hibiscus syriacus Linn.	July-September
Holoptelea integrifolia (Roxb.) Planch.	February-May
Humulus lupulus Linn.	June-July
Hydrangea macrophyla (Thunb.) DC.	June-July
Hymenocallis littoralis (Jacq.) Salisb.	July-August
Hyoscyamus niger Linn.	April-September
Hypericum cernuum Roxb.	April-October
Impatiens balsamina Linn.	July-September
Ipomoea cairica (L.) Sweet	June-November
Ipomoea purpurea (L.) Roth.	April-October
Ipomoea quamoclit Linn.	August-October
Iris germanica Linn.	June-July
Ixia chinensis Linn.	July-October
Jacaranda acutifolia Humb. and Bonpl.	March-June
Jasminum grandiflorum Linn.	April-September
Jasminum humile Linn.	May-September
Jasminum primulinum Hemsl. *ex* Baker	March-May
Jasminum sambac (L.) Ait	Summer-rainy seasons
Jatropha panduraefolia Andr.	Summer
Juglans regia Linn.	April-October
Juniperus chinensis Linn.	March-April
Juniperus macropoda Boiss.	September-March
Justicia gendarussa Linn. f.	January-March
Kochia trichophylla Voss.	Flowering after the rainy seasons
Lagerstroemia indica Linn.	June-October
Lannea coromandelica (Houtt.) Merrill.	February-June
Lawsonia alba Lam.	April-July
Lespedeza sericea (Thunb.) Miq.	July-August
Leucaena leucocephala (Lam.) de Wit.	September-November
Ligustrum lucidum Aiton.	July-November
Litchi chinensis Sonner.	April-June
Litsea polyantha Juss.	March-May
Litsea sebifera Pers.	May-July
Lonicera japonica Thunb.	May-August
Lycopersicum esculentum Mill.	May-June

Contd...

Table 2–*Contd...*

Name of the Plant	*Flowering and Fruiting Season*
Machilus gamblei King *ex* Hook. f.	March-June
Magnolia grandiflora Linn.	May-August
Mallotus philippinensis (Lam.) Muell.-Arg.	April-October
Malva sylvestris Linn.	May-October
Malva verticillata Linn.	May-September
Malvaviscus conzatii Greenm.	Major part of the year
Mangifera indica Linn.	March-July
Melia azadirachta Linn.	February-April
Melia azedarach Linn.	May-October
Mentha longifolia (Linn.) Huds.	November-February
Mentha piperita Linn.	July-September
Michelia champaca Linn.	April-September
Mimosa pudica Linn.	Flowers and fruits throughout the year
Mimusops elengi Linn.	March-May
Mirabilis jalapa Linn.	June-September
Momordica charantia Linn.	July-October
Morus alba Linn.	March-May
Morus serrata Roxb.	April-June
Murraya koenigii (Linn.) Spreng.	March-June
Murraya paniculata (Linn.) Jack.	March-September
Musa paradisiaca Linn.	Throughout the year
Narcissus tazetta Linn.	January-February
Nerium odorum Soland.	April-November
Nicandra physaloides (Linn.) Gaertn.	May-October
Nyctanthes arbor-tristis Linn.	July-October
Ocimum basilicum Linn.	June-September
Ocimum gratissimum Linn.	September-October
Ocimum kilimandscharicum Gurke	August-September
Ocimum sanctum Linn.	July-October
Oenothera drummondii Hook f.	August-September
Olea glandulifera Wall.	March-May
Papaver rhoeas L. var. *latifolia* Prain	Summer
Passiflora caerulea Linn.	June-September
Pedilanthus tithymaloides (L.) Poit.	Flowers mainly in summer, but intermittently throughout the year
Pelargonium hortorum Bailey	Summer season

Contd...

Table 2–*Contd...*

Name of the Plant	*Flowering and Fruiting Season*
Petunia hybrida Vilm.	Summer
Phaseolus vulgaris Linn.	June-September
Phoenix sylvestris (L.) Roxb.	May-June
Phytolacca acinosa Roxb.	June-September
Pinus gerardiana Wall. *ex* D. Don	June-July
Pinus roxburghii Sarg.	March-May
Pinus wallichiana A. B. Jackson	April-June
Pistacia integerrima J. L. Stewart *ex* Brand.	March-May
Pisum sativum Linn.	June-July
Platanus orientalis Linn.	April-May
Plumeria acutifolia Poiret	June-September
Polianthes tuberosa Linn.	Throughout the greater part of the year
Polyalthia longifolia (Sonn.) Thwaites	February-May
Polygonum alpinum All.	June-September
Pongamia pinnata (Linn.) Pierre	March-June
Populus ciliata Wall.	April-July
Populus nigra Linn.	April-July
Prinsepia utilis Royle	February-March
Prunus amygdalus Batsch.	May-September
Prunus armeniaca Linn.	May-August
Prunus domestica Linn.	February-March
Prunus persica (Linn.) Batsch	February-May
Prunus puddum (Roxb. *ex* Ser.) Brandis	November-May
Psidium guajava Linn.	July-September
Pterospermum acerifolium Willd.	April-June
Punica granatum Linn.	May-September
Putranjiva roxburghii Wall.	March-April
Pyrostegia venusta (Ker.-Gawl.) Miers	January-April
Pyrus communis Linn.	June-August
Pyrus malus Linn.	May-October
Pyrus pashia Buch.-Ham. *ex* D.Don	March-May
Quercus dilatata Lindl. *ex* A. DC.	April–May
Quercus incana Roxb.	April-October
Quisqualis indica Linn.	May-September
Ribes orientale Desf.	July-October
Ricinus communis Linn.	March-May

Contd...

Table 2–*Contd...*

Name of the Plant	*Flowering and Fruiting Season*
Robinia pseudoacacia Linn.	April-May
Rosa alba Linn.	June-August
Rosa banksiae Br.	May-June
Rosa foetida Herrm.	June-September
Rosa indica Linn.	June-September
Rosa jacquemontii Crep. *ex* Hook. f.	June-September
Rosa moschata Herrm.	April-October
Rosa multiflora Thunb.	June-July
Rosa webbiana Wall. *ex* Royle	June-September
Roylea cinerea (D. Don) Baill.	March-November
Ruellia tuberosa Linn.	March-September
Russelia equisetiformis Cham. and Schl.	Summer
Saccharum officinarum Linn.	January-April
Salix alba Linn.	March-May
Salix babylonica Linn.	February-May
Salix fragilis Linn.	April-June
Salix oxycarpa Anders.	June-September
Salix tetrasperma Roxb.	April-July
Salvia coccinia Juss.	July-August
Salvia splendens Sellow *ex* Roemer and J.A. Schultes	Summer onwards
Sambucus nigra Linn.	May-September
Sansevieria trifasciata Prain	June-July
Santalum album Linn.	February-May
Sapindus mukorossi Gaertn.	April-September
Sapium insigne (Royle) Benth.	May-August
Sapium sebiferum (L.) Roxb.	June-August
Schefflera arboricola (Hayata) Merr.	August-September
Sempervivum arborejum Linn.	January-March
Sesamum orientale Linn.	August-September
Shorea robusta Gaertn. f.	March-June
Solanum tuberosum Linn.	May-September
Syzigium cumini (Linn.) Skeels	April-August
Syzigium jambos (Linn.) Alston	February-June
Tabernaemontana coronaria (Jacq.)Willd.	May-December
Tagetes erecta Linn.	October-April
Tamarindus indica Linn.	June-July

Contd...

Table 2–*Contd...*

Name of the Plant	*Flowering and Fruiting Season*
Tecoma radicans Juss.	August-September
Tecomaria capensis (Thunberg) Spach.	May-September
Tecoma stans (Linn.) Juss *ex* H.B.K.	Flowers and fruits throughout the year
Tectona grandis Linn. f.	August-December
Terminalia bellirica (Gaertn.) Roxb.	May-September
Terminalia chebula (Gaertn.) Retz.	April-June
Thevetia neriifolia Juss *ex* Steud.	July-September
Thuja orientalis Linn.	July-October
Thuja plicata D. Don	June-September
Thunbergia erecta (Benth.) T. Anders	June-July
Tithonia tagetiflora Desf.	May-September
Tropaeolum majus Linn.	May-November
Ulmus villosa Brandis	February-March
Viola canescens Wall.	April-November
Viola tricolor Linn.	April-May
Wedelia calendulacea Less.	March-April
Wisteria sinensis (Sims) Sweet	April–May
Withania somnifera (L.) Dunal	November-February
Wrightia tomentosa (Roxb.) Roem and Schult.	March-July
Zanthoxylum armatum DC.	April-June
Zea mays Linn.	August-September
Zephyranthes candida (Lindl.) Herb.	September-October
Zephyranthes grandiflora Lindl.	June-July
Zinnia elegans Jacq.	Summer, Autumn
Zizyphus mauritiana Lamk.	April-September

Besides, first hand information on ethnobotanical aspects was collected through interviews from knowledgeable people (temple priests, members of temple committee, old experienced informants, medicinemen, village head, etc.) for which the proforma designed by Jain and Goel (1995) was modified to gather data for this purpose. Uses of plants narrated by a person were cross-checked and considered notable only if the author observed actual application, or similar application was reported by at least three informants. For the medicinal uses, information pertaining to method of preparation (powder, aq extr., infusion or decoction), dosage and combination with other plant, if any was also recorded. Status of courtyard plants in the surrounding areas was also noted in the field note book. Information pertaining to the nativity of the species was obtained from all possible sources (Bailey, 1949; Bentley and Trimen, 2008; Hajra and Das, 1982; Khuroo *et al.*, 2004; Maheshwari and Paul, 1975; Matthew, 1969; Murthy *et al.*, 2007; Nayar, 1977; Pandey and Parmar, 1994; Reddy, 2008; Reddy and Raju, 2002; Reddy and Reddy, 2004; Reddy *et al.*, 2000; Saxena, 1991; Sharma, 1984; Wasson, 2003; Weber, 2003).

The herbarium sheets of all the collected temple plants were dried, preserved and mounted as per standard herborizing practices outlined by Jain and Rao (1977). The specimens were identified using regional floras and various revisionary and monographic works (Bose and Chowdhury, 1991; Chauhan, 1999; Chowdhery and Wadhwa, 1984; Collett, 1902; Dhiman, 1976; Hooker, 1872-1897; Nair, 1977; Polunin and Stainton, 1984; Stainton, 1988; Warman, 1999) and later got authenticated by carefully matching with the specimens at the herbaria of Botanical Survey of India (BSI), Northern Circle and Forest Research Institute (FRI), Dehra Dun. Nomenclature of these taxa was confirmed from Bennet (1986) and Wielgorskaya (1995). Photographs of temples and plants in their natural habitats were also clicked. The voucher specimens of all the presently studied taxa were deposited in the herbarium section of Ethnobotanical Laboratory, Department of Biosciences, H.P. University, Shimla.

The collected information on temple plants of H.P. had been arranged district-wise. For each temple in a district, the compiled information pertains to the name of temple, main deity, other deities, total area of temple-courtyard, location (including place, approach and altitude), management, historical background, temple architect, offerings to God, offerings to people ('prasadam'), fairs and festivals, observations related to courtyard plants (protected as well as for landscaping), predominant plants of the courtyard, maintenance, ethnobotanical informations regarding sacred uses and associated myth, medicinal, edible, ornamental, material uses, status of courtyard plants in the surrounding areas, etc. Systematic enumeration of the plants is in alphabetical order of their botanical names, followed by synonyms (wherever available), family, vernaculars, English, Hindi and Sanskrit Names, distribution, habitat ecology, reproductive cycle, material examined and systematic account. Appendices include information regarding the families, local names and glossary of words.

Ethnobotanical Appraisal on the Temples of Himachal Pradesh

Temples of District Bilaspur

1. Baba Balak Nath Temple (Pl. 1A)

Main Deity : Baba Balak Nath.

Other Deities : Gugga Pir, Mata Ratno, Shiva, Radha-Krishana, Duttatrey.

Tot. Ar. of Tem. Courtyard : 4500 sq m.

Location

(i) Place

Town : Shah Talai

Block : Jhanduta

Tehsil : Jhanduta

District : Bilaspur

(ii) Approach : The place is 5 km from Deoth Sidh and 50 km from Hamirpur.

(iii) Altitude : 560 m.

Management : The State government has constituted Baba Balak Nath Temple Trust for its management.

Historical Background : The temple is believed to be about 500 years old. Prior to taking over by the management committee, it was reconstructed and repaired several times by the locals with donations given by different devotees. According to the prevalent legend, Baba Balak Nath used to work as a cowboy for a landlady at Talai village. For his services, he was daily served with maize bread and butter milk by the lady. One day the villagers complained to the landlady that their crops had been damaged oftenly by her wandering cattle under the command of cowboy. To show that they took her to their fields. To their surprise, the damaged fields were found to have better crops than the undamaged ones. On observing such miracles, Saint Gorakh Nath sought hime to become his disciple, for which he declined. Later, Baba Balak Nath resolved to proceed to the forests for which the landlady resented and reminded him how she served him with food and milk for his upbringing like her own child. On hearing this, he took her near a banyan tree, and asked to peep through the hollow of the tree. To her utmost surprise, she saw there the butter milk and the maize breads served to the boy for over twelve years. That Banyan tree is still present in the courtyard of the temple.

Temple Architect : Sikhara style.

Offerings to God

(i) By Priest : Milk and Sweets ('Pinni').

(ii) By People : 'Rot' (fried sweet hard wheat bread), 'Lachidana' (sugar balls), Puffed rice.

Off. to People (Prasadam) : 'Rot', 'Lachidana', Puffed rice.

Fairs and Festivals : Baba Balak Nath fair (held for one month during March-April), Shivratri.

Observations

(i) Courtyard Plants (Protected as well as for Landscaping : *Althaea rosea, Bryophyllum calycinum, Canna indica, Catharanthus roseus, Cestrum nocturnum, Chrysanthemum indicum, Duranta repens, Ficus benghalensis, Ficus religiosa, Hibiscus rosa-sinensis, Hibiscus syriacus, Impatiens balsamina, Ixia chinensis, Justicia gendarussa, Kochia trichophylla, Mangifera indica, Ocimum sanctum, Pelargonium hortorum, Rosa indica, Tagetes erecta, Thevetia neriifolia, Thuja orientalis, Tropaeolum majus.*

(ii) Predominant Plants of the Courtyard : *Tagetes erecta, Thuja orientalis.*

(iii) Maintenance : Excellent.

(iv) Ethnobotanical

Sacred and Associated Myth : Leaves and wood of *Mangifera indica* used in purification ceremonies. *Ficus benghalensis, Ficus religiosa* and

Ocimum sanctum worshipped as sacred plants by the people. *Ficus benghalensis* tree in sanctum sanctorum is believed to be as old as the temple and it is said that Baba Balak Nath used to meditate under this tree.

Medicinal : Decoction of leaves (250g leaves of *Bryophyllum calycinum* boiled in 4 litre of water) useful for kidney stone. Decoction of leaves of *Duranta repens* prescribed for joint pains. 4-5g powdered dried leaves of *Mangifera indica* prescribed daily for diabetes.

Edible : Fruits of *Mangifera indica* edible.

Ornamental : *Althaea rosea, Bryophyllum calycinum, Canna indica, Catharanthus roseus, Cestrum nocturnum, Chrysanthemum indicum, Duranta repens, Hibiscus rosa-sinensis, Hibiscus syriacus, Impatiens balsamina, Ixia chinensis, Justicia gendarussa, Kochia trichophylla, Ocimum sanctum, Pelargonium hortorum, Rosa indica, Tagetes erecta, Thevetia neriifolia, Thuja orientalis* and *Tropaeolum majus* grown as ornamentals.

Remarks : People whose wishes are fulfilled tend to offer goats but their sacrifice not permitted.

2. Baba Nahar Singh Temple or Dholra Temple (Pl. 1B)

Main Deity : Baba Nahar Singh.

Other Deity : Rani Kohli (there is a separate small temple of Rani Kohli near the main temple).

Tot. Ar. of Tem. Courtyard : About 3000 sq m.

Location

(i) *Place*

Town : Bilaspur

Block : Bilaspur

Tehsil : Bilaspur (Sadar)

District : Bilaspur

(ii) *Approach* : Temple is located in Dholra sector.

(iii) *Altitude* : 640 m.

Management : Temple committee.

Historical Background : Temple was built in 15th century by Raja Deep Chand (King of Bilaspur). Later, it was renovated by the temple committee.

It is believed that Rani Kohli (wife of Raja Deep Chand) brought Baba Nahar Singh at the time of her marriage from Kullu.

Temple Architect	:	Sikhara style.
Offerings to God		
(i) By Priest	:	Rice-Dal, Chapati-Vegetables.
(ii) By People	:	'Rot', 'Halva', Clarified butter, Jaggery, Gram, Clothes ('Dhoti-Kurta'), Wooden sleepers ('Khadawein'), Seasonal crops.
Off. to People (Prasadam)	:	Puffed rice, 'Rot', 'Halva'.
Fairs and Festivals	:	Fairs held on second and third Tuesday of Jyeshth month (May-June).
Observations		
(i) Courtyard Plants (Protected as well as for Landscaping	:	*Ficus benghalensis, Ocimum sanctum.*
(ii) Maintenance	:	Satisfactory.
(iii) Ethnobotanical		
Sacred	:	Both *Ficus benghalensis* (thought to be as old as the temple) and *Ocimum sanctum* worshipped as sacred plants.
Medicinal	:	Decoction of leaves of *Ocimum sanctum* prescribed for cough.
Remarks	:	Goats sacrificed as well as offered alive to the deity.

3. Garuna Jhadi Temple (Pl. 1C)

Main Deity	:	Baba Balak Nath.
Other Deities	:	Shiva, Parvati, Vishnu.
Tot. Ar. of Tem. Courtyard	:	1500 sq m.
Location		
(i) Place		
Town	:	Shah Talai
Block	:	Jhanduta
Tehsil	:	Jhanduta
District	:	Bilaspur
(ii) Approach	:	Located 5 km from Deoth Sidh.
(iii) Altitude	:	560 m.
Management	:	It is under Government constituted Baba Balak Nath Temple Trust, Shah Talai.
Historical Background	:	The temple is believed to be about 500 years old. Prior to its takeover by the State Government, it was reconstructed by the locals with donations given by different devotees.

		The temple is named as 'Garuna Jhadi' due to a bush of 'Garuna' (*Carissa spinarum*) in its courtyard. It is believed that Baba Balak Nath meditated under this bush.
Temple Architect	:	Sikhara style.
Offerings to God		
(i) By Priest	:	Milk, 'Khoya'.
(ii) By People	:	'Lachidana', Puffed rice.
Off. to People (Prasadam)	:	'Lachidana', Puffed rice.
Fairs and Festivals	:	Local fair held for one month during March-April.
Observations		
(i) Courtyard Plants (Protected as well as for Landscaping	:	*Carissa spinarum, Ficus benghalensis, Grevillea robusta, Rosa indica, Tagetes erecta, Wrightia tomentosa.*
(ii) Predominant Plants of the Courtyard	:	*Rosa indica, Tagetes erecta.*
(ii) Maintenance	:	Excellent.
(iv) Ethnobotanical		
Sacred and Associated Myth	:	*Carissa spinarum* and *Ficus benghalensis* worshipped as sacred plants. People tie sacred threads ('Dori') to *Carissa spinarum* for fulfilment of their wishes.
Medicinal	:	Fruits of *Carissa spinarum* considered good for anaemia.
Edible	:	Fruits of *Carissa spinarum* edible.
Ornamental	:	*Grevillea robusta, Rosa indica, Tagetes erecta* and *Wrightia tomentosa* grown as ornamental plants.
Status of Courtyard Plants in Surrounding Areas	:	Excepting *Wrightia tomentosa*, all other plants are common in the surrounding areas.
Remarks	:	No sacrifices are made. Plucking of fruits and/or leaves from the sacred plant of *Carissa spinarum* strictly prohibited.

4. Hari Devi Temple (Pl. 1D)

Main Deity	:	Hari Devi.
Other Deities	:	Shiva, Hanuman, Bhairav.
Tot. Ar. of Tem. Courtyard	:	About 3000 sq m.
Location		
(i) Place		
Village	:	Hari Devi
Block	:	Lehdi Sarel

Tehsil	:	Ghumarwin
District	:	Bilaspur
(ii) *Approach*	:	About 1 km from Dangar.
(iii) *Altitude*	:	900 m.
Management	:	Local temple committee.
Historical Background	:	The temple was built in Samvat 1831 by Rani Naagar Deyi (wife of King of Bilaspur, Maharaja Devi Chand).
Temple Architect	:	Sikhara style.
Offerings to God		
(i) *By Priest*	:	'Halva', Puffed rice, 'Kheer' (pudding), Sweets.
(ii) *By People*	:	'Halva', Puffed rice, Seasonal crops.
Off. to People (Prasadam)	:	'Halva', Puffed rice.
Fairs and Festivals	:	Local fair held in May.
Observations		
(i) *Courtyard Plants (Protected as well as for Landscaping*	:	*Bougainvillea spectabilis, Bryophyllum calycinum, Canna indica, Capsicum annuum, Cynodon dactylon, Ficus benghalensis, Ficus religiosa, Jasminum grandiflorum, Mangifera indica, Ocimum sanctum, Tagetes erecta.*
(ii) *Predominant Plants of the Courtyard*	:	*Jasminum grandiflorum.*
(iii) *Maintenance*	:	Satisfactory.
(iv) *Ethnobotanical*		
Sacred	:	*Ficus benghalensis, Ficus religiosa* and *Ocimum sanctum* worshipped. *Cynodon dactylon* considered sacred and employed in all religious ceremonies. Wood of *Ficus benghalensis, Ficus religiosa* and *Mangifera indica* used in purification ceremonies.
Medicinal	:	Leaf juice of *Ocimum sanctum* considered good for cold. Poultice of leaves of *Bryophyllum calycinum* applied for piles. Decoction of *Cynodon dactylon* (whole plant) anticancerous.
Edible	:	Fruits of *Mangifera indica* edible.
Ornamental	:	*Bougainvillea spectabilis, Bryophyllum calycinum, Canna indica, Jasminum grandiflorum, Ocimum sanctum, Tagetes erecta* grown as ornamentals.
Other Uses	:	*Bougainvillea spectabilis* planted for fencing purposes. A garland of fruits of *Capsicum annuum* alongwith lemon tied on the main door of houses to ward off evil spirits.

Remarks	:	Hari Devi treated as 'Kuldevi' by the villagers. However, no sacrifices made to appease deity.

5. Lakshmi Narayan Temple (Pl. 1E)

Main Deity	:	Lakshmi Narayan.
Other Deities	:	Shivlingam-called as Khanmukheshwar, Radha-Krishana, Durga. There is a separate temple of Shiva-Parvati, Shitla Mata, Gasiyan, Sati Parvati, Brahma, Vishnu, Mahesh, Kartik, Bhairav, Baba Zaahar Pir, Gugga and Nandi near the main temple.
Tot. Ar. of Tem. Courtyard	:	4500 sq m.
Location		
(i) Place		
Town	:	Bilaspur
Block	:	Bilaspur
Tehsil	:	Bilaspur (Sadar)
District	:	Bilaspur
(ii) Approach	:	The temple is located in the town.
(iii) Altitude	:	640 m.
Management	:	Temple committee.
Historical Background	:	The temple was built in 1958 by the locals.
Temple Architect	:	Sikhara style.
Offerings to God		
(i) By Priest		
Morning	:	Milk.
Noon/Evening	:	Rice-Dal, Chapati-Vegetables, 'Kheer' (Pudding).
(ii) By People	:	Puffed rice, 'Patasha', Fruits, Sweets.
Off. to People (Prasadam)	:	Puffed rice, Fruits, Sweets.
Fairs and Festivals	:	Janamashtmi, Shivratri and Navratra are commonly celebrated festivals. However, temple fair is held on every Tuesday of Jyeshth month (May–June).
Observations		
(i) Courtyard Plants (Protected as well as for Landscaping	:	*Aegle marmelos, Araucaria columnaris, Bougainvillea spectabilis, Calendula officinalis, Callistemon lanceolatus, Canna indica, Catharanthus roseus, Celosia argentea* var. *cristata, Cestrum nocturnum, Chlorophytum capense, Chrysanthemum indicum, Citrus limon, Citrus maxima, Cosmos sulphureus, Cynodon dactylon, Dahlia rosea,*

		Delphinium ajacis, Eriobotrya japonica, Euphorbia pulcherrima, Ficus benghalensis, Ficus elastica, Ficus religiosa, Gaillardia pulchella, Gomphrena globosa, Grevillea robusta, Hibiscus rosa-sinensis, Hibiscus syriacus, Impatiens balsamina, Ipomoea quamoclit, Jacaranda acutifolia, Jatropha panduraefolia, Malvaviscus conzatii, Mimosa pudica, Mirabilis jalapa, Morus alba, Musa paradisiaca, Ocimum sanctum, Papaver rhoeas var. *latifolia, Petunia hybrida, Pongamia pinnata, Psidium guajava, Punica granatum, Quisqualis indica, Rosa indica, Ruellia tuberosa, Salvia coccinia, Tabernaemontana coronaria, Tagetes erecta, Tecoma radicans, Tecoma stans, Thuja orientalis, Thunbergia erecta, Tropaeolum majus, Wedelia calendulacea, Zinnia elegans.*
(ii)	*Predominant Plants of the Courtyard*	: *Grevillea robusta, Mirabilis jalapa, Quisqualis indica.*
(ii)	*Maintenance*	: Excellent.
(iv)	*Ethnobotanical*	
	Sacred	: *Ficus benghalensis, Ficus religiosa, Musa paradisiaca* and *Ocimum sanctum* worshipped as sacred plants. *Cynodon dactylon* considered sacred plant and employed in all religious ceremonies. Leaves of *Aegle marmelos* specially offered to appease Lord Shiva.
	Medicinal	: Fruits of *Aegle marmelos* considered good for digestion. Pulp of the fruit given to cure leucorrhoea. One spoon leaf juice of *Citrus limon* mixed with leaf juice of Aru (*Prunus persica*) given regularly for 3 days for expelling intestinal worms. 5-10g powdered roots of *Mimosa pudica* prescribed twice daily for kidney stones.
	Edible	: Fruits of *Aegle marmelos, Citrus maxima, Eriobotrya japonica, Morus alba, Musa paradisiaca, Psidium guajava* and *Punica granatum* edible and that of *Citrus limon* for pickles. Fruit pulp of *Aegle marmelos* used for making cool drinks.
	Ornamental	: *Araucaria columnaris, Bougainvillea spectabilis, Calendula officinalis, Callistemon lanceolatus, Canna indica, Catharanthus roseus, Celosia argentea* var. *cristata, Cestrum nocturnum, Chlorophytum capense, Chrysanthemum indicum, Cosmos sulphureus, Dahlia rosea, Delphinium ajacis, Euphorbia pulcherrima, Ficus elastica, Gaillardia pulchella, Gomphrena globosa, Grevillea robusta, Hibiscus rosa-sinensis, Hibiscus syriacus, Impatiens balsamina, Ipomoea quamoclit, Jacaranda acutifolia, Jatropha panduraefolia, Malvaviscus conzatii, Mirabilis jalapa,*

		Ocimum sanctum, Papaver rhoeas var. *latifolia, Petunia hybrida, Pongamia pinnata, Quisqualis indica, Rosa indica, Ruellia tuberosa, Salvia coccinia, Tabernaemontana coronaria, Tagetes erecta, Tecoma radicans, Tecoma stans, Thuja orientalis, Thunbergia erecta, Tropaeolum majus, Wedelia calendulacea* and *Zinnia elegans* grown as ornamental plants.
Other Uses	:	Wood of *Morus alba* used for furniture; leaves as fodder and for rearing silk worms.
Remarks	:	No sacrifices offered.

6. Markanday Temple (Pl. 1F)

Main Deity	:	Markanday Rishi.
Other Deities	:	Yamraj, Shiv-Parvati, Ram-Lakshman-Sita, Radha-Krishana, Hanuman, Bhairav, Brahma, Vishnu, Mahesh.
Tot. Ar. of Tem. Courtyard	:	Of the about 5 hectare total area of the temple land, courtyard contains about 1 hectare.
Location		
(i) Place		
Village	:	Maakadi
Block	:	Bilaspur (Sadar)
Tehsil	:	Bilaspur (Sadar)
District	:	Bilaspur
(ii) Approach	:	About 5 km from Jukhala.
(iii) Altitude	:	590 m.
Management	:	Under Government control.
Temple Architect	:	Sikhara style.
Offerings to God		
(i) By Priest	:	Rice-Dal, Chapati-Vegetables and 'Halva' offered to deity in the noon.
(ii) By People	:	Puffed rice, 'Lachidana', Seasonal crops.
Off. to People (Prasadam)	:	Puffed rice, 'Lachidana'.
Fairs and Festivals	:	Baisakhi fair (April month).
Observations		
(i) Courtyard Plants (Protected as well as for Landscaping	:	*Albizia lebbeck, Boerhaavia diffusa, Cannabis sativa, Canna indica, Catharanthus roseus, Cedrela toona, Celtis australis, Cestrum nocturnum, Citrus jambhiri, Citrus limon,*

Clerodendrum philippinum, Colocasia esculenta, Cynodon dactylon, Datura fastuosa, Ficus benghalensis, Ficus glomerata, Ficus palmata, Ficus religiosa, Ficus roxburghii, Helianthus annuus, Hibiscus syriacus, Lycopersicum esculentum, Mangifera indica, Mentha piperita, Morus alba, Musa paradisiaca, Ocimum sanctum, Prunus puddum, Quisqualis indica, Roylea cinerea, Salix tetrasperma, Sapium insigne, Sesamum orientale, Syzygium cumini, Tagetes erecta, Thevetia neriifolia, Zinnia elegans.

(ii) Predominant Plants of the Courtyard : *Catharanthus roseus, Clerodendrum philippinum.*

(ii) Maintenance : Excellent.

(iv) Ethnobotanical

Sacred : Flowers of *Clerodendrum philippinum* used as special offering to the deity in the temple. *Ficus benghalensis, Ficus religiosa, Musa paradisiaca* and *Ocimum sanctum* worshipped as sacred plants. *Cynodon dactylon* employed in various religious ceremonies. Flowers of *Datura fastuosa* and leaves of *Cannabis sativa* considered sacred for offering to appease Lord Shiva. Wood of *Mangifera indica, Ficus benghalensis, Ficus religiosa*, and leaves of *Ficus glomerata* used in 'havan'.

Medicinal : Milky juice of *Ficus glomerata* applied to cure furuncles. Juice extracted from the leaves of *Roylea cinerea* used as nasal drops for headache and cold. Crushed seeds of *Syzygium cumini* taken with water daily to cure diabetes. Seed oil of *Sesamum orientale* used for body massage. Poultice of leaves and roots of *Boerhaavia diffusa* used to cure backache.

Edible : Fruits of *Celtis australis, Citrus jambhiri, Ficus palmata, Ficus roxburghii, Mangifera indica, Morus alba, Musa paradisiaca, Prunus puddum* and *Syzygium cumini* edible. Fruits of *Citrus limon* pickled. Leaves and underground corms of *Colocasia esculenta* and fruits of *Lycopersicum esculentum* used as vegetable. Leaves of *Mentha piperita* used as flavourants.

Ornamental : *Albizia lebbeck, Boerhaavia diffusa, Canna indica, Catharanthus roseus, Cestrum nocturnum, Datura fastuosa, Helianthus annuus, Hibiscus syriacus, Ocimum sanctum, Quisqualis indica, Sapium insigne, Tagetes erecta, Thevetia neriifolia* and *Zinnia elegans* grown as ornamental plants.

Other Uses : Wood of *Albizia lebbeck, Celtis australis, Salix tetrasperma* and *Syzygium cumini* used for making agricultural implements. Wood of *Cedrela toona* and *Morus alba* considered excellent for furniture. Leaves of *Albizia lebbeck, Celtis australis, Ficus palmata, Ficus roxburghii, Morus alba* and *Prunus puddum* used as fodder. Roughage obtained after the extraction of seed oil of *Sesamum orientale* used as excellent cattlefeed.

Remarks : No sacrifices made to deity. An old cave called 'Vyas cave' and 'Hawan Kund' exist in the temple courtyard.

7. Naina Devi Temple (Pl. 1G)

Main Deity : Naina Devi.

Other Deities : Kali, Bhagatad Dev, Bhairav, Bajrang Bali, Shiva.

Tot. Ar. of Tem. Courtyard : 800 sq m.

Location

(i) Place

Village : Kharota

Block : Brahmpukhar

Tehsil : Bilaspur

District : Bilaspur

(ii) Approach : 2 km from Namhol on Shimla-Hamirpur road.

(iii) Altitude : 880 m.

Management : Temple committee.

Historical Background : The temple is 30 years old and built by locals.

Temple Architect : Sikhara style.

Offerings to God

(i) By Priest : Rice, Dal, Chapati, Vegetables (prepared without the use of onion and garlic).

(ii) By People : 'Lachidana', Puffed rice, 'Halva'.

Off. to People (Prasadam) : 'Lachidana', Puffed rice.

Fairs and Festivals : Navratra.

Observations

(i) Courtyard Plants (Protected as well as for Landscaping : *Callistemon lanceolatus, Cestrum nocturnum, Citrus aurantifolia, Cynodon dactylon, Ficus religiosa, Jasminum primulinum, Lagerstroemia indica, Mangifera indica, Pinus roxburghii, Pistacia integerrima, Psidium guajava, Rosa indica, Salvia coccinia, Tagetes erecta, Tecoma stans, Thuja orientalis.*

(ii) Predominant Plants of the Courtyard	:	*Jasminum primulinum.*
(ii) Maintenance	:	Satisfactory.
(iv) Ethnobotanical		
Sacred	:	*Ficus religiosa* worshipped as incarnation of Lord Brahma. *Cynodon dactylon* considered sacred and employed in various religious ceremonies.
Medicinal	:	2-3g powdered dehusked seeds of *Citrus aurantifolia* given daily with warm water for typhoid. Fruit juice of *Mangifera indica* considered aphrodisiac.
Edible	:	Fruits of *Mangifera indica* and *Psidium guajava* edible and those of *Citrus aurantifolia* are pickled. Seeds of *Pinus roxburghii* are edible.
Ornamental	:	*Callistemon lanceolatus, Cestrum nocturnum, Jasminum primulinum, Lagerstroemia indica, Pinus roxburghii, Rosa indica, Salvia coccinia, Tagetes erecta, Tecoma stans* and *Thuja orientalis* grown as ornamentals.
Other Uses	:	Cones of *Pinus roxburghii* used for decorative purposes.
Remarks	:	Goats offered alive to the deity by the people whose wishes are fulfilled. Later, these are sold and the money obtained is saved in the account of the deity. 'Havan' ceremony is held on every 'Ashtmi'.

8. Naina Devi Temple (Pl. 2A, B)

Main Deity	:	Naina Devi.
Other Deities	:	Bhairav, Hanuman, Safed Batuk Bhairav, Kali, Shiva.
Tot. Ar. of Tem. Courtyard	:	About 3000 sq m.
Location		
(i) Place		
Village	:	Naina Devi
Block	:	Swar Ghat
Tehsil	:	Naina Devi
District	:	Bilaspur
(ii) Approach	:	The temple, situated on the top of Nainadhar, is about 57 km from Bilaspur and 25 km from Swarghat.
(iii) Altitude	:	1200 m.
Management	:	Temple trust (Government undertaken).
Historical Background	:	This temple though built by Pandvas was later renovated by Raja Bir Chand (King of Kehloor Riyasat)

and regarded the deity as his 'Kuldevi'. According to another prevalent legend Lord Shiva after the self-immolation of Sati carried the body of his consort over his shoulder roaming aimlessly. In order to pacify the Lord, Shani and Indra stealthily entered the body of Sati and dismembered it into pieces which fell at various places. One of the eyes of Sati was believed to fell at this place, earning it the name of Naina Devi.

Temple Architect : Sikhara style.

Offerings to God

(i) By Priest

Morning : (*i*) 'Mohan Bhog' (4 AM), comprising Sugar candy and 'Panchmeva' (five dry fruits like Cashewnut, Almond, Coconut, Dried palm fruit, currant, apricot, etc.; (*ii*) 'Bal Bhog' (6 AM)–'Halva', Sweets.

Noon : (*i*) 'Raj Bhog' (12:00 Noon)-'Kheer'.

Evening : (*i*) 'Shyam Bhog' (6:30 PM)-Gram, 'Puri'; (*ii*) 'Dugad Bhog' (9:30 PM)-Milk, Sweets.

(ii) By People : Puffed rice, 'Halva', Coconut, Sweets.

Off. to People (Prasadam) : Puffed rice, 'Halva'.

Fairs and Festivals : Navratra fairs held in Chaitra, Shravan and Ashwin. Shivratri is also celebrated enthusiastically.

Observations

(i) Courtyard Plants (Protected as well as for Landscaping : *Cynodon dactylon, Ficus religiosa, Hibiscus rosa-sinensis.*

(ii) Maintenance : Excellent.

(iii) Ethnobotanical

Sacred and Associated Myth : Tree of *Ficus religiosa,* flowers of *Hibiscus rosa-sinensis* and the plant of *Cynodon dactylon* considered sacred. People tie scarves and sacred thread to the branches of *Ficus religiosa* for the fulfilment of their wishes.

Medicinal : Decoction of leaves of *Ficus religiosa* prescribed for fever.

Ornamental : *Hibiscus rosa-sinensis.*

Remarks : Goats offered alive.

Long ago, Guru Gobind Singh established a 'havan kund' in the temple for religious offerings. Since then, 'havan' ceremony is held daily. Ash of this 'havan' is distributed as sacred ash ('vibhuti') to the devotees.

9. Naina Devi Temple (Pl. 2C)

Main Deity : Naina Devi.

Other Deities : Shiva, Parvati, Duttatrey, Ganesha, Baba Balak Nath, Markandey, Brahma (in the form of a Peepal tree), Bhairav, Hanuman.

Tot. Ar. of Tem. Courtyard : 3000 sq m.

Location

(i) Place

Village : Nihar

Block : Brahmpukhar

Tehsil : Bilaspur

District : Bilaspur

(ii) Approach : 1 km from Brahmpukhar.

(iii) Altitude : 850 m.

Management : Local committee.

Historical Background : Guru Rama Shanker Giri (Priest of the temple) meditated at this place and constructed the temple in 1988 with the help of locals.

Temple Architect : Sikhara style.

Offerings to God

(i) By Priest : 'Halva', 'Kheer'.

(ii) By People : 'Halva', Puffed rice, 'Lachidana', Sweets.

Off. to People (Prasadam) : 'Lachidana', Puffed rice.

Fairs and Festivals : Navratra, Shivratri.

Observations

(i) Courtyard Plants (Protected as well as for Landscaping : *Aegle marmelos, Bougainvillea spectabilis, Calendula officinalis, Cannabis sativa, Canna indica, Carica papaya, Catharanthus roseus, Cestrum nocturnum, Cynodon dactylon, Datura fastuosa, Ficus religiosa, Grevillea robusta, Hibiscus mutabilis, Hibiscus rosa-sinensis, Hibiscus syriacus, Jasminum grandiflorum, Lagerstroemia indica, Mangifera indica, Musa paradisiaca, Nerium odorum, Prunus persica, Psidium guajava, Rosa indica, Tecoma stans, Thevetia neriifolia.*

(ii) Predominant Plants of the Courtyard : *Mangifera indica.*

(ii) Maintenance : Excellent.

(iv) Ethnobotanical

Sacred : *Ficus religiosa* and *Cynodon dactylon* considered sacred. Leaves of *Aegle marmelos* ('Bael-Patri'), *Cannabis sativa* ('Bhang') and flowers of *Datura fastuosa* ('Dhatura') especially offered to appease Lord Shiva. Wood of *Mangifera indica* used in 'havan' alongwith seeds of *Sesamum orientale* ('Til'), grains of *Hordeum vulgare* ('Jou'), *Oryza sativa* (Rice) and wood of *Acacia catechu* ('Khair').

Medicinal : Candy ('Murabba') of fruits of *Aegle marmelos* given to cure dysentery. Seeds of *Cannabis sativa* prescribed to cure mental disorders.

Edible : Fruits of *Aegle marmelos, Carica papaya, Mangifera indica, Musa paradisiaca, Prunus persica* and *Psidium guajava* edible.

Ornamental : *Bougainvillea spectabilis, Calendula officinalis, Canna indica, Catharanthus roseus, Cestrum nocturnum, Grevillea robusta, Hibiscus mutabilis, Hibiscus rosa-sinensis, Hibiscus syriacus, Jasminum grandiflorum, Lagerstroemia indica, Nerium odorum, Rosa indica, Tecoma stans* and *Thevetia neriifolia* grown as ornamental plants.

10. Nav Durga Temple (Pl. 2D)

Main Deity : Nav Durga (9 forms of Durga).

Other Deities : Ganesha, Bajrangbali, Bhairav, Mahakali, Baba Balak Nath, Lakshmi Narayan, Radha-Krishana, Ram-Sita-Lakshman, Shiva, Shitla Mata and Shani Dev.

Tot. Ar. of Tem. Courtyard : 10000 sq m.

Location

(i) Place

Village : Kharota

Block : Jhanduta

Tehsil : Jhanduta

District : Bilaspur

(ii) Approach : 12 km from Ghumarwin and 20 km from Shah Talai.

(iii) Altitude : 600 m.

Management : Temple committee.

Historical Background : The temple was built in 1964 by Mahatma Mahinder ji Maharaj with the help of locals and different devotees.

Temple Architect : Sikhara style.

Offerings to God

(i) By Priest : 'Halva', Coconut, Currant, Dal-Rice, Chapati-Vegetable, 'Puri-Chana', Milk.

(ii) By People : Coconut, 'Halva', Fruits.

Off. to People (Prasadam) : 'Halva'.

Fairs and Festivals : 'Jagran' is held every year in April-May.

Observations

(i) Courtyard Plants (Protected as well as for Landscaping : *Aegle marmelos, Albizia lebbeck, Bougainvillea glabra, Broussonetia papyrifera, Bryophyllum calycinum, Cannabis sativa, Canna indica, Capsicum annuum, Catharanthus roseus, Cedrela toona, Cestrum nocturnum, Chrysanthemum indicum, Citrus aurantifolia, Citrus limon, Citrus reticulata, Cupressus sempervirens, Cynodon dactylon, Dalbergia sissoo, Datura fastuosa, Ficus religiosa, Grevillea robusta, Hibiscus rosa-sinensis, Ipomoea cairica, Jasminum primulinum, Jasminum sambac, Lagerstroemia indica, Leucaena leucocephala, Litchi chinensis, Mallotus philippinensis, Malvaviscus conzatii, Mangifera indica, Melia azadirachta, Mirabilis jalapa, Morus alba, Murraya koenigii, Murraya paniculata, Musa paradisiaca, Nerium odorum, Ocimum sanctum, Prunus persica, Psidium guajava, Quisqualis indica, Rosa indica, Tabernaemontana coronaria, Tagetes erecta, Tecoma radicans, Tectona grandis, Thevetia neriifolia, Zephyranthes candida.*

(ii) Predominant Plants of the Courtyard : *Tagetes erecta, Tabernaemontana coronaria, Mangifera indica, Grevillea robusta.*

(ii) Maintenance : Excellent.

(iv) Ethnobotanical

Sacred and Associated Myth : *Ficus religiosa, Ocimum sanctum* and *Musa paradisiaca* worshipped as sacred plants. Flowers of *Rosa indica* and *Tagetes erecta* specially offered to appease Durga. Leaves of *Aegle marmelos, Cannabis sativa* and flowers of *Datura fastuosa* and *Thevetia neriifolia* offered to Lord Shiva. Wood of *Mangifera indica* used in 'havan'. *Cynodon dactylon* employed in all religious ceremonies as a sacred plant. Newly married couples tie bangles to the mango tree (*Mangifera indica*) in the temple courtyard for prosperous married life.

Medicinal : Leaves of *Bryophyllum calycinum* tied to cure furuncle. seed powder of *Mallotus philippinensis* considered as blood purifier. 2-5g 'Ajvain' (*Trachyspermum ammi*) powdered with equal quantity of sugar candy ('mishri') prescribed with water in which 10-12 leaves of *Melia azadirachta* ('Neem') are kept overnight in a Copper vessel for gastric. Decoction of leaves of *Melia azadirachta* also given for diabetes.

Edible : Fruits of *Aegle marmelos, Citrus reticulata, Litchi chinensis, Mangifera indica, Morus alba, Musa paradisiaca, Prunus persica* and *Psidium guajava* are edible whereas those of *Citrus aurantifolia, Citrus limon* and *Capsicum annuum* are pickled.

Ornamental : *Albizia lebbeck, Bougainvillea glabra, Bryophyllum calycinum, Canna indica, Catharanthus roseus, Cestrum nocturnum, Chrysanthemum indicum, Cupressus sempervirens, Datura fastuosa, Grevillea robusta, Hibiscus rosa-sinensis, Ipomoea cairica, Jasminum primulinum, Jasminum sambac, Lagerstroemia indica, Malvaviscus conzatii, Melia azadirachta, Mirabilis jalapa, Murraya koenigii, Murraya paniculata, Nerium odorum, Ocimum sanctum, Quisqualis indica, Rosa indica, Tabernaemontana coronaria, Tagetes erecta, Tecoma radicans, Tectona grandis, Thevetia neriifolia* and *Zephyranthes grandiflora.*

Other Uses : Leaves of *Albizia lebbeck, Broussonetia papyrifera, Leucaena leucocephala* and *Morus alba* used as fodder. Wood of *Cedrela toona, Dalbergia sissoo* considered excellent for furniture and construction work, and that of *Mallotus philippinensis* for agricultural implements. Branches of *Melia azadirachta* and *Murraya koenigii* used as tooth brush.

Remarks : Use of onion (*Allium cepa*), garlic (*Allium sativum*) and masur (*Lens culinaris*) avoided in preparations offered to deity. No sacrifices made here. 'Havan' ceremony held daily. Holy fire ('akhand jyoti') burns continuously in the temple.

11. Shiv Temple or Shiv Baudi (Pl. 2E)

Main Deity : Lord Shiva.

Other Deities : Parvati, Ganesha.

Tot. Ar. of Tem. Courtyard : 2800 sq m.

Location

(i) Place

Village : Bachhretu

Block : Jhanduta

Tehsil : Jhanduta

District	:	Bilaspur
(ii) Approach	:	5 km from Shah Talai.
(iii) Altitude	:	700 m.
Management	:	Temple trust appointed by the Government.
Historical Background	:	In 9th century, Baba Visheshanand meditated at this place and it is believed that holy Ganges appeared here at that time and still present in the form of a small water reservoir ('baudi') in the temple courtyard. The temple was built by Raja Bir Chand.
Temple Architect	:	Sikhara style.
Offerings to God		
(i) By Priest	:	'Halva', Sweets.
(ii) By People	:	Puffed rice, 'Lachidana', 'Halva'.
Off. to People (Prasadam)	:	Puffed rice, 'Halva'.
Fairs and Festivals	:	Bachhretu fair, Shivratri.
Observations		
(i) Courtyard Plants (Protected as well as for Landscaping	:	*Cannabis sativa, Catharanthus roseus, Cynodon dactylon, Datura fastuosa, Ficus benghalensis, Mangifera indica, Ocimum sanctum, Tagetes erecta, Thuja orientalis.*
(ii) Predominant Plants of the Courtyard	:	*Catharanthus roseus, Mangifera indica.*
(ii) Maintenance	:	Satisfactory.
(iv) Ethnobotanical		
Sacred	:	*Ficus benghalensis* and *Ocimum sanctum* worshipped as sacred plants. *Cannabis sativa* and *Datura fastuosa* specially offered to Lord Shiva. Leaves of *Cannabis sativa* also employed to make 'Ghota' (a drink) during Shivratri. *Cynodon dactylon* considered sacred plant and employed in all religious ceremonies. Leaves of *Aegle marmelos* also offered to appease Lord Shiva.
Medicinal	:	Poultice of leaves of *Cannabis sativa* good for insect bites.
Edible	:	Fruits of *Mangifera indica* edible.
Ornamental	:	*Catharanthus roseus, Datura fastuosa, Ocimum sanctum, Tagetes erecta* and *Thuja orientalis* grown as ornamental plants.
Remarks	:	'Havan' for purification performed twice daily. A mixture of milk, curd, honey, clarified butter, loaf sugar and different flowers used for bathing Shivlingam everyday. No sacrifices made to deity.

Temples of District Chamba

1. Baba Lakhdata Pir Temple (Pl. 3A)

Main Deity : Lakhdata Pir.

Tot. Ar. of Tem. Courtyard : About 3000 sq m.

Location

(i) Place

Village : Dundiara

Block : Chawari

Tehsil : Dalhousie

District : Chamba

(ii) Approach : About 8 km from Banikhet on Chamba-Kangra road.

(iii) Altitude : 1500 m.

Management : Privately owned.

Historical Background : According to the local legend, about 100 years before, some foreigners came here and tried to misuse this place. Pir Baba appeared in the dream of one of them and warned to leave the place for which they did not pay any heed. Pir Baba then sent large number of snakes ('Nag') which ultimately forced them to leave the place. Since then the local people worship this place. The present temple structure is about 30 years old.

Temple Architect : Sikhara style.

Offerings to God

(i) *By Priest* : Sweet rice (cooked).

(ii) *By People* : 'Lachidana', 'Halva', Sweets.

Off. to People (Prasadam) : 'Lachidana'.

Fairs and Festivals : Two fairs are held in the month of September.

Observations

(i) *Courtyard Plants (Protected as well as for Landscaping* : *Bombax ceiba, Celtis australis, Cynodon dactylon, Phoenix sylvestris, Pyrus pashia.*

(ii) *Predominant Plants of the Courtyard* : *Phoenix sylvestris.*

(ii) *Maintenance* : Satisfactory.

(iv) *Ethnobotanical*

Sacred : *Phoenix sylvestris* considered sacred and worshipped.

Medicinal : Powdered roots of young plant of *Bombax ceiba* prescribed with milk to restore the sexual potency. Powdered fruits of *Pyrus pashia* prescribed for tonsils and other throat problems.

Edible : Fruits of *Celtis australis, Phoenix sylvestris* and *Pyrus pashia* edible.

Other Uses : Wood of *Bombax ceiba* used for making match-sticks. Cotton used for stuffing pillows and mattresses. *Cynodon dactylon* used as fodder.

Status of Courtyard Plants in Surrounding Areas : *Pyrus pashia* and *Celtis australis* are not common in the surrounding areas of the temple.

Remarks : Goats and Cocks sacrificed. All the courtyard trees considered sacred and hence protected.

2. Bansi Gopal Temple (Pl. 3B)

Main Deity : Lord Krishana.

Other Deity : Shiva.

Tot. Ar. of Tem. Courtyard : About 800 sq m.

Location

(i) *Place*

Town : Chamba

Block : Chamba

Tehsil : Chamba

District : Chamba

(ii) Approach	:	The temple is located in the vicinity of the Chamba palace.
(iii) Altitude	:	980 m.
Management	:	Temple falls under control of Archaeological Survey of India. Moreover, local temple committee also exists.
Historical Background	:	The temple dates back to 1595 A.D.
Temple Architect	:	Sikhara style.
Offerings to God		
(i) By Priest		
Noon	:	Dal-Rice.
Evening	:	'Halva-Puri'.
(ii) By People	:	'Halva', 'Lachidana', Puffed rice.
Off. to People (Prasadam)	:	'Halva', 'Lachidana', Puffed rice.
Fairs and Festivals	:	Janamashtmi, Shivratri.
Observations		
(i) Courtyard Plants (Protected as well as for Landscaping	:	*Ocimum sanctum.*
(ii) Maintenance	:	Satisfactory.
(iii) Ethnobotanical		
Sacred	:	*Ocimum sanctum* worshipped as sacred plant. Its leaves added for preparing 'Charnamat' (holy water possessing healing property).
Flavouring	:	Leaves of *Ocimum sanctum* added for flavouring tea.
Remark	:	No sacrifices are offered.

3. Bhalai Mata Temple (Pl. 3C)

Main Deity	:	Bhadrakali (Bhalai Mata).
Tot. Ar. of Tem. Courtyard	:	About 3000 sq m.
Location		
(i) Place		
Village	:	Bhalai
Block	:	Saluni
Tehsil	:	Saluni
District	:	Chamba
(ii) Approach	:	About 30 km from Banikhet.
(iii) Altitude	:	1030 m.

Management	:	Village temple committee.
Historical Background	:	The temple, believed to be about 500 years old, had been built by Raja Pratap Singh (King of Chamba) and named after the village. Long ago, Goddess Bhadrakali appeared in the dream of Raja Pratap Singh and gave him the direction for unearthing her idol lying buried at a place. Later it was unearthed and temple was built here by the king.
Temple Architect	:	Sikhara style.
Offerings to God		
(i) By Priest		
Morning	:	'Halva-Puri', Gram.
Evening and Noon	:	Dal-Rice, Chapati-Vegetables, 'Kheer'.
(ii) By People	:	Puffed rice, 'Lachidana', Coconut, Fruits, Sweets.
Off. to People (Prasadam)	:	Puffed rice, 'Lachidana'.
Fairs and Festivals	:	Navratra, and fair called 'Chhinj' organised in September.
Observations		
(i) Courtyard Plants (Protected as well as for Landscaping	:	*Bougainvillea spectabilis, Callistemon lanceolatus, Canna indica, Catharanthus roseus, Chrysanthemum indicum, Citrus aurantifolia, Citrus limon, Citrus reticulata, Cynodon dactylon, Dodonaea viscosa, Eucalyptus tereticornis, Ficus religiosa, Grevillea robusta, Hibiscus syriacus, Jasminum primulinum, Lagerstroemia indica, Mangifera indica, Melia azedarach, Morus alba, Narcissus tazetta, Pelargonium hortorum, Pinus roxburghii, Pistacia integerrima, Prunus armeniaca, Prunus persica, Psidium guajava, Robinia pseudoacacia, Rosa indica, Salix babylonica, Tabernaemontana coronaria, Tagetes erecta, Tecoma radicans, Thuja orientalis, Zephyranthes grandiflora.*
(ii) Predominant Plants of the Courtyard	:	*Callistemon lanceolatus, Thuja orientalis.*
(ii) Maintenance	:	Excellent.
(iv) Ethnobotanical		
Sacred and Associated Myth	:	Wood and leaves of *Ficus religiosa* and *Mangifera indica* used in 'havan' ceremony. *Ficus religiosa* worshipped as a sacred tree. It is believed that 84 deities reside on this tree. *Cynodon dactylon* considered sacred and employed in all religious ceremonies.

Medicinal : Fruits of *Pistacia integerrima* used for cough. 3-5g powdered seeds of *Melia azedarach* given daily on empty stomach to cure piles. Leaf juice of *Prunus persica* used for expelling intestinal worms.

Edible : Fruits of *Citrus aurantifolia* and *Citrus limon* pickled whereas those of *Citrus reticulata, Mangifera indica, Prunus armeniaca, Prunus persica, Psidium guajava* and *Morus alba* are edible. Seeds of *Pinus roxburghii* are also edible.

Ornamental : *Bougainvillea spectabilis, Callistemon lanceolatus, Canna indica, Catharanthus roseus, Chrysanthemum indicum, Dodonaea viscosa, Grevillea robusta, Hibiscus syriacus, Jasminum primulinum, Lagerstroemia indica, Melia azedarach, Narcissus tazetta, Pelargonium hortorum, Pinus roxburghii, Rosa indica, Salix babylonica, Tabernaemontana coronaria, Tagetes erecta, Tecoma radicans, Thuja orientalis, Zephyranthes grandiflora.*

Other Uses : Leaves of *Dodonaea viscosa, Morus alba* and *Robinia pseudoacacia* used as cattle feed. Fruits of *Citrus aurantifolia* used as facial cosmetics. Wood of *Eucalyptus tereticornis* and *Melia azedarach* employed for building works. Wood of *Melia azedarach* and *Robinia pseudoacacia* used for agricultural implements and that of *Eucalyptus tereticornis* and *Salix babylonica* for packing cases.

Remarks : No sacrifices made in the temple. Sheep and goats offered to the deity are taken back by the people to their places.

4. Bharmani Mata Temple (Pl. 3D)

Main Deity : Bharmani.

Tot. Ar. of Tem. Courtyard : About 1500 sq m.

Location

(i) Place

Village : Malkouta (Bharmani)

Block : Bharmour

Tehsil : Bharmour

District : Chamba

(ii) Approach : 3 km on foot from Bharmour.

(iii) Altitude : 2640 m.

Management	:	In the name of deity (Temple priest is the caretaker).
Historical Background	:	The original temple is very old but the new one was made by Mahant Shri Prem Narayan Giri in 1989.
Temple Architect	:	Pentroof style; 'Chir' (*Pinus roxburghii*) wood used in its construction.
Offerings to God		
(i) By Priest	:	Rice-Dal, Chapati-Vegetables.
(ii) By People	:	'Rot' (made of wheat or maize flour), 'Halva', Puffed rice.
Off. to People (Prasadam)	:	Puffed rice, 'Halva'.
Fairs and Festivals	:	Navratra and Local fair celebrated in September by women of the nearby villages.
Observations		
(i) Courtyard Plants (Protected as well as for Landscaping	:	*Mentha longifolia, Viola canescens.*
(ii) Maintenance	:	Satisfactory.
(iii) Ethnobotanical		
Sacred	:	Roots of *Viola canescens* used as incense ('Dhoop') in the temple.
Medicinal	:	Leaf juice of *Mentha longifolia* effective against stomach troubles.
Other Uses	:	Paste of *Mentha longifolia* applied inside the containers to protect grains from harmful insects.
Remarks	:	He-goats sacrificed by people on fulfilment of their desires.

5. Champawati Temple (Pl. 3E)

Main Deity	:	Champawati.
Other Deities	:	Shakti Mata, Nag Devta.
Tot. Ar. of Tem. Courtyard	:	About 800 sq m.
Location		
(i) Place		
Town	:	Chamba
Block	:	Chamba
Tehsil	:	Chamba
District	:	Chamba

(ii) Approach : The temple is located close to the Royal Palace.

(iii) Altitude : 950 m.

Management : The management of the temple is under Archaeological Survey of India. Besides, local temple committee also exists.

Historical Background : The temple was built in 10th century A.D. by King of Chamba. Champawati was the daughter of the king of Chamba (Sahil Varman). She used to visit the place of a saint for getting religious education. Local people suspected her relation with that saint for which the king decided to kill her. When she came to know about this, she committed suicide. Later, the king felt regretted and made this temple in her memory.

Temple Architect : Sikhara style.

Offerings to God

(i) By Priest : 'Halva-Puri'.

(ii) By People : Puffed rice, 'Lachidana'.

Off. to People (Prasadam) : Puffed rice, 'Lachidana'

Fairs and Festivals : Shivratri, Nag Panchami.

Observations

(i) Courtyard Plants (Protected as well as for Landscaping : *Tagetes erecta*.

(ii) Maintenance : Satisfactory.

(iii) Ethnobotanical

Sacred : Flowers of *Tagetes erecta* considered sacred for offering to the deity.

Medicinal : Fresh leaf juice of *Tagetes erecta* applied to cure pimples of face.

Remark : It is believed that the town Chamba is named after the name of deity, Champawati.

6. Chamunda Devi Temple (Pl. 3F)

Main Deity : Chamunda Devi.

Other Deity : Shiva.

Tot. Ar. of Tem. Courtyard : About 3000 sq m.

Location		
(i) Place		
Town	:	Chamba
Block	:	Chamba
Tehsil	:	Chamba
District	:	Chamba
(ii) Approach	:	About 3 km from the town.
(iii) Altitude	:	1080 m.
Management	:	The management of temple is under Archaeological Survey of India and the local temple committee.
Historical Background	:	The temple was built in 18th century A.D. It is said that the temple was in existence before the foundation of Chamba was laid by Raja Sahil Varman. Later, the new temple was built at its place.
Temple Architect	:	The temple architect is of Pentroof style. Deodar wood is used in its construction.
Offerings to God		
(i) By Priest	:	'Halva-Puri', Black gram offered in the morning.
(ii) By People	:	'Halva-puri', Gram, Puffed rice, 'Lachidana', Fruits.
Off. to People (Prasadam)	:	'Halva-Puri', Gram, Puffed rice, 'Lachidana' and Fruits.
Fairs and Festivals	:	Fair called 'Jaatar' is held in April (Baisakh).
Observations		
(i) Courtyard Plants (Protected as well as for Landscaping	:	*Eucalyptus tereticornis, Ficus religiosa, Ocimum sanctum, Tagetes erecta.*
(ii) Predominant Plants of the Courtyard	:	*Ficus religiosa.*
(ii) Maintenance	:	Excellent.
(iv) Ethnobotanical		
Sacred	:	Flowers of *Tagetes erecta* considered sacred for offering to the deity. *Ficus religiosa* and *Ocimum sanctum* worshipped as sacred plants.
Medicinal	:	Small branches of *Ficus religiosa* chewed to cure toothache. Leaf juice of *Ocimum sanctum* considered good for cold.
Ornamental	:	*Ocimum sanctum* and *Tagetes erecta* grown as ornamentals.

Other Use : Wood of *Eucalyptus tereticornis* used for construction work.

Remarks : The temple, dedicated to Goddess Chamunda, stands on the spur of a hill at a high point and commands a beautiful view of the river Ravi and Chamba town below. All the courtyard plants are strictly protected.

7. Chaurasi Temple (Pl. 3G, H)

Main Deities : Chaurasi Temple includes a group of 84 temples, having 84 Sidh and 9 Nath. These mainly include Manimahesh (Shiva) temple, Narsingh temple, Dharamraj temple, Lakhna Mata temple, Ganesh temple, Chamunda temple, Hanuman temple, Shitla Mata temple and several other small temples.

Tot. Ar. of Tem. Courtyard : About 4600 sq m.

Location

(i) Place

Town : Bharmour

Block : Bharmour

Tehsil : Bharmour

District : Chamba

(ii) Approach : Temple is located in the town.

(iii) Altitude : 2150 m.

Management : Ganesh temple, Manimahesh temple, Narsingh temple and Lakhna Mata temple are under Archaeological Survey of India. Dharamraj temple is under temple trust. Other temples are in the name of the deity.

Historical Background : Ganesha and Narsingh temples were erected during the 10th or 11th century A.D. whereas other temples were erected during the reign of Raja Meru Varman in the 7th century A.D.

Temple Architect : Most of the temples are in Pentroof style. Manimahesh temple is in Sikhara style.

Offerings to God

(i) By Priest

Morning : Rice with clarified butter.

Evening : Chapati-Vegetables.

(ii) By People : 'Halva-Puri', 'Rot' (made of maize flour, clarified butter, jaggery, dry fruits), Puffed rice, 'Lachidana'. Bil-patri

		(leaves of *Aegle marmelos*) offered to Lord Shiva and Dharamraj as special offering.
Off. to People (Prasadam)	:	'Halva-Puri', 'Rot', Puffed rice, 'Lachidana'.
Fairs and Festivals	:	Shivratri, Janamashtmi, Radha-ashtmi, Holi, and fair called 'Jaatar' held from August-September for 15 days during which journey for Manimahesh is organised.
Observations		
(i) Courtyard Plants (Protected as well as for Landscaping	:	*Althaea rosea, Cedrus deodara, Chrysanthemum indicum, Cupressus torulosa, Dahlia rosea, Juglans regia, Morus serrata, Ocimum sanctum, Prunus armeniaca, Pyrus malus, Tagetes erecta.*
(ii) Predominant Plants of the Courtyard	:	*Morus serrata.*
(iii) Maintenance	:	Excellent.
(iv) Ethnobotanical		
Sacred and Associated Myth	:	*Ocimum sanctum* and *Cedrus deodara* worshipped as sacred plants. A small piece of wood of *Morus serrata* growing in the temple courtyard, used symbolically for 'Chandan' (*Santalum album*) for burning dead bodies. One of the branches of 'Deodar' tree (*Cedrus deodara*) overarches the Manimahesh temple and shaped-like 'Nag' of Shiv, and believed to protect Lord Shiva from rain and sunshine.
Medicinal	:	Kernels of *Juglans regia* prescribed to lower the cholesterol level.
Edible	:	Fruits of *Morus serrata, Pyrus malus, Prunus armeniaca* and kernels of *Juglans regia* are edible.
Ornamental	:	*Althaea rosea, Chrysanthemum indicum, Cupressus torulosa, Dahlia rosea, Morus serrata, Ocimum sanctum, Tagetes erecta* grown as ornamental plants.
Other Uses	:	Wood of *Cedrus deodara* excellent for furniture and construction work.
Remarks	:	Being a group of 84 temples, this temple is only one of its kind in H.P. Sheep and goats sacrificed only in Lakhna temple. Use of onion and garlic avoided strictly in preparations for offerings to deities.

8. Hari Rai Temple

Main Deity	:	Lord Vishnu.
Other Deity	:	Gauri-Shanker.
Tot. Ar. of Tem. Courtyard	:	About 800 sq m.
Location		
(i) Place		
Town	:	Chamba
Block	:	Chamba
Tehsil	:	Chamba
District	:	Chamba
(ii) Approach	:	The temple is located in the town near Chougan.
(iii) Altitude	:	950 m.
Management	:	The temple is under Archeological Survey of India. Local temple committee also exists.
Historical Background	:	The temple was built in 11th century A.D. during the reign of Raja Ashtavarman by Prince Lakshman Varman.
Temple Architect	:	Sikhara style.
Offerings to God		
(i) By Priest	:	'Halva-Puri'.
(ii) By People	:	Puffed rice, 'Lachidana', Sweets, Fruits.
Off. to People (Prasadam)	:	Puffed rice, 'Lachidana'
Fairs and Festivals	:	Janamashtmi, Rath-Rathni fair held in September.
Observations		
(i) Courtyard Plants (Protected as well as for Landscaping	:	*Canna indica, Ficus religiosa, Jasminum primulinum, Lagerstroemia indica, Mirabilis jalapa, Nerium odorum, Ocimum sanctum, Rosa indica, Tagetes erecta, Tecoma radicans, Thuja orientalis.*
(ii) Predominant Plants of the Courtyard	:	*Tecoma radicans.*
(iii) Maintenance	:	Satisfactory.
(iv) Ethnobotanical		
Sacred	:	*Ficus religiosa* and *Ocimum sanctum* worshipped as sacred plants.
Medicinal	:	'Gulkand' prepared from petals of *Rosa indica* prescribed to cure constipation.

Ornamental : *Canna indica, Jasminum primulinum, Lagerstroemia indica, Mirabilis jalapa, Nerium odorum, Ocimum sanctum, Rosa indica, Tagetes erecta, Tecoma radicans* and *Thuja orientalis* grown as ornamentals.

9. Khajji Nag Temple (Pl. 3I)

Main Deity : Khajji Nag.

Other Deities : Pandavas, Hidimba Mata, Shiva.

Tot. Ar. of Tem. Courtyard : About 1500 sq m.

Location

(i) *Place*

Town : Khajiar

Block : Chamba

Tehsil : Chamba

District : Chamba

(ii) *Approach* : This fascinating place is 26 km from Dalhousie and 22 km from Chamba. Half way between Chamba and Dalhousie, there is a small lake with a floating island in the centre. The temple dedicated to Khajji Nag lies close to the lake.

(iii) *Altitude* : 1920 m.

Management : Temple committee.

Historical Background : The architecture of the original wooden temple of Khajji Nag dates back to the period earlier to twelfth century A.D. In the sixteenth century A.D., Raja Balbhadra Barman erected wooden statues of Pandavas in the temple. Later, it was renovated by Baltu, the religious nurse (Dai) of Raja Prithvi Singh. The place Khajiar is named after the deity Khajji Nag.

Temple Architect : Pentroof style.

Offerings to God

(i) *By Priest* : The offerings known as Sanj ('Halva-Puri') offered to deities in the morning.

(ii) *By People* : Puffed rice, 'Lachidana', Fruits.

Off. to People (Prasadam) : Puffed rice, 'Lachidana'.

Fairs and Festivals : Shivratri.

Observations

(i) *Courtyard Plants (Protected as well as for Landscaping* : *Canna indica, Cynodon dactylon, Dahlia rosea, Lonicera japonica, Narcissus tazetta, Tagetes erecta, Thuja orientalis.*

(ii) *Predominant Plants of the Courtyard* : *Lonicera japonica.*

(iii) *Maintenance* : Satisfactory.

(iv) *Ethnobotanical*

Sacred : Flowers of courtyard plants used for worshipping the deity. *Cynodon dactylon* considered sacred and employed in all religious ceremonies.

Medicinal : Leaf juice of *Tagetes erecta* applied to cure insect bites.

Remarks : Sheep and goats sacrificed to appease the deity.

10. Lakshmi Narayan Temple (Pl. 4A)

Main Deities : Lakshmi Narayan temple comprises a group of six main temples, namely Lakshmi Narayana, Radha-Krishana, Lakshmi Damodara, Chandergupta, Gauri Shanker and Trimukeshwara. Of these, three temples are devoted to Lord Vishnu and others to Lord Shiva.

Other Deities : Hanuman, Kali, Charpat Nath.

Tot. Ar. of Tem. Courtyard : About 3000 sq m.

Location

(i) *Place*

Town : Chamba

Block : Chamba

Tehsil : Chamba

District : Chamba

(ii) *Approach* : The temple is located in the town near Chougan.

(iii) *Altitude* : 950 m.

Management : The temple is under Archeological Survey of India. Local temple committee also exists.

Historical Background : The temples of Lakshmi Narayana and Chandergupta were built in the 10th-11th century A.D. by Raja Sahil Varma (Shail Varma), the founder of Chamba town. These temples were extensively repaired under the reign of Raja Pratap Singh Varma in the 16th century A.D. Later, the Radha-Krishana temple was constructed in 1828 A.D. by Rani Sharda, the widow of King Jit Singh of Chamba.

Temple Architect	:	Shikhar Style.
Offerings to God		
(i) By Priest		
Noon	:	Rice-Dal, 'Kheer'.
Evening	:	Puri-Vegetables.
(ii) By People	:	Puffed rice, 'Lachidana', 'Halva', Coconut. 'Bil-patri' (*Aegle marmelos*), 'Bhang' (*Cannabia sativa*) and 'Dhatura' (*Datura* sp.) offered to appease Lord Shiva.
Off. to People (Prasadam)	:	'Halva', 'Lachidana' Puffed rice.
Fairs and Festivals	:	Janamashtmi, Shivratri.
Observations		
(i) Courtyard Plants (Protected as well as for Landscaping	:	*Catharanthus roseus, Ficus religiosa, Hedychium coronarium, Jasminum grandiflorum, Jasminum humile, Mirabilis jalapa, Musa paradisiaca, Ocimum sanctum, Pelargonium hortorum, Rosa indica, Tagetes erecta, Thuja orientalis.*
(ii) Predominant Plants of the Courtyard	:	*Jasminum grandiflorum.*
(iii) Maintenance	:	Excellent.
(iv) Ethnobotanical		
Sacred	:	Flowers of *Catharanthus roseus, Jasminum grandiflorum, Jasminum humile, Mirabilis jalapa, Rosa indica* and *Tagetes erecta* considered sacred for offering to the deities. *Ficus religiosa, Ocimum sanctum* and *Musa paradisiaca* worshipped as sacred plants.
Medicinal	:	5g powdered roots of *Catharanthus roseus* given daily for diabetes.
Edible	:	Fruits of *Musa paradisiaca* are edible.
Ornamental	:	*Catharanthus roseus, Hedychium coronarium, Jasminum grandiflorum, Jasminum humile, Mirabilis jalapa, Ocimum sanctum, Pelargonium hortorum, Rosa indica, Tagetes erecta* and *Thuja orientalis* grown as ornamental plants.
Remark	:	No sacrifices are made.

11. Lakshmi Narayan Temple (Pl. 4B)

Main Deity	:	Lakshmi-Narayan.
Other Deities	:	Radha-Krishana, Durga Mata, Ram-Lakshman-Sita, Hanuman, Shiva.
Tot. Ar. of Tem. Courtyard	:	About 800 sq m.

Location	
(i) Place	
Town	: Dalhousie
Block	: Dalhousie
Tehsil	: Dalhousie
District	: Chamba
(ii) Approach	: The temple is located in Sadar Bazar.
(iii) Altitude	: 1950 m.
Management	: Local temple committee.
Historical Background	: The temple built by local people is about 150 years old.
Temple Architect	: Sikhara style.
Offerings to God	
(i) By Priest	: Dal-Rice.
(ii) By People	: Puffed rice, 'Lachidana', 'Halva-Puri', Fruits.
Off. to People (Prasadam)	: Puffed rice, 'Lachidana'.
Fairs and Festivals	: Baisakhi, Shivratri, Janamashtmi, Dussehra, Diwali, Ram-Navmi.
Observations	
(i) Courtyard Plants (Protected as well as for Landscaping	: A large-sized Akhrot tree (*Juglans regia*) is the only courtyard plant.
(ii) Maintenance	: Satisfactory.
(iii) Ethnobotanical	
Sacred	: The people worship Akhrot tree (*Juglans regia*) as sacred.
Medicinal	: Poultice of leaves of *Juglans regia* used for skin diseases.
Edible	: Kernels of *Juglans regia* are edible.
Other Uses	: Bark of *Juglans regia* used for scouring teeth and colouring the lips.
Status of Courtyard Plants in Surrounding Areas	: *Juglans regia* ('Akhrot') is absent in the surrounding areas of the temple.
Remarks	: The Akhrot tree is thought to be as old as the temple. No sacrifices are permitted.

12. Maha Kali Temple (Pl. 4C)

Main Deity	:	Kali Mata.
Other Deity	:	Hanuman.
Tot. Ar. of Tem. Courtyard	:	About 1500 sq m.
Location		
(i) *Place*		
Town	:	Chamba
Block	:	Chamba
Tehsil	:	Chamba
District	:	Chamba
(ii) *Approach*	:	Located at Police-line, Chamba.
(iii) *Altitude*	:	930 m.
Management	:	Police Department.
Historical Background	:	The temple was built by the employees of Police department about 20 years back.
Temple Architect	:	Sikhara style.
Offerings to God		
(i) *By Priest*	:	'Halva', Sweets, Fruits.
(ii) *By People*	:	Puffed rice, Coconut, Fruits.
Off. to People (Prasadam)	:	Puffed rice.
Fairs and Festivals	:	Navratra.
Observations		
(i) *Courtyard Plants (Protected as well as for Landscaping*	:	*Callistemon lanceolatus, Cedrus deodara, Grevillea robusta, Jacaranda acutifolia, Lagerstroemia indica, Ocimum basilicum, Pelargonium hortorum, Rosa indica, Tabernae-montana coronaria, Tagetes erecta, Tecomaria capensis.*
(ii) *Predominant Plants of the Courtyard*	:	*Tagetes erecta, Tecomeria capensis.*
(iii) *Maintenance*	:	Excellent.
(iv) *Ethnobotanical*		
Sacred	:	*Ocimum basilicum* worshipped as sacred plant.
Ornamental	:	*Callistemon lanceolatus, Grevillea robusta, Jacaranda acutifolia, Lagerstroemia indica, Ocimum basilicum, Pelargonium hortorum, Rosa indica, Tabernaemontana coronaria, Tagetes erecta* and *Tecomaria capensis* grown as ornamental plants.
Other Uses	:	Wood of *Cedrus deodara* excellent for furniture and construction work.

13. Narsingh Temple (Pl. 4D)

Main Deity : Narsingh.

Other Deities : Radha-Krishana, Ram-Sita, Sat Narayan, Gauri-Shanker, Ganesha, Hanuman.

Tot. Ar. of Tem. Courtyard : About 800 sq m.

Location

(i) Place

Town : Chamba

Block : Chamba

Tehsil : Chamba

District : Chamba

(ii) Approach : The temple is located in the town.

(iii) Altitude : 970 m.

Management : Privately managed.

Historical Background : The temple was built in 1832 by the astrologer (Jyotishi) of the King of Chamba.

Temple Architect : Sikhara style.

Offerings to God

(i) By Priest

Morning : 'Prantha' or Chapati, Vegetables.

Evening : Black Gram.

(ii) By People : Puffed rice, 'Lachidana', 'Halva', Fruits, Sweets.

Off. to People (Prasadam) : Puffed rice, 'Lachidana', 'Halva', Fruits, Sweets, Chestnut.

Fairs and Festivals : Janamashtmi, Shivratri.

Observations

(i) Courtyard Plants (Protected as well as for Landscaping : *Eriobotrya japonica, Ficus religiosa, Jasminum primulinum, Lonicera japonica, Melia azedarach, Mirabilis jalapa, Musa paradisiaca, Ricinus communis, Thuja orientalis.*

(ii) Predominant Plants of the Courtyard : *Mirabilis jalapa.*

(iii) Maintenance : Satisfactory.

(iv) Ethnobotanical

Sacred : *Ficus religiosa* and *Musa paradisiaca* worshipped as sacred plants.

Medicinal	: Seed oil of *Ricinus communis* given in milk to cure constipation. Leaf juice of *Melia azedarach* given for expelling intestinal worms.
Edible	: Fruits of *Eriobotrya japonica* and *Musa paradisiaca* are edible.
Ornamental	: *Jasminum primulinum, Lonicera japonica, Melia azedarach, Mirabilis jalapa* and *Thuja orientalis* grown as ornamentals.
Other Uses	: Wood of *Melia azedarach* used for building works and agricultural implements. Stem of *Ricinus communis* used for scouring teeth.

14. Raghunath Temple (Pl. 4E)

Main Deities	: Raghunath (Lord Rama), Sita, Lakshman.
Other Deities	: Hanuman, Shiva, Durga Mata.
Tot. Ar. of Tem. Courtyard	: About 800 sq m.
Location	
(i) Place	
Town	: Dalhousie
Block	: Dalhousie
Tehsil	: Dalhousie
District	: Chamba
(ii) Approach	: The temple is located in the town near Gandhi Chowk.
(iii) Altitude	: 2020 m.
Management	: Local temple committee.
Historical Background	: The temple, about 30 years old, had been built by locals with donations given by different devotees.
Temple Architect	: Sikhara style.
Offerings to God	
(i) By Priest	: Milk, Dal-Rice, Chapati-Vegetables.
(ii) By People	: Puffed rice, 'Lachidana', Fruits, Sweets.
Off. to People (Prasadam)	: Sugar candy, Puffed rice.
Fairs and Festivals	: Ram Navami, Shivratri, Navratra.
Observations	
(i) Courtyard Plants (Protected as well as for Landscaping	: *Bergenia ciliata, Cedrus deodara, Hydrangea macrophylla, Quercus incana, Tagetes erecta.*

(ii) Predominant Plants of the Courtyard	:	*Cedrus deodara.*
(iii) Maintenance	:	Satisfactory.
(iv) Ethnobotanical		
Sacred	:	Flowers of *Tagetes erecta* offered to the deity.
Medicinal	:	Paste of roots and leaves of *Bergenia ciliata* applied to heal wounds and also considered good against tooth pain. 2-5g powdered stem of *Bergenia ciliata* taken with milk after dinner for regaining lost sexual vigour.
Ornamental	:	*Bergenia ciliata, Hydrangea macrophylla,* and *Tagetes erecta* grown as ornamental plants.
Other Uses	:	Wood of *Cedrus deodara* excellent for furniture and construction works whereas that of *Quercus incana* used for agricultural implements and as fuel. Also the leaves of *Quercus incana* used as fodder.

15. Shitla Mata Temple (Pl. 4F)

Main Deity	:	Shitla Mata.
Other Deity	:	Durga Mata.
Tot. Ar. of Tem. Courtyard	:	About 750 sq m.
Location		
(i) Place		
Town	:	Chamba
Block	:	Chamba
Tehsil	:	Chamba
District	:	Chamba
(ii) Approach	:	Temple located near Shitla bridge.
(iii) Altitude	:	830 m.
Management	:	Privately owned.
Historical Background	:	The temple was built by Raja Bhuri Singh who donated it to the then priest of the temple.
Temple Architect	:	Sikhara style.
Offerings to God		
(i) By Priest	:	Sweets, Fruits.
(ii) By People	:	Wheat, Maize, Rice, Flour, 'Lachidana', Puffed rice, Fruits, Sweets.
Off. to People (Prasadam)	:	'Lachidana', Puffed rice.
Fairs and Festivals	:	Navratra.

Observations

(i) Courtyard Plants (Protected as well as for Landscaping	:	*Althaea rosea, Bougainvillea spectabilis, Callistemon lanceolatus, Canna indica, Cedrela toona, Celosia argentea* var. *cristata, Chrysanthemum indicum, Cupressus torulosa, Eriobotrya japonica, Ficus religiosa, Hibiscus mutabilis, Jasminum grandiflorum, Lagerstroemia indica, Mirabilis jalapa, Ocimum sanctum, Populus nigra, Salix babylonica, Tagetes erecta, Thuja orientalis.*
(ii) Predominant Plants of the Courtyard	:	*Mirabilis jalapa.*
(iii) Maintenance	:	Satisfactory.
(iv) Ethnobotanical		
Sacred	:	*Ficus religiosa* and *Ocimum sanctum* worshipped as sacred plants.
Edible	:	Fruits of *Eriobotrya japonica* are edible.
Ornamental	:	*Althaea rosea, Bougainvillea spectabilis, Callistemon lanceolatus, Canna indica, Celosia argentea* var. *cristata, Chrysanthemum indicum, Cupressus torulosa, Hibiscus mutabilis, Jasminum grandiflorum, Lagerstroemia indica, Mirabilis jalapa, Ocimum sanctum, Salix babylonica, Tagetes erecta* and *Thuja orientalis* grown as ornamentals.
Other Uses	:	Wood of *Cedrela toona* excellent for construction and furniture. Wood of *Salix babylonica* used for basketry and branches as fodder.
Remarks	:	Sheep and goats offered to appease deity.

16. Shiv-Shakti Temple or Chhatrari Temple (Pl. 4G)

Main Deity	:	Goddess Shiv-Shakti.
Other Deities	:	Shiva-Parvati (locally called Jogi-Jogan) and Mohar (person who built this temple).
Tot. Ar. of Tem. Courtyard	:	About 1500 sq m.
Location		
(i) Place		
Village	:	Chhatrari
Block	:	Mehla
Tehsil	:	Chamba
District	:	Chamba
(ii) Approach	:	8 km from Luna.
(iii) Altitude	:	1800 m.

Management : The management of temple is under Archaeological Survey of India. Village temple committee also exists.

Historical Background : The temple was built in 8th century A.D.

Temple Architect : Pentroof style. The temple is made of Deodar wood.

Offerings to God

(i) By Priest : Chapati, Clarified butter, Jaggery.

(ii) By People : Puffed rice, Sweets, Fruits.

Off. to People (Prasadam) : Puffed rice, Sweets, Fruits, Soaked gram.

Fairs and Festivals : Local fair starting from next day of Radha ashtmi held for 4 days.

Observations

(i) Courtyard Plants (Protected as well as for Landscaping : *Ipomoea purpurea, Morus serrata, Rosa indica, Tagetes erecta.*

(ii) Predominant Plants of the Courtyard : *Tagetes erecta.*

(iii) Maintenance : Excellent.

(iv) Ethnobotanical

Sacred : Flowers of *Rosa indica* and *Tagetes erecta* considered sacred for offering to the deity.

Medicinal : Leaf juice of *Ipomoea purpurea* prescribed for constipation.

Edible : Fruits of *Morus serrata* edible.

Ornamental : *Morus serrata, Rosa indica* and *Tagetes erecta* grown as ornamental plants.

Remarks : The entire temple is believed to be built from a single huge Deodar tree. Idol of deity (Shiv-Shakti) is made of Ashat dhatu (eight metals). Sheep and goats are regularly sacrificed to appease deities. Tree of *Morus serrata (*Kroon*)* in the courtyard is believed to be more than 100 years old and considered sacred.

17. Sita-Ram Temple (Pl. 4H)

Main Deity : Sita-Ram.

Other Deity : Hanuman.

Tot. Ar. of Tem. Courtyard : About 500 sq m.

Location		
(i) Place		
Town	:	Chamba
Block	:	Chamba
Tehsil	:	Chamba
District	:	Chamba
(ii) Approach	:	The temple is located in the town.
(iii) Altitude	:	980 m.
Management	:	The management of the temple falls under Archaeological Survey of India.
Historical Background	:	The temple is believed to be very old.
Temple Architect	:	It is a Sikhara type of temple built in stone.
Offerings to God		
(i) By Priest	:	'Halva-Puri', Rice-Dal.
(ii) By People	:	'Halva', 'Lachidana', Puffed rice.
Off. to People (Prasadam)	:	'Halva', 'Lachidana', Puffed rice.
Fairs and Festivals	:	Ram-Navami.
Observations		
(i) Courtyard Plants (Protected as well as for Landscaping	:	*Ocimum sanctum, Rosa indica.*
(ii) Maintenance	:	Satisfactory.
(iii) Ethnobotanical		
Sacred	:	*Ocimum sanctum* worshipped as sacred plant. Flowers of *Rosa indica* considered sacred for offering to the deity.
Medicinal	:	The entire plant of *Ocimum sanctum* considered anticancerous.
Ornamental	:	*Ocimum sanctum* and *Rosa indica* grown as ornamentals.

18. Suhi Mata Temple (Pl. 4I)

Main Deity	:	Suhi Mata.
Tot. Ar. of Tem. Courtyard	:	About 800 sq m.
Location		
(i) Place		
Town	:	Chamba
Block	:	Chamba

Tehsil	:	Chamba
District	:	Chamba
(ii) Approach	:	About 1km from Chamba town.
(iii) Altitude	:	1080 m.
Management	:	In the name of deity.
Historical Background	:	During the time of Raja Sahil Singh, there was once scarcity of water in the town. 'Kuldevi' appeared in the dream of king and told that this problem can be overcome if a member of his family sacrifices life. For this cause, the queen Sunaina took samadhi at Bhaloth in Chaitra month.
Temple Architect	:	Sikhara style.
Offerings to God		
(i) By Priest	:	'Halva-Puri'.
(ii) By People	:	'Halva-Puri', Puffed rice, 'Lachidana'.
Off. to People (Prasadam)	:	'Halva-Puri', Puffed rice, 'Lachidana'.
Fairs and Festivals	:	Fair called 'Fair of Women' is held from 15th of Chaitra to 30th of Chaitra in the memory of queen who sacrificed her life for the people of Chamba.
Observations		
(i) Courtyard Plants (Protected as well as for Landscaping	:	*Cestrum nocturnum, Mirabilis jalapa, Pinus roxburghii, Pistacia integerrima, Tagetes erecta.*
(ii) Predominant Plants of the Courtyard	:	*Mirabilis jalapa.*
(iii) Maintenance	:	Satisfactory.
(iv) Ethnobotanical		
Sacred	:	Flowers of *Mirabilis jalapa* and *Tagetes erecta* considered sacred for offering to the deity.
Medicinal	:	Juice of needles of *Pinus roxburghii* prescribed to cure tuberculosis.
Edible	:	Seeds of *Pinus roxburghii* are edible.
Ornamental	:	*Cestrum nocturnum, Mirabilis jalapa, Pinus roxburghii* and *Tagetes erecta* grown as ornamental plants.
Other Use	:	Cones of *Pinus roxburghii* used for decorative purposes.

Temples of District Hamirpur

1. Awah Devi Temple (Pl. 5A)

Main Deity	:	Awah Devi (Jalpa Mata).
Other Deities	:	Shiva, Gugga Pir, Baba Balak Nath, Hanuman, Ganesha.
Tot. Ar. of Tem. Courtyard	:	About 1500 sq m.
Location		
(i) Place		
Town	:	Awah Devi
Block	:	Bhoranj
Tehsil	:	Hamirpur
District	:	Hamirpur
(ii) Approach	:	24 km from Hamirpur and 14 km from Sarkaghat.
(iii) Altitude	:	1300 m.
Management	:	Local temple committee.
Historical Background	:	The original temple is believed to be very old. In 1971, it was renovated by Baba Sarvan Nath. Long ago, a farmer found a special type of stone ('Pindi') while ploughing his fields in Sangroh. The farmer was surprised to see the blood oozing out from it due to the impact of metal hull of plough. He decided to establish that 'pindi' there and started worshipping there. Later, it was shifted to

		the present temple. It is believed that Pandavas stayed here during their period of exile.
Temple Architect	:	Sikhara style.
Offerings to God		
(i) By Priest	:	Milk, Dry fruits, Seasonal fruits.
(ii) By People	:	Puffed rice, Fruits.
Off. to People (Prasadam)	:	Puffed rice.
Fairs and Festivals	:	Lohadi Fair, Gugga Navami (in August), Navratra.
Observations		
(i) Courtyard Plants (Protected as well as for Landscaping	:	*Bauhinia variegata, Bombax ceiba, Broussonetia papyrifera, Bryophyllum calycinum, Callistemon lanceolatus, Cannabis sativa, Canna indica, Catharanthus roseus, Cedrela toona, Chrysanthemum indicum, Citrus aurantifolia, Colocasia esculanta, Cynodon dactylon, Emblica officinalis, Ficus palmata, Ficus religiosa, Grevillea robusta, Grewia oppositifolia, Hibiscus rosa-sinensis, Hibiscus syriacus, Juglans regia, Litchi chinensis, Malvaviscus conzatii, Mangifera indica, Morus alba, Nerium odorum, Pinus roxburghii, Prinsepia utilis, Prunus amygdalus, Pyrus communis, Rosa indica, Syzygium cumini, Tagetes erecta, Thevetia neriifolia, Thuja orientalis, Zephyranthes candida, Zephyranthes grandiflora.*
(ii) Predominant Plants of the Courtyard	:	*Broussonetia papyrifera, Grewia oppositifolia.*
(iii) Maintenance	:	Excellent.
(iv) Ethnobotanical		
Sacred	:	*Ficus religiosa* worshipped as sacred tree. Leaves of *Cannabis sativa* offered especially to appease Lord Shiva. *Cynodon dactylon* employed in all religious ceremonies as sacred plant.
Medicinal	:	Decoction of bark of *Bauhinia variegata* given for urinary problems. Fruit candy of *Emblica officinalis* effective against heart ailments. 2-3g powdered seeds of *Syzigium cumini* taken in the morning with milk to cure diabetes.
Edible	:	Flower buds of *Bauhinia variegata* and leaves and underground corms of *Colocasia esculanta* cooked as vegetable. Fruits of *Broussonetia papyrifera, Emblica officinalis, Ficus palmata, Litchi chinensis, Mangifera indica, Morus alba, Prinsepia utilis, Pyrus communis, Syzigium cumini*; seeds of *Bombax ceiba, Pinus roxburghii, Prunus amygdalus* and kernels of *Juglans regia* are edible. Fruits of *Citrus aurantifolia* are pickled.

Ornamental : *Bombax ceiba, Bryophyllum calycinum, Callistemon lanceolatus, Canna indica, Catharanthus roseus, Chrysanthemum indicum, Emblica officinalis, Grevillea robusta, Hibiscus rosa-sinensis, Hibiscus syriacus, Malvaviscus conzatii, Nerium odorum, Pinus roxburghii, Rosa indica, Tagetes erecta, Thevetia neriifolia, Thuja orientalis, Zephyranthes candida, Zephyranthes grandiflora* grown as ornamentals.

Other Uses : Leaves of *Bauhinia variegata, Broussonetia papyrifera, Grewia oppositifolia* and *Morus alba* used as fodder. Seed cotton of *Bombax ceiba* used for filling pillows and mattresses. Wood of *Cedrela toona* and *Morus alba* excellent for construction and furniture, *Syzygium cumini* for agricultural implements and that of *Broussonetia papyrifera* for packing cases. Fibers from branches of *Grewia oppositifolia* used for making ropes.

Remarks : Coconuts symbolically sacrificed in place of goats.

2. Baba Balak Nath Temple (Pl. 5B, C)

Main Deity : Baba Balak Nath.

Other Deities : Shiva, Brahma, Vishnu, Duttatrey, Ganpati, Kali, Bhrit-hari.

Tot. Ar. of Tem. Courtyard : About 8000 sq m. (total area of the temple land is about 200 Kanal, where school, college, dharamshala, samadhi and hospitals are constructed).

Location

(i) *Place*

Village : Deoth Sidh

Block : Bijhari

Tehsil : Barsar

District : Hamirpur

(ii) *Approach* : 45 km from Hamirpur.

(iii) *Altitude* : 860 m.

Management : The temple is run by Baba Balak Nath Temple Trust, Deoth Sidh.

Historical Background : The temple was built in 17th–18th century A.D. Long ago, Baba Balak Nath came to Shahtalai from his native place Kathiabad (Gujarat). At that time, he was 14 years old. He stayed in the home of Ratno Mai and worked as a cowboy for 12 years. After that, he came to Deoth Sidh and meditated for several years in the cave in which, at present, his idol is kept.

Temple Architect : Architecture of the temple surrounding the cave is in Sikhara style.

Offerings to God

(i) *By Priest*

Morning : 'Rot' (at 5 AM daily), 'Halva' (on Sakranti).

Evening and Noon : Food prepared in 'langar' (Community meal).

(ii) *By People* : 'Rot', Puffed rice, 'Lachidana', 'Halva', Coconut and Dhoop (incense), Wooden sticks ('Bragan', 'Phodi'), Wooden sleepers ('Khdawan') and Scarves. People also offer wheat, rice, gram and wheat flour.

Off. to People (Prasadam) : 'Rot', Puffed rice, 'Lachidana', 'Dhoop', Sacred Thread, Sacred Ash ('vibhuti').

Fairs and Festivals : Baba Balak Nath fair held for one month from March 13 to April 13 (Chaitra Mass).

Observations

(i) *Courtyard Plants (Protected as well as for Landscaping* : *Aegle marmelos, Albizia lebbeck, Aloe vera, Antirrhinum majus, Araucaria columnaris, Bombax ceiba, Bougainvillea spectabilis, Callistemon lanceolatus, Canna indica, Catharanthus roseus, Celosia argentea* var. *cristata, Cestrum nocturnum, Chamaerops humilis, Chlorophytum capense, Chrysanthemum indicum, Cosmos sulphureus, Cupressus sempervirens, Cycas revoluta, Cynodon dactylon, Dahlia rosea, Duranta repens, Ficus benghalensis, Ficus elastica, Ficus religiosa, Grevillea robusta, Hibiscus rosa-sinensis, Impatiens balsamina, Jacaranda acutifolia, Jasminum grandiflorum, Jasminum primulinum, Juglans regia, Kochia trichophylla, Leucaena leucocephala, Malva sylvestris, Mangifera indica, Mimosa pudica, Nerium odorum, Ocimum sanctum, Pelargonium hortorum, Petunia hybrida, Pinus roxburghii, Polyalthia longifolia, Pterospermum acerifolium, Rosa indica, Salvia coccinia, Sambucus nigra, Tabernaemontana coronaria, Tagetes erecta, Thevetia neriifolia, Thuja orientalis, Tithonia tagetiflora, Viola tricolor* and *Zinnia elegans.*

(ii) *Predominant Plants of the Courtyard* : *Bougainvillea spectabilis, Canna indica, Grevillea robusta, Pinus roxburghii, Mangifera indica.*

(iii) *Maintenance* : Excellent.

(iv) *Ethnobotanical*

Sacred : *Ficus religiosa* and *Ocimum sanctum* worshipped as sacred plants. *Cynodon dactylon* employed in religious ceremonies as sacred plant.

Medicinal : Warm leaves of *Aloe vera* tied over the respective part of the body to cure body pain. 20ml decoction of bark of *Bombax ceiba* given twice daily for dysentery.

Edible : Fruits of *Aegle marmelos, Mangifera indica* and seeds of *Bombax ceiba* and *Pinus roxburghii* are edible.

Ornamental : *Albizia lebbeck, Aloe vera, Antirrhinum majus, Araucaria columnaris, Bombax ceiba, Bougainvillea spectabilis, Callistemon lanceolatus, Canna indica, Catharanthus roseus, Celosia argentea* var. *cristata, Cestrum nocturnum, Chamaerops humilis, Chlorophytum capense, Chrysanthemum indicum, Cosmos sulphureus, Cupressus sempervirens, Cycas revoluta, Dahlia rosea, Duranta repens, Ficus elastica, Grevillea robusta, Hibiscus rosa-sinensis, Impatiens balsamina, Jacaranda acutifolia, Jasminum grandiflorum, Jasminum primulinum, Juglans regia, Kochia trichophylla, Leucaena leucocephala, Malva sylvestris, Mimosa pudica, Nerium odorum, Ocimum sanctum, Pelargonium hortorum, Petunia hybrida, Pinus roxburghii, Polyalthia longifolia, Pterospermum acerifolium, Rosa indica, Salvia coccinia, Sambucus nigra, Tabernaemontana coronaria, Tagetes erecta, Thevetia neriifolia, Thuja orientalis, Tithonia tagetiflora, Viola tricolor* and *Zinnia elegans* are planted as ornamentals.

Other Uses : *Leucaena leucocephala* used as a source of excellent fodder. Wood of *Albizia lebbeck* used for agricultural implements and its leaves as fodder.

Remarks : The idol of the deity is kept inside a small natural cave overarched by a very old Banyan tree. Ladies are not allowed to enter the cave. 'Mitha Rot' is the special offering to appease Baba Balak Nath. It is a thick chapati prepared by mixing jaggery ('Gur') with wheat flour and fried in ghee. People whose wishes got fulfilled offer 'Rot' to the deity on Sunday. 'Vibhuti' resulted from continuous burning of log of wood in 'havan kund' offered to people as it is believed to possess magical power to cure various ailments.

3. Baba Sidh Chaino Temple (Pl. 5D)

Main Deity : Baba Sidh Chaino.

Other Deities : Nanga Bali, Bhairav, Hastbali, Mai Loona, Lal Khambli.

Tot. Ar. of Tem. Courtyard : About 1500 sq m.

Location	
(i) Place	
Village	: Smaila
Block	: Bijhari
Tehsil	: Barsar
District	: Hamirpur
(ii) Approach	: 9 km from Bijhari.
(iii) Altitude	: 660 m.
Management	: Privately owned.
Historical Background	: The temple is believed to be 500 years old, and reconstructed several times.
Temple Architect	: This is a flat roof temple.
Offerings to God	
(i) By Priest	: 'Rot Churma' (pieces of 'Rot'), 'Halva'.
(ii) By People	: Puffed rice, 'Lachidana', 'Rot', Seasonal crops.
Off. to People (Prasadam)	: 'Lachidana', Puffed rice.
Fairs and Festivals	: Dahaja (Saal) is celebrated on first Monday of December every year
Observations	
(i) Courtyard Plants (Protected as well as for Landscaping	: *Bauhinia vahlii, Bougainvillea spectabilis, Callistemon lanceolatus, Carica papaya, Catharanthus roseus, Chrysanthemum indicum, Cupressus sempervirens, Cynodon dactylon, Euphorbia pulcherrima, Grevillea robusta, Hibiscus rosa-sinensis, Hibiscus syriacus, Jacaranda acutifolia, Jasminum sambac, Lagerstroemia indica, Malvaviscus conzatii, Mangifera indica, Melia azedarach, Pelargonium hortorum, Prunus domestica, Prunus persica, Pterospermum acerifolium, Punica granatum, Rosa indica, Tabernaemontana coronaria, Tagetes erecta, Tecoma stans, Thuja orientalis.*
(ii) Predominant Plants of the Courtyard	: *Bauhinia vahlii, Mangifera indica, Rosa indica.*
(iii) Maintenance	: Satisfactory.
(iv) Ethnobotanical	
Sacred	: Wood of *Bauhinia vahlii* and *Mangifera indica* used for the preparation of holy ash ('Vibhuti'), and distributed to the people visiting the temple. *Cynodon dactylon* employed in all religious ceremonies as sacred plant.

Medicinal	:	Vegetable prepared from unripe fruits of *Carica papaya* considered good for jaundice. Fruit juice of *Punica granatum* effective against anaemia.
Edible	:	Fruits of *Carica papaya, Mangifera indica, Prunus domestica, Prunus persica* and *Punica granatum* are edible.
Ornamental	:	*Bougainvillea spectabilis, Callistemon lanceolatus, Catharanthus roseus, Chrysanthemum indicum, Cupressus sempervirens, Euphorbia pulcherrima, Grevillea robusta, Hibiscus rosa-sinensis, Hibiscus syriacus, Jacaranda acutifolia, Jasminum sambac, Lagerstroemia indica, Malvaviscus conzatii, Melia azedarach, Pelargonium hortorum, Pterospermum acerifolium, Rosa indica, Tabernaemontana coronaria, Tagetes erecta, Tecoma stans* and *Thuja orientalis* grown as ornamentals.
Other Uses	:	Leaves of *Bauhinia vahlii* used for making 'Pattal' and 'Duna' (meal plates). Wood of *Melia azedarach* good for building works and agricultural implements.
Remarks	:	Earlier, there used to be rich vegetation of 'Taur' plant (*Bauhinia vahlii*) in the temple courtyard, hence called 'Taur Wala Baba'. Over a period of time their number devindled. Goats, cocks and hens sacrificed to appease the deity.

4. Gauri Shanker Temple (Pl. 6A)

Main Deity	:	Silver image of Shiva and Parvati.
Other Deities	:	Nandigan, Kartikey, Brahma, Vishnu, Lakshmi.
Tot. Ar. of Tem. Courtyard	:	1200 sq m.
Location		
(i) Place		
Village	:	Tira (Sujanpur)
Block	:	Sujanpur
Tehsil	:	Sujanpur
District	:	Hamirpur
(ii) Approach	:	It is 2 km from Sujanpur and located inside the fort of Tira.
(iii) Altitude	:	900 m.
Management	:	Privately owned.
Historical Background	:	The temple was built in 1793 A.D. by Raja Sansar Chand and donated to the ancestors of the owner of the temple.
Temple Architect	:	This is a flat roof temple.

Offerings to God		
(i) By Priest	:	Rice, Dal, Chapati, Vegetables.
(ii) By People	:	'Halva', Puffed rice, 'Lachidana', 'Patasha'.
Off. to People (Prasadam)	:	'Halva', Puffed rice, 'Lachidana', 'Patasha'.
Fairs and Festivals	:	Basant Panchami, Shivratri.
Observations		
(i) Courtyard Plants (Protected as well as for Landscaping	:	*Aegle marmelos, Amaryllis reginae, Bougainvillea spectabilis, Bryophyllum calycinum, Callistemon lanceolatus, Catharanthus roseus, Cedrela toona, Chrysanthemum indicum, Citrus aurantifolia, Cynodon dactylon, Dalbergia sissoo, Eriobotrya japonica, Hibiscus syriacus, Jasminum humile, Jasminum sambac, Lagerstroemia indica, Mangifera indica, Mentha piperita, Morus alba, Musa paradisiaca, Nerium odorum, Ocimum sanctum, Prunus amygdalis, Psidium guajava, Rosa indica, Salvia coccinia, Sansevieria trifasciata, Tagetes erecta, Thuja orientalis.*
(ii) Predominant Plants of the Courtyard	:	*Catharanthus roseus, Bougainvillea spectabilis.*
(iii) Maintenance	:	Satisfactory.
(iv) Ethnobotanical		
Sacred	:	*Ocimum sanctum* worshipped as sacred plant. Leaves of *Aegle marmelos*, and flowers of *Nerium odorum* specially offered to the deity. *Cynodon dactylon* employed in all religious ceremonies.
Medicinal	:	2-3g powdered seeds of *Citrus aurantifolia* given daily with warm water for dysentery. Fruit pulp of *Aegle marmelos* given to cure pregnancy problems. Leaf juice of *Mentha piperita* effective against diarrhoea.
Edible	:	Leaves of *Mentha piperita* used for making 'chutney'. Fruits of *Aegle marmelos, Eriobotrya japonica, Mangifera indica, Morus alba, Musa paradisiaca, Psidium guajava* and seeds of *Prunus amygdalis* edible whereas those of *Citrus aurantifolia* are pickled.
Ornamental	:	*Amaryllis reginae, Bougainvillea spectabilis, Bryophyllum calycinum, Callistemon lanceolatus, Catharanthus roseus, Chrysanthemum indicum, Hibiscus syriacus, Jasminum humile, Jasminum sambac, Lagerstroemia indica, Nerium odorum, Ocimum sanctum, Rosa indica, Salvia coccinia, Sansevieria trifasciata, Tagetes erecta, Thuja orientalis* grown as ornamentals.
Other Uses	:	Wood of *Cedrela toona* and *Dalbergia sissoo* excellent for construction and furniture. *Cynodon dactylon* used for fodder.

5. Murli Manohar Temple (Pl. 6B)

Main Deity	:	Radha Krishana.
Other Deities	:	Garud, Ganesha, Annapurna, Chamunda, Jai-Vijay, Shiva, Kali, Durga, Brihaspati, Surya, Kartikey.
Tot. Ar. of Tem. Courtyard	:	About 800 sq m.
Location		
(i) Place		
Town	:	Sujanpur Tira
Block	:	Sujanpur
Tehsil	:	Sujanpur
District	:	Hamirpur
(ii) Approach	:	24 km from Hamirpur.
(iii) Altitude	:	700 m.
Management	:	Privately owned.
Historical Background	:	The temple is believed to be 250 years old and built by Maharaja Sansar Chand in memory of his mother.
Temple Architect	:	Sikhara style.
Offerings to God		
(i) By Priest		
Morning	:	Sugar candy.
Noon	:	'Bhojan Bhog' (Rice-Dal, Chapati-Vegetables), 'Halva'.
Evening	:	Black Gram.
(ii) By People	:	Black Gram, Sweets ('Bundi'), 'Halva'.
Off. to People (Prasadam)	:	Black Gram, Sweets ('Bundi'), 'Halva'.
Fairs and Festivals	:	Janamashtmi, Holi.
Observations		
(i) Courtyard Plants (Protected as well as for Landscaping	:	*Cynodon dactylon, Emblica officinalis, Ficus benghalensis, Ficus religiosa, Ocimum sanctum, Pelargonium hortorum, Tagetes erecta.*
(ii) Predominant Plants of the Courtyard	:	*Tagetes erecta.*
(iii) Maintenance	:	Excellent.
(iv) Ethnobotanical		
Sacred	:	*Ficus benghalensis, Ficus religiosa* and *Ocimum sanctum* worshipped as sacred plants. *Cynodon dactylon* employed in religious ceremonies as sacred plant.

Medicinal	:	Fruits of *Emblica officinalis* form one of the important constituents of 'Triphala'. Its powdered fruits kept in pitcher filled with water prescribed to wash head and mouth daily for 7 days to check nose bleeding.
Edible	:	Fruits of *Emblica officinalis* edible, also pickled and made into 'murabba' (candy–rich source of vitamin C).
Ornamental	:	*Emblica officinalis, Ocimum sanctum, Pelargonium hortorum,* and *Tagetes erecta* grown as ornamentals.

6. Narbadeshwar Temple (Pl. 6C, D)

Main Deity	:	Shiva.
Other Deities	:	Surya, Ganpati, Durga, Lakshmi Narayan, Ram-Sita, Nandi, Sheshnag, Ridhi-Sidhi
Tot. Ar. of Tem. Courtyard	:	About 3000 sq m.
Location		
(i) Place		
Village	:	Sujanpur Tira
Block	:	Sujanpur
Tehsil	:	Sujanpur
District	:	Hamirpur
(ii) Approach	:	It is 24 km from Hamirpur and situated on the bank of the river Beas.
(iii) Altitude	:	700 m.
Management	:	Government undertaken (under Archaeological Survey of India).
Historical Background	:	The temple was built in 1823 A.D. by Rani Suketan (Parsanni Devi-wife of Raja Sansar Chand).
Temple Architect	:	This is a flat roof temple built on Rajput-Mughal style of architecture. The temple walls are profusely decorated with paintings portraying scenes from Ramayana, Mahabharta, Bhagvata-Purana and other scriptures such as Shiva's marriage, Ganga Avatarna, etc.
Offerings to God		
(i) By Priest	:	Sweets ('Bundi').
(ii) By People	:	Sweets ('Bundi'), Fruits.
Off. to People (Prasadam)	:	Sweets ('Bundi').
Fairs and Festivals	:	Shivratri, Janamashtmi.

Observations

(i) Courtyard Plants (Protected as well as for Landscaping	:	*Aegle marmelos, Amaryllis reginae, Bauhinia variegata, Catharanthus roseus, Cedrela toona, Celosia argentea* var. *cristata, Chrysanthemum indicum, Cynodon dactylon, Dahlia rosea, Ficus religiosa, Gaillardia pulchella, Hedychium coronarium, Helianthus annuus, Hibiscus rosa-sinensis, Hibiscus syriacus, Jasminum sambac, Kochia trichophylla, Lagerstroemia indica, Mangifera indica, Nerium odorum, Pedilanthus tithymaloides, Plumeria acutifolia, Rosa indica, Tagetes erecta, Thuja orientalis, Zinnia elegans.*
(ii) Predominant Plants of the Courtyard	:	*Mangifera indica, Ficus religiosa, Helianthus annuus.*
(iii) Maintenance	:	Excellent.
(iv) Ethnobotanical		
Sacred	:	Leaves of *Aegle marmelos* and flowers of *Nerium odorum* especially offered to appease Lord Shiva. *Ficus religiosa* worshipped as sacred tree. *Cynodon dactylon* employed in all religious ceremonies as sacred plant.
Medicinal	:	Decoction of bark of *Bauhinia variegata* used to cure bile troubles.
Edible	:	Flower buds of *Bauhinia variegata* cooked as vegetable. Fruits of *Aegle marmelos* and *Mangifera indica* are edible.
Ornamental	:	*Amaryllis reginae, Catharanthus roseus, Celosia argentea* var. *cristata, Chrysanthemum indicum, Dahlia rosea, Gaillardia pulchella, Hedychium coronarium, Helianthus annuus, Hibiscus rosa-sinensis, Hibiscus syriacus, Jasminum sambac, Kochia trichophylla, Lagerstroemia indica, Nerium odorum, Pedilanthus tithymaloides, Plumeria acutifolia, Rosa indica, Tagetes erecta, Thuja orientalis* and *Zinnia elegans* are grown for beautification of temple.
Other Uses	:	Leaves of *Bauhinia variegata* used as fodder. Wood of *Cedrela toona* excellent for construction and furniture.

7. Santoshi Mata Temple (Pl. 6E)

Main Deity	:	Santoshi Mata.
Other Deities	:	Lakshmi Mata, Bajrangbali, Bhairav.
Tot. Ar. of Tem. Courtyard	:	About 3000 sq m.
Location		
(i) Place		
Village	:	Ladraur

Block : Bhoranj

Tehsil : Bhoranj

District : Hamirpur

(ii) *Approach* : 27 km from Hamirpur.

(iii) *Altitude* : 1080 m.

Management : Local temple committee.

Historical Background : The temple was built in 1969 by locals.

Temple Architect : Sikhara style.

Offerings to God

(i) *By Priest* : Rice, 'Chapati', Clarified butter, Sugar.

(ii) *By People* : 'Gur-Chana', Puffed rice, Coconut, Sweets.

Off. to People (Prasadam) : Puffed rice.

Fairs and Festivals : Navratra.

Observations

(i) *Courtyard Plants (Protected as well as for Landscaping* : *Callistemon lanceolatus, Carissa opaca, Citrus limon, Cynodon dactylon, Dalbergia sissoo, Eucalyptus tereticornis, Grevillea robusta, Grewia oppositifolia, Hibiscus rosa-sinensis, Hibiscus syriacus, Mangifera indica, Psidium guajava, Rosa indica, Tagetes erecta, Thuja orientalis.*

(ii) *Predominant Plants of the Courtyard* : *Eucalyptus tereticornis, Carissa opaca.*

(iii) *Maintenance* : Excellent.

(iv) *Ethnobotanical*

Sacred : *Cynodon dactylon* employed in all religious ceremonies as sacred plant.

Medicinal : Decoction of leaves of *Dalbergia sissoo* and tender branches of *Psidium guajava* prescribed for dysentery.

Edible : Fruits of *Carissa opaca, Mangifera indica* and *Psidium guajava* edible and those of *Citrus limon* for making candy and pickle.

Ornamental : *Callistemon lanceolatus, Grevillea robusta, Hibiscus rosa-sinensis, Hibiscus syriacus, Rosa indica, Tagetes erecta* and *Thuja orientalis* grown as ornamentals.

Other Uses : *Carissa opaca* planted as fence. Its leaves used as fodder for goats offered to the deity by the people. Leaves of *Grewia oppositifolia* also used as fodder. Its branches used for making ropes. Wood of *Dalbergia sissoo* excellent for furniture and building works, and that of *Eucalyptus tereticornis* for construction work.

Remarks	:	Goats offered to the deity by those people whose wishes are fulfilled. These later are sold by the temple committee.

8. Satyanarayan Temple

Main Deity	:	Satyanarayan.
Other Deities	:	Santoshi Mata, Shiv Panchayat (Shiva, Parvati, Ganesha and Kartik), Hanuman, Shaligdev.
Tot. Ar. of Tem. Courtyard	:	About 3000 sq m.
Location		
(i) Place		
Town	:	Hamirpur
Block	:	Hamirpur
Tehsil	:	Hamirpur
District	:	Hamirpur
(ii) Approach	:	The temple is located in the town.
(iii) Altitude	:	600 m.
Management	:	In the name of deity.
Historical Background	:	The temple is more than 100 years old.
Temple Architect	:	Sikhara style.
Offerings to God		
(i) By Priest		
Morning and Evening	:	Milk, Fruits.
Noon	:	Whatever being prepared in the home of temple priest.
(ii) By People	:	'Lachidana', Sweets, Flowers, Fruits.
Off. to People (Prasadam)	:	'Lachidana', Sweets, Flowers, Fruits.
Fairs and Festivals	:	Janamashtmi.
Observations		
(i) Courtyard Plants (Protected as well as for Landscaping	:	*Aegle marmelos, Bauhinia variegata, Broussonetia papyrifera, Cannabis sativa, Cestrum diurnum, Cupressus torulosa, Cynodon dactylon, Ficus palmata, Ficus religiosa, Hibiscus rosa-sinensis, Ipomoea cairica, Litsea polyantha, Mangifera indica, Morus alba, Melia azedarach, Musa paradisiaca, Psidium guajava, Rosa indica, Salvia coccinia.*
(ii) Predominant Plants of the Courtyard	:	*Ipomoea cairica* and *Salvia coccinia.*
(iii) Maintenance	:	Satisfactory.

(iv) Ethnobotanical

Sacred : *Ficus religiosa* and *Aegle marmelos* worshipped as sacred plants. Leaves of *Aegle marmelos* and *Cannabis sativa* offered to appease Lord Shiva. *Cynodon dactylon* employed to perform religious ceremonies.

Medicinal : Decoction of bark of *Bauhinia variegata* used to cure liver problems. Powdered seeds of *Cannabis sativa* given to cure joint pains. Mixture of 5g powdered bark of *Litsea polyantha* and 1 black pepper given twice daily for jaundice. Fruits of *Psidium guajava* used to recover iron deficiency in the body.

Edible : Flower buds of *Bauhinia variegata* cooked as vegetable. Fruits of *Ficus palmata, Mangifera indica, Morus alba, Musa paradisiaca, Psidium guajava, Aegle marmelos* and *Broussonetia papyrifera* are edible.

Ornamental : *Cestrum diurnum, Cupressus torulosa, Hibiscus rosa sinensis, Litsea polyantha, Melia azedarach, Rosa indica,* and *Salvia coccinia* planted as ornamentals.

Other Uses : Leaves of *Bauhinia variegata, Broussonetia papyrifera* and *Morus alba* used as fodder. Wood of *Broussonetia papyrifera* used for making plywood, packing cases and cheap furniture and that of *Melia azedarach* for building works and agricultural implements.

9. Shani Dev Temple (Pl. 6F)

Main Deity : Shani Dev.

Tot. Ar. of Tem. Courtyard : About 3000 sq m.

Location

(i) Place

Village : Sarlin (Lamblu)

Block : Tauni Devi

Tehsil : Hamirpur

District : Hamirpur

(ii) Approach : 10 km from Hamirpur.

(iii) Altitude : 940 m.

Management : Local temple committee.

Historical Background : The temple was built in 1996 by locals.

Temple Architect : Sikhara style.

Offerings to God	
(i) By Priest	
Morning	: 'Halva'.
Evening and Noon	: Rice, 'Dal', 'Chapati', Vegetables.
(ii) By People	: 'Mash', 'Til', 'Sarson', Salt, Iron, Mustard oil, Soap, Black piece of cloth, 'Ak', Wheat, Rice, 'Petha', 'Mungi Dal', 'Chana Dal'.
Off. to People (Prasadam)	: 'Halva', Puffed rice.
Fairs and Festivals	: Janamashtmi, Shivratri.
Observations	
(i) Courtyard Plants (Protected as well as for Landscaping	: *Acacia catechu, Cynodon dactylon, Ficus religiosa, Murraya koenigii, Pinus roxburghii, Pistacia integerrima, Tagetes erecta, Zizyphus mauritiana.*
(ii) Predominant Plants of the Courtyard	: *Pinus roxburghii.*
(iii) Maintenance	: Excellent.
(iv) Ethnobotanical	
Sacred	: *Ficus religiosa* worshipped as form of Lord Brahma. Flowers of *Tagetes erecta* considered sacred for offering to the deity. *Cynodon dactylon* employed in religious ceremonies as sacred plant.
Medicinal	: Decoction of bark of *Acacia catechu* used to cure throat infection. Powdered leaves of *Murraya koenigii* prescribed as blood purifier. Decoction of leaves of *Murraya koenigii* and *Psidiun guajava* used as gargle for gum-bleeding.
Edible	: Fruits of *Zizyphus mauritiana* and seeds of *Pinus roxburghii* are edible. Leaves of *Murraya koenigii* used as condiment for flavouring curries.
Ornamental	: *Tagetes erecta* grown as an ornamental plant.
Other Uses	: Stem of *Murraya koenigii* used as toothbrush. 'Katha' obtained from the heartwood of *Acacia catechu* is of great economic value as it is one of the main ingredients for betel chewing.
Remarks	: 'Ak' (Leaves and flowers of *Calotropis procera*) 'Mash' (seeds of *Vigna mungo*), mustard oil (seeds oil of *Brassica campestris*), black piece of cloth used as special offerings to appease Shani Dev. In order to protect themselves from the effect of evil spirits, people weigh themselves ('Tula Dan') in the temple with equal quantity of

different items including 'mash', 'til' (seeds of *Sesamum orientale*), 'sarson' (seeds of *Brassica campestris*), salt, iron piece, mustard oil, soap, black piece of cloth (at least 3 metre), wheat, rice, 'petha' (fruits of *Cucurbita hispida*), 'mungi dal' (seeds of *Vigna radiata*), 'chana dal' (split grains of *Cicer arietinum*) and 'chhaya patra' (a concave-shaped iron vessel) which afterwards are offered in the temple.

10. Shirdi Sai Baba Temple (Pl. 6G)

Main Deity : Shirdi Sai Baba.

Other Deities : Ganpati, Hanuman, Ashtbhuji Mata, Baba Balak Nath, Shiva, Ram, Sita, Lakshman, Radha–Krishana, Duttatrey.

Tot. Ar. of Tem. Courtyard : About 1600 sq m.

Location

(i) Place

Village : Kadriyana

Block : Bassi

Tehsil : Hamirpur

District : Hamirpur

(ii) Approach : 5 km from Bhota.

(iii) Altitude : 600 m.

Management : Temple committee.

Historical Background : The temple was built by locals about 12 years back.

Temple Architect : Sikhara style.

Offerings to God

(i) By Priest : 'Dal-Roti', Rice-Vegetables, 'Halva-Puri', Gram, Sweets, Fruits, Flowers.

(ii) By People : 'Lachidana', Puffed rice, Sweets, Fruits, Flowers.

Off. to People (Prasadam) : Sweets ('Mithi Bundi'), Puffed rice.

Fairs and Festivals : Ram-Navami, Nirwan-Diwas (15 October), Diwali, Janamashtmi, Shivratri.

Observations

(i) Courtyard Plants (Protected as well as for Landscaping : *Amaryllis reginae, Callistemon lanceolatus, Canna indica, Cestrum nocturnum, Chamaerops humilis, Chrysanthemum indicum, Citrus aurantifolia, Cynodon dactylon, Hemerocallis fulva, Hibiscus rosa-sinensis, Hibiscus syriacus,*

		Jasminum sambac, Lagerstroemia indica, Malvaviscus conzatii, Pelargonium hortorum, Prunus persica, Psidium guajava, Rosa indica, Tagetes erecta, Tecoma stans, Thuja orientalis, Zephyranthes grandiflora.
(ii) Predominant Plants of the Courtyard	:	*Lagerstroemia indica.*
(iii) Maintenance	:	Excellent.
(iv) Ethnobotanical		
Sacred	:	Flowers of *Chrysanthemum indicum, Hibiscus rosa-sinensis, Rosa indica, Tagetes erecta* and *Tecoma stans* offered to the deity. *Cynodon dactylon* employed in various religious ceremonies.
Edible	:	Fruits of *Prunus persica* and *Psidium guajava* edible and that of *Citrus aurantifolia* are pickled.
Medicinal	:	Fruits of *Psidium guajava* considered good to cure diabetes.
Ornamental	:	All the courtyard plants grown as ornamentals.

11. Tauni Devi Temple

Main Deity	:	Tauni Devi.
Other Deities	:	Hanuman, Shiva, Shani Dev, Baba Balak Nath, Sai Baba.
Tot. Ar. of Tem. Courtyard	:	About 1500 sq m.
Location		
(i) Place		
Village	:	Tauni Devi
Block	:	Tauni Devi
Tehsil	:	Hamirpur
District	:	Hamirpur
(ii) Approach	:	12 km from Hamirpur.
(iii) Altitude	:	1100 m.
Management	:	Local temple committee.
Temple Architect	:	Sikhara style.
Offerings to God		
(i) By Priest	:	Puffed rice, 'Halva', Fruits.
(ii) By People	:	Puffed rice, Fruits, Sweets, Coconut.
Off. to People (Prasadam)	:	Puffed rice, Fruits, Sweets.
Fairs and Festivals	:	Tauni Devi fair (held every year on 10th of Ashad month), Navratra.

Observations

(i) Courtyard Plants (Protected as well as for Landscaping	: *Bougainvillea spectabilis, Canna indica, Jasminum grandiflorum, Tagetes erecta, Thuja orientalis.*
(ii) Predominant Plants of the Courtyard	: *Bougainvillea spectabilis, Jasminum grandiflorum.*
(iii) Maintenance	: Satisfactory.
(iv) Ethnobotanical	
Sacred	: Flowers of *Bougainvillea spectabilis, Jasminum grandiflorum* and *Tagetes erecta* considered sacred for offerings to the deities.
Ornamental	: *Bougainvillea spectabilis, Canna indica, Jasminum grandiflorum, Tagetes erecta* and *Thuja orientalis* grown as ornamentals.

12. Trilokinath Temple (Pl. 6H)

Main Deity	: Trilokinath (Lord Shiva).
Other Deities	: Baba Balak Nath, Santoshi Mata.
Tot. Ar. of Tem. Courtyard	: About 800 sq m.
Location	
(i) Place	
Village	: Barah Mahino Ka Paron
Block	: Bijhari
Tehsil	: Barsar
District	: Hamirpur
(ii) Approach	: About 40 km from Hamirpur (5 km from Deoth Sidh).
(iii) Altitude	: 730 m.
Management	: Local temple committee.
Historical Background	: The temple was built by locals about 15 years back.
Temple Architect	: Sikhara style.
Offerings to God	
(i) By Priest	: 'Halva', Sweets.
(ii) By People	: 'Halva', Puffed rice, Sweets.
Off. to People (Prasadam)	: 'Halva', Puffed Rice, Sweets.
Fairs and Festivals	: Shivratri.

Observations

(i) Courtyard Plants (Protected as well as for Landscaping	:	*Aegle marmelos, Bougainvillea spectabilis, Callistemon lanceolatus, Catharanthus roseus, Cupressus sempervirens, Ficus elastica, Ficus religiosa, Hibiscus rosa-sinensis, Mangifera indica, Melia azadirachta, Psidium guajava, Tagetes erecta.*
(ii) Predominant Plants of the Courtyard	:	*Callistemon lanceolatus.*
(iii) Maintenance	:	Satisfactory.
(iv) Ethnobotanical		
Sacred	:	Leaves of *Aegle marmelos* offered to Lord Shiva. *Ficus religiosa* worshipped as sacred plant.
Medicinal	:	Powdered leaves of *Melia azadirachta* applied for quick healing of wounds, and its decoction considered good for diabetes and skin ailments.
Edible	:	Fruits of *Aegle marmelos, Mangifera indica* and *Psidium guajava* are edible.
Ornamental	:	*Bougainvillaea spectabilis, Callistemon lanceolatus, Catharanthus roseus, Cupressue sempervirens, Ficus elastica, Hibiscus rosa-sinensis, Melia azadirachta* and *Tagetes erecta* grown as ornamentals.
Other Uses	:	Leaves of *Mangifera indica* and *Psidium guajava* used for scouring teeth. Wood of *Mangifera indica* considered good for construction works.

13. Vaishno Mata Temple (Pl. 6I)

Main Deity	:	Vaishno Devi.
Other Deities	:	Shiva, Hanuman, Bhairav.
Tot. Ar. of Tem. Courtyard	:	About 1500 sq m.
Location		
(i) Place		
Village	:	Dhangota
Block	:	Bijhari
Tehsil	:	Barsar
District	:	Hamirpur
(ii) Approach	:	10 km from Bijhari.
(iii) Altitude	:	850 m.
Management	:	Local temple committee.

Historical Background	:	An old man of the village, Shri Chowdhary Ram had a strange dream in 1991. His niece also saw the same dream, in which he saw a cave having Vaishno Mata as deity having 'Bael' tree (*Aegle marmelos*) on one side and 'Amaltash' (*Cassia fistula*) on the other side. That place was found out with the help of villagers and the temple was built there.
Temple Architect	:	Sikhara style.
Offerings to God		
(i) By Priest		
Morning	:	'Halva'.
Noon and Evening	:	Rice, 'Chapati', 'Dal', Vegetables, Clarified butter, Loaf sugar.
(ii) By People	:	'Halva', 'Lachidana', Puffed rice. 'Rot' also offered by people on Saturday to Bhairav, Shiva and Hanuman.
Off. to People (Prasadam)	:	'Halva', 'Lachidana', Puffed rice.
Fairs and Festivals	:	Navratra.
Observations		
(i) Courtyard Plants (Protected as well as for Landscaping	:	*Acacia catechu, Aegle marmelos, Callistemon lanceolatus, Cassia fistula, Catharanthus roseus, Citrus limon, Cynodon dactylon, Ficus benghalensis, Ficus religiosa, Ficus roxburghii, Grevillea robusta, Grewia oppositifolia, Hibiscus rosa-sinensis, Jasminum sambac, Malvaviscus conzatii, Mangifera indica, Sansevieria trifasciata, Tagetes erecta, Thuja orientalis.*
(ii) Predominant Plants of the Courtyard	:	*Grevillea robusta.*
(iii) Maintenance	:	Excellent.
(iv) Ethnobotanical		
Sacred	:	Leaves of *Aegle marmelos* offered to Lord Shiva. Wood of *Aegle marmelos, Mangifera indica, Ficus religiosa, Ficus benghalensis, Acacia catechu* used in 'havan'. *Cynodon dactylon* considered sacred and employed in all religious ceremonies.
Medicinal	:	Fruit pulp of *Aegle marmelos* used for making candy which is considered very effective against dysentery and other stomach problems.
Edible	:	The young pods of *Cassia fistula* used for pickling and as vegetable. Fruits of *Aegle marmelos, Ficus roxburghii* and *Mangifera indica* are edible and that of *Citrus limon* pickled.

Ornamental	:	*Callistemon lanceolatus, Cassia fistula, Catharanthus roseus, Grevillea robusta, Hibiscus rosa-sinensis, Jasminum sambac, Malvaviscus conzatii, Sansevieria trifasciata, Tagetes erecta* and *Thuja orientalis* grown as ornamental.
Other Uses	:	'Katha' obtained from the heartwood of *Acacia catechu* used for betel chewing. Its branches used for fencing purposes. Leaves of *Grewia oppositifolia* and *Ficus roxburghii* used as fodder. Fibers from branches of *Grewia oppositifolia* used for making ropes. Wood of *Cassia fistula* good for agricultural implements.

Temples of District Kangra

1. Ashapuri Temple (Pl. 7A)

Main Deity	:	Goddess Ashapuri.
Other Deity	:	Shiva.
Tot. Ar. of Tem. Courtyard	:	3000 sq m.
Location		
(i) Place		
Village	:	Ashapuri
Block	:	Lamba Gaon
Tehsil	:	Jaisinghpur
District	:	Kangra
ii) Approach	:	18 km from Panchrukhi and 32 km from Palampur.
iii) Altitude	:	1500 m.
Management	:	Government undertaken (under Archaeological Survey of India). Local temple committee also exists.
Historical Background	:	The temple is believed to be built by Pandavas who stayed here during their exile. The present temple was built by Raja Vijaya Rama (the eldest son of Raja Chandrabhan) in 1687 A.D.
Temple Architect	:	Sikhara style.

Offerings to God

(i) By Priest

Morning : Black gram (boiled), 'Halva', 'Puffed rice, 'Patasha'.

Evening : Milk, Fruits.

(ii) By People : Puffed rice, 'Patasha', 'Halva', Seasonal crops. 'Bil-patri', 'Bhang' and 'Dhatura' are the special offerings to appease Lord Shiva.

Off. to People (Prasadam) : Puffed rice, 'Patasha', 'Halva'.

Fairs and Festivals : Navratra (twice in a year), Janamashtmi, Holi.

Observations

(i) Courtyard Plants (Protected as well as for Landscaping : *Cassia fistula, Cynodon dactylon, Phoenix sylvestris, Pistacia integerrima, Prunus puddum, Thuja orientalis.*

(ii) Predominant Plants of the Courtyard : *Phoenix sylvestris.*

(iii) Maintenance : Excellent.

(iv) Ethnobotanical

Sacred : *Cynodon dactylon* employed in all religious ceremonies as sacred plant.

Medicinal : 4-5g seed powder of *Cassia fistula* given daily to cure asthma. Powdered dried fruit of *Pistacia integerrima* used for cough.

Edible : Fruits of *Phoenix sylvestris* and *Prunus puddum* edible.

Ornamental : *Cassia fistula* and *Thuja orientalis* grown as ornamentals.

Other Uses : Leaves of *Phoenix sylvestris* used for making mats and brooms and that of *Prunus puddum* as good fodder for livestock.

2. Baba Balak Rupi Temple (Pl. 7B)

Main Deity : Baba Balak Rupi (form of Lord Shiva).

Other Deities : Hanuman, Ganesha, Kamakhya Devi, Nandigan.

Tot. Ar. of Tem. Courtyard : About 3000 sq m.

Location

(i) Place

Village : Balak Rupi

Block : Lamba Gaon

Tehsil : Jaisinghpur

District : Kangra

(ii) Approach : 4 km from Alampur.

(iii) Altitude : 700 m.

Management : Temple committee.

Historical Background : Temple was damaged during earthquake in 1905. Later, it was reconstructed by the local people.

Temple Architect : Sikhara style.

Offerings to God

(i) By Priest

Morning : Sugar Candy.

Noon : 'Rot'.

Evening : Black gram.

(ii) By People : Puffed rice, 'Lachidana', 'Halva', 'Rot', Sweets, Fruits, 'Bil-patri', 'Bhang', Rice, Wheat flour.

Off. to People (Prasadam) : Puffed rice, 'Lachidana', 'Halva', 'Rot'.

Fairs and Festivals : Fair is held on every Saturday of Jayeshth and Ashad month, Janamashtmi, Shivratri, Holi.

Observations

(i) Courtyard Plants (Protected as well as for Landscaping : *Euphorbia milii.*

(ii) Maintenance : Satisfactory.

(iii) Ethnobotanical

Ornamental : *Euphorbia milii* grown as an ornamental.

3. Baba Kalinath Temple (Pl. 7C)

Main Deity : Baba Kalinath.

Other Deities : Shiva-Parvati.

Tot. Ar. of Tem. Courtyard : About 1500 sq m.

Location

(i) Place

Town : Baroh

Block : Nagrota Bagwan

Tehsil : Baroh

District : Kangra

(ii) Approach : 14 km from Ranital.

(iii) Altitude : 700 m.

Management : Privately owned.

Historical background : The temple was built by King of Guler.

Temple Architect : Sikhara style.

Offerings to God

(i) By Priest : Rice with Clarified butter, Sugar.

(ii) By People : Puffed rice, 'Bundi', Sweets. 'Bil-patri', 'Bhang' offered to appease Lord Shiva.

Off. to People (Prasadam) : Puffed rice, 'Bundi', Sweets.

Fairs and Festivals : Local fairs are held in July-August. Besides, Shivratri and Janamashtmi are also celebrated.

Observations

(i) Courtyard Plants (Protected as well as for Landscaping : *Canna indica, Catharanthus roseus, Carica papaya, Cestrum nocturnum, Cynodon dactylon, Ficus religiosa, Holoptelea integrifolia, Hibiscus syriacus, Malvaviscus conzatii, Mangifera indica, Tabernaemontana coronaria, Tagetes erecta.*

(ii) Predominant Plants of the Courtyard : *Tagetes erecta.*

(iii) Maintenance : Satisfactory.

(iv) Ethnobotanical

Sacred : Wood of *Ficus religiosa* used for performing sacrificial fire ('havan') during religious ceremonies. *Holoptelea integrifolia* considered as sacred plant. *Cynodon dactylon* employed in various religious ceremonies.

Medicinal : Poultice of leaves of *Catharanthus roseus* appied to cure insect bites. Softwood of *Carica papaya* used for skin diseases.

Edible : Fruits of *Mangifera indica* edible. Its juice considered refreshing and cooling. Pulp of fruit used for making 'Am Papad'. Unripe fruits pickled and also used for making 'chutney'. Unripe fruits used for making 'Amchur' which is used for making various local dishes. Its fruits also used for making candy. Ripe fruits of *Carica papaya* edible whereas unripe ones cooked as vegetable.

Ornamental : *Canna indica, Catharanthus roseus, Cestrum nocturnum, Hibiscus syriacus, Holoptelea integrifolia, Malvaviscus conzatii, Tabernaemontana coronaria* and *Tagetes erecta* grown as ornamentals.

Other Uses	: Leaves of *Mangifera indica* used for scouring teeth and those of *Holoptelea integrifolia* used as fodder. Wood used for construction work.
Remark	: Goats sacrificed to appease deities.

4. Bagulamukhi Temple (Pl. 7D)

Main Deity	: Mata Bagulamukhi.
Other Deities	: Ganesha, Lakshmi, Saraswati, Hanuman, Annapurna Mata, Shiva.
Tot. Ar. of Tem. Courtyard	: About 1500 sq m.
Location	
(i) Place	
Village	: Vankhandi
Block	: Dehra
Tehsil	: Dehra
District	: Kangra
(ii) Approach	: About 10 km from Dehra and 10 km from Ranital.
(iii) Altitude	: 500 m.
Management	: Temple committee.
Historical Background	: It is believed that Lord Rama meditated at this place. Later, Meghnad (son of Ravana), meditated here for defeating Lord Rama in war. In his mission he was disturbed by Hanuman. Later, temple was made here by Pandavas.
Temple Architect	: Sikhara style.
Offerings to God	
(i) By Priest	: Whatever prepared in community meal ('Langar') is also offered to deity.
(ii) By People	: Puffed rice, 'Lachidana'. 'Bhang' offered to Lord Shiva.
Off. to People (Prasadam)	: Puffed rice, 'Lachidana'.
Fairs and Festivals	: Navratra, Shivratri.
Observations	
(i) Courtyard Plants (Protected as well as for Landscaping	: *Aegle marmelos, Bougainvillea spectabilis, Citrus aurantifolia, Cynodon dactylon, Ficus religiosa, Morus alba.*
(ii) Predominant Plants of the Courtyard	: *Bougainvillea spectabilis.*

(iii) Maintenance : Excellent.

(iv) Ethnobotanical

Sacred : Leaves of *Aegle marmelos* offered to appease Lord Shiva. *Ficus religiosa* worshipped as sacred tree. Its wood used in sacrificial fire ('havan') for performing religious rites. *Cynodon dactylon* considered sacred and employed in various religious ceremonies.

Medicinal : Decoction of leaves of *Bougainvillea spectabilis* given to cure diabetes.

Edible : Fruits of *Aegle marmelos, Morus alba* edible, and those of *Citrus aurantifolia* pickled.

Ornamental : *Bougainvillea spectabilis* grown as an ornamental plant.

Other Uses : Leaves of *Morus alba* used as fodder. Also, its dried branches used as firewood.

5. Bajreshwari Devi Temple (Pl. 7E)

Main Deity : Bajreshwari Devi (marble image of the goddess).

Other Deities : Shiva, Bhairav, Hanuman, Lakshmi, Kali, Rama, Lakshman, Sita.

Tot. Ar. of Tem. Courtyard : About 1500 sq m.

Location

(i) Place

Town : Kangra

Block : Kangra

Tehsil : Kangra

District : Kangra

(ii) Approach : The temple is located in the town.

(iii) Altitude : 700 m.

Management : Government controlled. Temple committee also exists.

Historical Background : The temple belongs to 8th-9th century A.D. This temple was damaged in the devastating earthquake of April 4, 1905. Later, it was rebuilt by the temple committee.

Temple Architect : Sikhara style.

Offerings to God

(i) By Priest

Morning : Dry Fruits like Chestnut, Cashewnut, Almond, Currant, Dried Palm Fruits, Coconut, etc.

Noon : Rice-Dal.

Evening : Puri-Chana, Milk.

(ii) By People	:	Coconut, Puffed rice, 'Lachidana', Sugar candy, Currant.
Off. to People (Prasadam)	:	Puffed rice, 'Lachidana', Sugar candy, Currant.
Fairs and Festivals	:	Minjar festival (for one week-at the end of July month), Makar Sakranti festival (13-14 January)-about 3 quintals clarified butter ('ghee') offered to Goddess, Navratra Fair.
Observations		
(i) Courtyard Plants (Protected as well as for Landscaping	:	*Amaryllis reginae, Araucaria columnaris, Ficus benghalensis, Ficus religiosa, Tagetes erecta.*
(ii) Maintenance	:	Excellent.
(iii) Ethnobotanical		
Sacred and Associated Myth	:	*Ficus benghalensis* and *Ficus religiosa* worshipped as sacred plants. Flowers of *Tagetes erecta* considered sacred for offering to the deity. People tie scarves to *Ficus benghalensis* and sacred threads to the *Ficus religiosa* for the fulfilment of their desires.
Ornamental	:	*Amaryllis reginae, Araucaria columnaris* and *Tagetes erecta* grown as ornamentals
Remarks	:	No sacrifices are offered.

6. Bhagsu Nath Temple (Pl. 7F, G)

Main Deity	:	Shiva, Nag Devta.
Other Deities	:	Radha, Krishana, Mira.
Tot. Ar. of Tem. Courtyard	:	About 3000 sq m including a pond in the courtyard of the temple.
Location		
(i) Place		
Town	:	Bhagsu
Block	:	Dharamshala
Tehsil	:	Dharamshala
District	:	Kangra
(ii) Approach	:	2 km from MacLeodganj.
(iii) Altitude	:	1750 m.
Management	:	Owned by Mahant Family.

Historical Background	:	In Dwaper Yug, there was a drought in Ajmer at that time ruled by Bhagsu-the king of devils. In search of the source of water, he reached at the top of a hill and brought water from Nag Dul with his supernatural powers. Nag followed him. At the place of present temple there was a big fight between them and Nag finally succeeded in killing Bhagsu. On his request Nag promised him that this place will be known by his name and both of them shall also be worshipped together. The temple dates back to the second half of the sixteenth century A.D. and built by Raja Dharam Chand.
Temple Architect	:	It is a Sikhara type of Structure.
Offerings to God		
(i) By Priest	:	Milk, Clarified butter, Rice, 'Dal', 'Chapati'.
(ii) By People	:	Puffed rice, 'Lachidana', Sugar candy, Fox nut ('Makhana'), Flowers, Fruits.
Off. to People (Prasadam)	:	Puffed rice, 'Lachidana', Fox nut, Currant, Sugar candy.
Fairs and Festivals	:	Radha Ashtmi, Krishana Ashtmi, Shivratri.
Observations		
(i) Courtyard Plants (Protected as well as for Landscaping	:	*Aesculus indica, Cedrus deodara, Cupressus torulosa, Cynodon dactylon, Jasminum primulinum, Lonicera japonica, Tagetes erecta.*
(ii) Predominant Plants of the Courtyard	:	*Tagetes erecta.*
(iii) Maintenance	:	Satisfactory.
(iv) Ethnobotanical		
Sacred	:	*Cynodon dactylon* considered sacred and employed in various religious ceremonies.
Edible	:	Fruits of *Aesculus indica* are edible.
Medicinal	:	Powdered fruits of *Aesculus indica* applied to cure skin problems.
Ornamental	:	*Aesculus indica, Cupressus torulosa, Jasminum primulinum, Lonicera japonica,* and *Tagetes erecta* grown as ornamental plants.
Other Uses	:	Wood of *Cedrus deodara* and *Aesculus indica* used for construction works. Leaves of *Aesculus indica* used as fodder.
Remarks	:	Goats sacrificed to appease deity.

7. Budhist Temple (Thek Chen Choiling) (Pl. 7H-K)

Main Deity	:	Gautam Buddha.
Other Deities	:	Avalokitesvara, Ajari Gusupema Sambhava.
Tot. Ar. of Tem. Courtyard	:	About 3000 sq m.
Location		
(i) Place		
Town	:	Dharamshala (Mc Leodganj)
Block	:	Dharamshala
Tehsil	:	Dharamshala
District	:	Kangra
(ii) Approach	:	15 km from Dharamshala.
(iii) Altitude	:	1730 m.
Management	:	Temple committee.
Historical Background	:	It was built in 1966.
Temple Architect	:	Buddhist Monastery.
Offerings to God		
By People	:	Biscuit, Fruits.
Off. to People (Prasadam)	:	'Lachidana', Biscuit, Fruits.
Fairs and Festivals	:	Budh Purnima.
Observations		
(i) Courtyard Plants (Protected as well as for Landscaping	:	*Aloe vera, Araucaria columnaris, Bougainvillea glabra, Cedrus deodara, Cestrum nocturnum, Cupressus torulosa, Diospyros lotus, Ehretia acuminata, Ficus religiosa, Ficus virens, Hydrangea macrophylla, Jacaranda acutifolia, Lagerstroemia indica, Pelargonium hortorum, Populus nigra, Pyrus pashia, Quercus incana, Robinia pseudoacacia, Salix tetrasperma, Tagetes erecta, Thuja orientalis.*
(ii) Predominant Plants of the Courtyard	:	*Cedrus deodara, Populus nigra, Thuja orientalis.*
(iii) Maintenance	:	Excellent.
(iv) Ethnobotanical		
Sacred	:	Leaves of *Cupressus torulosa* and *Thuja orientalis* burnt as incense in the temple. One of the Poplar trees (*Populus nigra*) in the courtyard treated as sacred and worshipped.
Medicinal	:	Leaf pulp of *Aloe vera* prescribed with milk to cure stomach problems.

Edible : *Aloe vera* used as vegetable. Fruits of *Ehretia acuminata* and *Pyrus pashia* eaten.

Ornamental : *Aloe vera, Araucaria columnaris, Bougainvillea glabra, Cestrum nocturnum, Cupressus torulosa, Diospyros lotus, Ficus religiosa, Ficus virens, Jacaranda acutifolia, Lagerstroemia indica, Pelargonium hortorum, Robinia pseudoacacia, Salix tetrasperma, Tagetes erecta, Thuja orientalis* grown as ornamental plants. *Jacaranda acutifolia* planted as an avenue plant.

Other Uses : Leaves of *Ehretia acuminata, Quercus incana* and *Robinia pseudoacaiaa* used as fodder. Wood of *Cedrus deodara* excellent for furniture and construction works. Wood of *Quercus incana* and *Robinia pseudoacaiaa* used for agricultural implements.

8. Buddha Temple (Pl. 8A)

Main Deity : Gautam Buddha.

Other Deities : Marpa, Milah, Thapo.

Tot. Ar. of Tem. Courtyard : About 1500 sq m.

Location

(i) *Place*

Village : Tilokpur

Block : Surian Nagrota

Tehsil : Jawali

District : Kangra

(ii) *Approach* : About 50 km from Kangra.

(iii) *Altitude* : 580 m.

Management : Temple committee.

Historical Background : The temple is about 37 years old.

Temple Architect : Buddhist Monastery.

Offerings to God : Biscuits, Fruits.

Off. to People (Prasadam) : Biscuits, Fruits.

Observations

(i) *Courtyard Plants (Protected as well as for Landscaping* : *Araucaria columnaris, Bougainvillea spectabilis, Callistephus chinensis, Citrus jambhiri, Dalbergia sissoo, Eriobotrya japonica, Ficus religiosa, Hibiscus syriacus, Jasminum sambac, Kochia trichophylla, Pelargonium hortorum, Psidium guajava, Rosa indica, Tagetes erecta, Thuja orientalis.*

(ii) *Predominant Plants of the Courtyard* : *Bougainvillea spectabilis, Tagetes erecta.*

(iii) *Maintenance* : Excellent.

(iv) *Ethnobotanical*

Sacred : Flowers of *Bougainvillea spectabilis, Hibiscus syriacus, Jasminum sambac, Rosa indica* and *Tagetes erecta* considered sacred for offering to the deity.

Edible : Fruits of *Citrus jambhiri, Eriobotrya japonica, Psidium guajava* edible.

Ornamental : *Araucaria columnaris, Bougainvillea spectabilis, Callistephus chinensis, Eriobotrya japonica, Hibiscus syriacus, Jasminum sambac, Kochia trichophylla, Pelargonium hortorum, Rosa indica, Tagetes erecta* and *Thuja orientalis* commonly grown as ornamentals.

Other Uses : Leaves and branches of *Psidium guajava* used for scouring teeth. Wood of *Dalbergia sissoo* excellent for furniture and building works.

9. Chamunda Nandikeshwar Temple (Pl. 8B)

Main Deity : Chamunda Devi, Shiva.

Tot. Ar. of Tem. Courtyard : About 7500 sq m including a lake and a park.

Location

(i) *Place*

Village : Jadrangal

Block : Nagrota Bagwan

Tehsil : Dharamshala

District : Kangra

(ii) *Approach* : The temple is situated in the lap of Dhauladhar range of mountains on the side of Baner Khadd (Baanganga), which is 15 km from Dharamshala on Palampur road.

(iii) *Altitude* : 980 m.

Management : Government owned (Shri Chamunda Nandikeshwar Dham Trust).

Historical Background : In the battle between the gods and the demons, Chandika emerged from an eyebrow of goddess 'Kaushiki' and assigned the task of eliminating the demons 'Chand' and 'Mund'. Chandika killed them after a fierce battle and took their slain heads to the goddess 'Kaushiki' who being immensely pleased,

		bestowed upon her the title of 'Chamunda'. The temple, about 750 years old, was built by Raja Chander Bhan.
Temple Architect	:	Sikhara style.
Offerings to God		
(i) By Priest		
Morning	:	Black gram, 'Halva', Sweets.
Noon/Evening	:	Rice-Dal, Chapati-Vegetables, 'Kheer' or Sweet rice.
(ii) By People	:	Puffed rice, Fox nut, 'Lachidana', Sugar candy, Currant. 'Bil-patri' (Leaves of Bil tree) specially offered to Lord Shiva.
Off. to People (Prasadam)	:	Puffed rice, Fox nut, 'Lachidana', Sugar candy, Currant.
Fairs and Festivals	:	Navratra Fair (Thrice in a year), Shivratri, Holi, Baisakhi, Dussehra.
Observations		
(i) Courtyard Plants (Protected as well as for Landscaping	:	*Albizia stipulata, Amaryllis reginae, Canna indica, Cynodon dactylon, Cyperus flabelliformis, Ficus religiosa, Litchi chinensis, Mangifera indica, Psidium guajava, Sansevieria trifasciata, Syzygium jambos, Tagetes erecta, Terminalia bellirica.*
(ii) Maintenance	:	Excellent.
(iii) Ethnobotanical		
Sacred	:	*Ficus religiosa* worshipped and its wood used in 'havan' for performing religious rites. *Cynodon dactylon* considered sacred and employed in all religious ceremonies.
Medicinal	:	Fruits of *Terminalia bellirica* used in the preparation of 'Triphala' which is prescribed to cure stomach problems. Its fruit pulp eaten raw for diarrhoea.
Edible	:	Fruits of *Litchi chinensis, Mangifera indica, Psidium guajava, Syzygium jambos* and *Terminalia bellirica* edible.
Ornamental	:	*Albizia stipulata, Amaryllis reginae, Canna indica, Cyperus flabelliformis, Sansevieria trifasciata, Tagetes erecta* and *Terminalia bellirica* grown as ornamental.
Other Uses	:	Leaves of *Psidium guajava* and *Mangifera indica* used for scouring teeth. Leaves of *Albizia stipulata* used as fodder. The plant of *Cyperus flabelliformis* grown in house courtyards to ward off snakes. Wood of *Albizia stipulata* and *Terminalia bellirica* used for agricultural implements. Wood of *Mangifera indica* used for construction works.
Status of Courtyard Plants in Surrounding Areas	:	Except *Syzygium jambos* all other plants are common outside.

10. Jwalamukhi Temple (Pl. 8C)

Main Deity	:	Jwalaji (Fire flames representing the deity).
Tot. Ar. of Tem. Courtyard	:	About 3000 sq m.
Location		
(i) Place		
Town	:	Jwalamukhi
Block	:	Dehra
Tehsil	:	Dehra
District	:	Kangra
(ii) Approach	:	55 km from Dharamshala.
(iii) Altitude	:	610 m.
Management	:	Government undertaken (Department of Language and Culture).
Historical Background	:	It is believed that the temple of Jwalamukhi must have existed as early as 650 A.D. A Yajna was arranged by Raja Daksh (father of Uma or Sati) for which Sati was also invited. There her husband, Lord Shiva was disgraced by Raja Daksh. On seeing this she ended her life in 'havan-kund'. Lord Shiva carried her body in burnt condition aimlessly. Lord Vishnu became afraid of the consequences of his anger. So her body was cut into pieces with the help of his 'Sudershan Chakra'. All the parts of her body fell at 51 places. These places are called as 'Shaktipeeth'. It is believed that tongue of Sati dropped at Jwalaji. Since then this place is worshipped as 'Shaktipeeth. There are 3 other 'Shaktipeeth' in Himachal Pradesh. These are Bajreshwari Devi at Kangra, Chintpurni at Una and Naina Devi at Bilaspur.
Temple Architect	:	Sikhara style.
Offerings to God		
(i) By Priest		
Morning	:	Sweets, Sugar Candy.
Noon	:	Rice-Dal, Sweets.
Evening	:	'Puri', Black gram, 'Halva'.
(ii) By People	:	Puffed rice, 'Lachidana', Sugar Candy.
Off. to People (Prasadam)	:	Puffed rice, 'Lachidana', Sugar Candy.
Fairs amd Festivals	:	Jwalaji Janamotsav celebrated on Ashtmi of Magh-Shukla Paksh; Katiyayini celebrated on the Ashtmi of Shravan Shukla Paksh and Navratra.

Observations

(i) Courtyard Plants (Protected as well as for Landscaping	: *Mimusops elengi.*
(ii) Maintenance	: Satisfactory.
(iii) Ethnobotanical	
Sacred	: *Mimusops elengi* considered as sacred plant. People tie threads to this tree for the fulfilment of their desires. Flowers of *Mimusops elengi* used for bedding of the deity.
Status of Courtyard Plants in Surrounding Areas	: *Mimusops elengi* is absent in the surrounding areas of the temple.
Remarks	: Goats are sacrificed. There are only 3 Moulsari trees in District Kangra: one in the Brijraj Swami temple, second at Shibbo Than and third at Jwalaji.

11. Kalinath Kaleshwar Mahadev Temple (Pl. 8D)

Main Deity	: Kaleshwar Mahadev (Shiva).
Other Deities	: Eleven Rudra (Eleven Forms of Lord Shiva), Ashtbhuji Mata, Radha-Krishana, Hanuman, Ganesh, Nandi.
Tot. Ar. of Tem. Courtyard	: Total area of the temple land is 7 hectare which includes cultivation land (about 2 Acre), an orchard of mango trees (1 hectare) and rest of the surrounding area of the temple courtyard of about 3000 sq m.
Location	
(i) Place	
Village	: Kaleshwar
Block	: Pragpur
Tehsil	: Dehra
District	: Kangra
(ii) Approach	: 15 km from Nadaun; temple is located on the side of Vyas River.
(iii) Altitude	: 465 m.
Management	: Temple committee.
Historical Background	: Long ago, Goddess Kali worshipped Lord Shiva at this place. Hence, it is named as Kalinath Kaleshwar Mahadev Temple. The present temple was built in 19th century A.D. by the King of Amb and Nadaun.
Temple Architect	: Sikhara style. The temple is a dressed stone structure.

Offerings to God		
(i) By Priest	:	Rice, Dal, Chapati, Vegetables.
(ii) By People	:	Puffed rice, Fruits, Sweets, Butter Milk ('Lassi'), Milk, 'Bil-patri', 'Bhang'.
Off. to People (Prasadam)	:	Puffed rice.
Fairs and Festivals	:	Shivratri, Baisakhi Fair (for 3 days), Local fair held on every Monday of Shravan Month.
Observations		
(i) Courtyard Plants (Protected as well as for Landscaping	:	*Aegle marmelos, Amaryllis reginae, Aloe vera, Cassia fistula, Cedrela toona, Citrus jambhiri, Cordia dichotoma, Cynodon dactylon, Emblica officinalis, Ficus benghalensis, Ficus religiosa, Grewia oppositifolia, Jasminum sambac, Lagerstroemia indica, Mallotus philippinensis, Mangifera indica, Mirabilis jalapa, Morus alba, Ocimum sanctum, Plumeria acutifolia, Polyalthia longifolia, Psidium guajava, Rosa indica, Tagetes erecta, Thevetia neriifolia, Zizyphus mauritiana.*
(ii) Predominant Plants of the Courtyard	:	*Ficus religiosa, Thevetia neriifolia.*
(iii) Maintenance	:	Satisfactory.
(iv) Ethnobotanical		
Sacred	:	Leaves of *Aegle marmelos* especially offered to Lord Shiva. *Ficus benghalensis, Ficus religiosa* and *Ocimum sanctum* worshipped as sacred plants. *Cynodon dactylon* employed in various religious ceremonies.
Medicinal	:	Powder removed from the fruits of *Mallotus philippinensis* consumed with butter milk ('Lassi') for stomachache and to expell worms. Fruits of *Emblica officinalis* used in preparation of 'Triphala' which is prescribed for stomach problems. Its powdered seeds effective for liver disorders. 50-100g of mixture containing equal quantity of *Aloe vera,* roasted black sesame ('Kala til') and jaggery ('Gur') taken daily to cure prostate and bed wetting.
Edible	:	*Aloe vera* used as vegetable. Fruits of *Aegle marmelos, Citrus jambhiri, Cordia dichotoma, Emblica officinalis, Mangifera indica, Morus alba, Psidium guajava* and *Zizyphus mauritiana* edible whereas those of *Mangifera indica, Cordia dichotoma, Emblica officinalis* used for making pickles. Fruits of *Mangifera indica* and *Emblica officinalis* also good for making candies.

Ornamental : *Amaryllis reginae, Aloe vera, Cassia fistula, Jasminum sambac, Lagerstroemia indica, Mirabilis jalapa, Ocimum sanctum, Plumeria acutifolia, Polyalthia longifolia, Rosa indica, Tagetes erecta* and *Thevetia neriifolia* planted as ornamentals.

Other Uses : Leaves of *Grewia oppositifolia* and *Morus alba* used as fodder. Wood of *Cedrela toona* excellent for construction and furniture; that of *Mallotus philippinensis* and *Cassia fistula* for agricultural implements.

Remarks : No sacrifices are offered. It is believed that Pandavas brought water from five sacred places ('Tirth Dham') and created a sacred water body near the temple, called 'Panchtirth'.

12. Lakshmi Narayan Temple or Thakurdwara (Pl. 8E)

Main Deities : Narayan (Lord Vishnu), Lakshmi.

Other Deities : Shiva, Vaishno Devi, Durga, Hanuman.

Tot. Ar. of Tem. Courtyard : About 1500 sq m.

Location

(i) Place

Village : Alampur

Block : Lamba Gaon

Tehsil : Jaisinghpur

District : Kangra

(ii) Approach : 3 km from Sujanpur.

(iii) Altitude : 680 m.

Management : Temple committee.

Historical Background : The temple is believed to be constructed in 1655 A.D. by Raja Abhay Chand-King of Sujanpur Tira.

Temple Architect : Sikhara style.

Offerings to God

(i) By Priest

Morning : Sugar, Sugar Candy, Sweet fennel ('Mithi saunf').

Noon : Cooked rice.

Evening : Sugar, Sugar Candy, Sweet fennel.

(ii) By People : 'Halva', Black gram, 'Kasaar', Sugar candy, Sugar, Sweet fennel. 'Bil-patri' and 'Bhang' offered to Lord Shiva as special offerings.

Off. to People (Prasadam) : Sugar candy, Sugar, Black gram.

Fairs and Festivals : Janamashtmi Fair.

Observations

(i) Courtyard Plants (Protected as well as for Landscaping : *Chrysanthemum indicum, Tagetes erecta.*

(ii) Maintenance : Excellent.

(iii) Ethnobotanical

Sacred : Flowers of *Chrysanthemum indicum* and *Tagetes erecta* offered to the deities.

Ornamental : *Chrysanthemum indicum* and *Tagetes erecta* grown as ornamental plants.

Remarks : No sacrifices are made.

13. Lakshmi Narayan Temple (Pl. 8F)

Main Deity : Lakshmi Narayan.

Other Deities : Durga Chandi, Rudra Bhagwan (form of Lord Shiva).

Tot. Ar. of Tem. Courtyard : About 3000 sq m (total area of the temple land is 9 hectare.

Location

(i) Place

Village : Chanor

Block : Pragpur

Tehsil : Dehra

District : Kangra

(ii) Approach : About 17 km from Dehra.

(iii) Altitude : 600 m.

Management : Temple committee.

Historical Background : The temple is said to be constructed by the Pandavas. Later it was renovated several times by the locals.

Temple Architect : Sikhara style.

Offerings to God

(i) By Priest

Morning : Milk-Bhog.

Noon and Evening : Food prepared in community meal.

(ii) By People : 'Kasaar', 'Lachidana', Puffed rice, 'Patasha'. 'Bil-patri', 'Bhang', 'Dhatura' offered to appease Lord Shiva.

Off. to People (Prasadam)	: 'Lachidana', Puffed rice, 'Patasha'.
Fairs and Festivals	: Sakranti celebrated every month, Janamashtmi.
Observations	
(i) Courtyard Plants (Protected as well as for Landscaping	: *Amaryllis reginae, Artocarpus lakoocha, Bougainvillea spectabilis, Bauhinia vahlii, Canna indica, Cassia fistula, Catharanthus roseus, Celosia argentea* var. *cristata, Citrus aurantifolia, Cynodon dactylon, Ficus religiosa, Ficus glomerata, Grevillea robusta, Hibiscus rosa-sinensis, Hibiscus syriacus, Lagerstroemia indica, Mangifera indica, Nerium odorum, Ocimum sanctum, Putranjiva roxburghii, Rosa indica, Sansevieria trifasciata, Syzygium cumini, Tagetes erecta, Thevetia neriifolia, Thuja orientalis.*
(ii) Predominant Plants of the Courtyard	: *Bougainvillea spectabilis, Canna indica.*
(iii) Maintenance	: Excellent.
(iv) Ethnobotanical	
Sacred and Associated Myth	: *Ficus religiosa* and *Ocimum sanctum* worshipped as sacred plants. Wood of *Ficus religiosa* used in 'havan' for performing religious rites. Newly married couples tie their 'bengals' (bride) and 'sehra' (bridegroom) to the tree of *Thuja orientalis* growing in the temple courtyard for their prosperous life. *Cynodon dactylon* employed in all religious ceremonies as sacred plant.
Medicinal	: Fruits of *Artocarpus lakoocha* considered good for cough. Seed powder of *Cassia fistula* given for constipation.
Edible	: Young pods of *Cassia fistula* used as vegetable. Ripe fruits of *Artocarpus lakoocha* generally used in preparing 'chutneys' whereas unripe ones cooked as vegetable or pickled. Fruits of *Ficus glomerata, Mangifera indica, Syzygium cumini* edible and that of *Citrus aurantifolia* pickled.
Ornamental	: *Amaryllis reginae, Bougainvillea spectabilis, Canna indica, Catharanthus roseus, Celosia argentea* var. *cristata, Grevillea robusta, Hibiscus rosa-sinensis, Hibiscus syriacus, Lagerstroemia indica, Nerium odorum, Ocimum sanctum, Putranjiva roxburghii, Rosa indica, Sansevieria trifasciata, Tagetes erecta, Thevetia neriifolia* and *Thuja orientalis* planted for beautification of the temple.
Other Uses	: Leaves of *Bauhinia vahlii* used for making meal plates. Wood of *Cassia fistula* good for making agricultural implements and that of *Syzygium cumini* for packing cases.
Remarks	: Thick forest (spreading to an area of 9 hectare) attached to the temple is protected on religious grounds.

14. Machhial Mahadev Temple (Pl. 8G)

Main Deity	: Shiva.
Other Deities	: Bhairav, Shani Dev, Hanuman, Kali, Durga.
Tot. Ar. of Tem. Courtyard	: About 3000 sq m.
Location	
(i) Place	
Village	: Andreta
Block	: Panchrukhi
Tehsil	: Palampur
District	: Kangra
(ii) Approach	: 2 km from Panchrukhi, on Palampur-Baijnath road.
(iii) Altitude	: 990 m.
Management	: In the name of Deity.
Historical Background	: Long ago, Baba Machhender Nath meditated here. Since then this place is known as 'Machhial Mahadev Temple'. It was reconstructed in 1999 by locals and the Mahant of the temple with donations given by devotees.
Temple Architect	: Sikhara style.
Offerings to God	
(i) By Priest	: 'Halva', Dry fruits like Almond, Currant, Cashewnut, Dried palm fruits, Coconut, etc.
(ii) By People	: 'Halva', Puffed rice, 'Lachidana', Fruits.
Off. to People (Prasadam)	: 'Halva', Puffed rice, 'Lachidana', Fruits.
Fairs and Festivals	: Machhial Ka Mela (14-16 September), Guru Purnima (in July), Shivratri.
Observation	
(i) Courtyard Plants (Protected as well as for Landscaping	: *Aegle marmelos, Calendula officinalis, Cannabis sativa, Canna indica, Carica papaya, Catharanthus roseus, Cedrela toona, Centella asiatica, Chrysanthemum indicum, Citrus reticulata, Colocasia esculenta, Cosmos sulphureus, Cynodon dactylon, Dalbergia sissoo, Datura fastuosa, Ehretia acuminata, Ficus glomerata, Ficus religiosa, Grewia oppositifolia, Hibiscus syriacus, Justicia gendarussa, Lagerstroemia indica, Machilus gamblei, Malvaviscus conzatii, Mentha piperita, Morus alba, Musa paradisiaca, Nerium odorum, Ocimum basilicum, Pisum sativum, Populus nigra, Quercus incana, Rosa indica, Salix*

			tetrasperma, Salvia splendens, Sapium sebiferum, Tagetes erecta, Zea mays, Zephyranthes grandiflora, Zinnia elegans.
(ii)	*Predominant Plants of the Courtyard*	:	*Cannabis sativa, Tagetes erecta.*
(iii)	*Maintenance*	:	Excellent.
(iv)	*Ethnobotanical*		
	Sacred	:	*Ficus religiosa, Ocimum basilicum, Aegle marmelos* and *Ficus glomerata* considered sacred plants. *Cynodon dactylon* employed in various religious ceremonies as sacred plant. Leaves of *Aegle marmelos, Cannabis sativa* and flowers of *Datura fastuosa* offered to appease Lord Shiva. Wood of *Ficus glomerata* and *Ficus religiosa* used as sacred wood in 'havan'.
	Medicinal	:	Leaves of *Mentha piperita* chewed for stomach troubles. One leaf of *Centella asiatica* and one black pepper (*Piper nigrum* Linn.) taken daily to cure stomach problems. 11 leaves of *Centella asiatica,* 3 black pepper and 1 cardamom powdered and then fried in cow's clarified butter and taken in morning with cow's milk to enhance memory. Powdered leaves of *Centella asiatica* used to cure liver problems. Tablets prepared by powdering 4-5 leaves of *Centella asiatica* alongwith 1 black pepper prescribed to cure jaundice. Poultice of leaves and bark of *Sapium sebiferum* applied to cure joint pains.
	Edible	:	Seeds of *Pisum sativum* and leaves and rhizome of *Colocasia esculenta* used as vegetable. Leaves of *Mentha piperita* used for making 'chutney'. Grains of *Zea mays* used in the form of flour for making 'chapatis' and its immature cobs eaten after roasting. Juice of fruits of *Citrus reticulata* considered refreshing and cooling. Fruits of *Aegle marmelos, Carica papaya, Litchi chinensis, Mangifera indica, Morus alba* and *Musa paradisiaca* edible.
	Ornamental	:	*Calendula officinalis, Canna indica, Catharanthus roseus, Chrysanthemum indicum, Cosmos sulphureus, Datura fastuosa, Hibiscus syriacus, Justicia gendarussa, Lagerstroemia indica, Malvaviscus conzatii, Nerium odorum, Ocimum basilicum, Rosa indica, Salix tetrasperma, Salvia splendens, Sapium sebiferum, Tagetes erecta, Zephyranthes grandiflora, Zinnia elegans* grown as ornamentals.
	Other Uses	:	Leaves of *Mangifera indica* used for scouring teeth and its wood for construction works. Leaves of *Morus alba, Ehretia acuminata, Quercus incana, Sapium sebiferum* used as fodder. Leaves of *Grewia oppositifolia* used as fodder

for enhancing lactation and its branches for making ropes. Wood of *Cedrela toona, Dalbergia sissoo* excellent for furniture and construction works. Wood of *Populus nigra* used for making cheap furniture and packing cases.

Status of Courtyard Plants in Surrounding Areas : *Ficus glomerata* is rare outside. All other plants are common outside.

Remarks : No sacrifices are made. The sacred water body on the backside of the temple contains numerous fish which are protected on religious grounds.

15. Mahakal Temple (Pl. 8H)

Main Deity : Shiva.

Other Deities : Goddess Durga, Hanuman, Ganpati, Kedarnath, Surya Narayan, Dharamraj.

Tot. Ar. of Tem. Courtyard : Of the total area (4) hectare of temple land, a large portion includes the forest area surrounding the temple courtyard and rest is cultivated. The forest trees are those of Jamun, Biyunsh, Poplar, Khajur and Bamboo plants, bearing religious protection.

Location

(i) Place

Village : Mahakal

Block : Baijnath

Tehsil : Baijnath

District : Kangra

(ii) Approach : 5 km from Baijnath.

(iii) Altitude : 860 m.

Management : Temple committee.

Historical Background : It is believed that Mahakal temple was originally made by Pandavas. Later, it was renovated by Adiguru Shankeracharaya.

Temple Architect : Sikhara style.

Offerings to God

(i) By Priest

Morning : Currant.

Noon : Rice, Dal, Chapati, Vegetables.

Evening : Chapati, Vegetables.

(ii) By People : Sweets. On Saturday people offer mustard oil, 'mash' (seeds of *Vigna mungo*), 'til' (seeds of *Sesamum orientale*), 'sarson' (seeds of *Brassica campestris*), a piece of black cloth, black ribbon, black bengals, etc. to protect themselves from the evil effect of Shani. 'Bil-patri' (leaves of *Aegle marmelos*) especially offered to Lord Shiva.

Off. to People (Prasadam) : Currant, 'Makhana'.

Fairs and Festivals : Fair held on every Saturday of Bhado month (August-September), Nirjal Ekadashi fair in June and Shivratri.

Observations

(i) Courtyard Plants (Protected as well as for Landscaping : *Bougainvillea spectabilis, Bryophyllum calycinum, Calendula officinalis, Celosia argentea* var. *cristata, Dahlia rosea, Ficus religiosa, Gaillardia pulchella, Hibiscus syriacus, Litchi chinensis, Rosa indica, Rosa multiflora, Tagetes erecta.*

(ii) Predominant Plants of the Courtyard : *Hibiscus syriacus, Tagetes erecta.*

(iii) Maintenance : Satisfactory.

(iv) Ethnobotanical

Sacred : *Ficus religiosa* worshipped as sacred plant. Its wood used in 'havan' for performing religious rites.

Medicinal : 3 fresh leaves of *Ficus religiosa* taken thrice a day to cure jaundice.

Edible : Fruits of *Litchi chinensis* edible.

Ornamental : *Bougainvillea spectabilis, Bryophyllum calycinum, Calendula officinalis, Celosia argentea* var. *cristata, Dahlia rosea, Gaillardia pulchella, Hibiscus syriacus, Rosa indica, Rosa multiflora* and *Tagetes erecta* grown as ornamentals.

Remarks : 'Rudra Abhishek' of Lord Shiva is done on special occasions like Shivratri in which Shivlingam is first washed with milk, curd, clarified butter, honey, loaf sugar, sugarcane juice and silver powder added to water and then with 'haldi' (*Curcuma longa*) and 'Kumkum' water. After the bathing ritual the deity is offered camphor ('Kapur'), saffron ('Kesar'), 'Janeu', leaves and fruits of 'Bael' tree ('Bael-Patri', 'Bael-Phal'), 'doob' (*Cynodon dactylon*) and dry fruits (Cashewnut, Almond, Sugar candy, Currant, Dried Palm Fruit etc.).

16. Mangla Mata Temple (Pl. 8I)

Main Deity : Mangla Mata.

Other Deities : Durga Asht Bhavani, Shiva.

Tot. Ar. of Tem. Courtyard : 1500 sq m.

Location

(i) *Place*

Village	:	Tilokpur
Block	:	Nagrota Suriyan
Tehsil	:	Jawali
District	:	Kangra
(ii) *Approach*	:	About 50 km from Kangra.
(iii) *Altitude*	:	650 m.
Management	:	Village temple committee.
Historical Background	:	The temple is about 400 years old. At that time, Sikh families used to live in this village who established the deity as their Kuldevi.
Temple Architect	:	Sikhara style.
Offerings to God		
(i) *By Priest*	:	'Halva', 'Khir', Currant, Flowers, Fruits.
(ii) *By People*	:	'Halva', Puffed rice, 'Patasha'.
Off. to People (Prasadam)	:	'Halva', Puffed rice, 'Patasha', Fruits.
Fairs and Festivals	:	Navratra.
Observations		
(i) *Courtyard Plants (Protected as well as for Landscaping*	:	*Bauhinia vahlii, Boerhaavia diffusa, Bombax ceiba, Callistemon lanceolatus, Cassia glauca, Catharanthus roseus, Cynodon dactylon, Emblica officinalis, Ficus religiosa, Grevillea robusta, Jacaranda acutifolia, Lannea coromandelica, Pinus roxburghii, Pistacia integerrima, Pinus roxburghii, Tectona grandis, Terminalia bellirica, Terminalia chebula, Thuja orientalis.*
(ii) *Predominant Plants of the Courtyard*	:	*Grevillea robusta.*
(iii) *Maintenance*	:	Satisfactory.
(iv) *Ethnobotanical*		
Sacred	:	*Ficus religiosa* worshipped as sacred plant. Its wood used in 'havan' for performing religious rites. *Cynodon dactylon* considered sacred plant and employed in religious ceremonies.
Medicinal	:	Fruits of *Emblica officinalis, Terminalia bellirica* and *Terminalia chebula* used in preparation of 'Triphala' which is prescribed for acidity, constipation and other stomach problems. Mixture of 1 kg fresh fruits of

Terminalia chebula boiled in water in an earthenware vessel alongwith 100g salt, 100g black pepper (*Piper nigrum*) and 50g Ajwain (*Trachyspermum ammi*) prescribed for constipation (one fruit daily till cure). Decoction of roots of *Boerhaavia diffusa* used to cure body swelling.

Edible : Seeds of *Bombax ceiba, Pinus roxburghii* and fruits of *Emblica officinalis, Terminalia bellirica, Terminalia chebula* are edible. Leaves of *Boerhaavia diffusa* cooked as vegetable.

Ornamental : *Bombax ceiba, Callistemon lanceolatus, Cassia glauca, Catharanthus roseus, Emblica officinalis, Grevillea robusta, Jacaranda acutifolia, Pinus roxburghii, Tectona grandis, Terminalia bellirica, Terminalia chebula* and *Thuja orientalis* grown as ornamentals.

Other Uses : Leaves of *Bauhinia vahlii* used for making meal plates. Cones of *Pinus roxburghii* used for decorative purposes. Wood of *Bombax ceiba* used for construction works. Its cotton used for stuffing pillows and mattresses. Wood of *Terminalia bellirica* used for making agricultural implements.

17. Mukut Nath Temple (Pl. 8J)

Main Deity : Mukut Nath (Shiva).

Other Deity : Goddess Parvati.

Tot. Ar. of Tem. Courtyard : Total area of the temple land is about 2 acre of which about 1500 sq m is earmarked for temple and its courtyard. The crops of wheat, rice etc. are grown in vacant land which also contains some trees of Akhrot, Orange, Aru, Kachnar, Khajur, Puna and Guggal.

Location

(i) *Place*

Village : Sansaal

Block : Baijnath

Tehsil : Baijnath

District : Kangra

(ii) *Approach* : 6 km from Baijnath.

(iii) *Altitude* : 1330 m.

Management : Temple committee.

Historical Background	:	It is believed that the devil Jalandhar when killed in 'Dwapar Yug' was dropped here.
Temple Architect	:	Sikhara style.
Offerings to God		
(i) By Priest		
Noon	:	Rice, Dal, Chapati, Vegetables.
Evening	:	Currant, Fruits.
(ii) By People	:	'Halva', Puffed rice. 'Bhang', 'Dhatura' and 'Bael-patri' offered specially to Lord Shiva.
Off. to People (Prasadam)	:	'Halva', Puffed rice, Currant.
Fairs and Festivals	:	Shivratri.
Observations		
(i) Courtyard Plants (Protected as well as for Landscaping	:	*Bauhinia variegata, Canna indica, Citrus aurantifolia, Citrus limon, Dahlia rosea, Ehretia acuminata, Hibiscus syriacus, Hydrangea macrophylla, Malvaviscus conzatii, Mangifera indica, Nerium odorum, Ocimum basilicum, Prunus armeniaca, Rosa indica, Tagetes erecta.*
(ii) Predominant Plants of the Courtyard	:	*Ocimum basilicum.*
(iii) Maintenance	:	Satisfactory.
(iv) Ethnobotanical		
Sacred	:	*Ocimum basilicum* worshipped as a sacred plant.
Medicinal	:	Decoction of bark of *Bauhinia variegata* used to dissolve blood clots in body. Fruit juice of *Citrus limon* used to cure respiratory problems.
Edible	:	Flower buds of *Bauhinia variegata* cooked as vegetable. Fruits of *Mangifera indica* and *Prunus armeniaca* are edible whereas those of *Citrus aurantifolia, Citrus limon* used for pickling
Ornamental	:	*Bauhinia variegata, Canna indica, Dahlia rosea, Hibiscus syriacus, Hydrangea macrophylla, Malvaviscus conzatii, Nerium odorum, Ocimum basilicum, Rosa indica* and *Tagetes erecta* grown as ornamentals.
Other Uses	:	Leaves of *Ehretia acuminata* used as fodder and its wood for furniture.
Remarks	:	No sacrifices are made.

18. Nag Devta Temple (Shri Sheshnag Temple) (Pl. 8K)

Main Deity	:	Nag Devta.
Tot. Ar. of Tem. Courtyard	:	About 800 sq m.
Location		
(i) Place		
Village	:	Chelian
Block	:	Dehra
Tehsil	:	Dehra
District	:	Kangra
(ii) Approach	:	2 km on foot from Ranital.
(iii) Altitude	:	550 m.
Management	:	Privately owned.
Historical Background	:	The present temple is around 40 years old. Earlier idols of Nag Devta were kept in open under the Banyan tree. Later, temple was constructed at the same spot after removing the tree.
Temple Architect	:	Sikhara style.
Offerings to God		
(i) By Priest	:	Whatever being prepared in the home of priest.
(ii) By People	:	'Lachidana', Puffed rice.
Off. to People (Prasadam)	:	'Lachidana', Puffed rice.
Fairs and Festivals	:	Fairs are held during July-September for 2 months.
Observations		
(i) Courtyard Plants (Protected as well as for Landscaping	:	*Cynodon dactylon, Ficus religiosa, Hibiscus rosa-sinensis, Ocimum sanctum, Tabernaemontana coronaria, Tagetes erecta, Zizyphus mauritiana.*
(ii) Predominant Plants of the Courtyard	:	*Ocimum sanctum.*
(iii) Maintenance	:	Satisfactory.
(iv) Ethnobotanical		
Sacred and Associated Myth	:	*Ficus religiosa, Ocimum sanctum* and *Zizyphus mauritiana* worshipped as sacred plants. Wood of *Ficus religiosa* used in 'havan' for performing religious rites. *Cynodon dactylon* considered sacred and employed in all religious ceremonies. *Zizyphus mauritiana,* generally present in Nag temples, believed to be associated with Nag Devta. Snake bites cured here.

Medicinal	:	Poultice of leaves of *Zizyphus mauritiana* applied to cure skin disorders.
Edible	:	Fruits of *Zizyphus mauritiana* edible.
Ornamental	:	*Hibiscus rosa-sinensis, Ocimum sanctum, Tabernaemontana coronaria* and *Tagetes erecta* grown as ornamentals.

19. Nagni Mata Temple (Pl. 9A)

Main Deity	:	Nagni Mata.
Tot. Ar. of Tem. Courtyard	:	About 1500 sq m.
Location		
(i) *Place*		
Village	:	Nagni
Block	:	Nurpur
Tehsil	:	Nurpur
District	:	Kangra
(ii) *Approach*	:	About 9 km from Nurpur, on the side of Nurpur-Kangra road.
(iii) *Altitude*	:	600 m.
Management	:	Local temple committee.
Historical Background	:	The temple is believed to be 200 years old with idols kept in open before its restructuring in 1968.
Temple Architect	:	Sikhara style.
Offerings to God		
(i) *By Priest*	:	'Halva-Puri'.
(ii) *By People*	:	'Lachidana', Puffed rice.
Off. to People (Prasadam)	:	'Lachidana', Puffed rice.
Fairs and Festivals	:	8-9 fairs held every year during July-September.
Observations		
(i) *Courtyard Plants (Protected as well as for Landscaping*	:	*Bischoffia javanica, Bougainvillea spectabilis, Ficus benghalensis, Ficus religiosa, Tagetes erecta.*
(ii) *Predominant Plants of the Courtyard*	:	*Bougainvillea spectabilis, Tagetes erecta.*
(iii) *Maintenance*	:	Excellent.

(iv) Ethnobotanical

Sacred and Associated Myth	:	*Ficus benghalensis* and *Ficus religiosa* worshipped as sacred plants. *Bischoffia javanica* also regarded as sacred tree. People tie sacred threads and scarves to the tree of *Ficus religiosa* for the fulfilment of their desires.
Medicinal	:	Decoction of leaves and bark (20 ml) of *Bischoffia javanica* given daily for cholera.
Ornamental	:	*Bougainvillea spectabilis* and *Tagetes erecta* grown as ornamentals.
Status of Courtyard Plants in Surrounding Areas	:	Excepting *Bischoffia javanica,* all other plants occur common outside.
Remarks	:	Goats sacrificed to appease the deity. People collect soil from this place for curing snake-bites.

20. Nagni Mata Temple (Pl. 9B)

Main Deity	:	Nagni Mata.
Other Deity	:	Shiva.
Tot. Ar. of Tem. Courtyard	:	About 8000 sq m.
Location		
(i) Place		
Village	:	Bhati
Block	:	Dehra
Tehsil	:	Dehra
District	:	Kangra
(ii) Approach	:	About 3 km from Jwalaji.
(iii) Altitude	:	550 m.
Management	:	Village temple committee.
Historical Background	:	The temple, about 50 years old, had been built by locals.
Temple Architect	:	Sikhara style.
Offerings to God		
(i) By Priest	:	Dal-Rice, Chapati-vegetable.
(ii) By People	:	'Halva', 'Lachidana', Puffed rice. 'Bil-patri' and 'Bhang' specially offered to appease Lord Shiva.
Off. to People (Prasadam)	:	'Halva', 'Lachidana', Puffed rice.
Fairs and Festivals	:	Shivratri, and local fairs held in July-September.

Observations

(i)	*Courtyard Plants (Protected as well as for Landscaping*	: *Acacia catechu, Araucaria columnaris, Asclepias curassavica, Bougainvillea spectabilis, Callistemon lanceolatus, Carica papaya, Catharanthus roseus, Cestrum nocturnum, Citrus aurantifolia, Citrus reticulata, Citrus jambhiri, Cupressus sempervirens, Cynodon dactylon, Emblica officinalis, Eucalyptus tereticornis, Ficus elastica, Ficus religiosa, Hibiscus rosa-sinensis, Litchi chinensis, Mangifera indica, Melia azedarach, Musa paradisiaca, Psidium guajava, Pyrus communis, Syzigium cumini, Tabernaemontana coronaria, Tecoma stans, Thevetia neriifolia, Thuja orientalis.*
(ii)	*Predominant Plants of the Courtyard*	: *Pyrus communis, Mangifera indica.*
(iii)	*Maintenance*	: Satisfactory.
(iv)	*Ethnobotanical*	
	Sacred and Associated Myth	: *Ficus religiosa* worshipped as sacred plant. Its wood used in 'havan' for performing religious rites. *Cynodon dactylon* employed in various religious ceremonies. Wood of *Acacia catechu* also used in 'havan'. For prosperous married life, bride and bridegroom tie their bengals and sehra, respectively to *Mangifera indica* trees in the temple courtyard.
	Medicinal	: Poultice of the rind of *Citrus reticulata* used to cure face pimples. Powdered dried fruits of *Emblica officinalis* mixed in mustard oil good for promoting hair growth. Juice of its fruit prescribed to lower the cholesterol level.
	Edible	: Unripe fruits of *Carica papaya* and *Musa paradisiaca* cooked as vegetable. Ripe fruits of *Carica papaya, Citrus jambhiri, Citrus reticulata, Litchi chinensis, Mangifera indica, Musa paradisiaca, Psidium guajava, Pyrus communis* and *Syzigium cumini* edible. Also, fruits of *Citrus aurantifolia* and *Emblica officinalis* are pickled.
	Ornamental	: *Araucaria columnaris, Asclepias curassavica, Bougainvillea spectabilis, Callistemon lanceolatus, Catharanthus roseus, Cestrum nocturnum, Cupressus sempervirens, Ficus elastica, Hibiscus rosa-sinensis, Melia azedarach, Tabernaemontana coronaria, Tecoma stans, Thevetia neriifolia,* and *Thuja orientalis* grown for ornamental purposes.
	Other Uses	: Leaves and branches of *Psidium guajava,* leaves of *Mangifera indica* and branches of *Acacia catechu* used for scouring teeth. Wood of *Syzigium cumini* and *Melia azedarach* used for making agricultural implements whereas that of *Eucalyptus tereticornis* used for construction works, household articles and packing cases.

21. Nag Temple (Pl. 9C)

Main Deity : Nag Devta.

Other Deities : Vaishno Mata, Baba Balak Nath, Hanuman, Ganesha, Saraswati, Durga.

Tot. Ar. of Tem. Courtyard : About 1536 sq m.

Location

(i) *Place*

Village : Kariyada

Block : Dehra

Tehsil : Dehra

District : Kangra

(ii) *Approach* : About 6 km from Dehra.

(iii) *Altitude* : 530 m.

Management : Privately owned.

Historical Background : The temple, about 25 years old, was got constructed by owner with donations from devotees.

Temple Architect : Sikhara style.

Offerings to God

(i) *By Priest*

Morning : 'Rot'.

Noon : Fruits.

Evening : Cardamom, Sugar candy, Currant.

(ii) *By People* : Puffed rice, 'Lachidana', 'Patasha', Fruits.

Off. to People (Prasadam) : Puffed rice, 'Lachidana'.

Fairs and Festivals : Janamashtmi, Shivratri, Nagpanchmi.

Observations

(i) *Courtyard Plants (Protected as well as for Landscaping* : *Aegle marmelos, Albizia lebbeck, Bauhinia variegata, Bougainvillea spectabilis, Cannabis sativa, Canna indica, Capsicum annuum, Carica papaya, Cassia fistula, Cedrela toona, Chlorophytum capense, Citrus reticulata, Citrus jambhiri, Cupressus sempervirens, Dahlia rosea, Euphorbia milii, Ficus religiosa, Gaillardia pulchella, Hibiscus rosa–sinensis, Hibiscus syriacus, Jasminum sambac, Kochia trichophylla, Malvaviscus conzatii, Mangifera indica, Mentha piperita, Morus alba, Pedilanthus tithymaloides, Psidium guajava, Punica granatum, Pyrus communis, Rosa indica, Russelia equisetiformis, Saccharum officinarum,*

Sansevieria trifasciata, Tabernaemontana coronaria, Tagetes erecta, Zaphyranthes grandiflora, Zinnia elegans, Zizyphus mauritiana.

(ii) *Predominant Plants of the Courtyard* : *Bougainvillea spectabilis, Mangifera indica, Tagetes erecta.*

(iii) *Maintenance* : Excellent.

(iv) *Ethnobotanical*

Sacred : *Morus alba* worshipped as the tree of Nag Devta. *Ficus religiosa* worshipped as sacred plant. Its wood used in 'havan' for performing religious rites. Leaves of *Aegle marmelos* and *Cannabis sativa* offered to appease Lord Shiva. Dried fruits of *Aegle marmelos* used in 'havan'.

Medicinal : Leaves of *Mentha piperita* consumed for indigestion. Vegetable of unripe fruits of *Carica papaya* considered good for jaundice.

Edible : Flower buds of *Bauhinia variegata* and unripe fruits of *Carica papaya* cooked as vegetable. Leaves of *Mentha piperita* used for flavouring chutney. Fruits of *Capsicum annuum* used for flavouring various dishes. Fruits of *Aegle marmelos, Carica papaya, Citrus jambhiri, Citrus reticulata, Mangifera indica, Morus alba, Psidium guajava, Punica granatum, Pyrus communis* and *Zizyphus mauritiana* edible. Fruit juice of *Mangifera indica* considered refreshing and cooling and its pulp used for making 'Am Papad'. Also, unripe fruits pickled and used for making 'Amchur' and candies. Cane juice of *Sachharum officinarum* considered refreshing.

Ornamental : *Bougainvillea spectabilis, Canna indica, Cassia fistula, Chlorophytum capense, Cupressus sempervirens, Dahlia rosea, Euphorbia milii, Gaillardia pulchella, Hibiscus rosa-sinensis, Hibiscus syriacus, Jasminum sambac, Kochia trichophylla, Malvaviscus conzatii, Pedilanthus tithymaloides, Rosa indica, Russelia equisetiformis, Sansevieria trifasciata, Tagetes erecta, Zaphyranthes grandiflora, Zinnia elegans* grown for ornamental purposes.

Other Uses : Leaves of *Mangifera indica* used for scouring teeth whereas leaves and branches of *Psidium guajava* used for scouring teeth. Leaves of *Albizia lebbeck, Bauhinia variegata, Morus alba* and *Zizyphus mauritiana* used as fodder. Wood of *Albizia lebbeck* and *Cassia fistula* used for agricultural implements, and that of *Cedrela toona* for construction and furniture.

22. Radha Krishana Temple (Pl. 9D)

Main Deity	:	Radha-Krishana.
Other Deity	:	Durga Mata.
Tot. Ar. of Tem. Courtyard	:	About 3000 sq m.
Location		
(i) Place		
Town	:	Baroh
Block	:	Nagrota Bagwan
Tehsil	:	Baroh
District	:	Kangra
(ii) Approach	:	14 km from Ranital.
(iii) Altitude	:	700 m.
Management	:	Privately owned.
Historical Background	:	The temple is 13 years old.
Temple Architect	:	Sikhara style.
Offerings to God		
(i) By Priest		
Noon	:	Rice-Dal, Pudding.
Evening	:	Chapati-Vegetable, Pudding.
(ii) By People	:	Puffed rice, 'Lachidana', Sugar candy.
Off. to People (Prasadam)	:	Puffed rice, 'Lachidana', Sugar candy.
Fairs and Festivals	:	Navratra, Janamashtmi.
Observations		
(i) Courtyard Plants (Protected as well as for Landscaping	:	*Asclepias physocarpa, Bougainvillea spectabilis, Callistemon lanceolatus, Canna indica, Cassia glauca, Catharanthus roseus, Cestrum nocturnum, Chrysanthemum indicum, Grevillea robusta, Hibiscus rosa-sinensis, Hibiscus syriacus, Malvaviscus conzatii, Musa paradisiaca, Polyalthia longifolia, Psidium guajava, Quisqualis indica, Tabernaemontana coronaria, Tagetes erecta, Thuja orientalis.*
(ii) Predominant Plants of the Courtyard	:	*Bougainvillea spectabilis, Grevillea robusta, Thuja orientalis.*
(iii) Maintenance	:	Excellent.
(iv) Ethnobotanical		
Sacred	:	Flowers of *Tagetes erecta* and *Hibiscus rosa-sinensis* considered sacred and offered in religious worshipping to the deities.

Medicinal	:	Decoction of tender leaves of *Psidium guajava* used to cure dysentery.
Edible	:	Fruits of *Psidium guajava, Musa paradisiaca* edible. Unripe fruits of *Musa paradisiaca* also cooked as vegetable.
Ornamental	:	*Asclepias physocarpa, Bougainvillea spectabilis, Callistemon lanceolatus, Canna indica, Cassia glauca, Catharanthus roseus, Cestrum nocturnum, Chrysanthemum indicum, Grevillea robusta, Hibiscus rosa-sinensis, Hibiscus syriacus, Malvaviscus conzatii, Polyalthia longifolia, Quisqualis indica, Tabernaemontana coronaria, Tagetes erecta, Thuja orientalis* grown as ornamentals.

23. Radha Krishana Temple (Pl. 9E)

Main Deity	:	Radha-Krishana.
Other Deities	:	Shiva, Hanuman, Ganesha.
Tot. Ar. of Tem. Courtyard	:	About 1500 sq m.
Location		
(i) *Place*		
Town	:	Dehra-Gopipur
Block	:	Pragpur
Tehsil	:	Dehra
District	:	Kangra
(ii) *Approach*	:	The temple is located in the town.
(iii) *Altitude*	:	650 m.
Management	:	Local temple committee.
Historical Background	:	The temple built by locals is about 100 years old.
Temple Architect	:	Sikhara style.
Offerings to God		
(i) *By Priest*	:	Sugar candy, Butter, Dal-Rice, Chapati, Vegetables.
(ii) *By People*	:	Puffed rice, 'Patasha'. 'Bil-patri' offered to Lord Shiva as special offering.
Off. to People (Prasadam)	:	Puffed rice, 'Patasha'.
Fairs and Festivals	:	Janamashtmi, Basant Panchmi, Shivratri.
Observations		
(i) *Courtyard Plants (Protected as well as for Landscaping*	:	*Amaryllis reginae, Canna indica, Capsicum annuum, Carica papaya, Catharanthus roseus, Chrysanthemum indicum, Citrus reticulata, Cynodon dactylon, Dahlia rosea, Hibiscus syriacus, Impatiens balsamina, Malvaviscus conzatii,*

Mangifera indica, Mentha piperita, Momordica charantia, Ocimum sanctum, Pisum sativum, Punica granatum, Quisqualis indica, Rosa indica, Sansevieria trifasciata, Tabernaemontana coronaria, Tagetes erecta, Tropaeolum majus.

(ii) Predominant Plants of the Courtyard : *Tagetes erecta, Quisqualis indica.*

(iii) Maintenance : Satisfactory.

(iv) Ethnobotanical

Sacred : *Ocimum sanctum* worshipped as a sacred plant. Leaves and stem of *Mangifera indica* used in 'havan'. *Cynodon dactylon* considered sacred and employed in various religious ceremonies.

Medicinal : Powdered burnt fruit peels of *Citrus reticulata* given to cure cough. Vegetable of fruits of *Momordica charantia* good for diabetes whereas its fruit juice prescribed as anthelmintic.

Edible : Fruits of *Carica papaya, Citrus reticulata, Mangifera indica* and *Punica granatum* edible. Leaves of *Mentha piperita* used as flavorant and those of *Capsicum annuum* used as spice. Seeds of *Pisum sativum* and unripe fruits of *Carica papaya* cooked as vegetable.

Ornamental : *Amaryllis reginae, Canna indica, Catharanthus roseus, Chrysanthemum indicum, Dahlia rosea, Hibiscus syriacus, Impatiens balsamina, Malvaviscus conzatii, Ocimum sanctum, Quisqualis indica, Rosa indica, Sansevieria trifasciata, Tabernaemontana coronaria, Tagetes erecta* and *Tropaeolum majus* grown as ornamentals.

24. Rock-Cut Temple (Masroor Temple) (Pl. 9F)

Main Deity : Ram-Lakshman-Sita.

Tot. Ar. of Tem. Courtyard : About 4000 sq m.

Location

(i) Place

Village : Masroor

Block : Nagrota Surian

Tehsil : Dehra (Gopipur)

District : Kangra

(ii) Approach : 11 km from Nagrota Surian.

(iii) Altitude : 770 m.

Management	:	Managed by Archaeological Survey of India.
Historical Background	:	Believed to be constructed during 8th-9th century A.D.
Temple Architect	:	The Rock-cut temple complex is a unique example of a monolithic temple. The entire complex comprises more than fifteen Sikhara temples. In the centre of complex stands the Thakurdwara, enshrining the black stone images of Rama, Sita and Lakshman facing East.
Offerings to God	:	'Halva', Puffed rice.
Fairs and Festivals	:	Ram-Navami.
Observations		
(i) Courtyard Plants (Protected as well as for Landscaping	:	*Acacia catechu, Albizia lebbeck, Amaryllis reginae, Boerhaavia diffusa, Canna indica, Catharanthus roseus, Celosia argentea* var. *cristata, Chrysanthemum indicum, Citrus aurantifolia, Cupressus sempervirens, Cynodon dactylon, Ficus benghalensis, Ficus glomerata, Ficus religiosa, Grewia oppositifolia, Hibiscus rosa-sinensis, Jasminum sambac, Mangifera indica, Ocimum sanctum, Rosa indica, Rosa multiflora, Syzygium cumini, Tagetes erecta, Thuja orientalis, Zephyranthes candida, Zephyranthes grandiflora, Zinnia elegans.*
(ii) Predominant Plants of the Courtyard	:	*Acacia catechu, Catharanthus roseus, Zinnia elegans.*
(iii) Maintenance	:	Excellent.
(iv) Ethnobotanical		
Sacred	:	*Ficus benghalensis, Ficus religiosa* and *Ocimum sanctum* considered sacred plants. *Cynodon dactylon* employed in religious ceremonies.
Medicinal	:	'Katha' obtained from the heartwood of *Acacia catechu* given to cure tonsils and mouth ulcers. 5-10g powdered roots of *Boerhaavia diffusa* prescribed twice daily for urinary problems.
Edible	:	Fruits of *Ficus glomerata, Syzygium cumini, Mangifera indica* edible and those of *Citrus aurantifolia* pickled.
Ornamental	:	*Amaryllis reginae, Canna indica, Catharanthus roseus, Celosia Argentea* var. *cristata, Chrysanthemum indicum, Cupressus sempervirens, Hibiscus rosa–sinensis, Jasminum sambac, Ocimum sanctum, Rosa indica, Rosa multiflora, Tagetes erecta, Thuja orientalis, Zephyranthes candida, Zephyranthes grandiflora, Zinnia elegans.*
Other Uses	:	Leaves of *Grewia oppositifolia* and *Albizia lebbeck* used as fodder. Fibers from branches of *Grewia oppositifolia* used

for making ropes. Wood of *Albizia lebbeck* used for agricultural implements. Branches of *Acacia catechu* used as toothbrush. Its pods relished by cattle, sheep and goats.

25. Shibbo Than Temple (Pl. 9G)

Main Deity : Baba Shibbo Nath.

Other Deities : Gugga Pir, Gorakh Nath, Bishander Nath, Mata Kashla, Gigdi, Kamdhenu Cow, Doodhia Nag, Lord Shiva.

Tot. Ar. of Tem. Courtyard : About 3000 sq m.

Location

(i) Place

Village : Bharmar

Block : Fatehpur

Tehsil : Jawali

District : Kangra

(ii) Approach : About 20 km from Jassur on Pathankot-Dehra road.

(iii) Altitude : 465 m.

Management : Privately owned.

Historical Background : Long back, Baba Shibbo Nath meditated at this place under the Ber tree. He worshipped Gugga Pir for grant of powers to save people from various ailments and evil spirits. Gugga Pir gave him the boon for the same. That Ber tree is still conserved in the courtyard of the temple. The original temple is about 850 years old and reconstructed about 260 years back by ancestors of present owners so called Mahant family.

Temple Architect : Sikhara style.

Offerings to God

(i) By Priest : Whatever prepared in the home of priest.

(ii) By People : Puffed rice, 'Lachidana', Flowers, Fruits.

Off. to People (Prasadam) : Puffed rice, 'Lachidana'.

Fairs and Festivals : Gugga Navami celebrated during July-August.

Observations

(i) Courtyard Plants (Protected as well as for Landscaping : *Aegle marmelos, Alstonia scholaris, Artocarpus lakoocha, Ficus religiosa, Hibiscus rosa-sinensis, Jasminum sambac, Mimusops elengi, Ocimum sanctum, Pterospermum acerifolium, Quisqualis indica, Tabernaemontana coronaria, Tagetes erecta, Zizyphus mauritiana.*

(ii) Predominant Plants of the Courtyard : *Tagetes erecta, Quisqualis indica.*

(iii) Maintenance : Satisfactory.

(iv) Ethnobotanical

Sacred and Associated Myth : *Ocimum sanctum* and *Ficus religiosa* considered sacred plants. Wood of *Ficus religiosa* used in 'havan' for performing religious rites. Tree of *Zizyphus mauritiana* in the courtyard of temple held in high esteem by the devotees as Baba Shibbo Nath meditated under this tree long ago.

Medicinal : Decoction of root-bark (5-10 ml) of *Alstonia scholaris* taken on empty stomach as an anthelmintic. Its milky latex applied on cuts and wounds.

Edible : Unripe fruits of *Artocarpus lakoocha* pickled or cooked as vegetable. Fruit pulp of *Aegle marmelos*, ripe fruits of *Artocarpus lakoocha* and *Zizyphus mauritiana* edible.

Ornamental : *Alstonia scholaris, Hibiscus rosa-sinensis, Jasminum sambac, Mimusops elengi, Ocimum sanctum, Pterospermum acerifolium, Quisqualis indica, Tabernaemontana coronaria* and *Tagetes erecta* grown as ornamentals.

Status of Courtyard Plants in Surrounding Areas : *Mimusops elengi* and *Pterospermum acerifolium* do not occur in the surrounding areas of the temple whereas *Artocarpus lakoocha* occurs infrequently. All other plants are common outside.

Remarks : It is commonly believed that people suffering from snake bites get cured here.

26. Shiva Temple (Pl. 9H)

Main Deity : Shivlingam.

Other Deities : Radha-Krishana, Garud, Shiva, Pashupati Nath, Bhairav, Neelkanth, Nandi.

Tot. Ar. of Tem. Courtyard : About 8000 sq m including a beautiful park in the surrounding area of the temple.

Location

(i) Place

Village : Baijnath

Block : Baijnath

Tehsil : Baijnath

District : Kangra

(ii) Approach	:	The temple is located in the town.
(iii) Altitude	:	1000 m.
Management	:	Temple managed by Archaeological Survey of India and local temple committee.
Historical Background	:	Temple built many centuries ago was got repaired by Raja Chand Katoch, who ruled Kangra from over half a century (1776 A.D. to 1824 A.D.) and later by Raja Sansar Chand. It is believed that Ravana worshipped Lord Shiva for 12 years at this place.
Temple Architect	:	Sikhara style.
Offerings to God		
(i) By Priest		
Morning	:	Currant, Coconut. Milk Bhog especially for Radha-Krishana.
Noon	:	Rice, Dal, Vegetables.
Evening	:	'Halva-Puri', Black gram.
(ii) By People	:	Currant, Coconut, Sweets, Fruits, Flowers. 'Bhang', 'Dhatura' offered to Lord Shiva as special offering.
Off. to People (Prasadam)	:	Sugar candy, Fox nut, Coconut, Milk (as 'charnamat'), Flowers.
Fairs and Festivals	:	Shivratri fair, Janamashtmi, and local fair held on every Monday of Shravan month (July-August).
Observations		
(i) Courtyard Plants (Protected as well as for Landscaping	:	*Aegle marmelos, Aloe vera, Bauhinia variegata, Bougainvillea spectabilis, Cedrus deodara, Cestrum nocturnum, Chlorophytum capense, Chrysanthemum indicum, Clerodendrum inerme, Cynodon dactylon, Duranta repens, Elaeocarpus ganitrus, Euphorbia milii, Ficus religiosa, Gaillardia pulchella, Hamelia patens, Hibiscus rosa-sinensis, Hibiscus syriacus, Hydrangea macrophylla, Hymenocallis littoralis, Jacaranda acutifolia, Jasminum sambac, Lagerstroemia indica, Nerium odorum, Nyctanthes arbor-tristis, Pelargonium hortorum, Petunia hybrida, Putranjiva roxburghii, Rosa indica, Rosa multiflora, Tabernaemontana coronaria, Tagetes erecta, Thevetia neriifolia, Thuja orientalis, Zanthoxylum armatum.*
(ii) Predominant Plants of the Courtyard	:	*Bougainvillea spectabilis, Clerodendrum inerme, Duranta repens, Tabernaemontana coronaria.*
(iii) Maintenance	:	Excellent.

(iv) Ethnobotanical

Sacred : Leaves of *Aegle marmelos* especially offered to Lord Shiva. *Ficus religiosa* worshipped as sacred tree. Its wood used in 'havan' for perfoming religious rites. *Cynodon dactylon* considered sacred and employed in various religious ceremonies. *Elaeocarpus ganitrus* considered symbolic to Lord Shiva. Its nuts are polished and made into rosaries and bracelets.

Medicinal : Decoction of leaves of *Zanthoxylum armatum* with black pepper used as gargle for gum-bleeding and removing bad smell from mouth.

Edible : Flower buds of *Bauhinia variegata* and leaves of *Aloe vera* cooked as vegetable. Fruits of *Aegle marmelos* edible.

Ornamental : *Aloe vera, Bauhinia variegata, Bougainvillea spectabilis, Cestrum nocturnum, Chlorophytum capense, Chrysanthemum indicum, Clerodendrum inerme, Duranta repens* (as hedge plant), *Elaeocarpus ganitrus, Euphorbia milii, Gaillardia pulchella, Hamelia patens, Hibiscus rosa-sinensis, Hibiscus syriacus, Hydrangea macrophylla, Hymenocallis littoralis, Jacaranda acutifolia, Jasminum sambac, Lagerstroemia indica, Nerium odorum, Nyctanthes arbor-tristis, Pelargonium hortorum, Petunia hybrida, Putranjiva roxburghii, Rosa indica, Rosa multiflora, Tabernaemontana coronaria, Tagetes erecta, Thevetia neriifolia, Thuja orientalis, Zanthoxylum armatum.*

Other Uses : Leaves of *Bauhinia variegata* used as fodder. Stem of *Zanthoxylum armatum* used as toothbrush. Wood of *Cedrus deodara* excellent for furniture and construction works.

Status of Courtyard Plants in Surrounding Areas : All other plants excepting *Cedrus deodara, Elaeocarpus ganitrus, Jacaranda acutifolia* and *Putranjiva roxburghii* occur commonly outside.

27. Shiva Temple or Chhota Haridwar (Pl. 9I)

Main Deity : Shiva.

Other Deities : Brahma, Vishnu, Ganesha, Hanuman, Kartik, Duttatrey, Saraswati, Durga.

Tot. Ar. of Tem. Courtyard : 768 sq m.

Location

(i) Place

Village : Amrella

Block : Nagrota Surian

Tehsil	:	Jawali
District	:	Kangra
(ii) Approach	:	2 km from Jawali on Jawali-Dehra road.
(iii) Altitude	:	460 m.
Management	:	In the name of Deity.
Historical Background	:	It is believed that Pandavas stayed here for some time during their exile. They decided to construct the stairs leading to heaven at this place in one night but could not do so, and accomplished the task at Haridwar. This place is, therefore, called as Chhota Haridwar. And people who can not afford to go to Haridwar perform the last rites of their near and dear ones at this place. This place is believed to be worshipped since 'Dwaper Yug' but the temple was constructed only about 20 years back by the locals.
Temple Architect	:	Sikhara style.
Offerings to God		
(i) By Priest	:	'Kasaar' (made of wheat flour).
(ii) By People	:	Puffed rice, 'Lachidana'.
Off. to People (Prasadam)	:	Puffed rice, 'Lachidana'.
Fairs and Festivals	:	Shivratri, Baisakhi.
Observations		
(i) Courtyard Plants (Protected as well as for Landscaping	:	*Callistemon lanceolatus, Calotropis procera, Catharanthus roseus, Cupressus sempervirens, Eriobotrya japonica, Gaillardia pulchella, Musa paradisiaca, Pelargoniun hortorum, Petunia hybrida, Psidium guajava, Tagetes erecta, Zinnia elegans.*
(ii) Predominant Plants of the Courtyard	:	*Tagetes erecta.*
(iii) Maintenance	:	Satisfactory.
(iv) Ethnobotanical		
Sacred	:	Leaves and flowers of *Calotropis procera*-'Shiv Ki Buti' offered to appease Lord Shiva. Its stem used in purification ceremonies. *Musa paradisiaca* also worshipped as sacred plant.
Medicinal	:	2-3g mixtire of powdered flowers of *Calotropis procera* and black pepper taken with honey to cure cough. Also, its leaf juice applied for rheumatism and latex for leucoderma.

Edible : Fruits of *Eriobotrya japonica, Musa paradisiaca* and *Psidium guajava* edible.

Ornamental : *Callistemon lanceolatus, Calotropis procera, Catharanthus roseus, Cupressus sempervirens, Gaillardia pulchella, Pelargoniun hortorum, Petunia hybrida, Tagetes erecta* and *Zinnia elegans* grown as ornamentals.

Other Uses : Leaves and branches of *Psidium guajava* used for scouring teeth.

Remarks : A small orchard attached to temple contains trees of mango, guava, litchi, loucat, lemon, banana, eucalyptus, tuni and shisham.

28. Shri Brijraj Swami Temple (Pl. 9J)

Main Deity : Shri Krishana.

Other Deity : Mira Bai.

Tot. Ar. of Tem. Courtyard : About 1500 sq m.

Location

(i) Place

Town : Nurpur

Block : Nurpur

Tehsil : Nurpur

District : Kangra

(ii) Approach : Temple is located amongst the old ruined fort of Nurpur.

(iii) Altitude : 700 m.

Management : Managed by Archaeological Survey of India.

Historical Background : The temple believed to be about 400 years old was built by Raja Jagat Singh who brought the idols of Lord Krishana and Mira along with a Moulsari plant (*Mimusops elingi*) as a gift presented by the king of Chittodgarh (Rajasthan). He built the temple at Nurpur and planted that Moulsari plant in the courtyard.

Temple Architect : Flat roof type.

Offerings to God

(i) By Priest

Morning : Rice, Dal, Chapati, Vegetables.

Evening : Milk.

(ii) By People : 'Lachidana'.

Off. to People (Prasadam)	:	'Lachidana'.
Fairs and Festivals	:	Janamashtmi, Govardhan Puja (Next day after Diwali).
Observations		
(i) Courtyard Plants (Protected as well as for Landscaping	:	*Bombax ceiba, Bougainvillea spectabilis, Citrus aurantifolia, Duranta repens, Tectona grandis, Dalbergia sissoo, Dodonea viscosa, Duranta repens, Eucalyptus tereticornis, Ficus benghalensis, Grewia oppositifolia, Jacaranda acutifolia, Litsea sebifera, Mangifera indica, Mimusops elingi, Morus alba, Murraya paniculata, Nyctanthes arbor-tristis, Tectona grandis, Terminalia chebula, Thevetia neriifolia, Thuja orientalis, Zizyphus mauritiana.*
(ii) Predominant Plants of the Courtyard	:	*Grewia oppositifolia.*
(iii) Maintenance	:	Satisfactory.
(iv) Ethnobotanical		
Sacred	:	*Ficus benghalensis* and *Mimusops elingi* considered sacred plants. *Mimusops elingi* is generally found at religious places in H.P.
Medicinal	:	Poultice of bark of *Bombax ceiba*, tender branches of *Vitex negundo*, leaves of *Adhatoda vasica* and *Cuscuta reflexa* applied to heel fractures or any kind of internal injury. Roasted dried fruits of *Terminalia chebula* chewed to cure old cough. 3-5g leaf powder of *Duranta repens* given daily to cure diabetes.
Edible	:	Seeds of *Bombax ceiba*, fruits of *Citrus aurantifolia, Mangifera indica, Morus alba, Terminalia chebula* and *Zizyphus mauritiana* are edible.
Ornamental	:	*Bombax ceiba, Bougainvillea spectabilis, Litsea sebifera, Mimusops elingi, Murraya paniculata, Nyctanthes arbor-tristis, Tectona grandis, Terminalia chebula, Thevetia neriifolia* and *Thuja orientalis* grown as ornamentals. *Jacaranda acutifolia* planted as an avenue plant whereas *Duranta repens* grown as hedges.
Other Uses	:	Leaves of *Zizyphus mauritiana, Morus alba* and young branches of *Grewia oppositifolia* used as fodder. Wood of *Bombax ceiba, Eucalyptus tereticornis, Mangifera indica, Dalbergia sissoo* and *Tectona grandis* used for construction work. Branches of *Zizyphus mauritiana* used for fencing whereas those of *Eucalyptus tereticornis* and *Morus alba* used as firewood.

Status of Courtyard Plants in Surrounding Areas	:	*Mimusops elingi* is absent outside the courtyard of the temple whereas *Litsea sebifera* and *Terminalia chebula* are rare outside.

29. Shri Raghunath Teda Temple (Trihayu Temple) (Pl. 9K)

Main Deities	:	Shri Rama (Raghunath), Lakshmana, Goddess Sita.
Other Deities	:	Mahavir, Vishnu, Brahma, Lakshmi, Garud, Panchmukhi Mahadev (Shiva).
Tot. Ar. of Tem. Courtyard	:	Total land in the name of deity covers an area of 7 hectare comprising of the forest land surrounding the temple and an orchard of mango, guava, banana, pear, lemon, papaya and jack fruit trees and about 1500 sq m of temple courtyard.
Location		
(i) Place		
Village	:	Chakvan Kalidhar
Block	:	Dehra
Tehsil	:	Dehra
District	:	Kangra
(ii) Approach	:	9 km by bus or 3 km on foot from Jwalaji.
(iii) Altitude	:	900 m.
Management	:	Managed by Shri Raghunath Teda Mandir Trihayu Trust.
Historical Background	:	It is believed that Lord Rama, Sita and Lakshmana came here during period of their exile and stayed at this place for some time. Since then this place is considered sacred.
Temple Architect	:	Sikhara style.
Offerings to God		
(i) By Priest		
Morning/Evening	:	Sugar candy, Currant, Fruits.
Noon (Rajbhog)	:	Dal-Rice, Chapati-Vegetables.
(ii) By People	:	Puffed rice, Sweets. 'Bil-patri' (leaves of Bil tree) offered to Lord Shiva in the temple alongwith different flowers, rice and 'dhoop'.
Off. to People (Prasadam)	:	Puffed rice, Sweets ('Mithi Bundi').
Fairs and Festivals	:	Ram Navami, Navratras celebrated thrice a year (Chait, Assu and Shravan).

Observations

(i) Courtyard Plants (Protected as well as for Landscaping	:	*Aristolochia elegans, Bryophyllum calycinum, Bougainvillea spectabilis, Carica papaya, Catharanthus roseus, Coix lachryma-jobi, Cynodon dactylon, Dolichos lablab, Ficus religiosa, Musa paradisiaca, Narcissus tazetta, Ocimum gratissimum, Rosa indica, Saccharum officinarum, Tagetes erecta, Tropaeolum majus.*
(ii) Predominant Plants of the Courtyard	:	*Carica papaya, Tagetes erecta, Bryophyllum calycinum.*
(iii) Maintenance	:	Satisfactory.
(iv) Ethnobotanical		
Sacred	:	Necklace made from the beads of *Coix lachryma-jobi* offered to the deity. *Ficus religiosa* considered sacred and its wood used for performing 'havan' during religious ceremonies. *Cynodon dactylon* employed in various religious ceremonies.
Medicinal	:	Slightly warmed leaves of *Bryophyllum calycinum* applied to cure furuncles. Decoction of leaves of *Ocimum gratissimum* considered aphrodisiac.
Edible	:	Pods of *Dolichos lablab* and unripe fruits of *Carica papaya* and *Musa paradisiaca* cooked as vegetable. Juice of *Saccharum officinarum* considered refreshing. Ripe fruits of *Carica papaya* and *Musa paradisiaca* edible.
Ornamental	:	*Aristolochia elegans, Bryophyllum calycinum, Bougainvillea spectabilis, Catharanthus roseus, Coix lachryma-jobi, Narcissus tazetta, Rosa indica, Tagetes erecta* and *Tropaeolum majus* planted as ornamental plants.
Other Uses	:	*Coix lachryma-jobi* and *Cynodon dactylon* used as fodder.
Status of Courtyard Plants in Surrounding Areas	:	*Aristolochia elegans, Coix lachryma-jobi,* and *Narcissus tazetta* occur infrequently as compared to others.
Remarks	:	'Havan' ceremony for purification is performed twice daily. During earthquake in 1905, the temple was partly tilted from one side and called as 'Teda Mandir'.

30. Sidh Chaino Temple (Pl. 9L)

Main Deity	:	Baba Sidh Chaino.
Other Deities	:	Baba Lal Khan, Bhairav.
Tot. Ar. of Tem. Courtyard	:	About 800 sq m.

Location

(i) Place

Village	:	Dangra
Block	:	Pragpur
Tehsil	:	Dehra
District	:	Kangra
(ii) Approach	:	3 km from Pragpur.
(iii) Altitude	:	550 m.
Management	:	Village temple committee.
Historical Background	:	It is believed that Baba Sidh Chaino appeared under 'Badaar' tree (*Ficus rumphi*) which is still there in the temple courtyard.
Temple Architect	:	Sikhara style.
Offerings to God		
(i) By Priest	:	'Rot'.
(ii) By People	:	'Halva', Puffed rice, 'Patasha'.
Off. to People (Prasadam)	:	'Halva', Puffed rice, 'Patasha'.
Fairs and Festivals	:	Baisakhi.
Observations		
(i) Courtyard Plants (Protected as well as for Landscaping	:	*Cedrela toona, Ficus benghalensis, Ficus rumphii.*
(ii) Maintenance	:	Satisfactory.
(iii) Ethnobotanical		
Sacred	:	*Ficus benghalensis* and *Ficus rumphii* worshipped as sacred trees.
Medicinal	:	Milky latex of *Ficus benghalensis* used for regaining lost sexual vigour.
Status of Courtyard Plants in Surrounding Areas	:	Excepting that of *Ficus rumphii*, other trees occur commonly outside.

31. Trilokinath Temple (Pl. 9M)

Main Deity	:	Trilokinath (Form of Lord Shiva).
Other Deities	:	Bhairav, Baba Balak Nath, Bhasmasur, Parvati, Hanuman.
Tot. Ar. of Tem. Courtyard	:	About 1500 sq m.

Location		
(i) Place		
Village	:	Tilokpur
Block	:	Surian Nagrota
Tehsil	:	Jawali
District	:	Kangra
(ii) Approach	:	Situated at Tilokpur on the side of Nurpur-Kangra road.
(iii) Altitude	:	600 m.
Management	:	Privately owned.
Historical Background	:	As per the prevalent legend, this cave having two pillars of gold used as a meditating hall by Lord Shiva with Sheshnag providing shadow to him. Once, a shepherd tried to remove some gold from the pillars. On seeing this, Lord Shiva turned him into stone along with his sheep which are still present as such inside the cave.
Temple Architect	:	The main temple is in the form of a cave. The surounding temples are in Sikhara style.
Offerings to God		
(i) By Priest	:	Puffed rice, 'Lachidana', 'Rot' (especially on Monday).
(ii) By People	:	Puffed rice, 'Lachidana', Flowers of Rose, Marigold, Seasonal Fruits.
Off. to People (Prasadam)	:	Puffed rice, 'Lachidana'.
Fairs and Festivals	:	Shivratri.
Observations		
(i) Courtyard Plants (Protected as well as for Landscaping	:	*Ficus benghalensis, Ficus religiosa.*
(ii) Maintenance	:	Satisfactory.
(iii) Ethnobotanical		
Sacred	:	*Ficus benghalensis* and *Ficus religiosa* (worshipped as form of Lord Brahma) considered sacred plants.
Remarks	:	Drops of water continuously dripping from the roof of cave shaped-like snake, ultimately fall on the idol of Shiva. This water believed to possess healing property.

32. Vindhyavasini Mata Temple (Bandla Mata Temple) (Pl. 9N)

Main Deity	:	Vindhyavasini Mata.
Other Deities	:	Bhairav, Shiva, Bajrangbali, Duttatrey.
Tot. Ar. of Tem. Courtyard	:	About 3000 sq m.
Location		
(i) Place		
Village	:	Bandla
Block	:	Bhavarna
Tehsil	:	Palampur
District	:	Kangra
(ii) Approach	:	About 5 km from Palampur.
(iii) Altitude	:	1050 m.
Management	:	In the name of Deity; managed by 'Mahantas'.
Historical Background	:	Once a shephered found 7 girls in the forest while grazing his animals. Out of these, six disappeared and the seventh one turned into stone which he brought with him while coming to home. On the way he stopped for rest, and found himself unable to carry the stone further as it became very heavy. Then deity appeared and told him to establish her at that place. Since then this place is worshipped by the people. The present temple is about 100 years old and built by 'Mahantas' and devotees.
Temple Architect	:	Sikhara style.
Offerings to God		
(i) By Priest	:	Rice-Dal, Chapati-Vegetable.
(ii) By People	:	Puffed rice, Coconut, 'Halva', Sweets. 'Bil-Patri' offered to appease Lord Shiva.
Off. to People (Prasadam)	:	Puffed rice, 'Halva'.
Fairs and Festivals	:	Navratra, Shivratri.
Observations		
(i) Courtyard Plants (Protected as well as for Landscaping	:	*Bougainvillea spectabilis, Cannabis sativa, Canna indica, Cedrus deodara, Celtis australis, Cestrum nocturnum, Citrus reticulata, Cynodon dactylon, Ehretia acuminata, Hemerocallis fulva, Hibiscus syriacus, Malvaviscus conzatii, Mangifera indica, Morus alba, Musa paradisiaca, Nerium odorum, Ocimum basilicum, Rosa indica, Tagetes erecta.*

(ii) Predominant Plants of the Courtyard : *Hibiscus syriacus.*

(iii) Maintenance : Excellent.

(iv) Ethnobotanical

Sacred : *Cannabis sativa* offered to appease Lord Shiva. *Ocimum basilicum* worshipped as sacred plant. *Cynodon dactylon* considered sacred and employed in all religious ceremonies.

Medicinal : Leaf paste of *Cannabis sativa* applied to cure insect bites.

Edible : Fruits of *Celtis australis, Citrus reticulata, Mangifera indica, Morus alba, Musa paradisiaca* edible.

Ornamental : *Bougainvillea spectabilis, Canna indica, Celtis australis, Cestrum nocturnum, Ehretia acuminata, Hemerocallis fulva, Hibiscus syriacus, Malvaviscus conzatii, Nerium odorum, Ocimum basilicum, Rosa indica, Tagetes erecta* grown as ornamentals.

Other Uses : Leaves of *Celtis australis, Morus alba* and *Ehretia acuminata* used as cattle feed. Wood of *Cedrus deodara* regarded excellent for furniture and construction works and that of *Celtis australis* for agricultural implements.

Temples of District Kinnaur

1. Badri Narayan Temple (Pl. 10A)

Main Deity	:	Badri Narayan.
Other Deity	:	Vishnu Narang.
Tot. Ar. of Tem. Courtyard	:	3000 sq m.
Location		
(i) Place		
Village	:	Batseri
Block	:	Kalpa
Tehsil	:	Sangla
District	:	Kinnaur
(ii) Approach	:	About 5 km from Sangla.
(iii) Altitude	:	2700 m.
Management	:	Temple committee.
Historical Background	:	Long ago, the king of devils, Banasur used to torture the people of this area. For protecting the people from his torture, the deity came here from Badrinath Dham and killed Banasur. The people constructed the temple for him. The previous temple, about 300 years old, was razed in fire in 1998. The present structure was constructed at the same spot.

Temple Architect	:	Pentroof type; Deodar and Qail wood used in its construction.
Offerings to God		
(i) By Priest	:	'Lachidana', Coconut.
(ii) By People	:	'Halva', Coconut, Seasonal Flowers and Fruits.
Off. to People (Prasadam)	:	'Lachidana', Coconut.
Fairs and Festivals	:	Bada Fulaich (celebrated for five days in September), Chhota Fulaich (held for three days in October), Dakhrain (celebrated during July-August), Bishu (held in April).
Observations		
(i) Courtyard Plants (Protected as well as for Landscaping	:	*Juniperus macropoda.*
(ii) Maintenance	:	Excellent.
(iii) Ethnobotanical		
Sacred	:	Leaves of *Juniperus macropoda* used as incense in the temple.

2. Baring Nag Temple (Pl. 10B)

Main Deity	:	Baring Nag.
Other Deity	:	Vaishno Devi.
Tot. Ar. of Tem. Courtyard	:	1536 sq m.
Location		
(i) Place		
Village	:	Sangla
Block	:	Kalpa
Tehsil	:	Sangla
District	:	Kinnaur
(ii) Approach	:	Located at Sangla.
(iii) Altitude	:	2640 m.
Management	:	Temple committee.
Historical Background	:	The present temple, about 56 years old, was got constructed by Temple committee from Deodar wood and stones.
Temple Architect	:	Pagoda type.

Offerings to God

(i) By Priest	:	'Halva', Sweets.
(ii) By People	:	'Halva', Sweets.
Off. to People (Prasadam)	:	'Halva', 'Poltu' ('Puri').
Fairs and Festivals	:	Bishu Festival-Baisakhi (celebrated in April), Fulaich (celebrated for five days in September in which flowers of *Saussurea obvallata* are collected, worshipped and distributed to be kept in houses for prosperity), Buddhi Diwali (celebrated exactly one month after Diwali.

Observations

(i) Courtyard Plants (Protected as well as for Landscaping	:	*Juniperus macropoda, Rosa indica, Thuja plicata.*
(ii) Maintenance	:	Excellent.
(iii) Ethnobotanical		
Ornamental	:	All courtyard plants are planted for beautification of the temple.
Remarks	:	No sacrifices are made to the deity.

3. Boirus Temple (Pl. 10C)

Main Deity	:	Bhairav.
Other Deities	:	Shiva, Pandavas.
Tot. Ar. of Tem. Courtyard	:	1536 sq m with three water bodies (ponds).
Location		
(i) Place		
Village	:	Kothi
Block	:	Kalpa
Tehsil	:	Kalpa
District	:	Kinnaur
(ii) Approach	:	2 km on foot from Reckong Peo (3 km from Peo by bus).
iii) Altitude	:	2400 m.
Management	:	Managed by Dev Sabha.
Historical Background	:	The temple is believed to be made by Pandavas, and later repaired by Raja Shamsher Singh (King of Bushehar). It is regarded that a devil called Bhairav was killed by Chandika Devi. Before his death he wished to have a place near Devi for which his temple was built by Pandavas.

Temple Architect	:	Pentroof style.
Offerings to God		
(i) By Priest	:	Flowers.
(ii) By People	:	People not allowed to enter the temple.
Off. to People (Prasadam)	:	Flowers.
Fairs and Festivals	:	Bhairav-Jalakha (Bhairav worshipped during this fair).
Observations		
(i) Courtyard Plants (Protected as well as for Landscaping	:	*Helianthus annuus, Oenothera drummondii, Salix babylonica, Thuja plicata.*
(ii) Predominant Plants of the Courtyard	:	*Thuja plicata.*
(iii) Maintenance	:	Excellent.
(iv) Ethnobotanical		
Sacred	:	*Salix babylonica* considered sacred and hence protected.
Ornamental	:	All the courtyard plants are considered for ornamental purposes.
Remarks	:	No body is permitted to enter the temple from the very beginning, excepting some selected members of the committee.

4. Chandika Devi Temple (Pl. 10D)

Main Deity	:	Chandika.
Tot. Ar. of Tem. Courtyard	:	3000 sq m.
Location		
(i) Place		
Village	:	Kothi
Block	:	Kalpa
Tehsil	:	Kalpa
District	:	Kinnaur
(ii) Approach	:	2 km on foot from Reckong Peo.
(iii) Altitude	:	2400 m.
Management	:	Temple committee.
Historical Background	:	Long ago, devils use to torture the people of this village. Chandika Devi killed all the devils. For that, temple was built for her.

Temple Architect : Pentroof type; dates back to the 16th-17th century A.D.

Offerings to God

(i) *By Priest* : 'Dhoop', Flowers, Wine of fruits (Apple, Grapes, Vaimi, Chuli) and Barley offered twice a day.

(ii) *By People* : 'Lachidana', Sweets.

Off. to People (Prasadam) : 'Lachidana', Fruits, Flowers.

Fairs and Festivals : Kashmir (on this day, Deity visits the nearby village, Kashmir), Holi (celebrated for 8 days), Bishu (Baisakhi), Lohdi, Dakhrain.

Observations

(i) *Courtyard Plants (Protected as well as for Landscaping* : *Robinia pseudoacacia, Salix alba, Tagetes erecta, Thuja plicata, Ulmus villosa.*

(ii) *Predominant Plants of the Courtyard* : *Thuja plicata.*

(iii) *Maintenance* : Excellent.

(iv) *Ethnobotanical*

Sacred : *Robinia pseudoacacia, Salix alba* and *Ulmus villosa* considered sacred and protected.

Ornamental : *Salix alba, Tagetes erecta* and *Thuja plicata* planted for ornamental purposes.

Status of Courtyard Plants in Surrounding Areas : Excepting *Ulmus villosa*, all other plants occur commonly outside.

Remarks : Locals hold the deity in great reverence and considered her to be one of the most powerful goddesses.

5. Kamakhaya Devi Temple (Kamru Kila) (Pl. 10E)

Main Deity : Kamakhya Devi, Thanapati.

Other Deities : Chhatra Singh, Badrinath, Parvorajas.

Tot. Ar. of Tem. Courtyard : About 3000 sq m.

Location

(i) *Place:*

Village : *Kamru*

Block : Kalpa

Tehsil : *Sangla*

District : *Kinnaur*

(ii) *Approach* : 1 km on foot from Sangla.

(iii) *Altitude* : 2800 m.

Management : Temple committee.

Historical Background : The temple is believed to be more than one thousand years old. It was initially built by ancestors of the king of Bushehar and later reconstructed by the temple committee.

Temple Architect : Pentroof type; it is a lofty square structure built of dressed stones held together at small intervals with wooden rafters (Deodar wood).

Offerings to God

(i) By Priest : Dry fruits like Coconut, Currant, Dried palm fruits, Almond, Chestnut, Sugar-candy.

(ii) By People : 'Lachidana', Coconut, Seasonal Flowers.

Off. to People (Prasadam) : *'Lachidana', Coconut.*

Fairs and Festivals : Fulaich (held in September), Phagli (held for 15 days in March), Holi (celebrated in March).

Observations

(i) Courtyard Plants (Protected as well as for Landscaping : *Dahlia rosea, Helianthus annuus, Oenothera drummondii, Phytolacca acinosa, Prunus persica, Pyrus malus, Rosa jacquemontii, Rosa moschata, Thuja plicata.*

(ii) Predominant Plants of the Courtyard : *Pyrus malus.*

(iii) Maintenance : Excellent.

(iv) Ethnobotanical

Sacred : Floral garlands of *Rosa jacquemontii* and *Rosa moschata* used as an offering to the deity.

Edible : *Phytolacca acinosa* used as vegetable (Sag). Fruits of *Prunus persica* and *Pyrus malus* edible.

Ornamental : *Dahlia rosea, Helianthus annuus, Phytolacca acinosa, Rosa jacquemontii, Rosa moschata, Thuja plicata* grown as ornamentals.

Remarks : Sheep and goats sacrificed in the temple. The temple is in the form of a fort. The coronation ceremony of the rulers of Bushehar state was used to be performed in this fort. Upto now 122 kings had been coronated here. The deity is never allowed to be touched except by the Priest and that too in Phagli fair for which he has to wear a particular dress after bathing.

6. Mahabodi Temple (Pl. 10F)

Main Deity	:	Gautam Budh.
Other Deities	:	Padam Sambhava, Tara Devi.
Tot. Ar. of Tem. Courtyard	:	6000 sq m.
Location		
(i) *Place*		
Village	:	Behringi (Kothi)
Block	:	Kalpa
Tehsil	:	Kalpa
District	:	Kinnaur
(ii) *Approach*	:	1 km on foot from Reckong Peo.
(iii) *Altitude*	:	2450 m.
Management	:	Temple committee.
Historical Background	:	The temple was built by Kinnaur Mahabodi Society under the supervision of respected Ga-Rinpoche about 12 years back.
Temple Architect	:	Buddhist temple; Deodar and Qail wood had been used in building the temple.
Offerings to God		
(i) *By Priest*	:	'Soor'-made of barley flour ('Sattu'), clarified butter.
(ii) *By People*	:	'Lachidana'.
Off. to People (Prasadam)	:	'Lachidana'.
Fairs and Festivals	:	Buddh Purnima.
Observations		
(i) *Courtyard Plants (Protected as well as for Landscaping*	:	*Betula utilis, Cupressus sempervirens, Dahlia rosea, Erigeron alpinus, Ficus cairica, Gaillardia pulchella, Juglans regia, Pinus gerardiana, Prunus armeniaca, Prunus persica, Pyrus malus, Robinia pseudoacacia, Rosa indica, Thuja plicata.*
(ii) *Predominant Plants of the Courtyard*	:	*Pyrus malus.*
(iii) *Maintenance*	:	Excellent.
(iv) *Ethnobotanical*		
Edible	:	Fruits of *Ficus cairica, Prunus armeniaca, Prunus persica, Pyrus malus,* kernels of *Juglans regia,* and seeds of *Pinus gerardiana* edible.
Medicinal	:	Kernel oil of *Juglans regia* used for massage to cure joint pains.

Ornamental	:	*Betula utilis, Cupressus sempervirens, Dahlia rosea, Erigeron alpinus, Ficus cairica, Gaillardia pulchella, Pinus gerardiana, Rosa indica* and *Thuja plicata* grown as ornamentals.
Status of Courtyard Plants in Surrounding Areas	:	Excepting *Betula utilis*, all plants occur commonly outside.
Remarks	:	There is an orchard of about 30 apple plants in the courtyard of temple.

7. Maheshwar Devta Temple (Pl. 10G)

Main Deity	:	Maheshwar Devta.
Other Deity	:	Narayana.
Tot. Ar. of Tem. Courtyard	:	1536 sq m.
Location		
(i) *Place*		
Village	:	Thanang
Block	:	Nichar
Tehsil	:	Nichar
District	:	Kinnaur
(ii) *Approach*	:	1 km on foot from Sungra.
(iii) *Altitude*	:	2150 m.
Management	:	Temple committee.
Historical Background	:	The temple belongs to 18th-19th century A.D.
Temple Architect	:	Pagoda type; Deodar wood used in its construction.
Offerings to God		
(i) *By Priest*	:	'Dhoop'.
(ii) *By People*	:	'Halva-Puri'.
Off. to People (Prasadam)	:	'Halva-Puri'.
Fairs and Festivals	:	Bishu, Fulaich, Dussehra, Janamashtmi.
Observations		
(i) *Courtyard Plants (Protected as well as for Landscaping*	:	*Althaea rosea, Berberis lycium, Clematis grata, Cupressus sempervirens, Dahlia rosea, Hydrangea macrophylla, Pelargonium hortorum, Pyrus malus, Tagetes erecta, Ulmus villosa.*
(ii) *Predominant Plants of the Courtyard*	:	*Cupressus sempervirens.*
(iii) *Maintenance*	:	Excellent.

(iv) Ethnobotanical

Sacred : *Ulmus villosa* and *Berberis lycium* in the temple courtyard are worshipped by locals as the deity resides in these plants. It is believed that *Ulmus villosa* grows at sacred places only.

Medicinal : 3-5g powdered roots of *Berberis lycium* prescribed on empty stomach for piles.

Edible : Fruits of *Berberis lycium* and *Pyrus malus* edible.

Ornamental : *Althaea rosea, Clematis grata, Cupressus sempervirens, Dahlia rosea, Hydrangea macrophylla, Pelargonium hortorum, Tagetes erecta* plants grown as ornamentals.

Status of Courtyard Plants in Surrounding Areas : *Ulmus villosa* is rare outside the temple courtyard.

Remarks : No sacrifices are made to God for the last 25-30 years.

8. Nag Devta Temple

Main Deity : Nag Devta.

Tot. Ar. of Tem. Courtyard : About 1500 sq m.

Location

(i) Place

Village : Bari

Block : Nichar

Tehsil : Nichar

District : Kinnaur

(ii) Approach : 14 km from Bhavanagar.

(iii) Altitude : 2300 m.

Management : Temple committee.

Historical Background : The temple is believed to be very old.

Temple Architect : Pentroof style with Deodar wood used for its construction.

Offerings to God

(i) By Priest : 'Dhoop' and Flowers collected from sacred places.

(ii) By People : 'Halva' prepared from wheat flour.

Off. to People (Prasadam) : 'Halva-Puri'.

Fairs and Festivals : Fulaich, Bishu (Baisakhi), Diwali.

Observations

(i) Courtyard Plants (Protected as well as for Landscaping	:	Nil.
(ii) Maintenance	:	Satisfactory.

9. Shamsheer Devta Temple (Shiv Temple) (Pl. 10H)

Main Deity	:	Shamsheer Devta.
Other Deities	:	Shiva, Bhagvati Devi, Nag Devta, Sonik Devi (Van Devi).
Tot. Ar. of Tem. Courtyard	:	3000 sq m.
Location		
(i) Place		
Village	:	Rakchham
Block	:	Kalpa
Tehsil	:	Sangla
District	:	Kinnaur
(ii) Approach	:	About 15 km from Sangla.
(iii) Altitude	:	2900 m.
Management	:	Temple committee.
Historical Background	:	The temple is believed to be very old.
Temple Architect	:	Pentroof style.
Offerings to God		
(i) By Priest	:	'Halva-Puri'.
(ii) By People	:	'Halva'.
Off. to People (Prasadam)	:	'Halva-Puri'.
Fairs and Festivals	:	Fulaich (festival of flowers, celebrated in September for five days), Shaji (fair held during February-March).
Observations		
(i) Courtyard Plants (Protected as well as for Landscaping	:	*Juniperus macropoda, Populus ciliata.*
(ii) Predominant Plants of the Courtyard	:	*Juniperus macropoda.*
(iii) Maintenance	:	Satisfactory.
(iv) Ethnobotanical		
Ornamental	:	*Juniperus macropoda* and *Populus ciliata* grown as ornamentals.

Remarks : The temple was completely burnt in a devastating fire on 18th November, 2002. At present, the idols are kept in the nearby Shiva temple which is considered to be the oldest Shiva temple in Kinnaur.

10. Usha Devi Temple (Pl. 10I)

Main Deity : Usha Devi.

Other Deities : Narayan Devta, Chelang Devta, Shiva.

Tot. Ar. of Tem. Courtyard : 1500 sq m.

Location

(i) *Place*

Village : Nichar

Block : Nichar

Tehsil : Nichar

District : Kinnaur

(ii) *Approach* : 4 km from Sungra (16 km from Bhavanagar).

(iii) *Altitude* : 2500 m.

Management : Temple committee.

Historical Background : Origin of the temple can be traced to 18th–19th century A.D.

Temple Architect : Pentroof type with Deodar wood used in its construction.

Offerings to God

(i) *By Priest* : 'Halva'.

(ii) *By People* : 'Halva', necklace of dry fruits like Chestnut and Apricot.

Off. to People (Prasadam) : 'Halva-Puri'.

Fairs and Festivals : Fulaich, Dakhrain (celebrated in July), Dussehra, Diwali.

Observations

(i) *Courtyard Plants (Protected as well as for Landscaping* : *Althaea rosea, Cupressus sempervirens, Dahlia rosea, Lonicera japonica, Thuja orientalis, Ulmus villosa.*

(ii) *Predominant Plants of the Courtyard* : *Althaea rosea.*

(iii) *Maintenance* : Excellent.

(iv) Ethnobotanical

Sacred	:	*Ulmus villosa* growing in the temple courtyard worshipped by the locals.
Ornamental	:	*Althaea rosea, Cupressus sempervirens, Dahlia rosea, Lonicera japonica, Thuja orientalis* grown as ornamentals.
Other Uses	:	Flowers of *Althaea rosea* used to reduce or kill termites.
Status of Courtyard Plants in Surrounding Areas	:	*Ulmus villosa* tree is absent outside the temple land.

Temples of District Kullu

1. Ambika Mata Temple (Pl. 11A)

Main Deity : Ambika (form of Durga).

Other Deities : Parshuram, Shiva.

Tot. Ar. of Tem. Courtyard : About 1500 sq m.

Location

(i) *Place*

Village : Nirmand

Block : Nirmand

Tehsil : Nirmand

District : Kullu

(ii) *Approach* : 17 km from Rampur.

(iii) *Altitude* : 1500 m.

Management : Local temple committee.

Historical Background : It is believed that Lord Parshuram established this temple here. It has been repaired and renovated many times in the past.

Temple Architect : Pentroof style.

Offerings to God

(i) By Priest

Morning : 'Halva-Puri'.

Evening : Salted Rice (cooked).

(ii) By People : 'Lachidana', 'Halva-Puri', 'Kasaar', Seasonal crops.

Off. to People (Prasadam) : 'Lachidana', 'Halva', 'Kasaar'.

Fairs and Festivals : Buddi Diwali, Navratra (twice a year).

Observations

(i) Courtyard Plants (Protected as well as for Landscaping : *Althaea rosea, Cannabis sativa, Canna indica, Clerodendrum philippinum, Cynodon dactylon, Dahlia rosea, Jasminum primulinum, Jasminum sambac, Lonicera japonica, Mirabilis jalapa, Musa paradisiaca, Nerium odorum, Ocimum sanctum, Prinsepia utilis, Prunus armeniaca, Rosa indica, Tagetes erecta, Tecoma stans* and *Thevetia neriifolia.*

(ii) Predominant Plants of the Courtyard : *Mirabilis jalapa.*

(iii) Maintenance : Excellent.

(iv) Ethnobotanical

Sacred : *Ocimum sanctum* considered sacred. *Cynodon dactylon* employed in all religious ceremonies.

Medicinal : Seeds of *Cannabis sativa* and stem of 'gloe' (*Tinospora cordifolia*) powdered. This powder is taken daily to reduce the accumulation of uric acid in body.

Edible : Fruis of *Musa paradisiaca* and *Prunus armeniaca* edible.

Ornamental : *Althaea rosea, Canna indica, Clerodendrum philippinum, Dahlia rosea, Jasminum primulinum, Jasminum sambac, Lonicera japonica, Mirabilis jalapa, Musa paradisiaca, Nerium odorum, Ocimum sanctum, Rosa indica, Tagetes erecta, Tecoma stans, Thevetia neriifolia.*

Remarks : Sheep and goats are sacrificed.

2. Bhekhali Mata Temple (Pl. 11B)

Main Deity : Bhuvneshwari Mata.

Other Deities : Narayana (form of Lord Krishana), Bhairav.

Tot. Ar. of Tem. Courtyard : 7220 sq m.

Location

(i) Place

Village : Bekhal

Block : Kullu

Tehsil	:	Kullu
District	:	Kullu
(ii) Approach	:	About 10 km from Kullu.
(iii) Altitude	:	1800 m.
Management	:	Temple committee appointed by Government.
Historical Background	:	Its origin traced back to medieval period. Later, it was repaired by the temple committee.
Temple Architect	:	Pentroof style.
Offerings to God		
(i) By Priest	:	'Chapati' alongwith clarified butter, Flowers from the temple courtyard.
(ii) By People	:	'Lachidana', 'Halva', Sweets, Flowers, Fruits.
Off. to People (Prasadam)	:	'Lachidana', Fruits, Sweets.
Fairs and Festivals	:	Baisakhi, Rath Yatra festival in June, Dussehra, Navratra.
Observations		
(i) Courtyard Plants (Protected as well as for Landscaping	:	*Buxus sempervirens, Cedrus deodara, Cestrum nocturnum, Narcissus tazetta, Rosa indica, Tagetes erecta, Ulmus villosa, Zinnia elegans.*
(ii) Predominant Plants of the Courtyard	:	*Tagetes erecta.*
(iii) Maintenance	:	Excellent.
(iv) Ethnobotanical		
Sacred	:	*Ulmus villosa* considered sacred and protected. Wood of *Cedrus deodara* employed for manufacturing palanquins of deity. Wood used in 'hawan'.
Ornamental	:	*Buxus sempervirens, Cestrum nocturnum, Narcissus tazetta, Rosa indica, Tagetes erecta, Zinnia elegans* grown as ornamental plants.
Other Uses	:	Wood of *Cedrus deodara* excellent for furniture and construction works.
Status of Courtyard Plants in Surrounding Areas	:	*Buxus sempervirens* is absent in the surrounding areas of the temple whereas *Ulmus villosa* occurs rarely.
Remarks	:	Buck and Ram regularly sacrificed for appeasing Bhairav. The temple is named after the name of village, *i.e.* Bekhal which several years ago was famous for having its large-sized shrubs.

3. Bijli Mahadev Temple (Pl. 11C)

Main Deity	: Bijli Mahadev (Lord Shiva) enshrined in the form of a 'Lingam'.
Other Deities	: Kuber, Mahuti Nag, Yogni Devi, Nandi, Bringi.
Tot. Ar. of Tem. Courtyard	: About 5000 sq m.
Location	
(i) Place	
Village	: Mathan
Block	: Kullu
Tehsil	: Kullu
District	: Kullu
(ii) Approach	: The temple located on the right bank of the Parbati river which is about 30 km from Kullu.
iii) Altitude	: 2400 m.
Management	: Temple committee.
Historical Background	: The temple, thought to be very old, was reconstructed and repaired from time to time by the temple committee.
Temple Architect	: Pentroof style; it is the most remarkable hill temple having a massive structure made of *Cedrus deodara* and *Pinus wallichiana* wood.
Offerings to God	
(i) By Priest	: Rice-Dal, 'Chapati' with clarified butter.
(ii) By People	: 'Halva', Sweets, Fruits.
Off. to People (Prasadam)	: 'Halwa', Fruits.
Fairs and Festivals	: Baisakhi, Shivratri, Phagli, Shaladra, Bijli mahadev fair, Dussehra.
Observations	
(i) Courtyard Plants (Protected as well as for Landscaping	: There is only one large tree of *Pyrus pashia* in the temple courtyard.
(ii) Maintenance	: Satisfactory.
(iii) Ethnobotanical	
Sacred and Associated Myth	: *Pyrus pashia* regarded as 'Kalpa Vriksha' by the villagers. It is believed that Lord Shiva takes rest under this tree and hence considered as sacred. Its wood used for religious activities in the temple.
Edible	: Fruit of *Pyrus pashia* edible.

Other Uses	:	Flag carrying post ('Dhwaja') is made from *Cedrus deodara* wood.
Status of Courtyard Plants in Surrounding Areas	:	Large-sized trees of *Pyrus pashia* are rare outside.
Remarks	:	Goats and Ram sacrificed to God. Every year, as a result of lightening Shivlingam breaks into pieces. Priest collects these pieces and then joins with butter to give it the original shape.

4. Chandi Temple (Pl. 11D)

Main Deity	:	Durga.
Other Deity	:	Shiva.
Tot. Ar. of Tem. Courtyard	:	About 750 sq m.
Location		
(i) Place		
Village	:	Nirmand
Block	:	Nirmand
Tehsil	:	Nirmand
District	:	Kullu
(ii) Approach	:	17 km from Rampur.
(iii) Altitude	:	1500 m.
Management	:	Local temple committee.
Historical Background	:	The temple belongs to the 7th century A.D.
Temple Architect	:	Pentroof style.
Offerings to God		
(i) By Priest	:	'Halva-Puri'.
(ii) By People	:	'Lachidana', 'Halva-Puri'.
Off. to People (Prasadam)	:	'Lachidana', 'Halva-Puri'.
Fairs and Festivals	:	Navratra.
Observations		
(i) Courtyard Plants (Protected as well as for Landscaping	:	*Jasminum sambac, Mirabilis jalapa, Musa paradisiaca, Nerium odorum, Thevetia neriifolia.*
(ii) Predominant Plants of the Courtyard	:	*Mirabilis jalapa.*
(iii) Maintenance	:	Satisfactory.

(iv) Ethnobotanical

Edible : Fruits of *Musa paradisiaca* edible.

Ornamental : *Jasminum sambac, Mirabilis jalapa, Nerium odorum* and *Thevetia neriifolia* grown as ornamentals.

5. Dakshini Mahadev Temple (Pl. 11E)

Main Deity : The main deity, a Shiv-lingam was brought from South India thus earning the temple the name of Dakshini Mahadev.

Other Deities : Brahma, Vishnu, Mahesh, Ganpati, Kali, Bhairav, Hanuman.

Tot. Ar. of Tem. Courtyard : About 1500 sq m.

Location

(i) Place

Village : Nirmand

Block : Nirmand

Tehsil : Nirmand

District : Kullu

(ii) Approach : 17 km from Rampur.

(iii) Altitude : 1500 m.

Management : In the name of Deity.

Historical Background : The temple belongs to the 7th-8th century A.D.

Temple Architect : Pentroof style.

Offerings to God

(i) By Priest

Morning : Milk.

Evening : Dry fruits like Almond, Chestnut, Coconut, Currant, etc.

(ii) By People : 'Lachidana', 'Halva-Puri'.

Off. to People (Prasadam) : 'Lachidana'.

Fairs and Festivals : Shivratri, Holi.

Observations

(i) Courtyard Plants (Protected as well as for Landscaping : *Aegle marmelos, Cannabis sativa, Canna indica, Celtis australis, Chrysanthemum indicum, Clerodendrum philippinum, Cynodon dactylon, Ficus palmata, Ficus religiosa, Impatiens balsamina, Jasminum grandiflorum, Mirabilis jalapa, Nerium odorum, Ocimum sanctum,*

Pelargonium hortorum, Prinsepia utilis, Ricinus communis, Rosa indica, Tagetes erecta.

(ii) Predominant Plants of the Courtyard : *Cannabis sativa, Tagetes erecta.*

(iii) Maintenance : Satisfactory.

(iv) Ethnobotanical

Sacred : *Aegle marmelos, Ficus religiosa* and *Ocimum sanctum* considered sacred by the locals. Leaves of *Aegle marmelos* and *Cannabis sativa* offered to appease Lord Shiva. *Cynodon dactylon* employed in various religious ceremonies.

Medicinal : Fruits of *Ficus palmata* good for constipation. Seed oil of *Ricinus communis* given with milk to cure stomach problems.

Edible : Fruits of *Aegle marmelos, Celtis australis* and *Ficus palmata* edible.

Ornamental : *Canna indica, Chrysanthemum indicum, Clerodendrum philippinum, Impatiens balsamina, Jasminum grandiflorum, Mirabilis jalapa, Nerium odorum, Ocimum sanctum, Pelargonium hortorum, Ricinus communis, Rosa indica, Tagetes erecta.*

Other Uses : Leaves of *Celtis australis* used as fodder for livestock.

6. Dhankeshwar Mahadev Temple (Dev Dhank) (Pl. 11F)

Main Deity : Shiva.

Tot. Ar. of Tem. Courtyard : About 1500 sq m.

Location

(i) Place

Village : Bial

Block : Nirmand

Tehsil : Nirmand

District : Kullu

(ii) Approach : About 3 km from Nirmand and 15 km from Rampur.

(iii) Altitude : 1350 m.

Management : Local temple committee.

Historical Background : Once, a goat belonging to Brahimin could not be traced despite all efforts to do so. In his dream Lord Shiva told him that the milk flowing from the udder of lost goat

		had been bathing his 'Lingam' in an inaccessible cave and further told to collect his lost animal from there. Thereafter, he dedicated his life to serve Lord Shiva.
Temple Architect	:	The temple is in the form of a cave.
Offerings to God		
(i) By Priest	:	'Halva', Loaf Sugar, Jaggery.
(ii) By People	:	'Halva', 'Lachidana', Fruits, Dry fruits.
Off. to People (Prasadam)	:	'Halva', 'Lachidana', Fruits, Dry fruits.
Fairs and Festivals	:	Shivratri, Saje (Sakranti of every month).
Observations		
(i) Courtyard Plants (Protected as well as for Landscaping	:	*Aegle marmelos, Canna indica Chrysanthemum indicum, Dahlia rosea, Datura fastuosa, Ficus benghalensis, Ficus glomerata, Ficus religiosa, Jasminum grandiflorum, Musa paradisiaca, Pelargonium hortorum, Pistacia integerrima, Rosa indica, Tagetes erecta, Thevetia neriifolia, Withania somnifera, Zephyranthes candida, Zinnia elegans.*
(ii) Predominant Plants of the Courtyard	:	*Tagetes erecta, Thevetia neriifolia.*
(iii) Maintenance	:	Satisfactory.
(iv) Ethnobotanical		
Sacred	:	Five sacred trees in the temple courtyard 'Panch Pallav', *viz. Aegle marmelos, Ficus benghalensis, Ficus glomerata, Ficus religiosa* and *Thevetia neriifolia* worshipped as sacred trees. Leaves of *Aegle marmelos*, flowers of *Datura fastuosa* and *Thevetia neriifolia* offered to appease Lord Shiva.
Medicinal	:	Minute quantity of seeds of *Datura fastuosa* given to cure mental disorders. Powdered dried fruits of *Pistacia integerrima* prescribed for asthma. Powdered roots of *Withania somnifera* given to cure mental disorders. Poultice of its leaves applied to cure skin disorders.
Edible	:	Fruits of *Aegle marmelos, Ficus glomerata* and *Musa paradisiaca* are edible.
Ornamental	:	*Canna indica Chrysanthemum indicum, Dahlia rosea, Datura fastuosa, Jasminum grandiflorum, Musa paradisiaca, Pelargonium hortorum, Rosa indica, Tagetes erecta, Thevetia neriifolia, Withania somnifera, Zephyranthes candida, Zinnia elegans.*
Status of Courtyard Plants in Surrounding Areas	:	Excepting *Ficus benghalensis, Ficus religiosa, Ficus glomerata, Aegle marmelos,* all other plants occur commonly outside.

7. Gurudwara Sahib Manikaran (Pl. 11G)

Main Deity : Guru Nanak Dev.

Tot. Ar. of Tem. Courtyard : About 7500 sq m. There is a separate land in the name of Gurudwara Sahib having an area of about 20 bigha where apples, seasonal vegetables and fodder are grown.

Location

(i) *Place*

Town : Manikaran

Block : Kullu

Tehsil : Kullu

District : Kullu

(ii) *Approach* : 35 km from Bhuntar.

(iii) *Altitude* : 1800 m.

Management : In the name of Deity.

Historical Background : Saint Shri Narayan Hari of Kamalpur (Pakistan) came to Manikaran with his family in 1947 A.D. The place of Gurudwara was neglected piece of land at that time and the present structure owes greatly to him and other followers.

Temple Architect : Dome-shaped.

Offerings to God

(i) *By Priest* : 'Halva'.

(ii) *By People* : Money.

Off. to People (Prasadam) : Sugar candy, 'Halva'.

Fairs and Festivals : Baisakhi, Budh Purnima, Dussehra, Diwali, Janamashtmi, Navratra (twice a year), Death Anniversary of Baba Narayan Hari (16-23 February).

Observations

(i) *Courtyard Plants (Protected as well as for Landscaping* : *Althaea rosea, Calendula officinalis, Callistemon lanceolatus, Cedrus deodara, Cosmos sulphureus, Cupressus sempervirens, Dahlia rosea, Ficus palmata, Helianthus annuus, Impatiens balsamina, Jasminum primulinum, Kochia trichophylla, Lagerstroemia indica, Mirabilis jalapa, Morus alba, Narcissus tazetta, Nerium odorum, Pelargonium hortorum, Populus ciliata, Prunus armeniaca, Pyrus malus, Rosa indica, Tagetes erecta, Thuja orientalis, Zephyranthes candida.*

(ii) Predominant Plants of the Courtyard	:	*Helianthus annuus, Tagetes erecta.*
(iii) Maintenance	:	Excellent.
(iv) Ethnobotanical		
Edible	:	Fruits of *Ficus palmata, Morus alba, Prunus armeniaca, Pyrus malus* edible.
Medicinal	:	Latex of *Ficus palmata* applied to cure cuts and wounds.
Ornamental	:	All the courtyard plants are also planted for ornamental purposes.
Other Uses	:	Leaves of *Morus alba* given as fodder. Wood of *Cedrus deodara* excellent for furniture and construction works.
Remarks	:	There are hot water springs in the Gurudwara Sahib, in which food to be served in community meal ('Langar') is cooked every day.

8. Hadimba Devi Temple (Pl. 11H, I)

Main Deity	:	Hadimba.
Tot. Ar. of Tem. Courtyard	:	About 4500 sq m.
Location		
(i) Place		
Town	:	Manali (at Dungri)
Block	:	Nagar
Tehsil	:	Manali
District	:	Kullu
(ii) Approach	:	About 3 km from Manali, situated in the forest of Devdar trees.
(iii) Altitude	:	2600 m.
Management	:	Managed by Archaeological Survey of India.
Historical Background	:	Pandavas spent a part of their exile in Kullu Himalaya. One day, Bhima saw Hadimba, only sister of a demon Hidimb, and fell in love with her. Bhima killed her monster brother in a combat and took her as his spouse. People worship Hadimba as their saviour in times of misery and natural calamities. The temple was built in 1553 A.D. by Raja Bahadur Singh.
Temple Architect	:	Pagoda type massive wooden structure.
Offerings to God		
(i) By Priest	:	Jaggery, Coconut, 'Halva', Flowers.
(ii) By People	:	Coconut, Scarves, Flowers.

Off. to People (Prasadam) : 'Lachidana'.

Fairs and Festivals : Dussehra, Hadimba fair on 15-16 May (for 2 days).

Observations

(i) Courtyard Plants (Protected as well as for Landscaping : *Berberis lycium, Cedrus deodara, Dahlia rosea, Hydrangea macrophylla, Tagetes erecta, Thuja orientalis.*

(ii) Predominant Plants of the Courtyard : *Cedrus deodara, Tagetes erecta.*

(iii) Maintenance : Excellent.

(iv) Ethnobotanical

Sacred : Flowers from the courtyard offered to the deity.

Ornamental : *Dahlia rosea, Hydrangea macrophylla, Tagetes erecta* and *Thuja orientalis* grown as ornamentals.

Remarks : Buck, ram and cocks sacrificed to deity by people whose wishes are fulfilled. It has been a custom for centuries that inaugural and closing ceremony of the 'Kullu Dussehra' is led by her chariot. Temple of Ghatotkach (son of Hadimba) is also situated at a distance of about 75 m from this temple where a tree of *Cedrus deodara* represent the deity.

9. Himalayan Buddhist Monastery (Pl. 11J, 12A)

Main Deity : Gautam Buddha.

Other Deities : Padamsambhava, Hyangrua.

Tot. Ar. of Tem. Courtyard : 7680 sq m.

Location

(i) Place

Town : Manali

Block : Nagar

Tehsil : Manali

District : Kullu

(ii) Approach : The temple is located in the town.

(iii) Altitude : 2200 m.

Management : Himalayan Buddhist Society.

Historical Background : The Monastery, about 80 years old, had been built by local Buddhist community.

Temple Architect : Buddhist Monastery. Deodar wood used in the construction of Monastery.

Offerings to God

(i) By Priest	:	Sweets, Fruits, Flowers.
(ii) By People	:	Sweets, Fruits, Flowers.
Off. to People (Prasadam)	:	Sweets.
Fairs and Festivals	:	Buddh Purnima.
Observations		
(i) Courtyard Plants (Protected as well as for Landscaping	:	*Althaea rosea, Calendula officinalis, Cannabis sativa, Chlorophytum capense, Cupressus torulosa, Dahlia rosea, Helianthus annuus, Hydrangea macrophylla, Morus alba, Populus ciliata, Prunus armeniaca, Pyrus malus, Rosa indica, Tagetes erecta, Tecoma radicans, Thuja orientalis, Tropaeolum majus, Ulmus villosa.*
(ii) Predominant Plants of the Courtyard	:	*Tagetes erecta.*
(iii) Maintenance	:	Excellent.
(iv) Ethnobotanical		
Sacred	:	*Cupressus torulosa* growing in the courtyard considered sacred. Its leaves and twigs burnt as incense ('Dhoop') in the Monastery.
Edible	:	Fruits of *Morus alba, Prunus armeniaca* and *Pyrus malus* edible.
Ornamental	:	*Althaea rosea, Calendula officinalis, Chlorophytum capense, Cupressus torulosa, Dahlia rosea, Helianthus annuus, Hydrangea macrophylla, Populus ciliata, Rosa indica, Tagetes erecta, Tecoma radicans, Thuja orientalis, Tropaeolum majus, Ulmus villosa* grown as ornamentals.

10. Maharishi Vashisht Temple (Pl. 12B)

Main Deity	:	Maharishi Vashisht.
Other Deities	:	Shri Rama, Shiva.
Tot. Ar. of Tem. Courtyard	:	About 3000 sq m.
Location		
(i) Place		
Village	:	Vashisht
Block	:	Nagar
Tehsil	:	Manali
District	:	Kullu

(ii) Approach	:	About 3 km from Manali on the side of Manali-Rohtang road.
(iii) Altitude	:	2650 m.
Management	:	Temple committee.
Historical Background	:	Long ago, Rishi Vashisht meditated at this place and blessed this place with natural hot water bodies. The old temple is believed to be 1800 years old. The present temple was constructed over the old temple in 1980 by Raja Takshpal.
Temple Architect	:	Pentroof style (made of Deodar wood and stone). Near the main temple, there is a temple of Sita-Rama-Lakshmana in Sikhara style.
Offerings to God		
(i) By Priest		
Morning	:	'Luchian', Rice-Dal, Chapati.
Evening	:	Dry fruits like Coconut, Almond, Currant, etc.
(ii) By People	:	'Lachidana', Fruits, Flowers.
Off. to People (Prasadam)	:	'Lachidana'.
Fairs and Festivals	:	Baisakhi (13-15 April).
Observations		
(i) Courtyard Plants (Protected as well as for Landscaping	:	*Juniperus chinensis, Lonicera japonica, Ocimum basilicum, Rosa indica, Tagetes erecta.*
(ii) Predominant Plants of the Courtyard	:	*Tagetes erecta.*
(iii) Maintenance	:	Satisfactory.
(iv) Ethnobotanical		
Sacred	:	Flowers from the temple courtyard offered twice daily to the deity in religious worshipping. *Ocimum basilicum* considered sacred.
Ornamental	:	All the courtyard plants grown for ornamental purposes.
Remarks	:	No sacrifices offered. There is a natural hot water spring in the courtyard of temple where people take holy bath to cure all types of skin ailments.

11. Manu Temple (Pl. 12C)

Main Deity	:	Manu Rishi.
Other Deities	:	Shiva, Brahama, Vishnu, Ganesh, Hanuman, Rama, Krishana, Kali, Durga.
Tot. Ar. of Tem. Courtyard	:	About 4500 sq m.

Location		
(i) Place		
Town	:	Manali (old Manali)
Block	:	Nagar
Tehsil	:	Manali
District	:	Kullu
(ii) Approach	:	2 km from Manali.
(iii) Altitude	:	2550 m.
Management	:	Temple committee.
Historical Background	:	Long ago, Rishi Manu came here and asked for milk from one of ladies in the village. The lady provided the same which the saint turned into curd. She was surprised to see all this. The saint suggested her to dig out her cowshed later. On doing so, the villagers found some idols for which temple was constructed.
Temple Architect	:	Pentroof type.
Offerings to God		
(i) By Priest	:	Dal-rice, Sweets.
(ii) By People	:	Sweets (especially 'Ladoo'), Flowers, Fruits.
Off. to People (Prasadam)	:	Puffed rice.
Fairs and Festivals	:	Phagli (in January), Diyai (celebrated for 11 days during January-February).
Observations		
(i) Courtyard Plants (Protected as well as for Landscaping	:	*Juglans regia, Ligustrum lucidum.*
(ii) Maintenance	:	Excellent.
(iii) Ethnobotanical		
Sacred	:	*Ligustrum lucidum,* planted near the entrance (gate) of temple, considered sacred and worshipped. People believe that Pataal Devta resides in this tree and regarded as gatekeeper of Manu Devta.
Medicinal	:	Kernels of *Juglans regia* effective against heart diseases.
Edible	:	Kernels of *Juglans regia* edible.
Ornamental	:	*Juglans regia* and *Ligustrum lucidum* grown as ornamentals.
Other Uses	:	Wood of *Juglans regia* excellent for construction works.

Status of Courtyard Plants in Surrounding Areas	:	Pataal tree (*Ligustrum lucidum*) is absent in the surrounding areas of the temple.
Remarks	:	The town Manali is named after the name of Rishi Manu.

12. Murlidhar Temple (Pl. 12D)

Main Deity	:	Radha-Krishana.
Tot. Ar. of Tem. Courtyard	:	384 sq m.
Location		
(i) Place		
Town	:	Manikaran
Block	:	Kullu
Tehsil	:	Kullu
District	:	Kullu
(ii) Approach	:	35 km from Bhuntar.
(iii) Altitude	:	1800 m.
Management	:	Privately owned.
Historical Background	:	The temple though constructed in 1600 A.D. was renovated in 1987 by locals.
Temple Architect	:	Sikhara style.
Offerings to God		
(i) By Priest		
Morning	:	Cooked rice.
Evening	:	'Babru', 'Rot' on Tuesday.
(ii) By People	:	Puffed rice, 'Lachidana'.
Off. to People (Prasadam)	:	Puffed rice, 'Lachidana'.
Fairs and Festivals	:	Janamashtmi, Ram navami.
Observations		
(i) Courtyard Plants (Protected as well as for Landscaping	:	*Canna indica, Citrus jambhiri, Cynodon dactylon, Dahlia rosea, Helianthus annuus, Jasminum primulinum, Lagerstroemia indica, Ocimum sanctum, Morus alba, Pelargonium hortorum, Prunus armeniaca, Punica granatum, Tagetes erecta.*
(ii) Predominant Plants of the Courtyard	:	*Tagetes erecta.*
(iii) Maintenance	:	Satisfactory.

(iv) Ethnobotanical

Sacred : *Ocimum sanctum* considered sacred. *Cynodon dactylon* employed in various religious ceremonies.

Edible : Fruits of *Citrus jambhiri, Morus alba, Prunus armeniaca* and *Punica granatum* edible.

Ornamental : *Canna indica, Dahlia rosea, Helianthus annuus, Jasminum primulinum, Lagerstroemia indica, Ocimum sanctum, Pelargonium hortorum, Tagetes erecta* and *Tecoma radicans* grown as ornamental plants.

Remarks : No sacrifices permitted.

13. Naina Devi Temple (Pl. 12E)

Main Deity : Naina Devi.

Tot. Ar. of Tem. Courtyard : 768 sq m.

Location

(i) Place

Town : Manikaran

Block : Kullu

Tehsil : Kullu

District : Kullu

(ii) Approach : 35 km from Bhuntar.

(iii) Altitude : 1800 m.

Management : In the name of Deity (Kardar from a particular family is the caretaker of the temple).

Historical Background : It is said that Lord Shiva opened his third eye here from which Naina Devi appeared.

Temple Architect : The temple had been built in Khash style.

Offerings to God

(i) By Priest : 'Halva'.

(ii) By People : 'Halva', 'Rot'.

Off. to People (Prasadam) : 'Halva'.

Fairs and Festivals : Faguli for 5 days during February-March, Saja (held on Sakranti of Baisakh), Navaratra, Dussehra.

Observations

(i) Courtyard Plants (Protected as well as for Landscaping : *Mirabilis jalapa, Morus serrata, Tagetes erecta, Tecoma radicans.*

(ii) Predominant Plants of the Courtyard : *Tagetes erecta.*

(iii) Maintenance : Satisfactory.

(iv) Ethnobotanical

Sacred : *Morus serrata* regarded as sacred plant.

Edible : Fruits of *Morus serrata* edible.

Ornamental : *Mirabilis jalapa, Tagetes erecta* and *Tecoma radicans* grown as ornamental plants.

Other Uses : Wood of *Morus serrata* used for furniture.

Remarks : Sheep and Goats sacrificed to the deity.

14. Parshuram Temple (Pl. 12F)

Main Deity : Parshuram (incarnation of Vishnu).

Tot. Ar. of Tem. Courtyard : About 1500 sq m.

Location

(i) Place

Village : Nirmand

Block : Nirmand

Tehsil : Nirmand

District : Kullu

(ii) Approach : 17 km from Rampur.

(iii) Altitude : 1500 m.

Management : Temple committee.

Historical Background : The temple is believed to be of the 7th century A.D. After destruction of Saraswati Sindhu region, Parshuram brought people from that place and provided shelter to 500 Brahmins, 6 Shilpi and 120 farmers. Therefore, it is also known as Parshuram Kothi.

Temple Architect : Pentroof structure (The temple has beautiful carvings depicting religious scenes and Hindu deities).

Fairs and Festivals : Bhunda Festival (last time held in 1981).

Observations

(i) Courtyard Plants (Protected as well as for Landscaping : *Clerodendrum philippinum, Mirabilis jalapa, Nerium odorum, Pelargonium hortorum, Tagetes erecta.*

(ii) Predominant Plants of the Courtyard : *Tagetes erecta.*

(iii) Maintenance	:	Satisfactory.
(iv) Ethnobotanical		
Ornamental	:	All the courtyard plants grown for ornamentals purposes.
Remarks	:	Temple opens only on the eve of Bhunda festival. Human sacrifices were common earlier on Bhunda festival but now it is completely banned and in lieu goats are sacrificed.

15. Raghunath Temple (Pl. 12G)

Main Deity	:	Sita-Ram.
Other Deities	:	Hanuman, Narsingh, Shaligram, Ganpati.
Tot. Ar. of Tem. Courtyard	:	About 3000 sq m.
Location		
(i) Place		
Town	:	Kullu (Sultanpur)
Block	:	Kullu
Tehsil	:	Kullu
District	:	Kullu
(ii) Approach	:	Located in the town at Sultanpur.
(iii) Altitude	:	1200 m.
Management	:	Privately owned.
Historical Background	:	Built by Raja Jagat Singh in 1660 A.D. Long ago, King of Kullu suffered from leprosy on killing an innocent Brahmin at Nagar (old Kullu). For its cure the king went to a saint for which he was advised to bring the idols of Rama and Sita from Treta Nath temple of Ayodhya. These idols ultimately were stolen from Ayodhya and established at Manikaran. The king worshipped the idols regularly, and completely got cured of the disease. Later, these idols were established at this temple and considered as the supreme deity of Kullu.
Temple Architect	:	Mixed look of Pahari and Pyramidal style; Deodar wood used in the construction of temple.
Offerings to God		
(i) By Priest	:	Milk-'Puri', Rice-Dal, 'Puri'-Vegetables.
(ii) By People	:	'Lachidana', Sweets, Flowers, Fruits.
Off. to People (Prasadam)	:	'Lachidana', Sugar candy.

Fairs and Festivals : Dussehra, Diwali, Basant panchami, Holi, Van-Vihar (on this day Diety goes to forest), Nirjala Ekadashi, Annkut (next day of Diwali).

Observations

(i) Courtyard Plants (Protected as well as for Landscaping : *Aegle marmelos, Bougainvillea spectabilis, Callistemon lanceolatus, Canna indica, Cynodon dactylon, Dahlia rosea, Ficus religiosa, Hedychium coronarium, Helianthus annuus, Jasminum grandiflorum, Jasminum humile, Narcissus tazetta, Nerium odorum, Polianthes tuberosa, Rosa indica, Tagetes eracta.*

(ii) Predominant Plants of the Courtyard : *Jasminum grandiflorum, Tagetes eracta.*

(iii) Maintenance : Satisfactory.

(iv) Ethnobotanical

Sacred : *Ficus religiosa* considered sacred. *Cynodon dactylon* employed in all religious ceremonies.

Medicinal : Leaves of *Jasminum humile* chewed to cure mouth ulcer. Also, its paste applied to cure skin disorders. 4-5g powdered rhizome of *Hedychium coronarium* prescribed twice daily for fever.

Edible : Fruits of *Aegle marmelos* edible.

Ornamental : *Bougainvillea spectabilis, Callistemon lanceolatus, Canna indica, Dahlia rosea, Hedychium coronarium, Helianthus annuus, Jasminum grandiflorum, Jasminum humile, Narcissus tazetta, Nerium odorum, Polianthes tuberosa, Rosa indica, Tagetes eracta* grown as ornamentals.

Remarks : Buck sacrificed to appease the deity.

16. Raghunath Temple (Pl. 12H)

Main Deity : Raghunath.

Other Deities : Vishnu, Shiv-Parvati.

Tot. Ar. of Tem. Courtyard : About 1500 sq m.

Location

(i) Place

Town : Manikaran

Block : Kullu

Tehsil : Kullu

District : Kullu

(ii) Approach	:	35 km from Bhuntar.
(iii) Altitude	:	1800 m.
Management	:	Local temple committee.
Historical Background	:	The temple could be traced to 18th century A.D.
Temple Architect	:	It is a pyramidal style of temple with a slate-roof at the top.
Offerings to God		
(i) By Priest		
Morning	:	Rice-Dal, 'Bhalle'.
Noon	:	Puffed rice.
Evening	:	Puffed rice.
(ii) By People	:	Puffed rice, 'Lachidana'.
Off. to People (Prasadam)	:	Puffed rice, 'Lachidana'.
Fairs and Festivals	:	Dussehra, Ram Navami, Janamashtmi, Holi, Basant Panchami, Annkut.
Observations		
(i) Courtyard Plants (Protected as well as for Landscaping	:	*Canna indica, Cynodon dactylon, Hibiscus syriacus, Ipomoea purpurea, Jasminum humile, Morus alba, Narcissus tazetta, Ocimum sanctum, Pelargonium hortorum, Populus ciliata, Pyrus malus, Rosa indica, Tagetes erecta, Tecoma radicans, Thuja orientalis.*
(ii) Predominant Plants of the Courtyard	:	*Tagetes erecta.*
(iii) Maintenance	:	Satisfactory.
(iv) Ethnobotanical		
Sacred	:	*Ocimum sanctum, Cynodon dactylon* and *Hordeum vulgare* specially offered to the deity.
Edible	:	Fruits of *Morus alba* and *Pyrus malus* edible.
Ornamental	:	*Canna indica, Hibiscus syriacus, Ipomoea purpurea, Jasminum humile, Narcissus tazetta, Ocimum sanctum, Pelargonium hortorum, Populus ciliata, Rosa indica, Tagetes erecta, Tecoma radicans* and *Thuja orientalis* grown as ornamentals purposes.
Remarks	:	'Kullu-ka-Dussehra' starts from this holy place. No sacrifices offered.

17. Rama Temple(Pl. 12I, J)

Main Deity	: Shri Rama.
Other Deities	: Shiva, Hanuman.
Tot. Ar. of Tem. Courtyard	: About 3000 sq m.
Location	
(i) Place	
Town	: Manikaran
Block	: Kullu
Tehsil	: Kullu
District	: Kullu
(ii) Approach	: 35 km from Bhuntar.
(iii) Altitude	: 1800 m.
Management	: Local temple committee.
Historical Background	: Rama and Lakshmana after completing their education in the hermitage of Saint Vashisht, worshipped Lord Shiva here and sought his blessings. Later, temple was built in 1653 A.D. by Raja Jagat Singh (King of Kullu).
Temple Architect	: Temple had been built in a combination of Pyramidal and Khash style with Deodar wood used in its construction.
Offerings to God	
(i) By Priest	
Morning	: Rice-Dal, 'Bhalle'.
Noon/Evening	: Puffed rice.
(ii) By People	: Puffed rice, 'Lachidana'. 'Bhang' and 'Dhatura' offered to Lord Shiva as special offering.
Off. to People (Prasadam)	: Puffed rice, 'Lachidana'.
Fairs and Festivals	: 'Bhadon-ki-Bees', Dussehra, Holi, Basant Panchmi, Shivratri.
Observations	
(i) Courtyard Plants (Protected as well as for Landscaping	: *Cynodon dactylon, Hibiscus syriacus, Narcissus tazetta, Ocimum sanctum, Pelargonium hortorum, Robinia pseudoacacia, Tagetes erecta.*
(ii) Predominant Plants of the Courtyard	: *Tagetes erecta.*
(iii) Maintenance	: Satisfactory.

(iv) Ethnobotanical

Sacred : Leaves of *Ocimum sanctum* used to make sanctified water given to devotees and possess curing properties. *Cynodon dactylon* considered sacred and employed in religious worshipping.

Ornamental : *Hibiscus syriacus, Narcissus tazetta, Ocimum sanctum, Pelargonium hortorum, Robinia pseudoacacia* and *Tagetes erecta* grown as ornamentals.

Other Uses : *Cynodon dactylon* used as fodder. Leaves of *Robinia pseudoacacia* used as fodder and its wood good for agricultural implements.

Remarks : No sacrifices made to deity. There is a boiling water source near the tank where rice can be cooked in small bags. There is also arrangement of holy bath within the temple premises, Rama-kund for males and Sita–kund for females.

18. Shitla Mata Temple (Pl. 12K)

Main Deity : Goddess Shitla.

Tot. Ar. of Tem. Courtyard : About 768 sq m.

Location

(i) Place

Town : Kullu

Block : Kullu

Tehsil : Kullu

District : Kullu

(ii) Approach : The temple is located in the town.

(iii) Altitude : 1180 m.

Management : Temple committee.

Temple Architect : Sikhara style.

Offerings to God

(i) By Priest : 'Halva-Puri'.

(ii) By People : 'Lachidana', Sweets, Flowers. 'Bhang' and 'Dhatura' specially offered to Lord Shiva.

Off. to People (Prasadam) : 'Lachidana'.

Fairs and Festivals : Navratra.

Observations

(i) *Courtyard Plants (Protected as well as for Landscaping*	:	*Araucaria columnaris, Cosmos sulphureus, Cynodon dactylon, Dahlia rosea, Debregeasia hypoleuca, Lycopersicum esculentum, Malvaviscus conzatii, Musa paradisiaca, Ocimum sanctum, Pelargonium hortorum, Psidium guajava, Pyrus malus, Ricinus communis, Rosa indica, Tagetes erecta.*
(ii) *Predominant Plants of the Courtyard*	:	*Ricinus communis.*
(iii) *Maintenance*	:	Excellent.
(iv) *Ethnobotanical*		
Sacred	:	*Ocimum sanctum, Musa paradisiaca* and *Cynodon dactylon* considered sacred plants.
Medicinal	:	Fruits of *Lycopersicum esculentum* effective against anaemia. Poultice of leaves of *Ricinus communis* and *Calotropis procera* applied to cure body swellings.
Edible	:	Ripe fruits of *Lycopersicum esculentum* cooked as vegetable and also used in salads whereas those of *Musa paradisiaca, Psidium guajava* and *Pyrus malus* are edible.
Ornamental	:	*Araucaria columnaris, Cosmos sulphureus, Dahlia rosea, Malvaviscus conzatii, Ocimum sanctum, Pelargonium hortorum, Rosa indica, Tagetes erecta* are grown as ornamentals.
Other Uses	:	Stem of *Ricinus communis* good for scouring teeth. Wood of *Debregeasia hypoleuca* used as fuel and leaves as fodder for livestock.

19. Shiva Temple (Pl. 12L)

Main Deity	:	Shiva.
Tot. Ar. of Tem. Courtyard	:	About 384 sq m.
Location		
(i) *Place*		
Town	:	Manikaran
Block	:	Kullu
Tehsil	:	Kullu
District	:	Kullu
(ii) *Approach*	:	Located 35 km from Bhuntar on the side of Parvati River.
(iii) *Altitude*	:	1800 m.
Management	:	Local temple committee.

Historical Background	: Once, the jewel of ear pendent of goddess Parvati was lost while taking bath, which men of Lord Shiva could not locate. Later, this was possessed by the Lord of jewels–Sheshnag. This enraged Lord Shiva. To save the world from the rage of Shiva, Sheshnag returned the jewel alongwith many other jewels. Therefore, this place has been named as Manikaran (Mani means jewel and Karan–the ear).
Temple Architect	: Sikhara style.
Offerings to God	
(i) By Priest	: Cooked rice.
(ii) By People	: Puffed rice, Fox nut. 'Bhang' and 'Dhatura' specially offered to appease Lord Shiva.
Off. to People (Prasadam)	: Fox nut, Puffed rice, Cooked rice.
Fairs and Festivals	: Shivratri.
Observations	
(i) Courtyard Plants (Protected as well as for Landscaping	: *Bougainvillea spectabilis, Canna indica, Dahlia rosea, Hibiscus rosa-sinensis, Jasminum sambac, Punica granatum, Rosa indica, Tagetes erecta.*
(ii) Predominant Plants of the Courtyard	: *Tagetes erecta.*
(iii) Maintenance	: Satisfactory.
(iv) Ethnobotanical	
Sacred	: Flowers of courtyard plants offered to the deity.
Edible	: Fruits of *Punica granatum* edible.
Ornamental	: All the courtyard plants grown as ornamental plants.
Remarks	: There are number of hot water springs in front of the temple. People love to cook rice in it in small bags and carry these as 'Prasadam'.

20. Shiva Temple (Pl. 12M)

Main Deity	: Shiva.
Tot. Ar. of Tem. Courtyard	: About 768 sq m.
Location	
(i) Place	
Village	: Nirmand
Block	: Nirmand
Tehsil	: Nirmand
District	: Kullu

(ii) Approach : 17 km from Rampur.

(iii) Altitude : 1550 m.

Management : Temple committee.

Historical Background : The temple belongs to the 10th century A.D.

Temple Architect : Pentroof style.

Offerings to God

(i) By Priest : 'Halva'.

(ii) By People : Seasonal crops. 'Bhang' and 'Dhatura' specially offered to appease Lord Shiva.

Off. to People (Prasadam) : 'Lachidana', 'Halva'.

Fairs and Festivals : Shivratri, Holi.

Observations

(i) Courtyard Plants (Protected as well as for Landscaping : *Cedrela toona, Jasminum grandiflorum, Nerium odorum, Tagetes erecta, Thevetia neriifolia.*

(ii) Predominant Plants of the Courtyard : *Tagetes erecta.*

(iii) Maintenance : Satisfactory.

(iv) Ethnobotanical

Sacred : Flowers of courtyard plants offered to the deity.

Ornamental : *Jasminum grandiflorum, Nerium odorum, Tagetes erecta* and *Thevetia neriifolia* grown as ornamental plants.

Other Uses : Wood of *Cedrela toona* excellent for construction and furniture.

Remarks : No sacrifices made to deity.

21. Siyali Mahadev Temple (Pl. 13A)

Main Deity : Shiva.

Other Deities : Parvati, Ganesha.

Tot. Ar. of Tem. Courtyard : About 6000 sq m.

Location

(i) Place

Town : Manali

Block : Nagar

Tehsil : Manali

District : Kullu

(ii) Approach	:	The temple is located in the town.
(iii) Altitude	:	2200 m.
Management	:	Temple committee.
Historical Background	:	The temple, about 25-30 years old, had been built by nearby villagers who regard the deity as their 'Kuldevta'.
Temple Architect	:	Sikhara Style.
Offerings to God		
(i) By Priest	:	'Lachidana', Flowers, Fruits.
(ii) By People	:	'Lachidana', Flowers, Fruits.
Off. to People (Prasadam)	:	'Lachidana'.
Fairs and Festivals	:	Shivratri, Phagu (held in March for 2 days).
Observations		
(i) Courtyard Plants (Protected as well as for Landscaping	:	*Cannabis sativa, Canna indica, Chrysanthemum indicum, Dahlia rosea, Helianthus annuus, Hydrangea macrophylla, Jasminum primulinum, Juglans regia, Lagerstroemia indica, Ocimum kilimandscharicum, Oenothera drummondii, Prunus armeniaca, Rosa indica, Tagetes erecta.*
(ii) Predominant Plants of the Courtyard	:	*Cannabis sativa, Helianthus annuus.*
(iii) Maintenance	:	Satisfactory.
(iv) Ethnobotanical		
Sacred	:	Leaves of *Cannabis sativa* especially offered to appease Lord Shiva. *Ocimum kilimandscharicum* worshipped as sacred plant.
Medicinal	:	40-50 ml decoction of roots of *Lagerstroemia indica* given, twice daily, to cure jaundice. Leaf paste of *Ocimum kilimandscharicum* applied to cure sprains.
Edible	:	Fruits of *Prunus armeniaca*, kernels of *Juglans regia* edible.
Ornamental	:	*Canna indica, Chrysanthemum indicum, Dahlia rosea, Helianthus annuus, Hydrangea macrophylla, Jasminum primulinum, Lagerstroemia indica, Ocimum kilimandscharicum, Oenothera drummondii, Rosa indica* and *Tagetes erecta* grown as ornamental plants.
Remarks	:	No sacrifices made to deity.

22. Vaishno Devi Temple (Shri Mahadevi Tirth) (Pl. 13B)

Main Deity	:	Vaishno Devi.
Other Deities	:	Shiva, Rama-Sita-Lakshmana, Hanuman, Radha Krishana, Shri Ramkrishan Paramhans, Mata Sharda, Vivekanand.

Tot. Ar. of Tem. Courtyard	:	About 7500 sq m.
Location		
(i) Place		
Village	:	Vashing
Block	:	Kullu
Tehsil	:	Kullu
District	:	Kullu
(ii) Approach	:	3 km from Kullu, on way to Manali.
(iii) Altitude	:	1250 m.
Management	:	Temple committee.
Historical Background	:	The temple was built in 1964 A.D. by Saint Sevak Das.
Temple Architect	:	Sikhara style; Deodar and Qail wood used in its construction.
Offerings to God		
(i) By Priest		
Morning	:	Sugar candy, Coconut, Puffed rice, Flowers, Fruits, Food prepared in community meal.
Evening	:	Sugar candy, Puffed rice, Coconut, Flowers, Fruits, Food prepared in community meal.
(ii) By People	:	Sugar candy, Coconut, Puffed rice, Sweets, Flowers, Fruits.
Off. to People (Prasadam)	:	Coconut, Puffed rice, Sugar candy, 'Lachidana'.
Fairs and Festivals	:	Navratra, Ram-Navami, Dussehra, Diwali, Shivratri, Janamashtmi.
Observations		
(i) Courtyard Plants (Protected as well as for Landscaping	:	*Aegle marmelos, Aloe vera, Alnus nitida, Celtis australis, Cestrum nocturnum, Chrysanthemum indicum, Cynodon dactylon, Datura innoxia, Ficus elastica, Ficus religiosa, Jasminum grandiflorum, Jasminum primulinum, Narcissus tazetta, Nerium odorum, Ocimum basilicum, Platanus orientalis, Prunus armeniaca, Rosa indica, Tagetes erecta.*
(ii) Predominant Plants of the Courtyard	:	*Chrysanthemum indicum.*
(iii) Maintenance	:	Excellent.
(iv) Ethnobotanical		
Sacred and Associated Myth	:	Leaves of *Aegle marmelos* offered to appease Lord Shiva. *Cynodon dactylon* employed in all religious ceremonies. *Aegle marmelos, Ficus religiosa* and *Ocimum basilicum*

		worshipped as sacred plants. For fulfillment of their desires, people tie sacred threads and bangles to the tree of *Ficus religiosa*.
Edible	:	Fruits of *Aegle marmelos, Celtis australis* and *Prunus armeniaca* edible.
Medicinal	:	Pulp of leaves of *Aloe vera* prescribed with milk to cure stomach problems.
Ornamental	:	*Alnus nitida, Aloe vera, Cestrum nocturnum, Chrysanthemum indicum, Datura innoxia, Ficus elastica, Jasminum grandiflorum, Jasminum primulinum, Narcissus tazetta, Nerium odorum, Ocimum basilicum, Platanus orientalis, Rosa indica* and *Tagetes erecta* grown as ornamentals.
Other Uses	:	Leaves of *Alnus nitida* and *Celtis australis* used as fodder. Wood of *Celtis australis* good for agricultural implements.
Status of Courtyard Plants in Surrounding Areas	:	Trees of *Platanus orientalis,* growing in the temple courtyard, are absent elsewhere. *Aegle marmelos* and *Ficus religiosa* are rare outside. All other plants are common outside.
Remarks	:	No sacrifices to deities.

23. Vishveshvar Mahadev Temple (Basheshar Mahadev Temple) (Pl. 13C)

Main Deity	:	The object of worship is a large stone 'Lingam' occupying most of the space in the sanctum.
Other Deities	:	Ganesha, Ganga-Yamuna, Vishnu, Kali.
Tot. Ar. of Tem. Courtyard	:	1000 sq m.
Location		
(i) *Place*		
Village	:	Bajaura
Block	:	Kullu
Tehsil	:	Kullu
District	:	Kullu
(ii) *Approach*	:	52 km from Mandi towards Kullu.
(iii) *Altitude*	:	1280 m.
Management	:	Managed by Archaeological Survey of India, local temple committee also exists.
Historical Background	:	The temple belongs to 8th century A.D.

Temple Architect	:	Sikhara style.
Offerings to God		
(i) By Priest	:	Coconut, Honey, Rice, Flowers.
(ii) By People	:	'Halva', Puffed rice, Fruits, Flowers. 'Bhang', 'Dhatura' and 'Doob' offered to appease the deity.
Off. to People (Prasadam)	:	'Halva', Puffed rice.
Fairs and Festivals	:	Shivratri.
Observations		
(i) Courtyard Plants (Protected as well as for Landscaping	:	*Althaea rosea, Canna indica, Chrysanthemum indicum, Cynodon dactylon, Dahlia rosea, Ficus religiosa, Helianthus annuus, Kochia trichophylla, Morus alba, Nerium odorum, Ocimum sanctum, Pelargonium hortorum, Prunus armeniaca, Rosa indica, Tagetes erecta, Tecoma radicans, Thuja orientalis, Zinnia elegans.*
(ii) Predominant Plants of the Courtyard	:	*Tagetes erecta.*
(iii) Maintenance	:	Excellent.
(iv) Ethnobotanical		
Sacred	:	*Ficus religiosa* and *Ocimum sanctum* worshipped as sacred plants. *Cynodon dactylon* employed in religious worshipping.
Edible	:	Fruits of *Morus alba* and *Prunus armeniaca* are edible.
Ornamental	:	*Althaea rosea, Canna indica, Chrysanthemum indicum, Dahlia rosea, Helianthus annuus, Kochia trichophylla, Nerium odorum, Ocimum sanctum, Pelargonium hortorum, Rosa indica, Tagetes erecta, Tecoma radicans, Thuja orientalis* and *Zinnia elegans* grown as ornamentals.
Other Uses	:	Leaves of *Morus alba* used as fodder.

Temples of District Lahul-Spiti

1. Kee Gompa (Pl. 14A)

Main Deity	:	Gautam Buddha.
Other Deities	:	Panjal Lama, Lochan Lama.
Tot. Ar. of Tem. Courtyard	:	About 8000 sq m.
Location		
(i) Place		
Village	:	Kee
Block	:	Kaza
Tehsil	:	Kaza
District	:	Lahul-Spiti
(ii) Approach	:	12 km from Kaza.
(iii) Altitude	:	3750 m.
Management	:	Temple committee.
Historical Background	:	The temple built by Sangsan Sherum Yangbo about 700 years back.
Temple Architect	:	Buddhist Monastery.
Offerings to God		
(i) By Priest	:	'Sow' (made of Barley), Sweets.
(ii) By People	:	'Sow', Sweets.

Off. to People (Prasadam)	:	'Sow', Sweets.
Fairs and Festivals	:	Molum chhupo (held in February for 15 days), Sagaldva (held in April for 7 days), Guitore (held in July for 7 days), Yarne (held in August-September for 45 days).
Observations		
(i) Courtyard Plants (Protected as well as for Landscaping	:	*Calendula officinalis, Helianthus annuus, Malva sylvestris, Ribes orientale, Rosa webbiana, Salix oxycarpa, Tagetes erecta.*
(ii) Predominant Plants of the Courtyard	:	*Tagetes erecta.*
(iii) Maintenance	:	Excellent.
(iv) Ethnobotanical		
Sacred	:	Flowers of *Tagetes erecta* predominantly used for religious activities in the monastery.
Edible	:	Fruits of *Rosa webbiana* and *Ribes orientale* edible.
Ornamental	:	*Calendula officinalis, Helianthus annuus, Malva sylvestris, Ribes orientale, Salix oxycarpa* and *Tagetes erecta* grown as ornamentals.
Status of Courtyard Plants in Surrounding Areas	:	Excepting *Rosa webbiana* and *Ribes orientale* all other plants occur commonly outside.

2. Labrang Monastery (Pl. 14B)

Main Deities	:	Gautam Buddha, Padam Sambhav, Kali, Charanji.
Tot. Ar. of Tem. Courtyard	:	About 1000 sq m.
Location		
(i) Place		
Village	:	Toche
Block	:	Keylong
Tehsil	:	Keylong
District	:	Lahul-Spiti
(ii) Approach	:	About 2 km on foot from Sissu.
(iii) Altitude	:	3050 m.
Management	:	Local committee.
Historical Background	:	The monastery is very old. It was reconstructed about 25 years before by local people.
Temple Architect	:	Buddhist Monastery.

Offerings to God		
(i) By Priest	:	'Sow' (made of Sattu/barley and Ghee)
(ii) By People	:	Sweets, 'Puri' 'Sattu' (used in monastery to make 'Sow')
Off. to People (Prasadam)	:	'Sow', Sweets.
Fairs and Festivals	:	Puranmashi.
Observations		
(i) Courtyard Plants (Protected as well as for Landscaping	:	*Brassica oleracea* var. *capitata, Convolvulus arvensis, Humulus lupulus, Hyoscyamus niger, Polygonum alpinum, Solanum tuberosum.*
(ii) Predominant Plants of the Courtyard	:	*Humulus lupulus.*
(iii) Maintenance	:	Satisfactory.
(iv) Ethnobotanical		
Medicinal	:	Burnt seeds of *Hyoscyamus niger* smoked to cure teeth pain.
Edible	:	Young stem of *Polygonum alpinum* edible. Leaves and stem of *Brassica oleracea* var. *capitata* cooked as vegetable. Tubers of *Solanum tuberosum* cooked as vegetable.
Other Uses	:	*Humulus lupulus* used in the preparation of Bear. *Convolvulus arvensis* used as fodder.
Remarks	:	Crops of cabbage (*Brassica oleracea* var. *capitata*) and potato (*Solanum tuberosum*) grown in small plots in the courtyard.

3. Losar Gompa

Main Deity	:	Jabru Devta (local deity).
Other Deities	:	Lord Buddha, Kali.
Tot. Ar. of Tem. Courtyard	:	7500 sq m.
Location		
(i) Place		
Village	:	Losar
Block	:	Kaza
Tehsil	:	Kaza
District	:	Lahul-Spiti
(ii) Approach	:	34 km from Kunzam Pass.
(iii) Altitude	:	About 3900 m.

Management : Village committee.

Historical Background : This Gompa is about 100 years old and built by the villagers.

Temple Architect : Buddhist Monastery.

Offerings to God

(i) *By Priest* : Sunflower, Wild flowers.

(ii) *By People* : Coins, Clarified butter, 'Sattu'.

Fairs and Festivals : Losar Festival, First day of new year according to Buddhist Calender.

Observations

(i) *Courtyard Plants (Protected as well as for Landscaping* : *Artemisia moorcroftiana, Helianthus annuus, Salix oxycarpa.*

(ii) *Predominant Plants of the Courtyard* : *Artemisia moorcroftiana.*

(iii) *Maintenance* : Satisfactory.

(iv) *Ethnobotanical*

Sacred : *Artemisia moorcroftiana* in the courtyard used as incense. Flowers of *Helianthus annuus* used as offering to the deity.

Ornamental : *Helianthus annuus* grown as an ornamental plant.

4. Mrikula Devi Temple (Pl. 14C, J)

Main Deity : Kali (Mahishasur Mardini).

Other Deity : Shiva.

Tot. Ar. of Tem. Courtyard : 1000 sq m.

Location

(i) *Place*

Town : Udaipur

Block : Keylong

Tehsil : Keylong

District : Lahul-Spiti

(ii) *Approach* : Located in the town.

(iii) *Altitude* : 2550 m.

Management : Managed by Archaeological Survey of India.

Historical Background	:	It is believed that Bhima (one of the Pandavas) brought wood of 'Devidiar' (*Juniperus macropoda*) from which Vishwakarma built the temple within 24 hours. Later, the present idol of Kali (made of Ashtdhatu) was established near the previous one by the king of Chamba (Raja Udai Singh).
Temple Architect	:	Pentroof-hill type; entirely made of wood of *Juniperus macropoda*.
Offerings to God		
(i) *By Priest*	:	'Dhoop'.
(ii) *By People*	:	Coconut, 'Halva', Clarified butter, Puffed rice, 'Lachidana', 'Shesh' (piece of cloth), flowers of Marigold.
Off. to People (Prasadam)	:	Puffed rice, 'Lachidana', 'Shesh' (piece of cloth tied on the wrists of devotees by priest for good luck.
Fairs and Festivals	:	Halda (January), Phagdi (February).
Observations		
(i) *Courtyard Plants (Protected as well as for Landscaping*	:	*Salix fragilis, Fraxinus xanthoxyloides.*
(ii) *Predominant Plants of the Courtyard*	:	*Fraxinus xanthoxyloides.*
(iii) *Maintenance*	:	Satisfactory.
(iv) *Ethnobotanical*		
Sacred and Associated Myth	:	Leaves of *Juniperus macropoda* used as incense in the temple. Flowers of *Tagetes erecta* are specially grown in fields for offering to the deity. These flowers are also used for making totems 'Par' (worn on chest) and 'Jholnu' (worn on cap). Cutting of courtyard trees is a taboo. Even these trees are not touched with iron for fear of punishment from the deity.
Other Uses	:	Young branches of *Salix fragilis* used as fodder. Wood of *Fraxinus xanthoxyloides* and *Salix fragilis* used for making agricultural implements and other household articles.
Remarks	:	Sheep and goats sacrificed to the deities.

5. Othang Gompa (Pl. 14D, E)

Main Deity	:	Gautam Buddha.
Tot. Ar. of Tem. Courtyard	:	About 1500 sq m.

Location		
(i) Place		
Village	:	Othang
Block	:	Keylong
Tehsil	:	Keylong
District	:	Lahul-Spiti
(ii) Approach	:	About 5 km on foot from Jahalma.
(iii) Altitude	:	3120 m.
Management	:	Local committee.
Temple Architect	:	Buddhist Monastery.
Offerings to God		
(i) By Priest	:	Fruits, Flowers, Sweets.
(ii) By People	:	Sweets, Fruits.
Off. to People (Prasadam)	:	Sweets, Fruits.
Fairs and Festivals	:	Halda.
Observations		
(i) Courtyard Plants (Protected as well as for Landscaping	:	*Althaea rosea, Brassica oleracea* var. *capitata, Calendula officinalis, Cosmos bipinnatus, Dahlia rosea, Daucus carota, Helianthus annuus, Hyoscyamus niger, Juniperus macropoda, Malva sylvestris, Pyrus malus, Rosa foetida, Salix fragilis, Solanum tuberosum, Tagetes erecta.*
(ii) Predominant Plants of the Courtyard	:	*Helianthus annuus.*
(iii) Maintenance	:	Satisfactory.
(iv) Ethnobotanical		
Sacred	:	Tree of *Juniperus macropoda* considered sacred. Floral garlands of *Tagetes erecta, Althaea rosea* and *Rosa foetida* offered to the deity by locals. Leaves of *Juniperus macropoda* used as incense in the monastry.
Medicinal	:	Burnt seeds of *Hyoscyamus niger* smoked for instant relief from toothache. Powdered dried petals of *Rosa foetida* given for curing the jaundice.
Edible	:	Fruits of *Pyrus malus* edible. *Brassica oleracea* var. *capitata* eaten raw or cooked as vegetable. Roots of *Daucus carota* and tubers of *Solanum tuberosum* cooked as vegetable.
Ornamental	:	*Althaea rosea, Calendula officinalis, Cosmos bipinnatus, Dahlia rosea, Helianthus annuus, Malva sylvestris, Rosa foetida, Salix fragilis, Tagetes erecta* are grown as ornamentals.

Other Uses	:	To keep off evil spirits, people tie plants of *Hyoscyamus niger* to the main door of the house. Wood of *Salix fragilis* used for making agricultural implements and household articles. Its leaves used as fodder.
Remarks	:	Crops of cabbage (*Brassica oleracea* var. *capitata*), carrot (*Daucus carota*) and potato (*Solanum tuberosum*) grown in small plots in the courtyard of temple.

6. Raja Ghepan Temple (Pl. 14F)

Main Deity	:	Raja Ghepan.
Other Deity	:	Devi Boti (Milaka).
Tot. Ar. of Tem. Courtyard	:	About 800 sq m.
Location		
(i) Place		
Village	:	Shashun
Block	:	Keylong
Tehsil	:	Keylong
District	:	Lahul-Spiti
(ii) Approach	:	14 km from Koksar on the side of Manali-Keylong road.
(iii) Altitude	:	2950 m.
Management	:	Local temple committee.
Historical Background	:	The present temple is about 80 years old and constructed by the locals.
Temple Architect	:	Sikhara style.
Offerings to God		
(i) By Priest	:	Butter.
(ii) By People	:	'Halva', Puffed rice, Sweets.
Off. to People (Prasadam)	:	'Halva', Puffed rice, Sweets.
Fairs and Festivals	:	Neuper (February), Phagdi (February).
Observations		
(i) Courtyard Plants (Protected as well as for Landscaping	:	*Calendula officinalis, Salix fragilis, Solanum tuberosum, Tagetes erecta.*
(ii) Predominant Plants of the Courtyard	:	*Tagetes erecta.*
(iii) Maintenance	:	Satisfactory.

(iv) Ethnobotanical

Sacred	: Flowers of *Tagetes erecta* considered sacred and offered to the deity. Totems made from it are worn on caps by locals as well as 'Gur' (temple priest).
Edible	: Tubers of *Solanum tuberosum* cooked as vegetable.
Ornamental	: *Calendula officinalis* and *Tagetes erecta* grown as ornamentals.
Other Uses	: Wood of *Salix fragilis* used for making agricultural implements and household articles. Its leaves used as fodder.
Remarks	: Ram and Goats sacrificed to appease the deity.

7. Sakya Monastery

Main Deity	: Gautam Buddha.
Other Deities	: Tara Devi, Padam Sambhav, Mahakal.
Tot. Ar. of Tem. Courtyard	: About 7500 sq m.
Location	
(i) Place	
Town	: Kaza
Block	: Kaza
Tehsil	: Kaza
District	: Lahul-Spiti
(ii) Approach	: The monastery is located in the town.
(iii) Altitude	: 3650 m.
Management	: Local committee called 'Sokpa'.
Historical Background	: Monastery is 30 years old and built by Lama Trop.
Temple Architect	: Buddhist Monastery.
Offerings to God	
(i) By Priest	: 'Sow' (made of 'Sattu', Cheese, Clarified butter, Cashewnut and dry fruits like Almond, Cashewnut, etc.).
(ii) By People	: Money, 'Ashi Khata' (a piece of cloth).
Off. to People (Prasadam)	: 'Sow', Sweets, Fruits.
Fairs and Festivals	: Buddh Purnima, Chham (in October), Birthday of Dalai Lama (6th July), Fair on 15th August.

Observations

(i) Courtyard Plants (Protected as well as for Landscaping	:	*Calendula officinalis, Malva sylvestris, Salix oxycarpa, Tagates erecta.*
(ii) Predominant Plants of the Courtyard	:	*Salix oxycarpa.*
(iii) Maintenance	:	Excellent.
(iv) Ethnobotanical		
Sacred	:	Flowers of courtyard plants considered sacred for offering to the deity.
Ornamental	:	*Calendula officinalis, Malva sylvestris, Tagetes erecta* are grown as ornamentals.

8. Shashur Monastery (Pl. 14G)

Main Deities	:	Guru Padam Sambhav, Deva Gyatso.
Other Deities	:	Manju Shri, Gialva, Gochhanva, Togdan Shakya Shri.
Tot. Ar. of Tem. Courtyard	:	About 3000 sq m.
Location		
(i) Place		
Village	:	Tingche
Block	:	Keylong
Tehsil	:	Keylong
District	:	Lahul-Spiti
(ii) Approach	:	About 5 km on foot from Keylong.
(iii) Altitude	:	3300 m.
Management	:	Local temple committee.
Historical Background	:	The Monastery founded by Lama Gyatso was built in 17 A.D.
Temple Architect	:	Buddhist Monastery.
Offerings to God		
(i) By Priest	:	'Nevet' (water, cereals, flowers), 'Disil' (made of rice, coconut, currant, clarified butter).
(ii) By People	:	Fruits, Sweets, 'Puri'.
Off. to People (Prasadam)	:	Fruits, Sweets, 'Puri'.
Fairs and Festivals	:	Sheshu Festival (held in May-June).

Observations

(i) Courtyard Plants (Protected as well as for Landscaping	:	*Juniperus macropoda, Populus ciliata, Ribes orientale, Rosa webbiana, Salix fragilis.*
(ii) Predominant Plants of the Courtyard	:	*Salix fragilis.*
(iii) Maintenance	:	Excellent.
(iv) Ethnobotanical		
Sacred	:	Leaves of *Juniperus macropoda* used as incense in the monastery. Floral garlands of *Rosa webbiana* offered to the deity. Two old courtyard trees of *Populus ciliata* considered sacred and worshipped by people.
Edible	:	Fruits of *Rosa webbiana* edible.
Other Uses	:	Young branches of *Salix fragilis* given as fodder. Its wood good for agricultural implements and household articles.
Status of Courtyard Plants in Surrounding Areas	:	*Populus ciliata* is not common in the surrounding areas of monastery.
Remarks	:	Plants growing in the courtyard are religiously protected.

9. Trilokinath Temple (Pl. 14H, I)

Main Deities	:	Trilokinath (Shiva), Gautam Buddha.
Other Deity	:	Tara Devi.
Tot. Ar. of Tem. Courtyard	:	About 1500 sq m.
Location		
(i) Place		
Village	:	Trilokinath
Block	:	Keylong
Tehsil	:	Keylong
District	:	Lahul Spiti
(ii) Approach	:	About 5 km on foot from Udaipur.
(iii) Altitude	:	2760 m.
Management	:	Temple committee constructed by State Government.
Historical Background	:	The previous temple was built by Pandavas and the new one by the local people about 50 years back.
Temple Architect	:	Sikhara style.

Offerings to God		
(i) By Priest	:	Puffed rice, 'Lachidana', 'Halva', Biscuit, Fruits.
(ii) By People	:	Puffed rice, 'Lachidana', Clarified butter, 'Halva', Groundnut, flowers of Marigold, ornaments of gold, silver and 'Khatak' (a piece of cloth).
Off. to People (Prasadam)	:	Puffed rice, 'Lachidana', 'Shesh'.
Fairs and Festivals	:	Halda (held in January), Phagdi (held in February), Yor (held in March for 3 days), Satdhara (held in July for 1 day), Pori (held in August for 3 days).
Observations		
(i) Courtyard Plants (Protected as well as for Landscaping	:	*Althaea rosea, Cosmos bipinnatus, Helianthus annuus, Humulus lupulus, Prunus armeniaca, Tagetes erecta.*
(ii) Predominant Plants of the Courtyard	:	*Althaea rosea.*
(iii) Maintenance	:	Excellent.
(iv) Ethnobotanical		
Sacred	:	Flowers of *Althaea rosea* offered to the deity. Floral garlands of *Tagetes erecta* offered to the deity by locals for which these are cultivated in their fields.
Edible	:	Fruits of *Prunus armeniaca* edible.
Ornamental	:	All the courtyard plants grown for ornamental purposes.
Other Uses	:	*Humulus lupulus* used in the preparation of Bear.
Remarks	:	No sacrifices made to deities. This temple is unique for both Hindus and Buddhists as both Lord Shiva and Gautam Buddha are worshipped together.

10. Yordang Gompa (Pl. 14K)

Main Deities	:	Padam Sambhav, Gautam Buddha.
Other Deities	:	Akhiyovia, Gosamba.
Tot. Ar. of Tem. Courtyard	:	About 1500 sq m.
Location		
(i) Place		
Village	:	Biling
Block	:	Keylong
Tehsil	:	Keylong
District	:	Lahul-Spiti

(ii) Approach	:	15 km from Tandi.
(iii) Altitude	:	3000 m.
Management	:	In the name of Deity.
Historical Background	:	The present Monastery (Gompa) is 20 years old built by locals.
Temple Architect	:	Buddhist Monastery.
Offerings to God		
(i) By Priest	:	'Sous'-made of barley.
(ii) By People	:	'Sous', Sweets, Fruits.
Off. to People (Prasadam)	:	'Sous', Sweets, Fruits.
Fairs and Festivals	:	Halda.
Observations		
(i) Courtyard Plants (Protected as well as for Landscaping	:	*Althaea rosea, Brassica oleracea* var. *capitata, Calendula officinalis, Cosmos bipinnatus, Dahlia rosea, Daucus carota, Helianthus annuus, Hemerocallis fulva, Mirabilis jalapa, Pyrus malus, Rosa foetida, Solanum tuberosum, Tagetes erecta.*
(ii) Predominant Plants of the Courtyard	:	*Althaea rosea.*
(iii) Maintenance	:	Satisfactory.
(iv) Ethnobotanical		
Sacred	:	Flowers of all the courtyard plants used for religious worshipping.
Medicinal	:	Dried petals of *Rosa foetida* given for curing jaundice.
Edible	:	Fruits of *Pyrus malus* edible. Leaves and stem of *Brassica oleracea* var. *capitata* cooked as vegetable. Roots of *Daucus carota* and tubers of *Solanum tuberosum* cooked as vegetable.
Ornamental	:	*Althaea rosea, Calendula officinalis, Cosmos bipinnatus, Dahlia rosea, Helianthus annuus, Hemerocallis fulva, Mirabilis jalapa, Pyrus malus, Rosa foetida, Tagetes erecta* are grown as ornamentals.
Remarks	:	Crops of cabbage (*Brassica oleracea* var. *capitata*), carrot (*Daucus carota*) and potato (*Solanum tuberosum*) cultivated in small plots in the temple courtyard.

Temples of District Mandi

1. Bhootnath Temple (Pl. 15A)

Main Deity	:	Shiva.
Other Deities	:	Radha-Krishana, Vishnu, Narsingh, Vaishno Devi.
Tot. Ar. of Tem. Courtyard	:	About 1500 sq m.
Location		
(i) Place		
Town	:	Mandi
Block	:	Mandi
Tehsil	:	Mandi
District	:	Mandi
(ii) Approach	:	Located in the town.
(iii) Altitude	:	880 m.
Management	:	Temple committee.
Historical Background	:	Temple was built by Raja Ajbar Sen (King of Mandi) in 1527 A.D.
Temple Architect	:	Sikhara style.
Offerings to God		
(i) By Priest	:	Rice-Dal, Chapati, Vegetables, Fruits.
(ii) By People	:	Puffed rice, Fruits, Sweets.

Off. to People (Prasadam)	:	Puffed rice, Fruits, Sweets.
Fairs and Festivals	:	Shivratri, Janamashtmi.
Observations		
(i) Courtyard Plants (Protected as well as for Landscaping	:	*Tagetes erecta.*
(ii) Maintenance	:	Excellent.
(iii) Ethnobotanical		
Sacred	:	Flowers of *Tagetes erecta* considered sacred for offering to the deity.
Medicinal	:	5-10g leaf juice of *Tagetes erecta* prescribed for urinary problems.
Remarks	:	No sacrifices.

2. Ekadash Rudra Temple (Pl. 15B)

Main Deity	:	Shiva.
Other Deities	:	Vishnu, Surya, Ganpati, Kartik, Bhagvati, Tripur Sundri, Hanuman.
Tot. Ar. of Tem. Courtyard	:	1200 sq m.
Location		
(i) Place		
Town	:	Mandi
Block	:	Mandi
Tehsil	:	Sadar Mandi
District	:	Mandi
(ii) Approach	:	Located in the town on the side of river Bias.
(iii) Altitude	:	870 m.
Management	:	State Government undertaken.
Historical Background	:	The temple was built by Raja Saheb Sen (King of Mandi) about 350 years back.
Temple Architect	:	Sikhara style.
Offerings to God		
(i) By Priest	:	Rice-Dal, Chapati-Vegetables.
(ii) By People	:	'Kasaar'. 'Bil patri' (Leaves of *Aegle marmelos*) offered to appease Lord Shiva.
Off. to People (Prasadam)	:	Sweets, Fruits.

Fairs and Festivals : Shivratri, Sawan-Mass (celebrated for one month from July 15-August 15).

Observations

(i) *Courtyard Plants (Protected as well as for Landscaping* : *Callistemon lanceolatus, Cynodon dactylon, Ficus roxburghii, Jasminum sambac, Mentha piperita, Nerium odorum, Rosa indica, Salvia coccinia.*

(ii) *Predominant Plants of the Courtyard* : *Jasminum sambac, Rosa indica.*

(iii) *Maintenance* : Excellent.

(iv) *Ethnobotanical*

Sacred : Flowers of the courtyard plants offered to the deity. *Cynodon dactylon* employed in various religious ceremonies as sacred plant.

Medicinal : Leaf juice of *Mentha piperita* effective against vomiting.

Edible : Leaves of *Mentha piperita* used for flavouring. Fruits of *Ficus roxburghii* edible.

Ornamental : *Callistemon lanceolatus, Jasminum sambac, Nerium odorum, Rosa indica, Salvia coccinia* are grown as ornamentals.

Remarks : No sacrifices offered.

3. Gurudwara Shri Guru Gobind Singh (Pl. 15C)

Main Deity : Guru Gobind Singh.

Tot. Ar. of Tem. Courtyard : About 9000 sq m.

Location

(i) *Place*

Town : Rewalsar

Block : Mandi

Tehsil : Sadar Mandi

District : Mandi

(ii) *Approach* : 24 km from Mandi.

(iii) *Altitude* : 1700 m.

Management : Temple committee.

Historical Background : In 1701, Guru Gobind Singh ji visited this place on 'Baisakhi'. He stayed here for a few months, and cajoled locals to fight against Aurangjeb. Since then this place has been worshipped, and the Gurudwara Sahib was built in 1930 by Raja Joginder Sen.

Temple Architect : Dome shaped.

Offerings to God

(i) By Priest : 'Halva'.

(ii) By People : 'Lachidana', Puffed rice.

Off. to People (Prasadam) : 'Halva', 'Lachidana', Puffed rice.

Fairs and Festivals : Baisakhi Fair.

Observations

(i) Courtyard Plants (Protected as well as for Landscaping : *Albizia stipulata, Bauhinia variegata, Debregeasia hypoleuca, Eriobotrya japonica, Eucalyptus tereticornis, Prunus puddum, Pyrus communis, Sapindus mukorossi, Tagetes erecta.*

(ii) Predominant Plants of the Courtyard : *Prunus puddum.*

(iii) Maintenance : Excellent.

(iv) Ethnobotanical

Sacred : Flowers of *Tagetes erecta* considered sacred for offering to the deity.

Medicinal : Ash obtained after burning seeds of *Sapindus mukorossi* consumed with hot water for headache. Its fruit used to cure piles. Infusion of bark of *Albizia stipulata* applied to cure skin problems.

Edible : Flower buds of *Bauhinia variegata* cooked as vegetable. Fruits of *Prunus puddum, Pyrus communis, Eriobotrya japonica, Debregeasia hypoleuca* are edible.

Ornamental : *Tagetes erecta* commonly grown as an ornamental plant.

Other Uses : Leaves of *Albizia stipulata, Debregeasia hypoleuca* and *Bauhinia variegata* given as fodder. Fruits of *Sapindus mukorossi* used for washing clothes.

Status of Courtyard Plants in Surrounding Areas : Excepting *Eriobotrya japonica* and *Sapindus mukorossi,* all other plants occur commonly in the surrounding areas.

4. Hanogi Mata Temple (Pl. 15D)

Main Deity : Durga.

Other Deities : Bhairav, Kali, Lakshmi, Saraswati.

Tot. Ar. of Tem. Courtyard : About 400 sq m.

Location

(i) Place

Village : Hanogi

Block : Mandi

Tehsil	:	Mandi
District	:	Mandi
(ii) Approach	:	30 km from Mandi on the side of Mandi-Kullu road.
(iii) Altitude	:	1100 m.
Management	:	Local temple committee.
Historical Background	:	The temple is about 30 years old and built by the temple committee.
Temple Architect	:	Sikhara style.
Offerings to God		
(i) By Priest	:	Fruits, Sweets.
(ii) By People	:	Fruits, Sweets.
Off. to People (Prasadam)	:	'Lachidana'
Fairs and Festivals	:	Navratra.
Observations		
(i) Courtyard Plants (Protected as well as for Landscaping	:	*Mirabilis jalapa, Tagetes erecta.*
(ii) Maintenance	:	Satisfactory.
(iii) Ethnobotanical		
Sacred	:	Flowers of *Mirabilis jalapa* and *Tagetes erecta* considered sacred for offering to the deity.
Ornamental	:	*Mirabilis jalapa* and *Tagetes erecta* commonly grown as ornamentals.
Remarks	:	No sacrifices made to the deity.

5. Kamaksha Devi Temple (Pl. 15E)

Main Deity	:	Kamaksha Devi.
Other Deities	:	Brahma, Vishnu, Surya, Shiva, Parshuram, Ganesha, Bhairav.
Tot. Ar. of Tem. Courtyard	:	About 3000 sq m.
Location		
(i) Place		
Village	:	Kao
Block	:	Karsog
Tehsil	:	Karsog
District	:	Mandi

(ii) Approach	:	7 km from Karsog.
(iii) Altitude	:	1240 m.
Management	:	Temple committee.
Historical Background	:	The temple dates back to 10th-11th century A.D. It is believed that Lord Parshuram worshipped goddess Kamaksha here. During exile Pandavas also visited this place and made this temple. Later, it was renovated several times.
Temple Architect	:	Pagoda type made of Devdar wood.
Offerings to God		
(i) By Priest		
Morning	:	Rice-Dal.
Evening	:	Chapati-Vegetables.
(ii) By People	:	'Halva', 'Lachidana', Seasonal crops.
Off. to People (Prasadam)	:	'Halva', 'Lachidana'.
Fairs and Festivals	:	Fair held on Ashtmi of Souj Navratra.
Observations		
(i) Courtyard Plants (Protected as well as for Landscaping	:	*Catharanthus roseus, Chrysanthemum indicum, Rosa indica.*
(ii) Maintenance	:	Excellent.
(iii) Ethnobotanical		
Sacred	:	Flowers of courtyard plants considered sacred for offering to the deity.
Medicinal	:	'Gulkand' prepared from petals of *Rosa indica* prescribed with milk to cure stomach problems.
Ornamental	:	All the courtyard plants are grown for ornamental purposes.
Remarks	:	Goats sacrificed.

6. Kamru Nag Temple (Pl. 15F)

Main Deity	:	Kamru Nag.
Tot. Ar. of Tem. Courtyard	:	About 6000 sq m.
Location		
(i) Place		
Village	:	Kamrah
Block	:	Gohr

Tehsil	:	Gohr
District	:	Mandi
(ii) Approach	:	5 km on foot from Rohanda.
(iii) Altitude	:	2720 m.
Management	:	Temple committee.
Historical Background	:	The present temple belongs to 14th-15th century A.D. It is regarded that Pandavas came here during their exile and established the deity.
Temple Architect	:	Pentroof style.
Offerings to God		
(i) By Priest	:	Uncooked rice, Clarified butter.
(ii) By People	:	'Halva', Sweets, Fruits. Newly harvested crops first offered to the deity by the locals before being used at their places.
Off. to People (Prasadam)	:	Unboiled rice, Sweets, 'Kaleya' (soaked rice mixed with clarified butter).
Fairs and Festivals	:	A local fair (held in June for 3 days).
Observations		
(i) Courtyard Plants (Protected as well as for Landscaping	:	Nil.
(ii) Maintenance	:	Excellent.
Remarks	:	Goats sacrificed to appease deity. The land surrounding the temple courtyard has a thick forest of Tosh (*Abies pindrow*), Rye (*Picea smithiana*), Kharshu (*Quercus semecarpifolia*) and Rakhaal (*Taxus baccata*). Offerings including ornaments of gold or silver and money are thrown in the lake by the devotees. 'Shesh' (rice grains) offered by the priest as 'prasadam' to the devotees. It is thought to be ominous if the grains are in odd number.

7. Mahamaya Durga Temple (Pangna Temple) (Pl. 15G)

Main Deity	:	Durga.
Other Deities	:	Bhairav, Lankada Danu, Shiva, Parvati, Ganesha.
Tot. Ar. of Tem. Courtyard	:	About 1500 sq m.
Location		
(i) Place		
Town	:	Pangna
Block	:	Mandi

Tehsil	:	Karsog
District	:	Mandi
(ii) Approach	:	About 12 km from Churag and 109 km from Mandi.
(iii) Altitude	:	1500 m.
Management	:	Local temple committee.
Historical Background	:	The temple is believed to be about 1300 years old.
Temple Architect	:	Pentroof style.
Offerings to God		
(i) By Priest		
Morning	:	Rice-Dal.
Evening	:	Chapati, Vegetables, Currant, Fruits, Sweets.
(ii) By People	:	'Halva', Fruits, Sweets.
Off. to People (Prasadam)	:	'Halva', Fruits, Sweets.
Fairs and Festivals	:	Durga Ashtmi in Souj Navratra, Local fair (held in April).
Observations		
(i) Courtyard Plants (Protected as well as for Landscaping	:	*Canna indica, Cestrum nocturnum, Chrysanthemum indicum, Citrus limon, Ficus palmata, Jasminum grandiflorum, Jasminum primulinum, Pistacia integerrima, Tagetes erecta, Thuja orientalis, Zephyranthes grandiflora.*
(ii) Predominant Plants of the Courtyard	:	*Tagetes erecta.*
(iii) Maintenance	:	Excellent.
(iv) Ethnobotanical		
Sacred and Associated Myth	:	Flowers of *Chrysanthemum indicum, Jasminum grandiflorum, Jasminum primulinum* and *Tagetes erecta* considered sacred for offering to the deity. Growing *Ficus palmata, Jasminum grandiflorum, Jasminum primulinum, Citrus limon, Pistacia integerrima* and *Cestrum nocturnum* considered sacred and worshipped by locals. Long ago, daughter of the king was blamed for some charges by the locals. In order to prove her innocence, she ended her life and told before death that her dead body would not decompose for next six months and regard it as a proof of her being innocent. It remained so and that spot in the temple courtyard is still worshipped. Any plant growing there is worshipped as sacred plant.
Edible	:	Fruits of *Ficus palmata* edible and those of *Citrus limon* pickled.

Ornamental	:	*Canna indica, Cestrum nocturnum, Chrysanthemum indicum, Jasminum grandiflorum, Jasminum primulinum, Tagetes erecta, Thuja orientalis* and *Zephyranthes grandiflora* grown as ornamentals.
Remarks	:	Goats sacrificed.

8. Mahunag Temple (Pl. 15H)

Main Deity	:	Mahunag (reincarnation of Karan).
Other Deities	:	Ragseta, Jad Devta, Mashanu, Surya Narayan.
Tot. Ar. of Tem. Courtyard	:	About 7500 sq m and it includes temple courtyard and an orchard of 70-80 apple trees.
Location		
(i) Place		
Village	:	Bakhaari
Block	:	Karsog
Tehsil	:	Karsog
District	:	Mandi
(ii) Approach	:	About 12 km from Churag and 105 km from Shimla via Tatapani.
(iii) Altitude	:	1940 m.
Management	:	Temple committee.
Historical Background	:	Long ago, a farmer found the idol of deity in his fields while ploughing in the nearby village (Shaidal). He brought the idol and temple was built.
Temple Architect	:	Pentroof style with temple made up of Devdar wood.
Offerings to God		
(i) By Priest	:	'Halva', 'Lachidana', Coconut.
(ii) By People	:	'Halva', 'Lachidana', Coconut, Rice grains, Fruits, Seasonal crops.
Off. to People (Prasadam)	:	'Lachidana', Fruits, Rice grains.
Fairs and Festivals	:	Mahunag fair (held in May for 5 days), Birthday of Mahunag (18 July).
Observations		
(i) Courtyard Plants (Protected as well as for Landscaping	:	*Thuja orientalis.*
(ii) Maintenance	:	Excellent.

(iii) Ethnobotanical

Sacred : Palanquin of the deity is made from Silver and wood of *Cedrus deodara*. Wood of 'Ban' (*Quercus* sp.) used for continuous burning of sacrificial fire ('Dhoona'). Its ash is believed to possess healing properties and given to devotees as 'prasadam'.

Ornamental : *Thuja orientalis* grown as an ornamental plant.

Remarks : Goats sacrificed.

9. Mamaleshwar Mahadeh Temple (Pl. 151)

Main Deity : Shiva.

Other Deities : Parvati, Kartikey, Ganesha, Surya Narayan, Brahma, Vishnu.

Tot. Ar. of Tem. Courtyard : About 3000 sq m.

Location

(i) Place

Village : Mamel

Block : Karsog

Tehsil : Karsog

District : Mandi

(ii) Approach : 2 km from Karsog.

(iii) Altitude : 1300 m.

Management : Temple committee.

Historical Background : It is believed that Pandavas during their exile made this temple in one night.

Temple Architect : Pagoda style with temple made up of Devdar wood.

Offerings to God

(i) By Priest

Morning : Rice-Dal, 'Kheer', Sweet-Rice, Vegetables.

Evening : 'Luchian', 'Halva', Loaf Sugar, Milk.

(ii) By People : 'Halva', Fruit, Cashewnut, Almond, Currant, Seasonal crops. Flowers of Ak, Dhatura offered to appease Lord Shiva.

Off. to People (Prasadam) : 'Halva', Fruits, Dry Fruits, Flowers, Sweets.

Fairs and Festivals : Shivratri, Lahul fair (held on 17-18 April), Buddhi Diwali (exactly one month after Diwali), Vijay Dashmi.

Observations

(i) Courtyard Plants (Protected as well as for Landscaping	: *Calendula officinalis, Cannabis sativa, Catharanthus roseus, Rosa indica, Tagetes erecta, Thevetia neriifolia.*
(ii) Predominant Plants of the Courtyard	: *Catharanthus roseus, Tagetes erecta.*
(iii) Maintenance	: Excellent.
(iv) Ethnobotanical	
Sacred and Associated Myth	: *Thevetia neriifolia* worshipped as sacred plant. Being very old, it is strictly protected. Flowers of *Catharanthus roseus, Rosa indica, Tagetes erecta* and *Thevetia neriifolia* considered sacred for offering to the deity. Flowers of *Thevetia neriifolia* and leaves of *Cannabis sativa* offered as special offering to Lord Shiva.
Medicinal	: Cotton plug dipped in latex of *Thevetia neriifolia* kept in tooth cavity for relief against pain.
Ornamental	: *Calendula officinalis, Catharanthus roseus, Rosa indica, Tagetes erecta, Thevetia neriifolia* are grown as ornamentals.
Status of Courtyard Plants in Surrounding Areas	: Occurance of *Thevetia neriifolia* outside the courtyard is rare.
Remarks	: No sacrifices made to deity in the temple.

10. Naina Devi Temple (Pl. 15J)

Main Deity	: Naina Devi.
Other Deity	: Shiva.
Tot. Ar. of Tem. Courtyard	: About 7500 sq m.
Location	
(i) Place	
Village	: Sarkidhar
Block	: Balh
Tehsil	: Sadar Mandi
District	: Mandi
(ii) Approach	: 11 km from Rewalsar.
(iii) Altitude	: 1900 m.
Management	: Village temple committee.
Historical Background	: The temple was built by Suraj Sen in the seventeenth century A.D. After the self-immolation of Sati, one of the eyes of Sati fell at this place, earning it the name of

		Naina Devi. The other eye was dropped at Naina Devi Temple (Bilaspur).
Temple Architect	:	Sikhara style.
Offerings to God		
(i) By Priest	:	'Halva', Sweets, Puffed rice.
(ii) By People	:	'Halva', Coconut, Puffed rice, Flowers, Fruits.
Off. to People (Prasadam)	:	'Halva', Puffed rice.
Fairs and Festivals	:	Baisakhi Fair, Navratra.
Observations		
(i) Courtyard Plants (Protected as well as for Landscaping	:	*Cynodon dactylon, Grevillea robusta, Prunus armeniaca, Prunus persica, Pyrus malus, Rosa indica, Tagetes erecta.*
(ii) Predominant Plants of the Courtyard	:	*Pyrus malus.*
(iii) Maintenance	:	Excellent.
(iv) Ethnobotanical		
Edible	:	Fruits of *Pyrus malus, Prunus persica* and *Prunus armeniaca* edible.
Ornamental	:	*Grevillea robusta, Rosa indica* and *Tagetes erecta* grown as ornamental plants.
Remarks	:	Goats sacrificed to appease deity. Religious purification ceremony ('havan') organised on Sakranti of each month.

11. Nawahi Devi Temple (Pl. 15K)

Main Deity	:	Nawahi Devi.
Other Deities	:	Mahakali, Ganpati, Satnarayan, Mahadev, Radha-Krishana, Santoshi Mata.
Tot. Ar. of Tem. Courtyard	:	About 1500 sq m.
Location		
(i) Place		
Village	:	Nawahi Devi
Block	:	Gopalpur
Tehsil	:	Sarkaghat
District	:	Mandi
(ii) Approach	:	6 km from Sarkaghat.
(iii) Altitude	:	930 m.

Management	:	Local temple committee.
Historical Background	:	The present temple dates back to seventeenth century A.D.
Temple Architect	:	Sikhara style.
Offerings to God		
(i) By Priest	:	Rice-Dal, Chapati-Vegetables, Sweets.
(ii) By People	:	Puffed rice, 'Halva', Seasonal crops.
Off. to People (Prasadam)	:	Puffed rice.
Fairs and Festivals	:	Local fair held on 14th June for a day, Navratra
Observations		
(i) Courtyard Plants (Protected as well as for Landscaping	:	*Bryophyllum calycinum, Catharanthus roseus, Cedrela toona, Cestrum nocturnum, Chrysanthemum indicum, Ficus benghalensis, Ficus religiosa, Hibiscus rosa-sinensis, Jasminum sambac, Mangifera indica, Mirabilis jalapa, Morus alba, Murraya koenigii, Ocimum sanctum, Rosa indica, Tagetes erecta.*
(ii) Predominant Plants of the Courtyard	:	*Tagetes erecta.*
(iii) Maintenance	:	Satisfactory.
(iv) Ethnobotanical		
Sacred	:	Plants of *Ficus benghalensis, Ficus religiosa* and *Ocimum sanctum* considered sacred.
Medicinal	:	Powdered leaves of *Murraya koenigii* prescribed to cure diabetes.
Edible	:	Fruits of *Mangifera indica* and *Morus alba* edible.
Ornamental	:	*Bryophyllum calycinum, Catharanthus roseus, Cestrum nocturnum, Chrysanthemum indicum, Hibiscus rosa–sinensis, Jasminum sambac, Mirabilis jalapa, Ocimum sanctum, Rosa indica, Tagetes erecta* grown as ornamentals.
Other Uses	:	Leaves of *Morus alba* considered as good fodder. Branches of *Murraya koenigii* used as toothbrush. Its leaves used for flavouring curries. Wood of *Cedrela toona* excellent for construction and furniture.

12. Ningmapa Budh Monastery (Pl. 15L)

Main Deity	:	Padamsambhav.
Tot. Ar. of Tem. Courtyard	:	3000 sq m.

Location		
(i) Place		
Town	:	Rewalsar
Block	:	Mandi
Tehsil	:	Sadar Mandi
District	:	Mandi
(ii) Approach	:	24 km from Mandi.
(iii) Altitude	:	1700 m.
Management	:	Gompa committee.
Historical Background	:	Long ago, Padamsambhav and the daughter of the king of Mandi fell in love with each other. She married him. When the king came to know about it, he got killed Padamsambhav by consigning him to flames. Next day, lake Rewalsar appeared at that place and the king saw Padamsambhav sitting on the flower of Punkesar (*Nelumbo nucifera*) in the lake. The king realised his mistake and built this monastery on one side of lake about 100 years back.
Temple Architect	:	Buddhist Monastery.
Offerings to God		
By People	:	'Lachidana', Puffed rice, Biscuit, Bread, Fruits.
Off. to People (Prasadam)	:	'Lachidana', Puffed rice, Biscuit, Bread, Fruits.
Fairs and Festivals	:	Chhishu Fair (held in February-March), Buddh Purnima.
Observations		
(i) Courtyard Plants (Protected as well as for Landscaping	:	*Cupressus sempervirens, Magnolia grandiflora.*
(ii) Maintenance	:	Excellent.
(iii) Ethnobotanical		
Sacred	:	Leaves of *Cupressus sempervirens* used as incense in the Monastery. One of the trees of *Magnolia grandiflora* was planted by Dalai Lama and other one by Pancham Lama on the same day in 1957. Both these trees are worshipped as sacred trees.
Ornamental	:	Both the plants grown as ornamentals.
Status of Courtyard Plants in Surrounding Areas	:	*Magnolia grandiflora* does not occur outside whereas *Cupressus sempervirens* commonly grows in the surrounding areas.

13. Panchvaktra Mahadev Temple (Pl. 15M)

Main Deity	:	Lord Shiva with five faces.
Other Deity	:	Batuk Bhairav.
Tot. Ar. of Tem. Courtyard	:	1536 sq m.
Location		
(i) Place		
Town	:	Mandi
Block	:	Mandi
Tehsil	:	Sadar Mandi
District	:	Mandi
(ii) Approach	:	Located on the side of river Bias in Mandi town.
(iii) Altitude	:	870 m.
Management	:	Managed by Archaeological Survey of India.
Historical Background	:	The temple dates back to 15th century A.D.
Temple Architect	:	Sikhara style.
Offerings to God		
(i) By Priest	:	Sugar-Candy, Currant.
(ii) By People	:	'Halva', 'Lachidana'. 'Bil patri' (Leaves of *Aegle marmelos*) offered to appease Lord Shiva.
Off. to People (Prasadam)	:	'Lachidana', Sugar-Candy, Currant.
Fairs and Festivals	:	Sawan Mass.
Observations		
(i) Courtyard Plants (Protected as well as for Landscaping	:	*Calendula officinalis, Callistemon lanceolatus, Canna indica, Catharanthus roseus, Celosia argentea* var. *cristata, Cynodon dactylon, Salvia coccinia, Tagetes erecta.*
(ii) Maintenance	:	Excellent.
(iii) Ethnobotanical		
Sacred	:	Flowers of *Tagetes erecta* considered sacred for offering to the deity. *Cynodon dactylon* considered sacred and employed in various religious ceremonies.
Ornamental	:	All courtyard plants are ornamentals.
Remarks	:	There are two temples in the sanctum sanctorum, the main temple for Lord Shiva and other one for Batuk Bhairav. Goats sacrificed for Batuk Bhairav only.

14. Shikari Devi Temple (Pl. 16A)

Main Deity : Shikari Devi (form of Kali).

Other Deities : Parshuram, Kamrunag, Shivlingam.

Tot. Ar. of Tem. Courtyard : About 1500 sq m.

Location

(i) Place

Village : Shikari Devi

Block : Thunag

Tehsil : Thunag

District : Mandi

(ii) Approach : 15 km from Janjaili.

(iii) Altitude : 3800 m.

Management : Temple committee.

Historical Background : According to the local legend, Pandavas visited this place during their exile. At one night, Sehdev saw a flame and raised his bow to shoot it. At that time the goddess appeared from the flame in the form of a girl. Later, Sehdev established this place as Shikari Jogni.

Temple Architect : The temple is without roof.

Offerings to God

(i) By Priest : 'Halva', 'Khir'.

(ii) By People : 'Luchian', 'Lachidana', Coconut, Currant, Dried fruits of Palm, Cashewnut, Almond.

Off. to People (Prasadam) : 'Lachidana', Puffed rice, Rice, Flowers.

Fairs and Festivals : Ganga Dashehra (1st week of June).

Observations

(i) Courtyard Plants (Protected as well as for Landscaping : None.

(ii) Maintenance : Satisfactory.

Remarks : Goats sacrificed to the deity.

15. Shri Lomash Rishi Temple (Pl. 16B)

Main Deity : Lomash Rishi.

Other Deities : Shiv-Parvati, Radha-Krishana.

Tot. Ar. of Tem. Courtyard : 768 sq m.

Location		
(i) Place		
Town	:	Rewalsar
Block	:	Mandi
Tehsil	:	Sadar Mandi
District	:	Mandi
(ii) Approach	:	24 km from Mandi.
(iii) Altitude	:	1700 m.
Management	:	Temple committee.
Historical Background	:	Long ago, Lomash Rishi worshipped Lord Shiva at this place for several years. Finally, Shiva was pleased and gave a boon that he would always be worshipped at this place by the people.
Temple Architect	:	Sikhara style.
Offerings to God		
(i) By Priest	:	'Rot', sweets.
(ii) By People	:	'Lachidana', Sweets, Flowers, Fruits. 'Bil patri' (Leaves of *Aegle marmelos*) offered to Lord Shiva as special offering.
Off. to People (Prasadam)	:	'Lachidana', sweets.
Fairs and Festivals	:	Baisakhi, Shivratri, Diwali.
Observations		
(i) Courtyard Plants (Protected as well as for Landscaping	:	*Ficus religiosa, Nerium odorum.*
(ii) Maintenance	:	Satisfactory.
(iii) Ethnobotanical		
Sacred and Associated Myth	:	Flowers of *Nerium odorum* considered sacred for offering to the deities and religious worshipping. *Ficus religiosa* regarded sacred and its wood used in 'havan' for performing religious rites. Newly married couples tie sacred threads around the tree of *Ficus religiosa* for prosperous married life.
Ornamental	:	*Nerium odorum* grown for ornamental purposes.
Remarks	:	No sacrifices made to deities.

16. Sidh Baba Kamlahiya Temple (Pl. 16C)

Main Deity	:	Baba Kamlahiya.
Other Deity	:	Jalpa Mata, Baba Balak Nath, Shiva, Hanuman.
Tot. Ar. of Tem. Courtyard	:	About 1000 sq m.
Location		
(i) Place		
Village	:	Kamlah
Block	:	Dharampur
Tehsil	:	Sarkaghat
District	:	Mandi
(ii) Approach	:	40 km from Sarkaghat.
(iii) Altitude	:	1420 m.
Management	:	Temple committee.
Historical Background	:	The temple belongs to 17th century A.D., and built by Raja Suraj Sen.
Temple Architect	:	Sikhara style.
Offerings to God		
(i) By Priest	:	Puffed rice, Flowers.
(ii) By People	:	Puffed rice, 'Lachidana', 'Patasha', Coconut, 'Rot', Seasonal crops.
Off. to People (Prasadam)	:	Puffed rice, 'Lachidana', 'Patasha', Fruits.
Fairs and Festivals	:	Local fair (held on first Sunday of May), Navratra.
Observations		
(i) Courtyard Plants (Protected as well as for Landscaping	:	*Bougainvillea spectabilis, Callistemon lanceolatus, Canna indica, Cestrum nocturnum, Chrysanthemum indicum, Cynodon dactylon, Ficus religiosa, Grevillea robusta, Hibiscus rosa–sinensis, Jasminum grandiflorum, Lagerstroemia indica, Mentha piperita, Phoenix sylvestris, Tagetes erecta.*
(ii) Predominant Plants of the Courtyard	:	*Bougainvillea spectabilis, Callistemon lanceolatus,*
(iii) Maintenance	:	Excellent.
(iv) Ethnobotanical		
Sacred	:	*Cynodon dactylon* employed in various religious ceremonies as sacred plant.
Edible	:	Fruits of *Phoenix sylvestris* edible.

Ornamental	:	All the courtyard plants also grown for ornamental purposes.
Remarks	:	No sacrifices permitted in the temple.

17. Tarna Devi Temple (Pl. 16D)

Main Deity	:	Goddess Kali.
Other Deities	:	Shiva, Bhairav.
Tot. Ar. of Tem. Courtyard	:	About 1000 sq m.
Location		
(i) Place		
Town	:	Mandi
Block	:	Mandi
Tehsil	:	Mandi
District	:	Mandi
(ii) Approach	:	Located about 2 km from the town at Tarna hill.
iii) Altitude	:	840 m.
Management	:	Privately owned.
Historical Background	:	The temple was built by Raja Shyam Sen in 17th century A.D.
Temple Architect	:	Dome type with a 'kalasa' at the top.
Offerings to God		
(i) By Priest		
Morning	:	'Bali-Bhog' which includes meat of Goat, Fish and Cock, Wine.
Evening	:	Chapati-Meat, Charnamat (Wine).
(ii) By People	:	Coconut, Fruits, Flowers, Sweets, Puffed rice.
Off. to People (Prasadam)	:	Puffed rice.
Fairs and Festivals	:	Navratra, Shivratri.
Observations		
(i) Courtyard Plants (Protected as well as for Landscaping	:	*Broussonetia papyrifera, Carica papaya, Celosia argentea* var. *cristata, Ficus religiosa, Hibiscus rosa-sinensis, Hibiscus syriacus, Jasminum grandiflorum, Ocimum sanctum, Rosa indica, Tagetes erecta, Zaphyranthes candida, Zaphyranthes grandiflora.*
(ii) Predominant Plants of the Courtyard	:	*Jasminum grandiflorum, Tagetes erecta.*

(iii) Maintenance : Excellent.

(iv) Ethnobotanical

Sacred and Associated Myth : Flowers of *Hibiscus-rosa sinensis* offered especially to appease the deity. Leaves of *Aegle marmelos* offered to Lord Shiva. *Cynodon dactylon* considered sacred and employed in religious worshipping. *Ficus religiosa* and *Ocimum sanctum* worshipped as sacred plants. A group of seven trees of *Ficus religiosa* in the courtyard considered sacred as it is believed that the goddess Mahakali appeared under this group of trees.

Edible : Fruits of *Carica papaya* edible.

Ornamental : *Celosia argentea* var. *cristata, Hibiscus rosa-sinensis, Hibiscus syriacus, Jasminum grandiflorum, Ocimum sanctum, Rosa indica, Tagetes erecta, Zaphyranthes candida* and *Zaphyranthes grandiflora* grown as ornamentals.

Other Uses : Leaves of *Broussonetia papyrifera* regarded as good fodder.

Remarks : In lieu of human sacrifices made earlier, goats regularly sacrificed to appease deity.

18. Trilokinath Temple (Pl. 16E)

Main Deity : Three-faced Shiva riding the bull Nandi, and consort Parvati to his left.

Other Deities : Narda-Sharda, Hanuman, Kali, Vishnu.

Tot. Ar. of Tem. Courtyard : About 1500 sq m.

Location

(i) Place

Village : Mandi

Block : Mandi

Tehsil : Mandi

District : Mandi

(ii) Approach : Located at old Mandi, on the bank of Beas River.

(iii) Altitude : 870 m.

Management : Managed by Archaeological Survey of India.

Historical Background : Dedicated to Lord Shiva, this beautiful temple was built by Sultana Devi, queen of Raja Ajbar Sen (King of Mandi) in 1520 A.D.

Temple Architect : Shikhar Style.

Offerings to God	
(i) By Priest	: Puffed rice, Flowers.
(ii) By People	: 'Halva', Puffed rice, Coconut, Flowers, Fruits. 'Bael-patri' (Leaves of *Aegle marmelos*), 'Bhang' (Leaves of *Cannabis sativa*) and 'Dhatura' (*Datura* sp.) offered to appease Lord Shiva.
Off. to People (Prasadam)	: Puffed rice, Flowers, Fruits.
Fairs and Festivals	: Shivratri.
Observations	
(i) Courtyard Plants (Protected as well as for Landscaping	: *Calendula officinalis, Catharanthus roseus, Chrysanthemum indicum, Cosmos sulphureus, Mangifera indica, Rosa indica, Tagetes erecta, Thuja orientalis.*
(ii) Predominant Plants of the Courtyard	: *Catharanthus roseus, Tagetes erecta.*
(iii) Maintenance	: Satisfactory.
(iv) Ethnobotanical	
Sacred	: Flowers of courtyard plants considered sacred for offering to the deity and employed in religious worshipping.
Edible	: Fruits of *Mangifera indica* edible.
Ornamental	: *Calendula officinalis, Catharanthus roseus, Chrysanthemum indicum, Cosmos sulphureus, Rosa indica, Tagetes erecta, Thuja orientalis* grown as ornamentals.
Remarks	: Goats sacrificed to goddess Kali.

Temples of District Shimla

1. Bhimakali Temple (Pl. 17A)

Main Deity	:	Bhimakali.
Other Deity	:	Lankadvir (Bhairav).
Tot. Ar. of Tem. Courtyard	:	About 5000 sq m.
Location		
(i) Place		
Village	:	Sarahan
Block	:	Rampur
Tehsil	:	Rampur
District	:	Shimla
(ii) Approach	:	180 km from Shimla.
(iii) Altitude	:	2150 m.
Management	:	Temple committee.
Historical Background	:	The temple dates back to 7th-8th century A.D.
Temple Architect	:	Pentroof style with Devdar and Qail wood used in its construction.
Offerings to God		
(i) By Priest	:	'Halva-Puri', Fruits, Flowers.
(ii) By People	:	'Lachidana', 'Halva', Coconut, Currant.

Off. to People (Prasadam)	:	'Lachidana', 'Halva', Coconut, Currant, Dried fruits of Palm.
Fairs and Festivals	:	Vijaydashmi.
Observations		
(i) Courtyard Plants (Protected as well as for Landscaping	:	*Calendula officinalis, Hemerocallis fulva, Hibiscus syriacus, Hydrangea macrophylla, Jasminum humile, Lonicera japonica, Tagetes erecta, Tropaeolum majus.*
(ii) Predominant Plants of the Courtyard	:	*Tagetes erecta, Tropaeolum majus.*
(iii) Maintenance	:	Excellent.
(iv) Ethnobotanical		
Sacred	:	Flowers of *Tagetes erecta* considered sacred for religious worshipping.
Medicinal	:	5-10 ml decoction of leaves of *Lonicera japonica* prescribed twice daily for diarrhoea.
Ornamental	:	All the courtyard plants considered good for ornamental purposes.
Remarks	:	Human sacrifices once prevalent in the temple are not permitted now.

2. Bhuteshwar Mahadev Temple (Pl. 17B)

Main Deity	:	Shiva.
Other Deities	:	Hanuman, Ganesha, Shani, Shitla Mata.
Tot. Ar. of Tem. Courtyard	:	About 1200 sq m.
Location		
(i) Place		
Town	:	Rampur
Block	:	Rampur
Tehsil	:	Rampur
District	:	Shimla
(ii) Approach	:	Temple is located in the town.
(iii) Altitude	:	900 m.
Management	:	In the name of Deity.
Historical Background	:	The temple is believed to be very old.
Temple Architect	:	Sikhara style.

Offerings to God		
(i) By Priest	:	Rice-Dal, Chapati, Vegetables.
(ii) By People	:	'Lachidana', Fruits, Flowers, Sweets. Flowers of Dhatura offered to appease Lord Shiva.
Off. to People (Prasadam)	:	'Lachidana', Fruits, Flowers, Sweets.
Fairs and Festivals	:	Shivratri.
Observations		
(i) Courtyard Plants (Protected as well as for Landscaping	:	*Aegle marmelos, Callistemon lanceolatus, Cannabis sativa, Canna indica, Catharanthus roseus, Cynodon dactylon, Ficus benghalensis, Ficus religiosa, Helianthus annuus, Hibiscus syriacus, Hydrangea macrophylla, Ipomoea purpurea, Jasminum primulinum, Jasminum sambac, Lagerstroemia indica, Mentha piperita, Mirabilis jalapa, Musa paradisiaca, Nerium odorum, Ocimum sanctum, Pelargonium hortorum, Punica granatum, Rosa indica, Tabernaemontana coronaria, Tagetes erecta, Thuja orientalis, Zinnia elegans.*
(ii) Predominant Plants of the Courtyard	:	*Helianthus annuus.*
(iii) Maintenance	:	Satisfactory.
(iv) Ethnobotanical		
Sacred	:	Leaves of *Aegle marmelos, Cannabis sativa* and *Ocimum sanctum* offered especially to appease Lord Shiva. *Ficus benghalensis, Ficus religiosa, Musa paradisiaca* and *Ocimum sanctum* worshipped as sacred plants. Fruits of *Punica granatum* offered to the deity. *Cynodon dactylon* employed in religious worshipping.
Medicinal	:	15-20 ml decoction of leaves of *Catharanthus roseus* prescribed twice daily for diabetes. 10-15 ml decoction of bark of *Lagerstroemia indica* prescribed twice daily for dysentery. Poultice of leaves of *Mirabilis jalapa* applied to cure boils. Paste of bark of *Tabernaemontana coronaria* applied to cure mouth ulcer.
Edible	:	Fruits of *Aegle marmelos, Musa paradisiaca, Punica granatum* edible. Leaves of *Mentha piperita* used for flavouring various preparations.
Ornamental	:	*Callistemon lanceolatus, Canna indica, Catharanthus roseus, Helianthus annuus, Hibiscus syriacus, Hydrangea macrophylla, Jasminum primulinum, Jasminum sambac, Lagerstroemia indica, Mirabilis jalapa, Nerium odorum, Pelargonium hortorum, Rosa indica, Tabernaemontana coronaria, Tagetes erecta, Thuja orientalis, Zinnia elegans* are grown as ornamentals.

Status of Courtyard Plants in Surrounding Areas	:	Excepting *Ficus religiosa, Ficus benghalensis, Aegle marmelos,* all other plants occur commonly.
Remarks	:	No sacrifices made to the deity.

3. Bijjat Maharaj Temple

Main Deity	:	Bijjat Maharaj.
Other Deity	:	Shirgul Devta.
Tot. Ar. of Tem. Courtyard	:	About 5000 sq m. About 6 hectare of sacred grove exists in the name of deity.
Location		
(i) Place		
Village	:	Saranh
Block	:	Chopal
Tehsil	:	Chopal
District	:	Shimla
(ii) Approach	:	28 km from Chopal.
iii) Altitude	:	2,350 m.
Management	:	Temple committee.
Historical Background	:	The deity is believed to be incarnation of Lord Indra. Bijjat Devta alongwith his elder brother Shirgul killed the demons at Choordhar. The temple was built in 1600 and renovated about 80 years later.
Temple Architect	:	Twin-Tower Pentroof style with Devdar wood used in its construction.
Offerings to God		
(i) By Priest	:	'Halva', Coconut, Seasonal fruits.
(ii) By People	:	'Halva', 'Lachidana'.
Off. to People (Prasadam)	:	'Lachidana', Coconut, Rice grains, Flowers
Fairs and Festivals	:	Bishu, Sakranti of Magh, Baisakh and Kartik months.
Observations		
(i) Courtyard Plants (Protected as well as for Landscaping	:	*Prunus armeniaca, Pyrus malus, Rosa indica.*
(ii) Predominant Plants of the Courtyard	:	*Pyrus malus.*
(iii) Maintenance	:	Excellent.

(iv) Ethnobotanical

Sacred : Flowers of *Rosa indica* considered sacred for offering to the deity.

Edible : Fruits of *Prunus armeniaca, Pyrus malus* edible.

Ornamental : *Rosa indica* grown as an ornamental.

Remarks : No sacrifices permitted in the temple.

4. Dingu Durga Temple (Pl. 17C)

Main Deity : Durga.

Other Deities : Satya Narayan, Shiva, Hanuman.

Tot. Ar. of Tem. Courtyard : About 3000 sq m.

Location

(i) Place

Town : Shimla (Sanjauli)

Block : Mashobra

Tehsil : Shimla

District : Shimla

(ii) Approach : About 8 km from Shimla.

(iii) Altitude : 2350 m.

Management : Sanaatan Dharam Prabandhak Committee.

Historical Background : The temple was originally built by the king of Qiar Koti about 150 years back after his desire of defeating his enemies in war was fulfilled.

Temple Architect : Sikhara style.

Offerings to God

(i) By Priest

Morning : 'Halva', Rice grains, Flowers, Betel-leaf, Dry Fruits.

Noon : 'Dal-Roti', Rice, Vegetables, Clarified butter with Loaf sugar

Evening : Same as at Noon alongwith milk.

(ii) By People : Coconut, Sweets, 'Lachidana', Ornaments of gold and silver.

Off. to People (Prasadam) : Coconut, 'Lachidana', Sweets.

Fairs and Festivals : Baisakhi, Navratra (twice in a year).

Observations

(i) *Courtyard Plants (Protected as well as for Landscaping*	:	*Althaea rosea, Antirrhinum majus, Callistemon lanceolatus, Cupressus torulosa, Dahlia rosea, Dianthus barbatus, Hemerocallis fulva, Iris germanica, Jasminum primulinum, Lonicera japonica, Papaver rhoeas, Passiflora caerulea, Petunia hybrida, Prunus armeniaca, Prunus domestica, Pyrus malus, Robinia pseudoacacia, Rosa multiflora, Salix tetrasperma, Thuja orientalis, Tropaeolum majus, Viola tricolor.*
(ii) *Predominant Plants of the Courtyard*	:	*Pyrus malus, Rosa multiflora.*
(iii) *Maintenance*	:	Excellent.
(iv) *Ethnobotanical*		
Sacred and Associated Myth	:	Fruits of *Pyrus malus* offered to the deity. Devotees tie scarves, sacred threads and bangles to courtyard trees for fulfillment of their desires.
Edible	:	Fruits of *Prunus armeniaca, Prunus domestica, Pyrus malus* are edible.
Ornamental	:	All courtyard plants also considered good for ornamental purposes.

5. Durga Temple (Pl. 17D)

Main Deity	:	Durga.
Other Deity	:	Hanuman.
Tot. Ar. of Tem. Courtyard	:	About 1500 sq m.
Location		
(i) *Place*		
Town	:	Shimla (Jatog Cant)
Block	:	Mashobra
Tehsil	:	Shimla
District	:	Shimla
(ii) *Approach*	:	About 2 km from Tutu.
(iii) *Altitude*	:	2250 m.
Management	:	Temple committee.
Historical Background	:	The temple was built in 1870 by Gorkhas living there at that time. It was renovated about 30 years back.
Temple Architect	:	Sikhara style.

Offerings to God

(i) By Priest	:	'Halva', Sweets, 'Lachidana'.
(ii) By People	:	'Halva', Sweets, 'Lachidana'.
Off. to People (Prasadam)	:	'Halva', Sweets, 'Lachidana'
Fairs and Festivals	:	Navratra fair.

Observations

(i) Courtyard Plants (Protected as well as for Landscaping	:	*Althaea rosea, Rosa indica, Tagetes erecta, Thuja orientalis.*
(ii) Predominant Plants of the Courtyard	:	*Tagetes erecta.*
(iii) Maintenance	:	Excellent.
(iv) Ethnobotanical		
Sacred and Associated Myth	:	Flowers of *Tagetes erecta* considered sacred for offering to the deity. For fulfilment of their desires, people tie sacred threads and scarves to the tree of *Thuja orientalis* growing in the temple courtyard.
Ornamental	:	All courtyard plants considered good for ornamental purposes.
Remarks	:	In earlier times, goats were sacrificed for goddess Kali. This practice has been stopped for the last 30 years. Goddess Durga is treated as Kuldevi by the people of surrounding 20-25 villages.

6. Duttatrey Temple (Pl. 17E)

Main Deities	:	Duttatrey (form of Lord Vishnu), Atri (Father of Duttatrey), Anusuya Mata (Mother of Duttatrey)
Other Deities	:	Bhuvaneshwari, Shivlingam, Vishnu-Lakshmi, Garud, Gorakh Nath, Parshuram.
Tot. Ar. of Tem. Courtyard	:	About 800 sq m.

Location

(i) Place		
Village	:	Dutt Nagar
Block	:	Rampur
Tehsil	:	Rampur
District	:	Shimla
(ii) Approach	:	12 km from Rampur.
(iii) Altitude	:	800 m.

Management	:	Local temple committee.
Historical Background	:	The temple dates back to 7th-8th century A.D.
Temple Architect	:	Pahadi style.
Offerings to God		
(i) By Priest	:	'Halva'.
(ii) By People	:	'Halva', Seasonal crops.
Off. to People (Prasadam)	:	'Halva'.
Fairs and Festivals	:	Duttatrey Jayanti, Buddi Diwali.
Observations		
(i) Courtyard Plants (Protected as well as for Landscaping	:	*Aegle marmelos, Cannabis sativa, Clerodendrum philippinum, Debregeasia hypoleuca, Ficus palmata, Ficus religiosa, Jasminum grandiflorum, Melia azedarach, Mirabilis jalapa, Morus alba, Pistacia integerrima, Prunus domestica.*
(ii) Predominant Plants of the Courtyard	:	*Mirabilis jalapa.*
(iii) Maintenance	:	Satisfactory.
(iv) Ethnobotanical		
Sacred	:	Leaves of *Aegle marmelos* and *Cannabis sativa* offered to Lord Shiva. *Ficus religiosa* worshipped as a sacred plant.
Medicinal	:	Powdered fruits of *Ficus palmata* alongwith black pepper and rock salt prescribed for dysentery. Its latex applied on cuts and wounds. Decoction of leaves of *Clerodendrum philippinum* prescribed for cold and cough.
Edible	:	Fruits of *Aegle marmelos, Ficus palmata Morus alba* and *Prunus domestica* edible.
Other Uses	:	Leaves of *Debregeasia hypoleuca* and *Ficus palmata* used as fodder.
Remarks	:	Sacrifices of sheep and goats for goddess Bhuvaneshwari permitted only from the outside of sanctum sanctorum.

7. Hatkeshwari Durga Mata Temple or Hatkoti Mata Temple (Pl. 17F)

Main Deity	:	Goddess Durga.
Other Deity	:	Shiva.
Tot. Ar. of Tem. Courtyard	:	About 5000 sq m.

Location	
(i) Place	
Village	: Hatkoti
Block	: Rohdu
Tehsil	: Jubbal
District	: Shimla
(ii) Approach	: 11 km from Jubbal towards Rohru.
(iii) Altitude	: 1500 m.
Management	: Managed by representative of the state Government.
Historical Background	: The temple dates back to 7th century A.D. It is believed that Pandavas stayed here and built the temple.
Temple Architect	: The original Sikhara type of construction was replaced with a Pagoda type of roof by Raja Padam Chand. Devdar wood used predominantly in its construction.
Offerings to God	
(i) By Priest	: Rice-Dal.
(ii) By People	: Coconut, Currant, 'Halva', 'Lachidana'.
Off. to People (Prasadam)	: 'Lachidana', Currant, Coconut, 'Halva'.
Fairs and Festivals	: Diwali, Shivratri, Navratra (twice a year), Baisakhi.
Observations	
(i) Courtyard Plants (Protected as well as for Landscaping	: *Canna indica, Celosia argentea* var. *cristata, Ficus religiosa, Hydrangea macrophylla, Impatiens balsamina, Ocimum sanctum, Pelargonium hortorum, Rosa alba, Tagetes erecta, Thuja orientalis.*
(ii) Predominant Plants of the Courtyard	: *Tagetes erecta.*
(iii) Maintenance	: Excellent.
(iv) Ethnobotanical	
Sacred and Associated Myth	: *Ficus religiosa* and *Ocimum sanctum* worshipped as sacred plants. Flowers of *Tagetes erecta* considered sacred for religious worshipping. People tie sacred threads to *Ficus religiosa* for fulfilment of their desires.
Medicinal	: Decoction of leaves of *Ocimum sanctum* given to cure cold.
Ornamental	: All courtyard plants considered good for ornamental purposes.

Status of Courtyard Plants in Surrounding Areas	:	Excepting *Ficus religiosa*, all plants occur commonly in the surrounding areas.
Remarks	:	Goats sacrificed to the deity.

8. Jakhoo Hanuman Temple (Pl. 17G)

Main Deity	:	Hanuman.
Other Deities	:	Ram-Lakshman, Durga.
Tot. Ar. of Tem. Courtyard	:	About 3000 sq m.
Location		
(i) Place		
Town	:	Shimla
Block	:	Mashobra
Tehsil	:	Shimla
District	:	Shimla
(ii) Approach	:	Located on the top of Jakhoo hill about 2 km from The Mall, Shimla.
(iii) Altitude	:	2500 m.
Management	:	Government undertaken.
Historical Background	:	The temple believed to be very old and built by Saint Yaksha. As per the prevalent legend, Lord Hanuman stopped here while searching for 'Sanjivani' required to cure Lakshman, who lay mortally wounded in battlefield in Lanka for which practically he was helped by a sage meditating here at that time. Later, the image of Lord Hanuman appeared at this place and this famous temple came to its existence.
Temple Architect	:	Sikhara style with Devdar wood predominantly used in its construction.
Offerings to God		
(i) By Priest		
Morning	:	Gram, Jaggery, Currant.
Evening	:	'Chapati', 'Halva', Clarified butter, Loaf-Sugar, Sweets.
(ii) By People	:	Gram, 'Lachidana'.
Off. to People (Prasadam)	:	Gram, 'Lachidana'
Fairs and Festivals	:	Hanuman Jayanti, Dussehra.

Observations

(i) Courtyard Plants (Protected as well as for Landscaping	:	*Antirrhinum majus, Calendula officinalis, Cedrus deodara, Chrysanthemum indicum, Pelargonium hortorum, Rosa indica, Tagetes erecta.*
(ii) Predominant Plants of the Courtyard	:	*Cedrus deodara, Tagetes erecta, Chrysanthemum indicum.*
(iii) Maintenance	:	Excellent.
(iv) Ethnobotanical		
Sacred	:	Flowers of *Chrysanthemum indicum, Rosa indica* and *Tagetes erecta* considered sacred for offering to the deity.
Ornamental	:	All courtyard plants considered good for ornamental purposes.
Remarks	:	No sacrifices permitted here.

9. Jathiya Devi Temple

Main Deity	:	Jathiya Devi.
Other Deities	:	Kali, Badmoo, Shitla Mata, Shivlingam.
Tot. Ar. of Tem. Courtyard	:	About 2500 sq m.
Location		
(i) Place		
Town	:	Dhonkhri
Block	:	Mashobra
Tehsil	:	Shimla
District	:	Shimla
(ii) Approach	:	18 km from Shimla.
(iii) Altitude	:	1500 m.
Management	:	Village temple committee.
Historical Background	:	The temple is believed to be very old.
Temple Architect	:	Sikhara style.
Offerings to God		
(i) By Priest	:	'Halva'.
(ii) By People	:	Coconut, 'Halva', 'Lachidana'.
Off. to People (Prasadam)	:	Coconut, 'Halva', 'Lachidana'.
Fairs and Festivals	:	Fair called Jathiya Mela held in April.

Observations

(i) Courtyard Plants (Protected as well as for Landscaping	:	*Acer oblongum, Bauhinia variegata, Callistemon lanceolatus, Canna indica, Cedrela toona, Cupressus torulosa, Cynodon dactylon, Ficus religiosa, Grevillea robusta, Machilus gamblei, Melia azedarach, Narcissus tazetta, Pyrus pashia, Salvia coccinia, Sapindus mukorossi, Tagetes erecta, Zephyranthes grandiflora.*
(ii) Predominant Plants of the Courtyard	:	*Cupressus torulosa, Tagetes erecta.*
(iii) Maintenance	:	Excellent.
(iv) Ethnobotanical:		
Sacred and Associated Myth	:	Flowers of *Tagetes erecta* considered sacred for offering to the deity. *Ficus religiosa* worshipped as a sacred tree. People tie bengals, ribbons and scarves to *Pyrus pashia* for fulfilment of their wishes. *Machilus gamblei* considered sacred tree. People tie flags to this tree after offering to the deity.
Medicinal	:	Mixture of 10g powdered seeds of *Melia azedarach* and 10g sugar candy prescribed for piles.
Edible	:	Flower buds of *Bauhinia variegata* cooked as vegetable. Fruits of *Pyrus pashia* edible.
Ornamental	:	*Acer oblongum, Callistemon lanceolatus, Canna indica, Cupressus torulosa, Grevillea robusta, Machilus gamblei, Narcissus tazetta, Salvia coccinia, Tagetes erecta, Zephyranthes grandiflora* are grown for ornamental purposes.
Other Uses	:	Leaves of *Bauhinia variegata* used as fodder. Wood of *Acer oblongum, Cedrela toona* and *Melia azedarach* used for furniture and construction purposes. Wood of *Melia azedarach* also good for agricultural implements. Fruits of *Sapindus mukorossi* used as detergent to wash clothes.
Remarks	:	No sacrifices permitted.

10. Kalibari Temple (Pl. 18A)

Main Deity	:	Kali.
Other Deities	:	Shyamala Durga, Mangal Chandi, Shiva.
Tot. Ar. of Tem. Courtyard	:	About 1500 sq m.
Location		
(i) Place		
Town	:	Shimla (Kalibari)

Block	:	Mashobra
Tehsil	:	Shimla
District	:	Shimla
(ii) Approach	:	The temple is located in the town.
(iii) Altitude	:	2200 m.
Management	:	Temple committee.
Historical Background	:	The temple was built in 1823 by Ram Charan Bhramachari, a Bengali Brahmin, who worshipped goddess Shyamala here in a cave. Later the temple was made at the place of cave. After the name of deity this city was named as Shimla.
Temple Architect	:	Sikhara style.
Offerings to God		
(i) By Priest		
Morning	:	Fruits, Sweets.
Noon	:	Rice-Vegetables.
Night	:	Puri-Vegetables, Milk.
(ii) By People	:	Coconut, 'Lachidana', Sweets, Fruits.
Off. to People (Prasadam)	:	'Lachidana'.
Fairs and Festivals	:	Durga Pooja, Diwali, Janamashtmi, Shivratri.
Observations		
(i) Courtyard Plants (Protected as well as for Landscaping	:	*Calendula officinalis, Cedrus deodara, Dianthus barbatus, Hydrangea macrophylla, Tagetes erecta, Thuja orientalis.*
(ii) Predominant Plants of the Courtyard	:	*Tagetes erecta.*
(iii) Maintenance	:	Excellent.
(iv) Ethnobotanical		
Ornamental	:	*Calendula officinalis, Dianthus barbatus, Hydrangea macrophylla, Tagetes erecta* and *Thuja orientalis* are grown for ornamental purposes.
Remarks	:	No sacrifices made to the deity.

11. Kali Mata Temple (Pl. 18B)

Main Deity	:	Goddess Kali.
Other Deities	:	Mahalakshmi, Mahasaraswati, Shanker, Mahavir, Vishnu.
Tot. Ar. of Tem. Courtyard	:	About 3000 sq m.

Location		
(i) Place		
Village	:	Padhai
Block	:	Mashobra
Tehsil	:	Shimla
District	:	Shimla
(ii) Approach	:	Located at Padhai about 5 km from Lakkar Bazar (Shimla).
(iii) Altitude	:	2000 m.
Management	:	Local temple committee.
Historical Background	:	The temple is believed to be 150-200 years old.
Temple Architect	:	Sikhara style.
Offerings to God		
(i) By Priest	:	'Halva', Coconut, 'Lachidana'.
(ii) By People	:	'Lachidana'
Off. to People (Prasadam)	:	'Lachidana', Coconut, Sweets, Flowers.
Fairs and Festivals	:	Navratra.
Observations		
(i) Courtyard Plants (Protected as well as for Landscaping	:	*Buxus sempervirens, Cedrus deodara, Cupressus torulosa, Ficus religiosa, Jasminum primulinum, Pinus roxburghii, Quercus incana.*
(ii) Predominant Plants of the Courtyard	:	*Jasminum primulinum.*
(iii) Maintenance	:	Excellent.
(iv) Ethnobotanical		
Sacred	:	Flowers of *Jasminum primulinum* especially offered to the deity and also given to the visitors alongwith Parsadam. *Ficus religiosa* worshipped as a sacred tree.
Edible	:	Seeds of *Pinus roxburghii* edible.
Ornamental	:	*Buxus sempervirens, Cupressus torulosa, Jasminum primulinum, Pinus roxburghii* and *Quercus incana* grown for ornamental purposes.
Other Uses	:	Wood of *Cedrus deodara* excellent for furniture and construction works. Wood of *Quercus incana* used for making agricultural implements and as fuel. Its leaves used as fodder.

Status of Courtyard Plants in Surrounding Areas	:	Excepting *Ficus religiosa*, all other plants occur commonly in the surrounding areas.
Remarks	:	No sacrifices.

12. Kamna Devi Temple (Pl. 18C)

Main Deity	:	Goddess Durga.
Tot. Ar. of Tem. Courtyard	:	1536 sq m.
Location		
(i) Place		
Town	:	Shimla (Kamna Devi)
Block	:	Mashobra
Tehsil	:	Shimla
District	:	Shimla
(ii) Approach	:	About 1 km from Boileauganj.
(iii) Altitude	:	2200 m.
Management	:	Privately owned.
Historical Background	:	The temple is believed to be more than 200 years old.
Temple Architect	:	Sikhara style.
Offerings to God		
(i) By Priest	:	Sweets and Fruits.
(ii) By People	:	'Lachidana', Sweets.
Off. to People (Prasadam)	:	'Bundi'.
Fairs and Festivals	:	Ashtmi (Navratra).
Observations		
(i) Courtyard Plants (Protected as well as for Landscaping	:	*Cedrus deodara, Chlorophytum capense, Chrysanthemum indicum, Hydrangea macrophylla, Juglans regia, Quercus dilatata.*
(ii) Predominant Plants of the Courtyard	:	*Cedrus deodara.*
(iii) Maintenance	:	Excellent.
(iv) Ethnobotanical		
Sacred and Associated Myth	:	People tie sacred threads and scarves to the tree of *Juglans regia* for fulfilment of their wishes.
Edible	:	Kernels of *Juglans regia* edible.
Ornamental	:	*Chlorophytum capense, Chrysanthemum indicum, Hydrangea macrophylla, Quercus dilatata* are grown as ornamentals.

Status of Courtyard Plants in Surrounding Areas : Excepting *Juglans regia,* all other plants are common outside.

Remarks : The deity regarded as 'Kuldevi' by the people of all surrounding villages.

13. Lankadvir Temple (Pl. 18D, E)

Main Deities : Lankadvir (Bhairav), Brijeshwari.

Other Deities : Shiva, Hanuman.

Tot. Ar. of Tem. Courtyard : About 3000 sq m.

Location

(i) Place

Town : Chopal

Block : Chopal

Tehsil : Chopal

District : Shimla

(ii) Approach : Located in the town.

(iii) Altitude : 2300 m.

Management : Temple committee.

Historical Background : The temple is believed to be more than 500 years old, and renovated about 20 years back.

Temple Architect : Pentroof type.

Offerings to God

(i) By Priest : 'Lachidana', Coconut, Currant. 'Rot' (made of wheat-flour) offered to Lankadvir on every Tuesday and Sunday.

(ii) By People : 'Rot', 'Halva', Coconut, 'Lachidana'.

Off. to People (Prasadam) : 'Lachidana', Sweets.

Fairs and Festivals : Navratra, Rishi Panchmi.

Observations

(i) Courtyard Plants (Protected as well as for Landscaping : *Cedrus deodara, Pinus wallichiana, Rosa indica.*

(ii) Predominant Plants of the Courtyard : *Cedrus deodara.*

(iii) Maintenance : Satisfactory.

(iv) Ethnobotanical

Sacred : Flowers of *Rosa indica* considered sacred for offering to the deity. Wood of *Cedrus deodara* employed for making articles used in various religious rites; and that of *Pinus wallichiana* for making musical instruments.

Ornamental : *Cedrus deodara, Pinus wallichiana* and *Rosa indica* grown for ornamental purposes.

Other Uses : Wood of *Cedrus deodara* and *Pinus wallichiana* employed for construction works.

Remarks : 'Lankadvir' considered as body guard of goddess Brijeshwari Devi. Lamb sacrificed by the people on fulfilment of their desires. Trees in the courtyard can be cut only for temple use. Flags also offered to the deity by people visiting the temple.

14. Mahasu Devta Temple

Main Deity : Mahasu Devta (Lord Shiva).

Tot. Ar. of Tem. Courtyard : About 1500 sq m.

Location

(i) *Place*

Village : Gajedi

Block : Theog

Tehsil : Theog

District : Shimla

(ii) *Approach* : 5 km from Theog towards Rohru.

(iii) *Altitude* : 2400 m.

Management : Village temple committee.

Historical Background : The temple, about 200 years old, was built by the king of Theog.

Temple Architect : Sikhara style.

Offerings to God

(i) *By Priest* : Rice-grains, Clarified butter.

(ii) *By People* : Coconut, 'Halva'.

Off. to People (Prasadam) : 'Halva', 'Lachidana'.

Fairs and Festivals : Rishi Panchmi, Saji celebrated four times in a year on the eve of Diwali, Baisakhi, Lohdi and in Sawan (August).

Observations

(i) *Courtyard Plants (Protected as well as for Landscaping* : *Canna indica, Nicandra physaloides, Prinsepia utilis.*

(ii) Predominant Plants of the Courtyard	:	*Nicandra physaloides.*
(iii) Maintenance	:	Satisfactory.
(iv) Ethnobotanical		
Sacred and Associated Myth	:	Wood of *Prinsepia utilis* considered good for 'havan'. People tie flags, scarves and sacred threads to *Prinsepia utilis* for fulfilment of their wishes. Flowers of *Nicandra physaloides* offered to the deity.
Ornamental	:	*Canna indica* grown as an ornamental plant.
Remarks	:	Sheep and goats sacrificed to appease the deity.

15. Mahish Mardini Mahalakshmi Temple (Pl. 18F)

Main Deity	:	Maha Lakshmi.
Other Deity	:	Hanuman.
Tot. Ar. of Tem. Courtyard	:	About 2000 sq m.
Location		
(i) Place		
Town	:	Rampur
Block	:	Rampur
Tehsil	:	Rampur
District	:	Rampur
(ii) Approach	:	Located at Khopdi, about 1 km from Rampur.
(iii) Altitude	:	940 m.
Management	:	Local temple committee.
Historical Background	:	The temple was built in 2000 by the locals.
Temple Architect	:	Sikhara style.
Offerings to God		
(i) By Priest		
Morning	:	Milk, Dry fruits.
Noon	:	Dal-Rice, Chapati, Vegetables.
Evening	:	Dal-Rice, Chapati, Vegetables.
(ii) By People	:	'Halva', Gram, 'Lachidana' Puffed rice.
Off. to People (Prasadam)	:	Sugar candy, 'Lachidana'.
Fairs and Festivals	:	Navratra.

Observations

(i) Courtyard Plants (Protected as well as for Landscaping	:	*Aegle marmelos, Araucaria columnaris, Bougainvillea spectabilis, Bryophyllum calycinum, Cestrum nocturnum, Chrysanthemum indicum, Ficus benghalensis, Ficus religiosa, Hibiscus rosa-sinensis, Hibiscus syriacus, Hydrangea macrophylla, Ipomoea quamoclit, Jasminum primulinum, Jasminum sambac, Mangifera indica, Mentha piperita, Musa paradisiaca, Narcissus tazetta, Nerium odorum, Ocimum sanctum, Pelargonium hortorum, Psidium guajava, Rosa indica, Saccharum officinarum, Tagetes erecta, Tecoma radicans, Tecoma stans, Thuja orientalis, Zinnia elegans.*
(ii) Predominant Plants of the Courtyard	:	*Tagetes erecta.*
(iii) Maintenance	:	Excellent.
(iv) Ethnobotanical		
Sacred	:	*Ficus benghalensis, Ficus religiosa, Mangifera indica, Ocimum sanctum* and *Musa paradisiaca* worshipped as sacred plants. Leaves *of Aegle marmelos* (Bael-patri) offered to the deity.
Medicinal	:	Juice of *Saccharum officinarum* given to cure jaundice.
Edible	:	Fruits of *Mangifera indica, Musa paradisiaca* and *Psidium guajava* edible. Juice of *Saccharum officinarum* considered refreshing.
Ornamental	:	*Araucaria columnaris, Bougainvillea spectabilis, Bryophyllum calycinum, Cestrum nocturnum, Chrysanthemum indicum, Hibiscus rosa-sinensis, Hibiscus syriacus, Hydrangea macrophylla, Ipomoea quamoclit, Jasminum primulinum, Jasaminum sambac, Narcissus tazetta, Nerium odorum, Pelargonium hortorum, Rosa indica, Tagetes erecta, Tecoma radicans, Tecoma stans, Thuja orientalis, Zinnia elegans.*
Other Uses	:	Leaves of *Saccharum officinarum* used as fodder.
Status of Courtyard Plants in Surrounding Areas	:	Excepting *Ficus benghlensis, Ficus religiosa* and *Mangifera indica,* all other plants are common in surrounding areas.
Remarks	:	Coconut sacrificed in lieu of sheep and goats.

16. Narsingh Temple (Pl. 18G)

Main Deity	:	Nar Singh.
Tot. Ar. of Tem. Courtyard	:	About 800 sq m.

Location		
(i) Place		
Village	:	Shingla
Block	:	Rampur
Tehsil	:	Rampur
District	:	Shimla
(ii) Approach	:	7 km from Rampur.
(iii) Altitude	:	1200 m.
Management	:	Temple committee.
Historical background	:	The temple is believed to be very old and built by Parshuram.
Temple Architect	:	Sikhara style.
Offerings to God		
(i) By Priest	:	'Halva'.
(ii) By People	:	'Halva'.
Off. to People (Prasadam)	:	'Halva'.
Fairs and Festivals	:	Narsingh Chaturdashi, Janamashtmi.
Observations		
(i) Courtyard Plants (Protected as well as for Landscaping	:	*Callistemon lanceolatus, Canna indica, Ficus palmata, Ocimum sanctum, Tagetes erecta.*
(ii) Predominant Plants of the Courtyard	:	*Tagetes erecta.*
(iii) Maintenance	:	Satisfactory.
(iv) Ethnobotanical		
Sacred	:	*Ocimum sanctum* worshipped as a sacred plant.
Edible	:	Fruits of *Ficus palmata* edible.
Ornamental	:	*Callistemon lanceolatus, Canna indica,* and *Tagetes erecta* grown as ornamentals.
Remarks	:	No sacrifices to the deity.

17. Shikhdu Devta Temple

Main Deity	:	Shikhdu Devta.
Other Deity	:	Mahasu Devta.
Tot. Ar. of Tem. Courtyard	:	3000 sq m.

Location		
(i) Place		
Town	:	Rohru
Block	:	Rohru
Tehsil	:	Rohru
District	:	Shimla
(ii) Approach	:	Located in the town.
(iii) Altitude	:	1540 m.
Management	:	Temple committee.
Historical Background	:	The temple is believed to be very old but renovated about 50 years back.
Temple Architect	:	Sikhara style.
Offerings to God		
(i) By Priest	:	'Halva', Flowers, Fruits.
(ii) By People	:	'Halva', Flowers, Fruits.
Off. to People (Prasadam)	:	'Halva'.
Observations		
(i) Courtyard Plants (Protected as well as for Landscaping	:	*Cyperus flabelliformis, Dahlia rosea, Ficus palmata, Jasminum grandiflorum, Robinia pseudoacacia, Rosa indica.*
(ii) Predominant Plants of the Courtyard	:	*Rosa indica.*
(iii) Maintenance	:	Satisfactory.
(iv) Ethnobotanical		
Sacred	:	*Robinia pseudoacacia* considered a sacred plant.
Edible	:	Fruits of *Ficus palmata* edible.
Ornamental	:	*Cyperus flabelliformis, Dahlia rosea, Jasminum grandiflorum* and *Rosa indica* are grown as ornamental.
Other Uses	:	*Robinia pseudoacacia* considered good for fodder and fuel, and its wood for agricultural implements.
Remarks	:	Goats sacrificed to appease the deity. People also use coconut and pumpkin as a substitute for sheep and goats.

18. Shri Ayodhya Nath Temple (Pl. 18H)

Main Deities	:	Ram-Sita-Lakshman.
Other Deities	:	Shiv-Parvati, Hanuman, Radha-Krishana, Badri-Narayan, Narsingh, Saraswati, Ganesh.
Tot. Ar. of Tem. Courtyard	:	About 800 sq m.
Location		
(i) Place		
Town	:	Rampur
Block	:	Rampur
Tehsil	:	Rampur
District	:	Shimla
(ii) Approach	:	It is 130 km from Shimla.
(iii) Altitude	:	850 m.
Management	:	Temple Trust (Government undertaken).
Historical Background	:	The temple was built by Raja Kahari Singh in 17th century A.D.
Temple Architect	:	Sikhara style.
Offerings to God		
(i) By Priest	:	'Halva'.
(ii) By People	:	'Halva' ('Rot' offered to Lord Hanuman on every Tuesday).
Off. to People (Prasadam)	:	'Halva', Jaggery.
Fairs and Festivals	:	Dussehra, Basant Panchami.
Observations		
(i) Courtyard Plants (Protected as well as for Landscaping	:	*Debregeasia hypoleuca, Jasminum primulinum, Nerium odorum, Ocimum sanctum, Roylea cinerea, Tagetes erecta, Thevetia neriifolia.*
(ii) Predominant Plants of the Courtyard	:	*Tagetes erecta, Jasminum primulinum.*
(iii) Maintenance	:	Satisfactory.
(iv) Ethnobotanical		
Sacred and Associated Myth	:	*Ocimum sanctum* worshipped as well as offered to appease Lord Vishnu. Traditionally, Tulsi plant is married to Saligram (Lord Vishnu).
Ornamental	:	*Jasminum primulinum, Nerium odorum, Tagetes erecta* and *Thevetia neriifolia* grown as ornamental.

Other Uses	:	Wood of *Debregeasia hypoleuca* used as fuel, and leaves as fodder for domestic animals.
Remarks	:	No sacrifices offered.

19. Shri Sankatmochan Hanuman Temple (Pl. 18I)

Main Deities	:	Ram-Sita, Hanuman.
Other Deities	:	Shiva, Ganesha, Baba Neeb Karouri.
Tot. Ar. of Tem. Courtyard	:	About 4000 sq m.
Location		
(i) *Place*		
Town	:	Shimla
Block	:	Mashobra
Tehsil	:	Shimla
District	:	Shimla
(ii) *Approach*	:	Located in the town.
(iii) *Altitude*	:	2100 m.
Management	:	Temple Trust.
Historical Background	:	The temple is 40 years old built by Shri Bajrang Bahadur (Goverener of H.P.) under the guidance of Baba Neeb Karouri.
Temple Architect	:	Sikhara style.
Offerings to God		
(i) *By Priest*	:	Dal-Rice, Chapati, Sweets, 'Halva'. Milk offered in the morning daily.
(ii) *By People*	:	'Lachidana', Gram.
Off. to People (Prasadam)	:	'Lachidana', Gram.
Fairs and Festivals	:	Ram-Navami, Dussehra, Hanuman-Jayanti (Chhoti Diwali), Janamashtmi, Shivratri, Navratra.
Observations		
(i) *Courtyard Plants (Protected as well as for Landscaping*	:	*Acacia dealbata, Calendula officinalis, Callistemon lanceolatus, Cedrus deodara, Centaurea cyanus, Cupressus torulosa, Ficus religiosa, Grevillea robusta, Hibiscus syriacus, Hydrangea macrophylla, Hypericum cernuum, Jasminum grandiflorum, Lagerstroemia indica, Lonicera japonica, Nerium odorum, Pelargonium hortorum, Pinus roxburghii, Prunus puddum, Pyrus malus, Robinia pseudoacacia, Rosa alba, Rosa banksiae, Rosa multiflora, Tagetes erecta, Thuja orientalis, Wisteria sinensis.*

(ii) Predominant Plants of the Courtyard	:	*Pinus roxburghii, Tagetes erecta, Pelargonium hortorum.*
(iii) Maintenance	:	Excellent.
(iv) Ethnobotanical		
Sacred and Associated Myth	:	Flowers of *Nerium odorum, Rosa alba, Rosa banksiae, Rosa multiflora* and *Tagetes erecta* considered sacred for offering to the deity. *Ficus religiosa* worshipped on every saturday and monday. People tie scarves and sacred threads to *Ficus religiosa* and *Rosa banksiae* for fulfilment of their wishes.
Ornamental	:	*Acacia dealbata, Calendula officinalis, Callistemon lanceolatus, Centaurea cyanus, Cupressus torulosa, Ficus religiosa, Grevillea robusta, Hibiscus syriacus, Hydrangea macrophylla, Hypericum cernuum, Jasminum grandiflorum, Lagerstroemia indica, Lonicera japonica, Nerium odorum, Pelargonium hortorum, Pinus roxburghii, Prunus puddum, Pyrus malus, Robinia pseudoacacia, Rosa alba, Rosa banksiae, Rosa multiflora, Tagetes erecta, Thuja orientalis, Wisteria sinensis.*
Other Uses	:	Cones of *Pinus roxburghii* used for decoration purposes. Wood of *Cedrus deodara* excellent for furniture and construction works.
Status of Courtyard Plants in Surrounding Areas	:	*Ficus religiosa* not found in the surrounding vicinity.
Remarks	:	No sacrifices permitted.

20. Shri Sat Narayan Temple (Pl. 18J)

Main Deity	:	Vishnu.
Other Deities	:	Radha-Krishana, Duttatrey, Ram-Lakshman-Sita, Brahma, Mata Durga, Ganesha, Shiv-Parvati, Garud, Tulsi, Hanuman, Shivlingam.
Tot. Ar. of Tem. Courtyard	:	About 1200 sq m.
Location		
(i) Place		
Village	:	Rampur
Block	:	Rampur
Tehsil	:	Rampur
District	:	Shimla
(ii) Approach	:	Located in the town.
(iii) Altitude	:	850 m.

Management	:	Temple Trust built in 1998.
Historical Background	:	The temple was built by Lala Kapur Chand Seth in memory of his father Lala Gobinda Mal in 1926.
Temple Architect	:	Sikhara style.
Offerings to God		
(i) By Priest		
Morning	:	Dal-Rice, Chapati, Vegetables, Milk, Fruits, Sweets.
Evening	:	Dry Fruits.
(ii) By People	:	'Lachidana', Puffed rice, Fruits offered generally. However, 'Bundi'-Sweets offered to Hanuman on Tuesday whereas gram, banana, yellow flowers and yellow sweets ('Besan') offered to Lord Vishnu on Thursday.
Off. to People (Prasadam)	:	'Lachidana', 'Charnamat' (sacred water comprising Tulsi leaves).
Fairs and Festivals	:	Janamashtmi, Ram-Navami.
Observations		
(i) Courtyard Plants (Protected as well as for Landscaping	:	*Musa paradisiaca, Ocimum sanctum, Punica granatum, Rosa indica, Tagetes erecta.*
(ii) Predominant Plants of the Courtyard	:	*Tagetes erecta.*
(iii) Maintenance	:	Excellent.
(iv) Ethnobotanical		
Sacred	:	*Musa pradisiaca* and *Ocimum sanctum* worshipped as sacred plants.
Medicinal	:	Decoction of leaves of *Ocimum sanctum* given to cure fever and cold.
Edible	:	Fruits of *Musa paradisiaca* and *Punica granatum* edible.
Ornamental	:	*Rosa indica* and *Tagetes erecta* grown as ornamentals.
Other Uses	:	Leaves of *Ocimum sanctum* used for flavouring tea and 'Charanamat'.
Remarks	:	No sacrifices permitted.

21. Surya Temple (Pl. 18K)

Main Deity	:	Surya Devta.
Other Deities	:	Durga, Lakshmi-Narayan, Shiv-Parvati, Ganesh, Shivlingam.
Tot. Ar. of Tem. Courtyard	:	About 1500 sq m.

Location		
(i) Place		
Village	:	Nirth
Block	:	Rampur
District	:	Shimla
(ii) Approach	:	18 km from Rampur.
(iii) Altitude	:	800 m.
Management	:	Local temple committee.
Historical Background	:	It is one of the rare temples of Lord Surya, which dates back to the 8th century A.D. and constructed by Lord Parshuram.
Temple Architect	:	Sikhara type.
Offerings to God		
(i) By Priest	:	'Halva'.
(ii) By People	:	'Halva', 'Rot', Seasonal crops.
Off. to People (Prasadam)	:	'Lachidana', 'Halva'.
Fairs and Festivals	:	Buddi Diwali (November-December).
Observations		
(i) Courtyard Plants (Protected as well as for Landscaping	:	*Callistemon lanceolatus, Canna indica, Ficus palmata, Ficus religiosa, Grevillea robusta, Jasminum primulinum, Jasminum sambac, Melia azedarach, Nerium odorum, Ocimum sanctum, Prunus puddum, Rosa indica, Tagetes erecta, Thevetia neriifolia, Withania somnifera.*
(ii) Predominant Plants of the Courtyard	:	*Tagetes erecta.*
(iii) Maintenance	:	Satisfactory.
(iv) Ethnobotanical		
Sacred	:	Flowers of *Tagetes erecta, Jasminum primulinum, Rosa indica, Jasminum sambac, Nerium odorum, Thevetia neriifolia* considered sacred for religious ceremonies. *Ocimum sanctum* and *Ficus religiosa* worshipped as sacred plants.
Medicinal	:	Poultice of leaves of *Withania somnifera* applied to cure joint pains. Its powdered roots given to promote sexual vigour.
Edible	:	Fruits of *Ficus palmata* and *Prunus puddum* edible.
Ornamental	:	*Callistemon lanceolatus, Canna indica, Grevillea robusta, Jasminum primulinum, Jasminum sambac, Nerium odorum,*

	Rosa indica, Tagetes erecta, Thevetia neriifolia grown for ornamental purposes.
Remarks	: No sacrifices permitted.

22. Tara Devi Temple (Pl. 18L, M)

Main Deity	: Tara Devi.
Other Deities	: Mahakali, Mahalakshmi, Mahasaraswati, Dwarpal Devta, Bhairav.
Tot. Ar. of Tem. Courtyard	: 3000 sq m.
Location	
(i) *Place*	
Village	: Patyud
Block	: Mashobra
Tehsil	: Shimla
District	: Shimla
(ii) *Approach*	: 18 km from Shimla.
(iii) *Altitude*	: 2400 m.
Management	: Managed by State Government since 1988.
Historical Background	: The temple is believed to be more than 200 years old, and built by Bhupender Sen (King of Junga) at Juggar. Later, the deity was shifted to hill top at Tara Devi by Raja Balbir Sen (king of Junga) on the advice of saint Taradhinath. This place came to be known as Taarab after the name of this saint, and subsequently as Tara Devi.
Temple Architect	: Sikhara style.
Offerings to God	
(i) *By Priest*	
Morning	: Dry Fruits, Milk, Fruits, 'Halva' made of Ogle flour.
Evening	: Dry Fruits, Milk, Fruits, Sweets.
(ii) *By People*	: Coconut, 'Lachidana', Puffed rice, Sugar candy.
Off. to People (Prasadam)	: 'Lachidana', Sugar candy, Fruits.
Fairs and Festivals	: Navratra, Janamashtmi.
Observations	
(i) *Courtyard Plants (Protected as well as for Landscaping*	: *Althaea rosea, Antirrhinum majus, Asclepias physocarpa, Calendula officinalis, Callistemon lanceolatus, Canna indica, Chrysanthemum indicum, Chrysanthemum leucanthemum,*

		Cosmos bipinnatus, Dahlia rosea, Grevillea robusta, Hemerocallis fulva, Hibiscus syriacus, Impatiens balsamina, Juglans regia, Kochia trichophylla, Lonicera japonica, Nicandra physaloides, Ocimum sanctum, Pelargonium hortorum, Petunia hybrida, Prunus puddum, Punica granatum, Pyrus communis, Pyrus pashia, Quercus incana, Rosa indica, Tagetes erecta, Thuja orientalis, Tropaeolum majus.
(ii)	*Predominant Plants of the Courtyard*	: *Hibiscus syriacus, Tagetes erecta.*
(iii)	*Maintenance*	: Excellent.
(iv)	*Ethnobotanical*	
	Sacred and Associated Myth	: Flowers and wood of *Prunus puddum* used for religious ceremonies in the temple. Fruits of *Pyrus communis* offered to the deity. *Ocimum sanctum* worshipped as sacred plant. People tie scarves, sacred threads and bangles to the *Pyrus communis* for the fulfilment of their wishes.
	Medicinal	: Powdered fruits of *Pyrus pashia* prescribed for the treatment of dysentery.
	Edible	: Fruits of *Pyrus communis, Prunus puddum, Pyrus pashia* and *Punica granatum* and kernels of *Juglans regia* edible.
	Ornamental	: All the courtyard plants considered good for ornamental purposes.
Remarks		: The idol of the deity is made of 'Ashat Dhatu' (eight metals). No sacrifices made to the deity.

Temples of District Sirmaur

1. Balasundari Temple (Pl. 19A, B)

Main Deity	:	Goddess Balasundari.
Other Deities	:	Kali, Bajrangbali, Ashtbhuji Mata, Santoshi Mata, Shiva.
Tot. Ar. of Tem. Courtyard	:	About 1000 sq m.
Location		
(i) Place		
Village	:	Trilokpur
Block	:	Nahan
Tehsil	:	Nahan
District	:	Sirmaur
(ii) Approach	:	7 km from Kala Amb.
(iii) Altitude	:	500 m.
Management	:	Trilokpur Temple Trust (under the supervision of Language and Culture Department).
Historical Background	:	As per the prevalent legend, a 'Pindi' that emerged due to uprooting of a tree due to storm was brought by Lala Ramdas to his home for worshipping. Later, the deity appeared in his dream and asked him to establish 'Pindi' in temple which he got done with help of the king Raja Pradeep Prakash of Sirmaur.

Temple Architect : Sikhara type.

Offering to God:

(i) By Priest : 'Lachidana', Puffed rice, Fox nut, Flowers.

(ii) By People : 'Lachidana', Puffed rice.

Off. to People (Prasadam) : 'Lachidana', Puffed rice, Flowers.

Fairs and Festivals : Local fair (held in March-April and in October), Navratra.

Observations

(i) Courtyard Plants (Protected as well as for Landscaping : *Asparagus densiflorus, Bougainvillea spectabilis, Chamaerops humilis, Elaeocarpus ganitrus, Ficus religiosa, Hibiscus rosa-sinensis, Mimusops elengi, Ocimum basilicum.*

(ii) Predominant Plants of the Courtyard : *Bougainvillea spectabilis.*

(iii) Maintenance : Excellent.

(iv) Ethnobotanical

Sacred and Associated Myth : *Ficus religiosa* and *Ocimum basilicum* worshipped as sacred plants. People tie sacred threads to *Bougainvillea spectabilis* present in the temple courtyard for getting their wishes fulfilled.

Edible : Fruits of *Mimusops elengi* edible.

Ornamental : *Asparagus densiflorus, Bougainvillea spectabilis, Chamaerops humilis, Elaeocarpus ganitrus, Hibiscus rosa–sinensis* and *Mimusops elengi* grown as ornamentals.

Status of Courtyard Plants in Surrounding Areas : *Mimusops elengi* and *Elaeocarpus ganitrus* do not occur in the surrounding areas.

Remarks : No sacrifices permitted.

2. Brahmchari Ashram (Pl. 19C)

Main Deity : Gaytri Mata.

Other Deities : Yamdagni (father of Parshuram), Parshuram, Vashisht, Ganesha, Durga, Vishnu, Krishana, Shiva.

Tot. Ar. of Tem. Courtyard : About 800 sq m.

Location

(i) Place

Town : Renuka ji

Block : Sangrah

Tehsil : Sangrah

District : Sirmaur

(ii) Approach	:	36 km from Nahan.
(iii) Altitude	:	650 m.
Management	:	In the name of Deity.
Historical Background	:	The temple is 50 years old, built by Swami Dyanand Bharti and Brahmchari Pyare Nand.
Temple Architect	:	Sikhara style.
Offerings to God		
(i) By Priest	:	Rice, Chapati with milk.
(ii) By People	:	Coconut, 'Lachidana', Puffed rice.
Off. to People (Prasadam)	:	'Lachidana', 'Kasaar'
Fairs and Festivals	:	Renuka fair, Parshuram Jayanti, Guru Purnima (2nd June).
Observations		
(i) Courtyard Plants (Protected as well as for Landscaping	:	*Aegle marmelos, Aloe vera, Cannabis sativa, Cedrela toona, Citrus aurantifolia, Colocasia esculenta, Datura fastuosa, Ficus religiosa, Juglans regia, Lagerstroemia indica, Litchi chinensis, Mangifera indica, Melia azedarach, Morus alba, Ocimum sanctum, Prunus persica, Psidium guajava, Punica granatum, Syzygium cumini, Tabernaemontana coronaria.*
(ii) Predominant Plants of the Courtyard	:	*Cannabis sativa.*
(iii) Maintenance	:	Satisfactory.
(iv) Ethnobotanical		
Sacred	:	Leaves of *Aegle marmelos, Cannabis sativa* and flowers of *Datura fastuosa* especially offered to appease Lord Shiva. Leaves of *Aegle marmelos* also offered to Gayatri Mata. *Ficus religiosa* and *Ocimum sanctum* worshipped as sacred plants.
Medicinal	:	Poultice of jelly of *Aloe vera* mixed with roasted latex of *Ferula asafoetida* applied around the navel of infants to cure constipation or other stomach problems. Fruits of *Syzygium cumini* effective against anaemia. 2-3g powdered seeds of *Citrus aurantifolia* given daily with warm water for spermatorrhoea.
Edible	:	Leaves and tubers of *Colocasia esculenta* cooked as vegetable. Fruits of *Aegle marmelos, Citrus aurantifolia, Litchi chinensis, Mangifera indica, Morus alba, Prunus persica, Psidium guajava, Punica granatum, Syzygium cumini* and kernels of *Juglans regia* edible. Fruits of *Citrus aurantifolia* also pickled.

Ornamental : *Lagerstroemia indica, Melia azedarach,* and *Tabernaemontana coronaria* grown as ornamentals.

Other Uses : Leaves of *Morus alba* given as fodder. Wood of *Melia azedarach* used for building works and agricultural implements. Wood of *Cedrela toona* excellent for furniture and construction works.

Remarks : No sacrifices permitted.

3. Gurudwara Shri Paonta Sahib (Pl. 19D)

Main Deity : Guru Gobind Singh.

Tot. Ar. of Tem. Courtyard : About 8000 sq m.

Location

(i) *Place*

Town : Paonta Sahib

Block : Paonta Sahib

Tehsil : Paonta Sahib

District : Sirmaur

(ii) *Approach* : Located in the town on the bank of Yamuna river.

(iii) *Altitude* : 550 m.

Management : Gurudwara Parbandhak Committee.

Historical Background : The Gurudwara built by Guru Gobind Singh in 1685 got renovated in 1978 by Gurudwara Parbandhak Committee which came into existence in 1964.

Temple Architect : Dome-shaped.

Offerings to God

(i) *By Priest* : 'Halva'.

(ii) *By People* : 'Halva'.

Off. to People (Prasadam) : 'Halva'.

Fairs and Festivals : Hola-Mohalla, Guru Gobind Singh Jayanti, Guru Nanak Dev Jayanti, Purnima of every month.

Observations

(i) *Courtyard Plants (Protected as well as for Landscaping* : *Acalypha wilkesiana, Alpinia nutans, Amaryllis reginae, Araucaria columnaris, Asparagus densiflorus, Beloperone guttata, Bougainvillea spectabilis, Callistemon lanceolatus, Caryota urens, Catharanthus roseus, Cestrum nocturnum, Chamaerops humilis, Chrysanthemum indicum, Clerodendrum thomsonae, Coleus blumei, Cycas revoluta, Duranta repens, Grevillea robusta, Hibiscus rosa-sinensis,*

		Impatiens balsamina, Jasminum sambac, Juniperus chinensis, Michelia champaca, Ocimum sanctum, Pedilanthus tithymaloides, Petunia hybrida, Plumeria acutifolia, Polyalthia longifolia, Quisqualis indica, Rosa indica, Tabernaemontana coronaria, Tagetes erecta, Thuja orientalis, Tropaeolum majus, Zephyranthes grandiflora.
(ii) Predominant Plants of the Courtyard	:	*Bougainvillea spectabilis, Caryota urens, Polyalthia longifolia, Tagetes erecta.*
(iii) Maintenance	:	Excellent.
(iv) Ethnobotanical		
Medicinal	:	Decoction of bark of *Michelia champaca* prescribed for fever.
Ornamental	:	All courtyard plants considered good for ornamental purposes.
Remarks	:	All courtyard plants considered sacred and protected with all reverence.

4. Jagan Nath Temple (Pl. 19E)

Main Deity	:	Jagan Nath.
Other Deities	:	Radha-Krishana, Badrinath, Ladoo-Gopal, Hanuman.
Tot. Ar. of Tem. Courtyard	:	About 3000 sq m.
Location		
(i) Place		
Town	:	Nahan
Block	:	Nahan
Tehsil	:	Nahan
District	:	Sirmaur
(ii) Approach	:	Located in the town.
iii) Altitude	:	600 m.
Management	:	Temple committee.
Historical Background	:	The temple was built by Raja Mahi Singh (King of Sirmaur) in 1681 A.D. on the request of Banwari baba who meditated at this place.
Temple Architect	:	Sikhara style.
Offerings to God		
(i) By Priest		
Morning	:	Sugar Candy, Milk.
Noon	:	Dal-Chapati, Rice, Vegetables, Fruits.
(ii) By People	:	Sweets.

Off. to People (Prasadam) : Sweets.

Fairs and Festivals : Janamashtmi, Bawan-Dwadshi (held in August-September).

Observations

(i) *Courtyard Plants (Protected as well as for Landscaping* : *Aegle marmelos, Artocarpus heterophyllus, Cynodon dactylon, Deeringia amaranthoides, Ficus benghalensis, Ficus glomerata, Ficus palmata, Mimusops elengi, Musa paradisiaca* and *Ocimum sanctum.*

(ii) *Predominant Plants of the Courtyard* : *Ficus palmata.*

(iii) *Maintenance* : Satisfactory.

(iv) *Ethnobotanical*

Sacred and Associated Myth : *Musa paradisiaca* and *Ocimum sanctum* worshipped as sacred plants. The plant of *Musa paradisiaca* worshipped on every thursday by the people for fulfilment of their wishes. *Cynodon dactylon* employed in religious worshipping.

Edible : Fruits of *Aegle marmelos, Ficus glomerata, Ficus palmata, Mimusops elengi, Musa paradisiaca* edible whereas those of *Artocarpus heterophyllus* cooked as vegetable and also pickled.

Ornamental : *Deeringia amaranthoides, Mimusops elengi, Musa paradisiaca, Ocimum sanctum* grown as ornamental plants.

Status of Courtyard Plants in Surrounding Areas : Except that of *Mimusops elengi*, all other plants occur commonly outside.

Remarks : No sacrifices to the deity. The idol of Jagan Nath made of Neem wood brought from Orissa (Jagannath Puri).

5. Kalisthan Temple (Pl. 19F)

Main Deity : Kali.

Other Deities : Shiva, Bhairav, Hanuman.

Tot. Ar. of Tem. Courtyard : About 5000 sq m.

Location

(i) *Place*

Town : Nahan

Block : Nahan

Tehsil : Nahan

District : Sirmaur

(ii) Approach	:	Located in the town.
(iii) Altitude	:	600 m.
Management	:	In the name of Deity.
Historical Background	:	The temple was built in 1730 A.D. Later, it was renovated several times.
Temple Architect	:	Sikhara style.
Offerings to God		
(i) By Priest	:	'Halva', milk.
(ii) By People	:	'Lachidana', Coconut, Betel leaf and Betel nut ('Pan-Supari'), Sweets.
Off. to People (Prasadam)	:	'Lachidana', Flowers, Sweets.
Fairs and Festivals	:	Navratra.
Observations		
(i) Courtyard Plants (Protected as well as for Landscaping	:	*Artocarpus lakoocha, Bougainvillea spectabilis, Cynodon dactylon, Emblica officinalis, Ficus benghalensis, Ficus religiosa, Mangifera indica, Mimusops elengi, Musa paradisiaca, Ocimum sanctum, Plumeria acutifolia, Psidium guajava, Roylea cinerea, Terminalia bellirica.*
(ii) Predominant Plants of the Courtyard	:	*Mangifera indica.*
(iii) Maintenance	:	Excellent.
(iv) Ethnobotanical		
Sacred	:	*Ocimum sanctum, Musa paradisiaca, Ficus benghalensis* and *Ficus religiosa* worshipped on all religious ceremonies. *Cynodon dactylon* employed in religious worshipping.
Medicinal	:	Fruits of *Emblica officinalis* and *Terminalia bellirica* used in preparation of 'Triphala' for stomach ailments. Aqueous extract of the fruits of *Emblica officinalis* applied to promote hair growth.
Edible	:	Unripe fruits of *Artocarpus lakoocha* and *Musa paradisiaca* cooked as vegetable. Fruits of *Emblica officinalis, Mangifera indica, Mimusops elengi, Musa paradisiaca, Psidium guajava* and *Terminalia bellirica* edible.
Ornamental	:	*Bougainvillea spectabilis, Plumeria acutifolia* and *Roylea cinerea* grown as ornamental plants.
Other Uses	:	Leaves of *Mangifera indica* used for scouring teeth. Wood of *Mangifera indica* used for construction works whereas that of *Terminalia bellirica* for agricultural implements.

Status of Courtyard Plants in Surrounding Areas	:	*Mimusops elengi* does not occur in the surrounding areas.
Remarks	:	No sacrifices made to the deity.

6. Katasan Devi Temple (Pl. 19G)

Main Deity	:	Katasan Devi.
Other Deity	:	Shiva.
Tot. Ar. of Tem. Courtyard	:	About 3000 sq m.
Location		
(i) Place		
Village	:	Bada Van (Katasan Devi)
Block	:	Nahan
Tehsil	:	Nahan
District	:	Sirmaur
(ii) Approach	:	About 17 km from Nahan, on way to Paonta Sahib.
(iii) Altitude	:	550 m.
Management	:	Privately owned.
Historical Background	:	The temple built by King of Sirmaur belongs to 18th-19th century A.D.
Temple Architect	:	Sikhara style.
Offerings to God		
(i) By Priest	:	Milk, Fruits.
(ii) By People	:	'Lachidana', Puffed rice.
Off. to People (Prasadam)	:	'Lachidana', Puffed rice, Fox nut, Flowers.
Fairs and Festivals	:	Navratra (twice a year).
Observations		
(i) Courtyard Plants (Protected as well as for Landscaping	:	*Aegle marmelos, Bauhinia vahlii, Ficus benghalensis, Ficus religiosa, Mallotus philippinensis, Mangifera indica, Shorea robusta, Thevetia neriifolia.*
(ii) Predominant Plants of the Courtyard	:	*Shorea robusta.*
(iii) Maintenance	:	Satisfactory.
(iv) Ethnobotanical		
Sacred and Associated Myth	:	Leaves of *Aegle marmelos* and flowers of *Thevetia neriifolia* offered to appease Lord Shiva. *Aegle marmelos, Ficus benghalensis,* and *Ficus religiosa* worshipped for all

religious activities. People tie scarves and sacred threads to *Ficus religiosa* and *Thevetia neriifolia* for getting their wishes fulfilled.

Medicinal : Seed powder of *Mallotus philippinensis* mixed in ointments used to cure skin problems.

Edible : Fruits of *Aegle marmelos* and *Mangifera indica* edible.

Other Uses : Leaves of *Bauhinia vahlii* used for making meal plates. Wood of *Mangifera indica* and *Shorea robusta* used for construction works whereas that of *Mallotus philippinensis* for agricultural implements.

Remarks : Goats sacrificed to appease the deity.

7. Narsingh Temple (Pl. 19H)

Main Deity : Nar Singh.

Other Deity : Shiva.

Tot. Ar. of Tem. Courtyard : About 1500 sq m.

Location

(i) *Place*

Town : Nahan

Block : Nahan

Tehsil : Nahan

District : Sirmaur

(ii) *Approach* : Located near the town.

(iii) *Altitude* : 600 m.

Management : Temple committee.

Historical Background : The temple is 300 years old and built by King of Sirmaur.

Temple Architect : Sikhara style.

Offerings to God

(i) *By Priest* : 'Lachidana', Fruits.

(ii) *By People* : 'Lachidana'.

Off. to People (Prasadam) : 'Lachidana'.

Fairs and Festivals : Narsingh Jayanti, Bawan Dwadshi, Shivratri.

Observations

(i) *Courtyard Plants (Protected as well as for Landscaping* : *Artocarpus heterophyllus, Bombax ceiba, Bougainvillea spectabilis, Canna indica, Catharanthus roseus, Cupressus sempervirens, Emblica officinalis, Mangifera indica, Mimusops elengi, Ocimum sanctum, Tamarindus indica.*

(ii) Predominant Plants of the Courtyard	:	*Mangifera indica.*
(iii) Maintenance	:	Satisfactory.
(iv) Ethnobotanical		
Sacred	:	*Ocimum sanctum* worshipped as sacred plant. *Cupressus sempervirens* also considered sacred.
Medicinal	:	Poultice of bark of *Bombax ceiba* used to cure furuncles. Aqueous extract of fruits of *Emblica officinalis* applied for promoting hair growth. Fruits of *Emblica officinalis* employed in the preparation of 'Triphala'.
Edible	:	Fruits of *Artocarpus heterophyllus* cooked as vegetable. Fruits of *Emblica officinalis, Mangifera indica, Mimusops elengi, Tamarindus indica* edible.
Ornamental	:	*Bombax ceiba, Bougainvillea spectabilis, Canna indica, Catharanthus roseus, Cupressus sempervirens* and *Mimusops elengi* grown as ornamentals.
Other Uses	:	Leaves of *Mangifera indica* employed for scouring teeth. Wood of *Bombax ceiba* and *Mangifera indica* used for construction works.
Status of Courtyard Plants in Surrounding Areas	:	Occurrence of *Tamarindus indica* is rare whereas *Cupressus sempervirens* and *Mimusops elengi* do not occur in the surrounding areas.
Remarks	:	No sacrifices permitted. There is a thick sacred forest around the temple covering an area of about 10 hectare.

8. Nirwan Ashram (Pl. 19I)

Main Deity	:	Shri Guru Sunder Muni Nirwan.
Other Deities	:	Renuka Mata, Parshuram, Vishnu, Hanuman, Ganesha.
Tot. Ar. of Tem. Courtyard	:	About 1500 sq m.
Location		
(i) Place		
Town	:	Renuka ji
Block	:	Sangrah
Tehsil	:	Sangrah
District	:	Sirmaur
(ii) Approach	:	About 36 km from Nahan.
(iii) Altitude	:	650 m.

Management	:	In the name of Deity.
Historical Background	:	The temple was built in 1972 by Shri Guru Sunder Muni who meditated here for several years.
Temple Architect	:	Sikhara style.
Offerings to God		
(i) By Priest	:	Roti-Dal, Vegetables, Tea, 'Rot-Parsad', 'Kasaar'.
(ii) By People	:	Rice, Flour, Clarified butter, Fruits.
Off. to People (Prasadam)	:	'Rot-Parsad', 'Kasaar', Sweets, Fruits.
Fairs and Festivals	:	Guru Purnima (held in July), Renuka Fair.
Observations		
(i) Courtyard Plants (Protected as well as for Landscaping	:	*Araucaria columnaris, Canna indica, Catharanthus roseus, Chlorophytum capense, Cupressus sempervirens, Datura fastuosa, Elaeocarpus ganitrus, Ficus religiosa, Mangifera indica, Ocimum sanctum, Pelargonium hortorum, Punica granatum, Rosa indica, Tagetes erecta, Thuja orientalis.*
(ii) Predominant Plants of the Courtyard	:	*Thuja orientalis.*
(iii) Maintenance	:	Satisfactory.
(iv) Ethnobotanical		
Sacred	:	*Ficus religiosa, Mangifera indica* and *Ocimum sanctum* considered sacred plants.
Edible	:	Fruits of *Mangifera indica, Punica granatum* edible.
Ornamental	:	*Araucaria columnaris, Canna indica, Catharanthus roseus, Chlorophytum capense, Cupressus sempervirens, Elaeocarpus ganitrus, Pelargonium hortorum, Rosa indica, Tagetes erecta* and *Thuja orientalis* grown as ornamentals.
Other Uses	:	The hard seeds of *Elaeocarpus ganitrus* used for making prayer beads (rosaries).
Status of Courtyard Plants in Surrounding Areas	:	*Elaeocarpus ganitrus* does not occur in the surrounding areas.
Remarks	:	No sacrifices made to the deity.

9. Pujari Ghat Rama Temple (Pl. 19J)

Main Deities	:	Rama, Sita, Lakshman.
Other Deities	:	Shiva, Ganesha, Hanuman.
Tot. Ar. of Tem. Courtyard	:	About 1500 sq m.

Location		
(i) Place		
Town	:	Paonta Sahib (Devi Nagar)
Block	:	Paonta Sahib
Tehsil	:	Paonta Sahib
District	:	Sirmaur
(ii) Approach	:	Located in the town on the side of Yamuna river.
(iii) Altitude	:	550 m.
Management	:	Government undertaken.
Historical Background	:	The temple was built in 1889 by Rani Dei Sahiba (Queen of Kangra) in memory of her deceased husband, Raja Pratap Chand Bahadur (King of Kangra) with the help of her brother Maharaja Shamsher Prakash Bahadur.
Temple Architect	:	Sikhara style.
Offerings to God		
(i) By Priest	:	'Lachidana', Puffed rice, 'Halva', Fruits, Flowers.
(ii) By People	:	'Lachidana', Fruits, Flowers, Sweets. 'Bhang' and 'Dhatura' offered to appease Lord Shiva.
Off. to People (Prasadam)	:	'Lachidana', Puffed rice.
Fairs and Festivals	:	Ram Navami, Shivratri, Janamashtmi.
Observations		
(i) Courtyard Plants (Protected as well as for Landscaping	:	*Bougainvillea spectabilis, Carica papaya, Canna indica, Catharanthus roseus, Citrus aurantifolia, Cynodon dactylon, Ficus religiosa, Grevillea robusta, Mangifera indica, Melia azadirachta, Musa paradisiaca, Ocimum sanctum, Psidium guajava, Tabernaemontana coronaria, Tagetes erecta, Tectona grandis, Thevetia neriifolia.*
(ii) Predominant Plants of the Courtyard	:	*Thevetia neriifolia.*
(iii) Maintenance	:	Satisfactory.
(iv) Ethnobotanical		
Sacred	:	Leaves of *Aegle marmelos* offered to Lord Shiva on Monday on special occasions. *Ficus religiosa, Musa paradisiaca,* and *Ocimum sanctum* worshipped as sacred plants. *Cynodon dactylon* employed in various religious ceremonies.
Medicinal	:	Tablets made by taking equal quantity of powdered seeds of *Melia azadirachta* and Drek (*Melia azedarach*),

		fruit of Ritha (*Sapindus mukorossi*) prescribed (one tablet taken daily) to cure piles. Leaf powder of *Melia azadirachta* applied for skin disorders. Decoction of leaves of *Melia azadirachta* prescribed for diabetes and as blood purifier.
Edible	:	Fruits of *Carica papaya, Mangifera indica, Musa paradisiaca* and *Psidium guajava* edible whereas those of *Citrus aurantifolia* used in preparation of refreshing drinks and pickling.
Ornamental	:	*Bougainvillea spectabilis, Canna indica, Catharanthus roseus, Grevillea robusta, Melia azadirachta, Tabernaemontana coronaria, Tagetes erecta* and *Thevetia neriifolia* grown as ornamentals.
Other Uses	:	Branches of *Melia azadirachta* used as tooth brush for scouring teeth.
Remarks	:	No sacrifices made to the deity.

10. Radha Krishan Temple (Pl. 19K)

Main Deity	:	Radha-Krishana.
Other Deities	:	Shiva, Hanuman, Ram Chandra.
Tot. Ar. of Tem. Courtyard	:	About 400 sq m.
Location		
(i) Place		
Village	:	Naina Tikker
Block	:	Saranh
Tehsil	:	Pachhad
District	:	Sirmaur
(ii) Approach	:	About 15 km from Kumarhatti.
(iii) Altitude	:	550 m.
Management	:	In the name of Deity.
Historical Background	:	The temple though quite old was renovated about 5 years back by locals.
Temple Architect	:	Sikhara style.
Offerings to God		
(i) By Priest	:	Nothing special.
(ii) By People	:	Sweets.
Off. to People (Prasadam)	:	Sweets.
Fairs and Festivals	:	Janamashtmi, Shivratri, Ram Navmi.

Observations

(i) Courtyard Plants (Protected as well as for Landscaping	:	*Dalbergia sissoo, Lagerstroemia indica, Musa paradisiaca, Zephyranthes grandiflora.*
(ii) Predominant Plants of the Courtyard	:	*Lagerstroemia indica.*
(iii) Maintenance	:	Satisfactory.
(iv) Ethnobotanical		
Sacred	:	*Musa paradisiaca* considered sacred and worshipped for all religious purposes.
Edible	:	Fruits of *Musa paradisiaca* edible.
Ornamental	:	All courtyard plants considered good for beautification of the temple.
Other Uses	:	Wood of *Dalbergia sissoo* excellent for furniture and building works.
Remarks	:	No sacrifice permitted.

11. Renuka Mata Temple (Pl. 19L)

Main Deities	:	Renuka Mata, Parshuram.
Other Deities	:	Ganesha, Vishnu, Shiva, Goddess Durga, Saraswati.
Tot. Ar. of Tem. Courtyard	:	About 3000 sq m.
Location		
(i) Place		
Town	:	Renuka ji
Block	:	Sangrah
Tehsil	:	Sangrah
District	:	Sirmaur
(ii) Approach	:	About 36 km from Nahan.
(iii) Altitude	:	650 m.
Management	:	Renuka Development Board.
Historical Background	:	Earlier, the lake was worshipped as Goddess Renuka. About 100 years back, the temple was constructed with the idol of the deity.
Temple Architect	:	Sikhara style.
Offerings to God		
(i) By Priest	:	Dry Fruits, 'Halva'.
(ii) By People	:	'Lachidana', Puffed rice, Flowers

Off. to People (Prasadam) : 'Lachidana', Puffed rice, Flowers.

Fairs and Festivals : Local fair (held every year in the month of November for 5 days), Parshuram Jayanti (held in May), Lohdi, Janamashtmi, Shivratri.

Obervations:

(i) *Courtyard Plants (Protected as well as for Landscaping* : *Bombax ceiba, Bougainvillea spectabilis, Callistemon lanceolatus, Canna indica, Cedrela toona, Citrus limon, Ficus glomerata, Ficus religiosa, Grevillea robusta, Populus nigra, Psidium guajava, Rosa indica, Rosa multiflora, Tabernaemontana coronaria, Tagetes erecta, Thevetia neriifolia, Thuja orientalis.*

(ii) *Predominant Plants of the Courtyard* : *Thuja orientalis.*

(iii) *Maintenance* : Excellent.

(iv) *Ethnobotanical*

Sacred and Associated Myth : Flowers of *Bougainvillea spectabilis, Rosa indica, Rosa multiflora, Tabernaemontana coronaria, Tagetes erecta* and *Thevetia neriifolia* considered sacred for offering to the deity. *Ficus religiosa* and *Ficus glomerata* worshipped as sacred trees. It is believed that deities reside on the roots of *Ficus glomerata* (Gullar).

Medicinal : Fruits of *Ficus glomerata* considered good for eyesight.

Edible : Fruits of *Citrus limon, Ficus glomerata* and *Psidium guajava* edible

Ornamental : *Bombax ceiba, Bougainvillea spectabilis, Callistemon lanceolatus, Canna indica, Grevillea robusta, Rosa indica, Rosa multiflora, Tabernaemontana coronaria, Tagetes erecta, Thevetia neriifolia* and *Thuja orientalis* grown as ornamentals.

Other Uses : Leaves of *Populus nigra* used as fodder. Its wood used for cheap furniture. Wood of *Bombax ceiba* and *Cedrela toona* excellent for construction works. Seed cotton of *Bombax ceiba* used for filling pillows and mattresses.

Remarks : No sacrifices permitted to the deity.

12. Rishi Markanda Dham (Pl. 19M)

Main Deity : Markanda Rishi.

Other Deities : Goddess Jwala, Kali Saraswati, Shiva, Hanuman, Bhairav, Narsingh.

Tot. Ar. of Tem. Courtyard : About 1000 sq m.

Location		
(i) Place		
Village	:	Jogivan (Bohliyon)
Block	:	Nahan
Tehsil	:	Nahan
District	:	Sirmaur
(ii) Approach	:	16 km from Nahan.
(iii) Altitude	:	520 m.
Management	:	Local temple committee.
Historical Background	:	The temple, about 25 years old, had been built by locals.
Temple Architect	:	Sikhara style.
Offerings to God		
(i) By Priest	:	Gruel ('Daliya').
(ii) By People	:	'Lachidana', Puffed rice.
Off. to People (Prasadam)	:	'Lachidana', Puffed rice.
Fairs and Festivals	:	Baisakhi.
Observations		
(i) Courtyard Plants (Protected as well as for Landscaping	:	*Acacia catechu, Aegle marmelos, Artocarpus lakoocha, Ficus glomerata, Ficus religiosa, Ficus roxburghii, Mangifera indica, Psidium guajava, Shorea robusta, Syzygium cumini.*
(iii) Maintenance	:	Satisfactory.
(iii) Ethnobotanical		
Sacred and Associated Myth	:	Wood of *Ficus glomerata* and *Ficus roxburghii* used in purification ceremonies. *Ficus glomerata* worshipped as a sacred tree. It is believed that the river Saraswati originated from the roots of the 'Gullar' tree (*Ficus glomerata*) present still in the temple courtyard under which Rishi Markanda meditated long ago. The river ultimately flows toward the west.
Medicinal	:	Latex of *Ficus glomerata* prescribed with milk to increase sexual potentiality. Seeds of *Syzygium cumini* used to cure diabetes and fruits prescribed as blood purifier.
Edible	:	Fruits of *Aegle marmelos, Ficus glomerata, Ficus roxburghii, Mangifera indica, Psidium guajava* and *Syzygium cumini* edible whereas unripe ones of *Artocarpus lakoocha* and *Mangifera indica* pickled.
Other Uses	:	Branches of *Acacia catechu* used as toothbrush. Wood of *Shorea robusta, Syzygium cumini* and *Mangifera indica* used for construction works.

Status of Courtyard Plants in Surrounding Areas : Excepting *Ficus glomerata* and *Ficus roxburghii*, all plants are common outside.

Remarks : No sacrifices permitted in the sanctum sanctorum.

13. Sahibji Maharaj Gurudwara (Pl. 20A)

Main Deity : Sahibji Maharaj.

Tot. Ar. of Tem. Courtyard : About 1500 sq m.

Location

(i) Place

Village : Shambhuwala

Block : Nahan

Tehsil : Nahan

District : Sirmaur

(ii) Approach : About 10 km from Nahan.

(iii) Altitude : 550 m.

Management : In the name of Deity.

Historical Background : It is about 30 years old.

Temple Architect : Dome-shaped.

Offerings to God

(i) By Priest : 'Halva-Puri', Dal-Chapati.

(ii) By People : 'Lachidana', 'Patasha'.

Off. to People (Prasadam) : 'Lachidana'.

Fairs and Festivals : Purnima.

Observations

(i) Courtyard Plants (Protected as well as for Landscaping : *Hibiscus rosa-sinensis, Lawsonia alba, Mangifera indica, Psidium guajava, Tagetes erecta, Thuja orientalis.*

(ii) Predominant Plants of the Courtyard : *Mangifera indica.*

(iii) Maintenance : Satisfactory.

(iv) Ethnobotanical

Sacred : Flowers of *Hibiscus rosa-sinensis* and *Tagetes erecta* considered sacred for offering to the deity.

Medicinal : Poultice of leaves of *Lawsonia alba* with alum ('fatkari') and rock salt applied to cure bone dislocation. Its leaf paste applied to relieve the burning sensation in the sole of feet and fungal infection. Also, poultice of leaves of *Lawsonia alba*, tender leaves of 'Kainth' (*Pyrus pashia*) and tea applied for hair growth and hair colouring.

Edible : Fruits of *Mangifera indica* and *Psidium guajava* edible.

Ornamental : *Hibiscus rosa-sinensis, Lawsonia alba, Tagetes erecta* and *Thuja orientalis* grown for ornamental purposes.

Other Uses : Leaf paste of *Lawsonia alba* extensively used for colouring hands, hair and beard.

Remarks : No sacrifices made to the deity.

14. Sanyas Ashram (Pl. 20B)

Main Deity : Shiva.

Other Deities : Parshuram, Renuka Mata.

Tot. Ar. of Tem. Courtyard : About 1500 sq m.

Location

(i) Place

Town : Renuka ji

Block : Sangrah

Tehsil : Sangrah

District : Sirmaur

(ii) Approach : 36 km from Nahan.

(iii) Altitude : 650 m.

Management : In the name of Deity.

Historical Background : The temple was built by Guru Ganga Puri, Devki Nandan and Bodhanand about 50 years back.

Temple Architect : Sikhara style.

Offerings to God

(i) By Priest : 'Rot', Jaggery, Rice, 'Khir', Clarified butter.

(ii) By People : 'Lachidana', Puffed rice.

Off. to People (Prasadam) : 'Lachidana', Puffed rice.

Fairs and Festivals : Shivratri, Janamashtmi, Renuka Fair.

Observations

(i) Courtyard Plants (Protected as well as for Landscaping : *Aegle marmelos, Amaryllis reginae, Callistemon lanceolatus, Cannabis sativa, Canna indica, Catharanthus roseus, Datura fastuosa, Hibiscus rosa-sinensis, Juglans regia, Lagerstroemia indica, Mangifera indica, Musa paradisiaca, Ocimum*

		sanctum, Quisqualis indica, Syzigium cumini, Tabernaemontana coronaria, Thuja orientalis.
(ii) Predominant Plants of the Courtyard	:	*Thuja orientalis.*
(iii) Maintenance	:	Satisfactory.
(iv) Ethnobotanical		
Sacred	:	Leaves of *Cannabis sativa, Aegle marmelos* and flowers of *Datura fastuosa* epecially offered to Lord Shiva. *Musa paradisiaca* and *Ocimum sanctum* worshipped as sacred plants. Extract prepared by mixing *Cannabis sativa* ('Bhang'), dry fruits and milk given as special offering to people on Shivratri. Wood of *Mangifera indica* used in purification ceremonies.
Medicinal	:	2-3g powdered dried seeds of *Syzygium cumini* taken daily with water to cure diabetes.
Edible	:	Fruits of *Aegle marmelos, Mangifera indica, Musa paradisiaca* and *Syzygium cumini* and kernels of *Juglans regia* edible. Unripe fruits of *Mangifera indica* pickled and also used for making 'chutney'.
Ornamental	:	*Amaryllis reginae, Callistemon lanceolatus, Canna indica, Catharanthus roseus, Hibiscus rosa-sinensis, Lagerstroemia indica, Quisqualis indica, Tabernaemontana coronaria* and *Thuja orientalis* grown as ornamentals.
Other Uses	:	Wood of *Syzygium cumini* used for agricultural implements and that of *Mangifera indica* for construction works.
Remarks	:	No sacrifices permitted.

15. Shirgul Devta Temple (Chudeshwar Maharaj Temple)

Main Deity	:	Shirgul Devta.
Other Deity	:	Shiva.
Tot. Ar. of Tem. Courtyard	:	About 3000 sq m.
Location		
(i) Place		
Locality	:	Churdhar
Block	:	Sangrah
Tehsil	:	Sangrah
District	:	Sirmaur
(ii) Approach	:	About 10 km on foot from Saranh.
(iii) Altitude	:	3700 m.

Management	:	Temple committee.
Historical Background	:	The temple is thought to be 1000-1200 years old.
Temple Architect	:	Pentroof style with Devdar wood used in its construction.
Offerings to God		
(i) *By Priest*	:	Currant, Coconut, 'Halva'.
(ii) *By People*	:	'Lachidana', Coconut, 'Halva'.
Off. to People (Prasadam)	:	Rice grains, 'Lachidana', Sugar candy, Flowers.
Fairs and Festivals	:	Samavati (whenever Amavas falls on Monday), Ekadashi of Kartik.
Observations		
(i) *Courtyard Plants (Protected as well as for Landscaping*	:	Nil.
(ii) *Maintenance*	:	Satisfactory.
Remarks	:	No sacrifices permitted. Rice grains offered as 'prasadam' are generally kept in granary for prosperity. People offer wheat, maize or barley (seasonal crops) to the deity as it is believed to eradicate mice from their homes and fields. For fulfilment of their wishes, people keep coins near the front wall of the temple.

16. Shiva Temple (Pl. 20C)

Main Deity	:	Shiva.
Other Deities	:	Goddess Durga, Hanuman.
Tot. Ar. of Tem. Courtyard	:	About 1500 sq m.
Location		
(i) *Place*		
Village	:	Rukhdi
Block	:	Nahan
Tehsil	:	Nahan
District	:	Sirmaur
(ii) *Approach*	:	14 km from Nahan.
(iii) *Altitude*	:	525 m.
Management	:	In the name of Deity.
Historical Background	:	Temple is believed to be 200 years old.
Temple Architect	:	Sikhara style.

Offerings to God		
(i) By Priest	:	Sweets, Fruits, Milk.
(ii) By People	:	'Lachidana', Puffed rice. 'Bil-patri' (Leaves of *Aegle marmelos*) offered to appease Lord Shiva.
Off. to People (Prasadam)	:	'Lachidana', Puffed rice, Fruits.
Fairs and Festivals	:	Shivratri, Ram Navami.
Observations		
(i) Courtyard Plants (Protected as well as for Landscaping	:	*Aloe vera, Araucaria columnaris, Bombax ceiba, Carica papaya, Cannabis sativa, Canna indica, Catharanthus roseus, Cedrela toona, Chamaerops humilis, Cupressus sempervirens, Datura fastuosa, Ficus religiosa, Juniperus chinensis, Malvaviscus conzatii, Mangifera indica, Morus alba, Prunus persica, Psidium guajava, Punica granatum, Saccharum officinarum, Syzygium cumini, Tagetes erecta.*
(ii) Predominant Plants of the Courtyard	:	*Mangifera indica.*
(iii) Maintenance	:	Satisfactory.
(iv) Ethnobotanical		
Sacred	:	Leaves of *Cannabis sativa* and flowers of *Datura fastuosa* offered as special offerings to Lord Shiva. *Ficus religiosa* worshipped for religious purposes.
Medicinal	:	Pieces of *Aloe vera* sprinkled with powdered black pepper and rock salt prescribed for curing cough. 2-3g powdered seeds of *Syzygium cumini* taken twice daily with water to cure diabetes. Juice of *Saccharum officinarum* effective against jaundice. Powdered young plant roots of *Bombax ceiba* prescribed with milk to restore the sexual potency. Poultice of its bark applied to cure furuncles.
Edible	:	Leaves of *Aloe vera* cooked as vegetable. Fruits of *Carica papaya, Mangifera indica, Morus alba, Prunus persica, Psidium guajava, Punica granatum* and *Syzygium cumini* edible.
Ornamental	:	*Araucaria columnaris, Canna indica, Catharanthus roseus, Chamaerops humilis, Cupressus sempervirens, Malvaviscus conzatii* and *Tagetes erecta* grown as ornamentals.
Other Uses	:	Leaves of *Morus alba* considered as good fodder. Wood of *Cedrela toona* employed for construction works. Seed cotton of *Bombax ceiba* employed for filling pillows and mattresses.
Remarks	:	No sacrifices permitted. 'Akhand Jyoti' is kept burning in the temple for the last several years.

17. Shiva Temple

Main Deity	:	Shiva.
Tot. Ar. of Tem. Courtyard	:	About 3000 sq m.
Location		
(i) *Place*		
Village	:	Trilokpur
Block	:	Nahan
Tehsil	:	Nahan
District	:	*Sirmaur*
(ii) *Approach*	:	7 km from Kala Amb.
(iii) *Altitude*	:	500 m.
Management	:	Committee under Trilokpur Temple Trust.
Historical Background	:	The temple, about 20 years old, had been built by Temple Trust.
Temple Architect	:	Sikhara style.
Offerings to God		
(i) *By Priest*	:	'Lachidana', Puffed rice.
(ii) *By People*	:	'Lachidana', Puffed rice. 'Bil-Patri' (Leaves of *Aegle marmelos*) offered to appease Lord Shiva.
Off. to People (Prasadam)	:	'Lachidana', Puffed rice.
Fairs and Festivals	:	Shivratri, Trilokpur fair (held in the month of April and October).
Observations		
(i) *Courtyard Plants (Protected as well as for Landscaping*	:	*Araucaria columnaris, Bougainvillea spectabilis, Callistemon lanceolatus, Canna indica, Catharanthus roseus, Cordia dichotoma, Duranta repens, Ficus benghalensis, Hamelia patens, Hibiscus rosa-sinensis, Jatropha panduraefolia, Mangifera indica, Melia azadirachta, Mimusops elengi, Nerium odorum, Petunia hybrida, Polyalthia longifolia, Rosa indica, Syzygium cumini, Tabernaemontana coronaria, Tecomaria capensis, Thuja orientalis.*
(ii) *Predominant Plants of the Courtyard*	:	*Polyalthia longifolia, Thuja orientalis.*
(iii) *Maintenance*	:	Excellent.
(iv) *Ethnobotanical*		
Sacred	:	*Ficus benghalensis* worshipped as a sacred tree.
Medicinal	:	Vegetable of fruits of *Cordia dichotoma* good for stomach disorders. 2-3g powdered seeds of *Syzygium cumini* taken

		twice daily with water as an effective remedy against diabetes.
Edible	:	Fruits of *Cordia dichotoma* pickled as well as cooked as vegetable. Fruits of *Mangifera indica, Mimusops elengi* and *Syzygium cumini* edible.
Ornamental	:	*Araucaria columnaris, Bougainvillea spectabilis, Callistemon lanceolatus, Canna indica, Catharanthus roseus, Duranta repens, Hamelia patens, Hibiscus rosa-sinensis, Jatropha panduraefolia, Nerium odorum, Petunia hybrida, Polyalthia longifolia, Rosa indica, Tabernaemontana coronaria, Tecomaria capensis* and *Thuja orientalis* grown as ornamental plants.
Other Uses	:	*Duranta repens* considered good for fencing purposes. Wood of *Mangifera indica* employed for construction works, and that of *Syzygium cumini* for agricultural implements and construction works.
Status of Courtyard Plants in Surrounding Areas	:	Excepting *Mimusops elengi*, all other plants occur commonly in the surrounding areas.
Remarks	:	No sacrifices permitted.

18. Vishwakarma Temple (Pl. 20D)

Main Deity	:	Vishwakarma.
Other Deities	:	Shiva, Radha-Krishana, Ganesha, Rama, Goddess Bhagwati.
Tot. Ar. of Tem. Courtyard	:	About 1500 sq m.
Location		
(i) Place		
Town	:	Paonta Sahib
Block	:	Paonta Sahib
Tehsil	:	Paonta Sahib
District	:	Sirmaur
(ii) Approach	:	Located in the town.
iii) Altitude	:	550 m.
Management	:	Local temple committee.
Historical Background	:	The temple was constructed in 1978 by the Temple Committee with help from villagers and devotees.
Temple Architect	:	Sikhara style.

Offerings to God

(i) By Priest : 'Dal-Chapati', Rice, Vegetables.

(ii) By People : 'Lachidana', Fruits, Sweets.

Off. to People (Prasadam) : 'Lachidana'.

Fairs and Festivals : Vishwakarma Day, Sthapana Diwas, Janamashtmi.

Observations

(i) Courtyard Plants (Protected as well as for Landscaping : *Callistemon lanceolatus, Canna indica, Cassia glauca, Catharanthus roseus, Cedrela toona, Ficus religiosa, Grevillea robusta, Hymenocallis littoralis, Lagerstroemia indica, Musa paradisiaca, Nerium odorum, Ocimum sanctum, Petunia hybrida, Polyalthia longifolia, Psidium guajava, Rosa multiflora, Tabernaemontana coronaria, Tagetes erecta, Tectona grandis, Thevetia neriifolia, Thuja orientalis.*

(ii) Predominant Plants of the Courtyard : *Thuja orientalis.*

(iii) Maintenance : Satisfactory.

(iv) Ethnobotanical

Sacred : *Ficus religiosa* and *Ocimum sanctum* worshipped as sacred plants.

Edible : Fruits of *Musa paradisiaca* and *Psidium guajava* edible.

Ornamental : *Callistemon lanceolatus, Canna indica, Cassia glauca, Catharanthus roseus, Grevillea robusta, Hymenocallis littoralis, Lagerstroemia indica, Nerium odorum, Petunia hybrida, Polyalthia longifolia, Rosa multiflora, Tabernaemontana coronaria, Tagetes erecta, Thevetia neriifolia* and *Thuja orientalis* grown as ornamentals.

Other Uses : Wood of *Cedrela toona* and *Tectona grandis* excellent for construction and furniture purposes.

Remarks : No sacrifices allowed in the sanctorum.

19. Yamuna Temple (Pl. 20E)

Main Deity : Radha-Krishana.

Other Deities : Shiva, Durga, Hanuman.

Tot. Ar. of Tem. Courtyard : About 1500 sq m.

Location

(i) Place

Town : Paonta Sahib

Block : Paonta Sahib

Tehsil	:	Paonta Sahib
District	:	Sirmaur
(ii) Approach	:	Located on the bank of river Yamuna in the town.
(iii) Altitude	:	550 m.
Management	:	Temple committee.
Historical Background	:	The temple was built several hundred years ago. It is believed that the war of Kurukshetra, led by Lord Krishana, started from this place.
Temple Architect	:	Sikhara style.
Offerings to God		
(i) By Priest	:	Milk, 'Halva'.
(ii) By People	:	'Halva', Puffed rice, 'Lachidana', Fruits, Flowers.
Off. to People (Prasadam)	:	Puffed rice, 'Lachidana', Flowers, Fruits.
Fairs and Festivals	:	Janamashtmi, Shivratri, Navratra, Bawan Dwadshi.
Observations		
(i) Courtyard Plants (Protected as well as for Landscaping	:	*Aegle marmelos, Ficus religiosa, Musa paradisiaca, Ocimum sanctum.*
(ii) Predominant Plants of the Courtyard	:	*Ocimum sanctum.*
(iii) Maintenance	:	Excellent.
(iv) Ethnobotanical		
Sacred and Associated Myth	:	Leaves of *Aegle marmelos* especially offered to appease Lord Shiva. It is believed that 33 crore deities reside on the Peepal tree. Therefore it is worshipped and considered sacred.
Edible	:	Fruits of *Aegle marmelos* and *Musa paradisiaca* edible.
Remarks	:	No sacrifices permitted inside.

Temples of District Solan

1. Baduwada Temple (Pl. 21A)

Main Deity	:	Sehdev.
Tot. Ar. of Tem. Courtyard	:	About 1000 sq m.
Location		
(i) Place		
Town	:	Dadlaghat
Block	:	Kunihar
Tehsil	:	Arki
District	:	Solan
(ii) Approach	:	Located in the town.
(iii) Altitude	:	1500 m.
Management	:	Temple committee.
Historical Background	:	The temple was built about 5 years back though this place worshipped for the last 200 years
Temple Architect	:	Sikhara style.
Offerings to God		
(i) By Priest	:	'Halva'.
(ii) By People	:	'Halva', Coconut.
Off. to People (Prasadam)	:	'Halva', Wheat grains.
Fairs and Festivals	:	Sakranti.

Observations

(i) Courtyard Plants (Protected as well as for Landscaping	:	*Eucalyptus tereticornis, Dendrophthoe falcata.*
(ii) Maintenance	:	Satisfactory.
(iii) Ethnobotanical		
Sacred	:	*Dendrophthoe falcata* regarded as sacred and worshipped by the locals.
Medicinal	:	Decoction comprising of a leaf of *Eucalyptus tereticornis* with one-fifth part of fruit of 'Kakadan' (*Pistacia integerrima*), 3-5g 'Adrak' (*Zingiber officinale*), 2 black pepper, rock salt and jaggery in 2 glasses of water prescribed for cold and cough.
Remarks	:	No sacrifices permitted. The locals consider the deity as 'Kuldevta'. Whenever wishes are fulfilled, people invite the deity to their homes and worship it with all reverence.

2. Bhadrakali Temple (Jakholi Mata Temple) (Pl. 21B)

Main Deities	:	Goddess Bhadrakali, Durga, Goddess Mansa.
Other Deities	:	Shiva, Ganesha, Bhairav.
Tot. Ar. of Tem. Courtyard	:	About 800 sq m.
Location		
(i) Place		
Village	:	Jakholi
Block	:	Kunihar
Tehsil	:	Arki
District	:	Solan
(ii) Approach	:	6 km from Arki.
(iii) Altitude	:	890 m.
Management	:	In the name of Deity.
Historical Background	:	The inner (older) temple and the idols are from the time of Pandavas but the outer one was constructed in 1991 by locals.
Temple Architect	:	Sikhara style.
Offerings to God		
(i) By Priest	:	'Halva'.
(ii) By People	:	'Halva', Seasonal crops.

Off. to People (Prasadam)	:	'Halva, 'Lachidana'.
Fairs and Festivals	:	Navratra (Durga Ashtmi).
Observations		
(i) Courtyard Plants (Protected as well as for Landscaping	:	*Acacia catechu, Callistemon lanceolatus, Cupressus sempervirens, Ficus religiosa, Sapindus mukorossi.*
(ii) Maintenance	:	Satisfactory.
(iii) Ethnobotanical		
Sacred	:	*Ficus religiosa* worshipped as sacred tree.
Ornamental	:	*Callistemon lanceolatus, Cupressus sempervirens* and *Sapindus mukorossi* grown as ornamental plants.
Other Uses	:	Fruits of *Sapindus mukorossi* used as detergent to wash clothes. 'Katha' obtained from *Acacia catechu* is the main ingredient of 'Pan' prepared with betel leaves.
Remarks	:	Goats sacrificed by the people. Coconuts also offered to the deity.

3. Dev Dhar Temple (Pl. 21C)

Main Deity	:	Dev Dhar Devta.
Other Deity	:	Nar Singh.
Tot. Ar. of Tem. Courtyard	:	About 1600 sq m.
Location		
(i) Place		
Town	:	Arki
Block	:	Kunihar
Tehsil	:	Arki
District	:	Solan
(ii) Approach	:	40 km from Shimla.
(iii) Altitude	:	1050 m.
Management	:	Temple committee.
Historical Background	:	This place though worshipped for the last more than 100 years, the temple was built about 30 years back.
Temple Architect	:	Sikhara style.
Offerings to God		
(i) By Priest	:	'Halva-Rot'.
(ii) By People	:	'Halva-Rot', Seasonal crops.

Off. to People (Prasadam) : 'Halva'.

Fairs and Festivals : Chhamahi Fair (held after every six months) to celebrate the harvesting of crops.

Observations

(i) *Courtyard Plants (Protected as well as for Landscaping* : *Acacia catechu, Callistemon lanceolatus, Cynodon dactylon, Ficus religiosa, Grewia oppositifolia, Olea glandulifera, Pinus roxburghii, Syzygium cumini, Thuja orientalis.*

(ii) *Predominant Plants of the Courtyard* : *Pinus roxburghii.*

(iii) *Maintenance* : Satisfactory.

(iv) *Ethnobotanical*

Sacred : *Ficus religiosa* worshipped as sacred plant. *Cynodon dactylon* employed in various religious ceremonies.

Medicinal : Juice of Pine needles (*Pinus roxburghii*) prescribed to cure menstrual problems. Decoction of bark of *Syzygium cumini* given for purifying blood.

Edible : Fruits of *Syzygium cumini* and seed of *Pinus roxburghii* edible.

Ornamental : *Callistemon lanceolatus, Olea glandulifera, Pinus roxburghii,* and *Thuja orientalis* grown as ornamentals.

Other Uses : Leaves and branches of *Grewia oppositifolia* used as fodder. Wood of *Syzygium cumini* used for agricultural implements and packing cases. 'Katha' obtained from *Acacia catechu* used as the main ingredient for betel chewing.

Remarks : No sacrifices made.

4. Lutru Mahadev Temple (Pl. 21D)

Main Deity : Shiva.

Tot. Ar. of Tem. Courtyard : 1536 sq m.

Location

(i) *Place*

Town : Arki

Block : Kunihar

Tehsil : Arki

District : Solan

(ii) *Approach* : 40 km from Shimla.

(iii) *Altitude* : 1250 m.

Management : Local temple committee.

Historical Background : The temple believed to be very old. It is said that Pandavas meditated here at the time of their exile. Later, saint Lutru worshipped Lord Shiva at this place; and hence called as Lutru Mahadev temple.

Temple Architect : Cave temple.

Offerings to God

(i) By Priest : 'Halva'.

(ii) By People : Milk, Clarified butter, Flowers, Seasonal crops.

Off. to People (Prasadam) : 'Lachidana', Flowers, Fruits.

Fairs and Festivals : Shivratri.

Observations

(i) Courtyard Plants (Protected as well as for Landscaping : *Ficus virens, Sempervivum arboreum, Thevetia neriifolia.*

(ii) Maintenance : Satisfactory.

(iii) Ethnobotanical

Sacred : Flowers of *Thevetia neriifolia* considered sacred for offering to the deities (particularly to Lord Shiva). *Ficus virens* worshipped as sacred tree.

Ornamental : *Sempervivum arboreum* and *Thevetia neriifolia* commonly grown as ornamentals.

Remarks : Inside the cave there is a natural Shivlingam over which water drips continuously from the roof of cave.

5. Mutru Mahadev Temple (Pl. 21E)

Main Deity : Shiva.

Tot. Ar. of Tem. Courtyard : About 3000 sq m.

Location

(i) Place

Town : Arki

Block : Kunihar

Tehsil : Arki

District : Solan

(ii) Approach : 40 km from Shimla.

(iii) Altitude : 950 m.

Management : Local temple committee.

Historical Background	:	The Shiva temple is about 80 years old and built by locals whereas the caves containing the natural Shivlingam are more than one thousand years old. Long ago, saint Mutru meditated here for Lord Shiva, and the temple accordingly is known as Mutru Mahadev Temple.
Temple Architect	:	Sikhara style.
Offerings to God		
(i) By Priest	:	'Bhang', 'Datura', 'Ak' flowers, 'Bil Patri', Fruits.
(ii) By People	:	'Lachidana', Fruits, Flowers, Seasonal crops.
Off. to People (Prasadam)	:	'Lachidana', Fruits, Flowers.
Fairs and Festivals	:	Shivratri, Navratra.
Observations		
(i) Courtyard Plants (Protected as well as for Landscaping	:	*Aegle marmelos, Berberis lycium, Bombax ceiba, Bougainvillea spectabilis, Broussonetia papyrifera, Callistemon lanceolatus, Cannabis sativa, Canna indica, Clerodendrum philippinum, Cupressus sempervirens, Datura stramonium, Ficus glomerata, Ficus religiosa, Grevillea robusta, Leucaena leucocephala, Mangifera indica, Musa paradisiaca, Pistacia integerrima, Psidium guajava, Ricinus communis, Rosa indica, Roylea cinerea, Tagetes erecta, Thevetia neriifolia.*
(ii) Predominant Plants of the Courtyard	:	*Mangifera indica.*
(iii) Maintenance	:	Satisfactory.
(iv) Ethnobotanical		
Sacred	:	Leaves of *Aegle marmelos, Cannabis sativa,* and flowers of *Datura stramonium* offered to appease Lord Shiva. Flowers of *Clerodendrum philippinum* also offered to the deity. Wood and leaves of *Ficus glomerata* and leaves of *Mangifera indica* used in religious worshipping.
Medicinal	:	Thorns of *Bombax ceiba* kept over eyes to cure eye infection ('Tandali'). Crushed roots of *Berberis lycium* applied on eyes to cure eye disorders. Fruits of *Pistacia integerrima* considered good to cure cough. Latex of *Ficus glomerata* applied on cuts and wounds.
Edible	:	Fruits of *Aegle marmelos, Broussonetia papyrifera, Mangifera indica, Musa paradisiaca, Psidium guajava* and seeds of *Bombax ceiba* edible.
Ornamental	:	*Bombax ceiba, Bougainvillea spectabilis, Callistemon lanceolatus, Canna indica, Cupressus sempervirens, Grevillea*

		robusta, Rosa indica, Tagetes erecta and *Thevetia neriifolia* grown as ornamentals.
Other Uses	:	Stem of *Ricinus communis* used for scouring teeth. Leaves of *Broussonetia papyrifera* and *Leucaena leucocephala* used as an excellent fodder. Seed cotton of *Bombax ceiba* used for filling pillows and mattresses. Wood of *Broussonetia papyrifera* used for making cheap furniture.
Remarks	:	There is a natural Shivlingam inside the cave in the temple courtyard. Water dripping from the roof of cave fall continuously on the Shivlingam, said to possess healing properties.

6. Sai Baba Temple (Pl. 21F)

Main Deity	:	Shiridi Sai Baba.
Other Deities	:	Ganpati, Shiv-Parvati, Sita-Ram, Hanuman, Duttatrey, Durga, Radha-Krishana, Lakshmi-Narayan, Maha Lakshmi, Saraswati, Satya Narayan, Satya Sai Baba, Guru Nanak, Swami Vivekanand, Mahatma Buddh, Mahavir Swami, Swami Dayanand, Yishu Masih, Baba Balak Nath.
Tot. Ar. of Tem. Courtyard	:	About 5000 sq m.
Location		
(i) *Place*		
Village	:	Garkhal
Block	:	Dharampur
Tehsil	:	Kasauli
District	:	Solan
(ii) *Approach*	:	About 10 km from Dharampur.
(iii) *Altitude*	:	1550 m.
Management	:	Temple committee.
Historical Background	:	Earlier, there was only a Shiva temple at this place which was more than 100 years old. In 1978, the temple committee decided to construct a temple for Shiridi Sai Baba by the side of Shiva temple.
Temple Architect	:	Sikhara style.
Offerings to God		
(i) *By Priest*	:	Dal-Chapati, Rice-Vegetable (Milk Bhog at 9 A.M. and 9 P.M.).
(ii) *By People*	:	'Lachidana', Sugar Candy, Fruits, Sweets. 'Bil-patri' and 'Bhang' offered to appease Lord Shiva.

Off. to People (Prasadam)	:	'Lachidana', Sugar Candy.
Fairs and Festivals	:	Ram Navami (April), Guru Purnima (July), Death Anniversary of Shiridi Sai Baba (15 October), Shivratri.
Observations		
(i) Courtyard Plants (Protected as well as for Landscaping	:	*Aesculus indica, Asparagus densiflorus, Calendula officinalis, Callistemon lanceolatus, Canna indica, Catharanthus roseus, Cestrum nocturnum, Chrysanthemum indicum, Cynodon dactylon, Dahlia rosea, Ficus elastica, Ficus religiosa, Gomphrena globosa, Helianthus annuus, Hydrangea macrophylla, Malva sylvestris, Musa paradisiaca, Ocimum sanctum, Pelargonium hortorum, Petunia hybrida, Phaseolus vulgaris, Pisum sativum, Populus nigra, Prunus amygdalus, Prunus armeniaca, Prunus persica, Rosa indica, Salvia splendens, Tagetes erecta, Thuja orientalis, Zea mays.*
(ii) Predominant Plants of the Courtyard	:	*Tagetes erecta.*
(iii) Maintenance	:	Excellent.
(iv) Ethnobotanical		
Sacred and Associated Myth	:	*Ficus religiosa, Musa paradisiaca* and *Ocimum sanctum* worshipped as sacred plants. *Musa paradisiaca* considered symbol of Lord Brihaspati and *Ficus religiosa* as that of Lord Brahma, and worshipped on thursday and saturday, respectively. Flowers of the courtyard plants, particularly those of *Catharanthus roseus, Chrysanthemum indicum, Dahlia rosea, Rosa indica* and *Tagetes erecta* offered to the deity. *Cynodon dactylon* considered sacred and employed in all religious ceremonies.
Medicinal	:	Seeds oil of *Prunus amygdalus* used for body massage and as hair oil. Seeds considered good to enhance memory.
Edible	:	Pods of *Phaseolus vulgaris* and seeds of *Pisum sativum* cooked as vegetable. Grains of *Zea mays* used in the form of flour for making 'chapatis' and its immature cobs eaten after roasting. Fruits of *Aesculus indica, Musa paradisiaca, Prunus amygdalus, Prunus armeniaca, Prunus persica* edible.
Ornamental	:	*Aesculus indica, Asparagus densiflorus, Calendula officinalis, Callistemon lanceolatus, Canna indica, Catharanthus roseus, Cestrum nocturnum, Chrysanthemum indicum, Dahlia rosea, Ficus elastica, Gomphrena globosa, Helianthus annuus, Hydrangea macrophylla, Malva*

		sylvestris, Ocimum sanctum, Pelargonium hortorum, Petunia hybrida, Populus nigra, Rosa indica, Salvia splendens, Tagetes erecta and *Thuja orientalis* grown as ornamentals
Other Uses	:	*Zea mays* as a whole plant and leaves of *Aesculus indica* and *Populus nigra* employed as fodder. Wood of *Populus nigra* used for cheap furniture and as fuel.
Status of Courtyard Plants in Surrounding Areas	:	*Aesculus indica* is rare whereas other plants occur commonly in the surrounding areas.
Remarks	:	No sacrifices made to the deity.

7. Shakani Ganga Temple (Gangeshwar Mahadev Temple) (Pl. 21G)

Main Deity	:	Goddess Shakani.
Other Deities	:	Shiva, Hanuman, Santoshi Mata.
Tot. Ar. of Tem. Courtyard	:	About 5000 sq m.
Location		
(i) Place		
Town	:	Arki
Block	:	Kunihar
Tehsil	:	Arki
District	:	Solan
(ii) Approach	:	Located near the town.
(iii) Altitude	:	1100 m.
Management	:	Local temple committee.
Historical Background	:	About 50 years ago, Baba Sheelnath came and meditated here for some time. After his death, temple was constructed here in his memory about 35 years back by Baba Sukhdevanand Giri.
Temple Architect	:	Sikhara style.
Offerings to God		
(i) By Priest	:	'Halva-Rot'.
(ii) By People	:	'Halva', 'Lachidana', Flowers, Fruits.
Off. to People (Prasadam)	:	'Lachidana', Fruits.
Fairs and Festivals	:	Shivratri, Navratra, Birthday of Baba Sheelnath.
Observations		
(i) Courtyard Plants (Protected as well as for Landscaping	:	*Acacia catechu, Aloe vera, Callistemon lanceolatus, Casuarina equisetifolia, Cedrela toona, Citrus aurantifolia, Cupressus sempervirens, Emblica officinalis, Ficus religiosa, Mangifera*

indica, Melia azedarach, Olea glandulifera, Pinus roxburghii, Psidium guajava, Punica granatum, Pyrus communis, Salix babylonica, Salix tetrasperma, Syzygium cumini.

(ii) Predominant Plants of the Courtyard : *Acacia catechu.*

(iii) Maintenance : Excellent.

(iv) Ethnobotanical

Sacred : *Ficus religiosa* worshipped as sacred.

Medicinal : Cake prepared by mixing jelly of *Aloe vera* with wheat flour considered good for curing waist pain. 2-3g powdered seeds of *Syzygium cumini* taken twice daily with water for cough, stomachache and diabetes. Fruits of *Emblica officinalis* form one of the important constituents of 'Triphala' which is prescribed for stomach disorders. Seeds of *Punica granatum* used to prepare chutney which is good for diarrhoea.

Edible : Fruits of *Emblica officinalis, Mangifera indica, Psidium guajava, Punica granatum, Pyrus communis* and *Syzygium cumini* edible whereas that of *Citrus aurantifolia* pickled.

Ornamental : *Aloe vera, Callistemon lanceolatus, Casuarina equisetifolia, Cupressus sempervirens, Emblica officinalis, Pinus roxburghii, Salix babylonica* and *Salix tetrasperma* grown as ornamental plants.

Other Uses : Branches of *Acacia catechu* used as toothbrush. Its pods relished by sheep and goats. 'Katha' used for betel chewing. Cones of *Pinus roxburghii* used for decorative purposes. Wood of *Cedrela toona* excellent for construction and furniture whereas that of *Melia azedarach* used for building works and agricultural implements.

Status of Courtyard Plants in Surrounding Areas : *Casuarina equisetifolia* is absent in the surrounding areas whereas other plants are common.

Remarks : No sacrifices made to the deity.

8. Shiva Temple (Pl. 22A)

Main Deity : Shiva.

Other Deities : Hanuman, Vaishno Devi, Bhairav.

Tot. Ar. of Tem. Courtyard : 3000 sq m.

Location		
(i) Place		
Town	:	Dadlaghat
Block	:	Kunihar
Tehsil	:	Arki
District	:	Solan
(ii) Approach	:	About 50 km from Shimla.
(iii) Altitude	:	1500 m.
Management	:	In the name of Deity.
Historical Background	:	The temple is about 100 years old, and built by locals.
Temple Architect	:	Sikhara style.
Offerings to God		
(i) By Priest	:	'Lachidana', Fruits, Sweets.
(ii) By People	:	'Lachidana', Fruits, Sweets. Bil-patri, Bhang and Dhatura offered to Lord Shiva as special offering.
Off. to People (Prasadam)	:	'Lachidana', Fruits, Sweets.
Fairs and Festivals	:	Shivratri, Janamashtmi, Vijay Dashmi.
Observations		
(i) Courtyard Plants (Protected as well as for Landscaping	:	*Acer oblongum, Bauhinia variegata, Bougainvillea spectabilis, Callistemon lanceolatus, Canna indica, Cedrela toona, Citrus limon, Cupressus sempervirens, Cynodon dactylon, Datura stramonium, Eriobotrya japonica, Euphorbia pulcherrima, Ficus elastica, Ficus palmata, Ficus roxburghii, Jasminum primulinum, Juglans regia, Lagerstroemia indica, Leucaena leucocephala, Machilus gamblei, Malva sylvestris, Malva verticillata, Morus alba, Pinus roxburghii, Populus nigra, Prunus domestica, Pyrostegia venusta, Pyrus communis, Ricinus communis, Salix tetrasperma, Tropaeolum majus.*
(ii) Predominant Plants of the Courtyard	:	*Jasminum primulinum, Pyrostegia venusta.*
(iii) Maintenance	:	Satisfactory.
(iv) Ethnobotanical		
Sacred	:	Flowers of *Datura stramonium* offered to appease Lord Shiva. Fruits of *Pyrus communis* offered to the deity. *Cynodon dactylon* considered sacred and employed in religious ceremonies.

Medicinal	:	Decoction of leaves of *Lagerstroemia indica* effective against allergic disorders. Seed oil of *Ricinus communis* used for massage to cure joint pains. 5-10 ml decoction of bark of *Cedrela toona* prescribed twice daily for diarrhoea.
Edible	:	Flower buds of *Bauhinia variegata* cooked as vegetable. Fruits of *Citrus limon* used for pickling whereas those of *Eriobotrya japonica, Ficus palmata, Ficus roxburghii, Juglans regia, Morus alba, Prunus domestica* and *Pyrus communis* edible.
Ornamental	:	*Acer oblongum, Bauhinia variegata, Bougainvillea spectabilis, Callistemon lanceolatus, Canna indica, Cupressus sempervirens, Euphorbia pulcherrima, Ficus elastica, Ficus roxburghii, Jasminum primulinum, Lagerstroemia indica, Machilus gamblei, Malva sylvestris, Malva verticillata, Pinus roxburghii, Pyrostegia venusta, Ricinus communis* and *Tropaeolum majus* grown for ornamental purposes.
Other Uses	:	Leaves of *Bauhinia variegata, Leucaena leucocephala* and *Morus alba* used as fodder. Wood of *Salix tetrasperma* specially used for making 'Jungda' (used for ploughing the fields) and household articles whereas those of *Acer oblongum, Cedrela toona, Pinus roxburghii, Populus nigra* are used for furniture and construction purposes.
Status of Courtyard Plants in Surrounding Areas	:	Occurrence of *Acer oblongum* in the surrounding areas is rare whereas other plants occur commonly.
Remarks	:	No sacrifices permitted.

9. Shiva Temple (Pl. 22B)

Main Deities	:	Shiva, Parvati, Ganesha, Duttatrey.
Tot. Ar. of Tem. Courtyard	:	1536 sq m.
Location		
(i) *Place*		
Village	:	Jakholi
Block	:	Kunihar
Tehsil	:	Arki
District	:	Solan
(ii) *Approach*	:	6 km from Arki.
(iii) *Altitude*	:	890 m.
Management	:	In the name of Deity.

Historical Background	:	Though the place is worshipped since long back, the present temple was built by locals about 60 years back.
Temple Architect	:	Sikhara style.
Offerings to God		
(i) By Priest	:	'Lachidana', Flowers, Fruits.
(ii) By People	:	'Lachidana', Flowers, Fruits. 'Bil-patri', 'Bhang' and 'Dhatura' offered to appease Lord Shiva.
Off. to People (Prasadam)	:	'Lachidana'.
Fairs and Festivals	:	Shivratri, Durga Ashtmi.
Observations		
(i) Courtyard Plants (Protected as well as for Landscaping	:	*Artocarpus lakoocha, Cannabis sativa, Canna indica, Cedrela toona, Celtis australis, Dalbergia sissoo, Datura fastuosa, Ficus religiosa, Mangifera indica, Prunus domestica, Rosa indica, Salvia coccinia, Thevetia neriifolia, Thuja orientalis.*
(ii) Predominant Plants of the Courtyard	:	*Cedrela toona, Rosa indica.*
(iii) Maintenance	:	Satisfactory.
(iv) Ethnobotanical		
Sacred	:	Leaves and wood of *Mangifera indica* used in religious ceremonies. Leaves of *Cannabis sativa* and flowers of *Datura fastuosa* offered to appease Lord Shiva. *Ficus religiosa* considered sacred for worshipping.
Edible	:	Fruits of *Artocarpus lakoocha, Mangifera indica, Prunus domestica* and *Celtis australis* edible.
Ornamental	:	*Canna indica, Rosa indica, Salvia coccinia, Thevetia neriifolia* and *Thuja orientalis* grown as ornamentals.
Other Uses	:	Wood of *Cedrela toona* and *Dalbergia sissoo* excellent for construction and furniture.
Status of Courtyard Plants in Surrounding Areas	:	Occurrence of *Artocarpus lakoocha* (Diyun) is rare in the surrounding areas.
Remarks	:	No sacrifices permitted inside.

10. Shiva Temple (Pl. 22C)

Main Deity	:	Shiva.
Other Deity	:	Swami Krishananand Paramhans.
Tot. Ar. of Tem. Courtyard	:	About 5000 sq m.

Location

(i) Place

Village : Jatoli

Block : Solan

Tehsil : Solan

District : Solan

(ii) Approach : About 7 km from Solan.

(iii) Altitude : 1420 m.

Management : Temple committee.

Historical Background : The temple about 25 years old was built by Swami Krishananand Paramhans. He came here in 1950 and believed to possess supernatural powers by which he used to cure various ailments of people. It was his keen desire to build a Shiva temple at this sacred place for which he requested people to donate money. Consequently, the work started on 10th July, 1983.

Temple Architect : Sikhara style.

Offerings to God

(i) By Priest : Dal-Rice, Chapati.

(ii) By People : Fruits, Flowers, Sweets.

Off. to People (Prasadam) : Puffed rice, 'Lachidana'.

Fairs and Festivals : Janamashtmi, Shivratri.

Observations

(i) Courtyard Plants (Protected as well as for Landscaping : *Bougainvillea spectabilis, Callistemon lanceolatus, Cedrela toona, Celtis australis, Cestrum nocturnum, Chrysanthemum indicum, Cupressus torulosa, Dahlia rosea, Ficus roxburghii, Ficus virens, Grevillea robusta, Hemerocallis fulva, Hibiscus syriacus, Impatiens balsamina, Jacaranda acutifolia, Jasminum primulinum, Lespedeza sericea, Melia azedarach, Musa paradisiaca, Tagetes erecta, Tithonia tagetiflora, Zinnia elegans.*

(ii) Predominant Plants of the Courtyard : *Callistemon lanceolatus, Cupressus torulosa, Tagetes erecta.*

(iii) Maintenance : Excellent.

(iv) Ethnobotanical

Sacred : Flowers of *Bougainvillea spectabilis, Chrysanthemum indicum, Dahlia rosea, Hemerocallis fulva, Hibiscus syriacus, Jasminum primulinum, Tagetes erecta* considered sacred

		and offered to the deity. *Musa paradisiaca* worshipped for religious purposes.
Medicinal	:	Vegetable prepared from unripe fruits of *Musa paradisiaca* considered good for dysentery.
Edible	:	Fruits of *Celtis australis, Ficus roxburghii* and *Musa paradisiaca* are edible.
Ornamental	:	*Bougainvillea spectabilis, Callistemon lanceolatus, Cestrum nocturnum, Chrysanthemum indicum, Cupressus torulosa, Dahlia rosea, Grevillea robusta, Hemerocallis fulva, Hibiscus syriacus, Impatiens balsamina, Jacaranda acutifolia, Jasminum primulinum, Lespedeza sericea, Musa paradisiaca, Tagetes erecta, Tithonia tagetiflora* and *Zinnia elegans* grown as ornamentals.
Other Uses	:	Leaves of *Celtis australis* used as fodder. Wood of *Celtis australis* and *Melia azedarach* considered good for agricultural implements whereas that of *Cedrela toona* excellent for construction and furniture.
Status of Courtyard Plants in Surrounding Areas	:	Excepting *Ficus virens*, all other plants occur commonly in the surrounding areas.
Remarks	:	No sacrifices permitted.

11. Shri Sidh Baba Balak Nath Temple (Pl. 22D)

Main Deity	:	Baba Balak Nath.
Other Deities	:	Shiva, Kali, Vaishno Devi, Hanuman.
Tot. Ar. of Tem. Courtyard	:	2300 sq m.
Location		
(i) Place		
Town	:	Garkhal
Block	:	Dharampur
Tehsil	:	Kasauli
District	:	Solan
(ii) Approach	:	About 10 km from Dharampur.
(iii) Altitude	:	1650 m.
Management	:	Temple committee.
Historical Background	:	Long ago, Baba Balak Nath stayed here to take rest for a while during his journey to Tirth Dham. Since then, this place is considered sacred.
Temple Architect	:	Sikhara style.

Offerings to God

(i) *By Priest* : 'Rot', 'Halva', Coconut, Milk, Dry Fruits.

(ii) *By People* : 'Rot', 'Halva', Coconut, 'Lachidana'.

Off. to People (Prasadam) : 'Lachidana'.

Fairs and Festivals : Navratra (Twice in a year), Janamashtmi, Shivratri.

Observations

(i) *Courtyard Plants (Protected as well as for Landscaping* : *Aloe vera, Canna indica, Catharanthus roseus, Cupressus sempervirens, Dahlia rosea, Duranta repens, Ficus benghalensis, Ficus religiosa, Jasminum primulinum, Musa paradisiaca, Nerium odorum, Populus nigra, Salvia coccinia, Syzygium cumini, Tagetes erecta, Thuja orientalis.*

(ii) *Predominant Plants of the Courtyard* : *Tagetes erecta.*

(iii) *Maintenance* : Excellent.

(iv) *Ethnobotanical*

Sacred : *Ficus religiosa* and *Musa paradisiaca* worshipped for all religious purposes.

Medicinal : Pulp of leaves of *Aloe vera* consumed daily to cure diabetes. Decoction of leaves of *Duranta repens* prescribed for diabetes.

Edible : Fruits of *Syzygium cumini* edible.

Ornamental : *Aloe vera, Canna indica, Catharanthus roseus, Cupressus sempervirens, Dahlia rosea, Duranta repens, Jasminum primulinum, Nerium odorum, Populus nigra, Salvia coccinia, Tagetes erecta* and *Thuja orientalis* grown as ornamentals.

Other Uses : Leaves of *Populus nigra* employed as fodder and its wood for cheap furniture and as fuel.

Status of Courtyard Plants in Surrounding Areas : *Ficus benghalensis* and *Ficus religiosa* are rare outside.

Remarks : No sacrifices permitted.

12. Shulini Mata Temple (Pl. 22E)

Main Deity : Shulini Devi (Durga).

Other Deity : Shirgul Dev.

Tot. Ar. of Tem. Courtyard : 768 sq m.

Location

(i) *Place*

Town : Solan

Block : Solan

Tehsil	: Solan
District	: Solan
(ii) Approach	: Located in the town.
(iii) Altitude	: 1430 m.
Management	: Temple committee.
Historical Background	: About 200 years back, King of Bhaghaat (old name of Solan) built the temple for Shulini Devi at this place. Recently, the temple has been renovated.
Temple Architect	: Sikhara style.
Offerings to God	
(i) By Priest	: Sweets, Fruits.
(ii) By People	: 'Lachidana', Sweets.
Off. to People (Prasadam)	: 'Lachidana', Sweets.
Fairs and Festivals	: Local Fair (held in June), Navratra (twice a year).
Observations	
(i) Courtyard Plants (Protected as well as for Landscaping	: *Canna indica, Jasminum primulinum.*
(ii) Predominant Plants of the Courtyard	: *Canna indica.*
(iii) Maintenance	: Satisfactory.
(iv) Ethnobotanical	
Sacred	: Flowers of *Canna indica* and *Jasminum primulinum* considered sacred and offered to the deity.
Ornamental	: *Canna indica* and *Jasminum primulinum* grown as ornamentals.
Remarks	: The town Solan got its name after the deity Shulini Devi. Goats sacrificed rarely.

13. Sidheshwari Temple (Pl. 22F)

Main Deity	: Sidheshwari Devi.
Other Deity	: Shiva.
Tot. Ar. of Tem. Courtyard	: About 1500 sq m.
Location	
(i) Place	
Town	: Solan
Block	: Solan

Tehsil	:	Solan
District	:	Solan
(ii) Approach	:	Located in the outskirts of the town.
(iii) Altitude	:	1370 m.
Management	:	Managed by the Police Department.
Historical Background	:	The temple was built in 1990 by Brigadier Kapil Mohan with the help of Police Department.
Temple Architect	:	Sikhara style.
Offerings to God		
(i) By Priest	:	Puffed rice, 'Lachidana', Fruits, Sweets, Flowers.
(ii) By People	:	Puffed rice, 'Lachidana', Fruits, Sweets, Flowers.
Off. to People (Prasadam)	:	Puffed rice, 'Lachidana', Fruits, Flowers.
Fairs and Festivals	:	Navratra (twice a year).
Observations		
(i) Courtyard Plants (Protected as well as for Landscaping	:	*Callistemon lanceolatus, Cedrela toona, Chrysanthemum indicum, Cosmos sulphureus, Cynodon dactylon, Dahlia rosea, Ficus religiosa, Jasminum primulinum, Musa paradisiaca, Pinus roxburghii, Populus nigra, Rosa indica, Tagetes ereca, Thuja orientalis, Zinnia elegans.*
(ii) Predominant Plants of the Courtyard	:	*Cosmos sulphureus, Pinus roxburghii, Tagetes erecta, Thuja orientalis.*
(iii) Maintenance	:	Excellent.
(iv) Ethnobotanical		
Sacred and Associated Myth	:	Flowers of *Chrysanthemum indicum, Cosmos sulphureus, Dahlia rosea, Jasminum primulinum, Rosa indica, Tagetes ereca* and *Zinnia elegans* offered to appease the deity. *Ficus religiosa* and *Musa paradisiaca* worshipped on saturday and thursday, respectively. *Cynodon dactylon* considered sacred and employed in religious ceremonies.
Edible	:	Fruits of *Musa paradisiaca* edible whereas unripe ones cooked as vegetable.
Ornamental	:	*Callistemon lanceolatus, Chrysanthemum indicum, Cosmos sulphureus, Dahlia rosea, Jasminum primulinum, Rosa indica, Tagetes ereca, Thuja orientalis* and *Zinnia elegans* grown as ornamentals.

Other Uses	:	Leaves of *Populus nigra* employed as fodder. Wood of *Cedrela toona* excellent for construction and furniture.
Status of Courtyard Plants in Surrounding Areas	:	*Ficus religiosa* is rare in the surrounding areas.
Remarks	:	Coconut (*Cocos nucifera*) and 'Petha' (*Cucurbita hispida*) sacrificed in place of animals.

Temples of District Una

1. Baba Achutanand Thakurdwara (Pl. 23A)

Main Deity	:	Swami Achutanand.
Other Deities	:	Shivlingam, Hanuman, Durga, Vishnu, Devi Bhagvati, Radha-Krishana, Ganesha, Saligram.
Tot. Ar. of Tem. Courtyard	:	About 8000 sq m.
Location		
(i) Place		
Village	:	Amlehad
Block	:	Gagret
Tehsil	:	Amb
District	:	Una
(ii) Approach	:	6 km from Mubarkpur.
(iii) Altitude	:	650 m.
Management	:	Temple trust.
Historical Background	:	About 250 years back, Swami Achutanand came to this place from Bengal and meditated here. Later, the temple was constructed.
Temple Architect	:	Sikhara style.

Offerings to God		
(i) By Priest		
Noon	:	Rice-Dal, Chapati-Vegetables.
Evening	:	Fruits, Sweets.
(ii) By People	:	Fruits, Sweets, Flowers.
Off. to People (Prasadam)	:	Puffed rice.
Fairs and Festivals	:	Sakranti, Shivratri, Janamashtmi, Rakhar Punya, Kusha Amavasya.
Observations		
(i) Courtyard Plants (Protected as well as for Landscaping	:	*Aegle marmelos, Araucaria columnaris, Capsicum annuum, Catharanthus roseus, Cestrum nocturnum, Chrysanthemum indicum, Citrus aurantifolia, Citrus limon, Colocacia esculenta, Cynodon dactylon, Dahlia rosea, Duranta repens, Ficus benghalensis, Ficus religiosa, Hibiscus rosa-sinensis, Jasminum sambac, Mangifera indica, Melia azadirachta, Mentha piperita, Mirabilis jalapa, Murraya paniculata, Ocimum sanctum, Plumeria acutifolia, Psidium guajava, Pyrus communis, Rosa indica, Tabernaemontana coronaria, Tagetes erecta, Tamarindus indica, Thevetia neriifolia, Thuja orientalis, Zinnia elegans.*
(ii) Predominant Plants of the Courtyard	:	*Mangifera indica, Tabernaemontana coronaria, Tagetes erecta.*
(iii) Maintenance	:	Satisfactory.
(iv) Ethnobotanical		
Sacred	:	*Ficus religiosa* and *Ocimum sanctum* worshipped for all religious purposes. *Tamarindus indica* also considered sacred and as old as the temple. *Cynodon dactylon* employed in religious worshipping.
Medicinal	:	Fruit pulp of *Aegle marmelos* prescribed for gastrointestinal ailments. Leaves of *Jasminum sambac* chewed to cure mouth ulcers. 5-10 ml of leaf juice of *Melia azadirachta* considered good against blood disorders, skin ailments and as anthelmintic. Fruit pulp of *Tamarindus indica* boiled in water given for fever.
Edible	:	Leaves and underground corms of *Colocasia esculenta* cooked as vegetable. Raw pods of *Tamarindus indica* used for making Chutney for which leaves of *Mentha piperita* and fruits of *Capsicum annuum* used as flavorant. Fruits of *Aegle marmelos, Mangifera indica, Psidium guajava* and *Pyrus communis* edible whereas those of *Citrus aurantifolia* and *Citrus limon* pickled.

Ornamental : *Araucaria columnaris, Catharanthus roseus, Cestrum nocturnum, Chrysanthemum indicum, Colocacia esculenta, Dahlia rosea, Hibiscus rosa–sinensis, Jasminum sambac, Mirabilis jalapa, Murraya paniculata, Ocimum sanctum, Plumeria acutifolia, Rosa indica, Tabernaemontana coronaria, Tagetes erecta, Tamarindus indica, Thevetia neriifolia, Thuja orientalis* and *Zinnia elegans* considered as ornamentals whereas *Duranta repens* grown as hedges.

Other Uses : Leaves of *Mangifera indica* used for scouring teeth whereas its wood used for construction works. Wood of *Tamarindus indica* used for making domestic articles. *Cynodon dactylon* used as fodder.

2. Baba Barbhag Singh Gurudwara (Pl. 23B)

Main Deity : Baba Barbhag Singh.

Other Deity : Nahar Singh.

Tot. Ar. of Tem. Courtyard : About 6000 sq m.

Location

(i) *Place*

Village : Maidi

Block : Amb

Tehsil : Amb

District : Una

(ii) *Approach* : 10 km from Amb.

(iii) *Altitude* : 670 m.

Management : Managed by Intezamia Sewa Committee.

Historical Background : Long ago, Baba Bharbhag Singh meditated here under a 'Beri' tree for 18 years which is still present in the courtyard. The Gurudwara was later renovated in 1970 by Gurudwara Committee.

Temple Architect : Dome-shaped.

Offerings to God

(i) *By Priest* : 'Halva'.

(ii) *By People* : 'Halva', Puffed rice, 'Lachidana'.

Off. to People (Prasadam) : 'Halva', 'Lachidana'.

Fairs and Festivals : Hola-Mohalla (held in March for 15 days), Rakhad Puniya, Baisakhi.

Observations

(i) Courtyard Plants (Protected as well as for Landscaping	:	*Amaryllis reginae, Araucaria columnaris, Broussonetia papyrifera, Catharanthus roseus, Dahlia rosea, Ficus religiosa, Hibiscus rosa-sinensis, Hibiscus syriacus, Jasminum sambac, Jatropha panduraefolia, Lagerstroemia indica, Lannea coromandelica, Mangifera indica, Mirabilis jalapa, Nerium odorum, Rosa indica, Tabernaemontana coronaria, Thevetia neriifolia, Thuja orientalis, Zizyphus mauritiana.*
(ii) Predominant Plants of the Courtyard	:	*Hibiscus rosa-sinensis, Mirabilis jalapa.*
(iii) Maintenance	:	Excellent.
(iv) Ethnobotanical		
Sacred	:	*Ficus religiosa* and *Zizyphus mauritiana* worshipped as sacred plants.
Medicinal	:	Leaf paste of *Catharanthus roseus* applied to cure insect bites. Leaf juice of *Lagerstroemia indica* prescribed for fever. Poultice of leaves of *Lannea coromandelica* applied for sprain.
Edible	:	Fruits of *Broussonetia papyrifera, Mangifera indica* and *Zizyphus mauritiana* edible. Unripe fruits of *Mangifera indica* pickled.
Ornamental	:	*Amaryllis reginae, Araucaria columnaris, Catharanthus roseus, Dahlia rosea, Hibiscus rosa-sinensis, Hibiscus syriacus, Jasminum sambac, Jatropha panduraefolia, Lagerstroemia indica, Lannea coromandelica, Mirabilis jalapa, Nerium odorum, Rosa indica, Tabernaemontana coronaria, Thevetia neriifolia, Thuja orientalis* grown as ornamentals.
Other Uses	:	Leaves of *Broussonetia papyrifera* and *Zizyphus mauritiana* employed as cattle feed. Wood of *Mangifera indica* used for construction works.

3. Bagula Mukhi Temple (Pitambar Dham) (Pl. 23C)

Main Deity	:	Bagula Mukhi.
Other Deities	:	Saligram (form of Lord Vishnu), Ganesha, Durga. There is a separate small Shiva temple in front of the main temple.
Tot. Ar. of Tem. Courtyard	:	About 3000 sq m.

Location

(i) Place

Village	:	Kaloh
Block	:	Gagret
Tehsil	:	Amb
District	:	Una
(ii) Approach	:	2 km from Gagret towards Hoshiarpur.
(iii) Altitude	:	560 m.
Management	:	Privately owned.
Historical Background	:	The previous temple was very old but the new one is 5 years old, built by the owner of the temple, local people and different devotees.
Temple Architect	:	Sikhara style.

Offerings to God

(i) By Priest

Morning	:	'Halva', Sweets ('Besan', 'Ladoo'), Yellow Flowers
Noon	:	Food preparations devoid of onion and garlic.
(ii) By People	:	Sweets, Yellow clothes, Coconut, Puffed rice, 'Patasha', Yellow flowers, Fruits, 'Halva-Puri'. Bhang, 'Bil-patri' and Dhatura offered to Lord Shiva as special offering.
Off. to People (Prasadam)	:	Sweets, Puffed rice, 'Patasha'.
Fairs and Festivals	:	Navratra.

Observations

(i) Courtyard Plants (Protected as well as for Landscaping	:	*Bougainvillea spectabilis, Cannabis sativa, Celosia argentea* var. *cristata, Cynodon dactylon, Ficus benghalensis, Mirabilis jalapa, Ocimum sanctum, Pinus roxburghii, Punica granatum, Rosa indica, Tagetes erecta, Tecoma stans, Thevetia neriifolia, Thuja orientalis.*
(ii) Predominant Plants of the Courtyard	:	*Thuja orientalis.*
(iii) Maintenance	:	Excellent.

(iv) Ethnobotanical

Sacred	:	Yellow coloured flowers of *Tecoma stans, Thevetia neriifolia* and *Tagetes erecta* considered sacred for offering to the deity. *Ficus benghalensis* and *Ocimum sanctum* worshipped for all religious purposes. Leaves of *Cannabis sativa* offered to appease Lord Shiva. *Cynodon dactylon* employed in religious ceremonies as sacred plant.

Medicinal	: 5-10 ml leaf juice of *Ocimum sanctum* taken twice daily for high fever.
Edible	: Fruits of *Punica granatum* and seeds of *Pinus roxburghii* edible. Dried seeds of *Punica granatum* used in preparing Chutney.
Ornamental	: *Bougainvillea spectabilis, Celosia argentea* var. *cristata, Mirabilis jalapa, Pinus roxburghii, Rosa indica, Tagetes erecta, Tecoma stans, Thevetia neriifolia* and *Thuja orientalis* grown as ornamentals.
Remarks	: 'Petha' (*Cucurbita hispida*) and lemon (*Citrus aurantifolia*) offered in lieu of animals in the temple.

4. Bhadrakali Temple (Pl. 23D)

Main Deity	: Bhadrakali.
Other Deity	: Saligram.
Tot. Ar. of Tem. Courtyard	: About 1500 sq m.
Location	
(i) Place	
Village	: Bhadrakali
Block	: Gagret
Tehsil	: Amb
District	: Una
(ii) Approach	: 4 km from Doulatpur.
(iii) Altitude	: 630 m.
Management	: In the name of Deity
Historical Background	: The temple is believed to be constructed by Pandavas.
Temple Architect	: Sikhara style.
Offerings to God	
(i) By Priest	
Morning	: Sweets, Dry fruits.
Evening	: Dry fruits, Fruits, Sweets, Puffed rice.
(ii) By People	: Sweets, 'Halva', Puffed rice, 'Lachidana', Seasonal crops.
Off. to People (Prasadam)	: 'Halva', Puffed rice.
Fairs and Festivals	: Birthday festival of the deity (held in June), Navratra.

Observations

(i) Courtyard Plants (Protected as well as for Landscaping	:	*Capsicum annuum, Cedrela toona, Cestrum nocturnum, Diospyros Montana* var. *cordifolia, Mangifera indica, Momordica charantia, Morus alba, Tagetes erecta, Thevetia neriifolia.*
(ii) Predominant Plants of the Courtyard	:	*Diospyros montana* var. *cordifolia.*
(iii) Maintenance	:	Satisfactory.
(iv) Ethnobotanical		
Sacred	:	Flowers of *Tagetes erecta* and *Thevetia neriifolia* considered sacred for offerings.
Medicinal	:	Fruit juice of *Momordica charantia* effective against diabetes.
Edible	:	Fruits of *Mangifera indica* and *Morus alba* edible. Fruits of *Capsicum annuum* pickled and also used for flavouring.
Ornamental	:	*Cestrum nocturnum, Diospyros montana* var. *cordifolia, Tagetes erecta* and *Thevetia neriifolia* grown as ornamentals.
Other Uses	:	Leaves of *Morus alba* employed as fodder. Wood of *Cedrela toona* excellent for construction and furniture.
Remarks	:	No sacrifices permitted.

5. Dera Baba Rudranand (Pl. 23E)

Main Deities	:	Baba Rudranand, Lakshmi-Narayan.
Tot. Ar. of Tem. Courtyard	:	About 6000 sq m including a garden attached to the temple.
Location		
(i) Place		
Village	:	Nari
Block	:	Una
Tehsil	:	Una
District	:	Una
(ii) Approach	:	About 10 km from Una.
(iii) Altitude	:	500 m.
Management	:	In the name of deity.
Historical Background	:	The temple is about 150 years old.
Temple Architect	:	Dome-shaped.

Offerings to God

(i) By Priest

Morning : Fruits, Sweets.

Noon and Evening : 'Manni' (large Chapati made of wheat flour), Vegetarian food preparations.

(ii) By People : Puffed rice, Flowers, Fruits.

Off. to People (Prasadam) : 'Lachidana', Puffed rice, Fruits, 'Vibhuti' (sacred ash left after burning of wood in 'Dhoona') and 'Peepal Churi' (made of pieces of 'Rot' mixed in clarified butter and offered to five Peepel trees growing in the courtyard).

Fairs and Festivals : Panch Bhikh Puniya (in November), Guru Purnima, Holi.

Observations

(i) Courtyard Plants (Protected as well as for Landscaping : *Acalypha wilkesiana, Aegle marmelos, Aloe vera, Amaryllis reginae, Araucaria columnaris, Artocarpus heterophyllus, Beloperone guttata, Bryophyllum calycinum, Carica papaya, Canna indica, Capsicum annuum, Carissa carandas, Catharanthus roseus, Cedrela toona, Celosia argentea* var. *cristata, Cestrum diurnum, Chrysanthemum indicum, Cinnamomum camphora, Citrus limon, Citrus reticulata, Coleus blumei, Colocasia esculenta, Cupressus sempervirens, Cynodon dactylon, Dahlia rosea, Duranta repens, Emblica officinalis, Eucalyptus tereticornis, Ficus elastica, Ficus religiosa, Hibiscus rosa-sinensis, Ipomoea quamoclit, Jasminum humile, Jasminum sambac, Juniperus chinensis, Kochia trichophylla, Lagerstroemia indica, Malvaviscus conzatii, Mangifera indica, Melia azadirachta, Melia azedarach, Mimusops elengi, Mirabilis jalapa, Musa paradisiaca, Nerium odorum, Ocimum sanctum, Pedilanthus tithymaloides, Plumeria acutifolia, Psidium guajava, Pterospermum acerifolium, Punica granatum, Pyrus communis, Quisqualis indica, Rosa indica, Sansevieria trifasciata, Tabernaemontana coronaria, Tagetes erecta, Tecoma radicans, Tecoma stans, Thevetia neriifolia, Thuja orientalis, Tropaeolum majus, Zephyranthes candida, Zephyranthes grandiflora, Zinnia elegans.*

(ii) Predominant Plants of the Courtyard : *Tagetes erecta.*

(iii) Maintenance : Excellent.

(*iv*) *Ethnobotanical*

Sacred and Associated Myth : *Ficus religiosa* and *Ocimum sanctum* worshipped as sacred plants. Wood of *Ficus religiosa* and *Cinnamomum camphora* used in ceremonial fires for performing religious rites. *Cynodon dactylon* employed in religious worshipping. Five Peepal trees (*Ficus religiosa*) in the temple courtyard considered sacred. It is believed that these trees emerged from the spots where five holy men took 'samadhi' about 500 years ago.

Medicinal : Juice of the leaves of *Melia azadirachta* prescribed as blood purifier. 3-4g powdered fruits of *Emblica officinalis* taken with water twice a day to check acidity. Poultice of leaves of *Bryophyllum calycinum* applied to cure furuncles. 2-3 g powdered stem bark of *Cinnamomum camphora* with banana prescribed for piles. 10g powdered stem bark of *Cinnamomum camphora* mixed with 100g coconut oil used for removing hair dandruff.

Edible : Fruits of *Aegle marmelos, Carica papaya, Carissa carandas, Citrus limon, Citrus reticulata, Emblica officinalis, Mangifera indica, Mimusops elengi, Musa paradisiaca, Psidium guajava, Punica granatum, Pyrus communis* edible. Leaves and underground corms of *Colocasia esculenta* used as vegetable. Leaves of *Aloe vera* and fruits of *Musa paradisiaca* (unripe ones) cooked as vegetable whereas those of *Artocarpus heterophyllus, Citrus limon* pickled.

Ornamental : *Acalypha wilkesiana, Aloe vera, Amaryllis reginae, Araucaria columnaris, Beloperone guttata, Bryophyllum calycinum, Canna indica, Carissa carandas, Catharanthus roseus, Celosia argentea* var. *cristata, Cestrum diurnum, Chrysanthemum indicum, Cinnamomum camphora, Coleus blumei, Colocasia esculenta, Cupressus sempervirens, Dahlia rosea, Duranta repens, Emblica officinalis, Ficus elastica, Hibiscus rosa-sinensis, Ipomoea quamoclit, Jasminum humile, Jasminum sambac, Juniperus chinensis, Kochia trichophylla, Lagerstroemia indica, Malvaviscus conzatii, Mirabilis jalapa, Nerium odorum, Ocimum sanctum, Pedilanthus tithymaloides, Plumeria acutifolia, Pterospermum acerifolium, Quisqualis indica, Rosa indica, Sansevieria trifasciata, Tabernaemontana coronaria, Tagetes erecta, Tecoma radicans, Tecoma stans, Thevetia neriifolia, Thuja orientalis, Tropaeolum majus, Zephyranthes candida, Zephyranthes grandiflora* and *Zinnia elegans* grown for ornamental purposes.

Other Uses : Leaves of *Pterospermum acerifolium* employed for making meal plates. Wood of *Cedrela toona* considered excellent for construction and furniture whereas that of *Melia azedarach* for building works and agricultural implements. *Cynodon dactylon* used as fodder. Wood of *Eucalyptus tereticornis* used for construction works.

Status of Courtyard Plants in Surrounding Areas : *Carissa carandas* and *Mimusops elengi* do not occur in the surrounding areas.

Remarks : Food preparations bereft of onion and garlic offered to the deity.

6. Dron Shiva Temple (Shiva Badi) (Pl. 23F)

Main Deity : Shiva.

Tot. Ar. of Tem. Courtyard : About 3000 sq m.

Location

(i) Place

Village : Ambota

Block : Gagret

Tehsil : Amb

District : Una

(ii) Approach : 2 km from Gagret towards Hoshiarpur.

(iii) Altitude : 500 m.

Management : In the name of deity.

Historical Background : Yajogita was the daughter of Guru Dronacharya. Long ago, she meditated at this place for Lord Shiva who finally was pleased with her devotion. Later, Guru Drona established 'Pindi' here.

Temple Architect : Sikhara style.

Offerings to God

(i) By Priest : 'Kasaar' (wheat flour, clarified butter and Sugar), Fruits.

(ii) By People : Fruits, Flowers. 'Ak', 'Bhang', 'Bil-patri', 'Dhatura' offered to appease Lord Shiva

Off. to People (Prasadam) : Fruits (generally Banana and Apple).

Fairs and Festivals : Shivratri, Loal Fair (held on 2nd Saturday after Baisakhi in April and celebrated as the day of arrival of Shiva), Sakranti.

Observations

(i) Courtyard Plants (Protected as well as for Landscaping	:	*Aegle marmelos, Broussonetia papyrifera, Carica papaya, Citrus jambhiri, Datura fastuosa, Datura innoxia, Ficus palmata, Ficus religiosa, Hibiscus rosa-sinensis, Kochia trichophylla, Lagerstroemia indica, Mallotus philippinensis, Mangifera indica, Mirabilis jalapa, Morus alba, Nyctanthes arbor-tristis, Ocimum sanctum, Quisqualis indica, Psidium guajava, Putranjiva roxburghii, Rosa indica, Tabernaemontana coronaria, Thevetia neriifolia, Zizyphus mauritiana.*
(ii) Predominant Plants of the Courtyard	:	*Mangifera indica, Putranjiva roxburghii, Broussonetia papyrifera.*
(iii) Maintenance	:	Excellent.
(iv) Ethnobotanical		
Sacred	:	Flowers of *Mirabilis jalapa, Rosa indica* and *Thevetia neriifolia* considered sacred for offering to the deity. Leaves of *Aegle marmelos* ('Bil-patri') and flowers of *Datura fastuosa* and *Datura innoxia* offered to appease Lord Shiva. *Ficus religiosa, Ocimum sanctum* and *Musa paradisiaca* worshipped as sacred plants.
Medicinal	:	Paste of seeds of *Putranjiva roxburghii* applied on forehead to check pain. Also, its 5 seeds given daily for one month to women for conception. Powdered fruits of *Mallotus philippinensis* added to 'Churan' to cure acidity.
Edible	:	Fruits of *Aegle marmelos, Broussonetia papyrifera, Carica papaya, Citrus jambhiri, Ficus palmata, Mangifera indica, Morus alba, Psidium guajava* and *Zizyphus mauritiana* edible. Unripe fruits of *Carica papaya* cooked as vegetable whereas those of *Mangifera indica* pickled and used for making chutney.
Ornamental	:	*Hibiscus rosa-sinensis, Kochia trichophylla, Lagerstroemia indica, Mirabilis jalapa, Nyctanthes arbor-tristis, Quisqualis indica, Rosa indica, Tabernaemontana coronaria* and *Thevetia neriifolia* grown as ornamentals.
Other Uses	:	Leaves of *Mangifera indica* and *Psidium guajava* used for scouring teeth whereas that of *Broussonetia papyrifera, Morus alba* and *Zizyphus mauritiana* given as fodder. Branches of *Zizyphus mauritiana* used for fencing. Wood of *Mallotus philippinensis* used for agricultural implements.
Remarks	:	Trees of *Ficus religiosa, Mangifera indica, Putranjiva roxburghii* and *Zizyphus mauritiana* are more than 100 years old.

7. Gugga Jahar Pir Temple (Pl. 23G)

Main Deity	:	Gugga Pir.
Tot. Ar. of Tem. Courtyard	:	About 600 sq m.
Location		
(i) Place		
Village	:	Chak Sarai
Block	:	Amb
Tehsil	:	Amb
District	:	Una
(ii) Approach	:	About 15 km from Amb.
(iii) Altitude	:	620 m.
Management	:	Managed by private owners.
Historical Background	:	The temple is believed to be very old.
Temple Architect	:	Pentroof style.
Offerings to God		
(i) By Priest	:	Puffed rice, 'Rot'.
(ii) By People	:	Puffed rice, 'Rot', Seasonal crops, Salt.
Off. to People (Prasadam)	:	Puffed rice.
Fairs and Festivals	:	Gugga Navami.
Observations		
(i) Courtyard Plants (Protected as well as for Landscaping	:	Nil.
(ii) Maintenance	:	Satisfactory.

8. Lakshmi Narayan Temple (Pl. 23H)

Main Deity	:	Lakshmi Narayan.
Other Deities	:	Shiva, Hanuman, Chamunda.
Tot. Ar. of Tem. Courtyard	:	About 3000 sq m.
Location		
(i) Place		
Village	:	Chowki Maniar
Block	:	Bangana
Tehsil	:	Bangana
District	:	Una

(ii) Approach	:	About 15 km from Amb.
(iii) Altitude	:	580 m.
Management	:	Managed by 'Jai Maa Mission'.
Historical Background	:	Built by locals about 8 years back.
Temple Architect	:	Flatroof type.
Offerings to God		
(i) By Priest		
Morning	:	Rice-Dal, Chapati-Vegetables, Milk, Clarified butter, Sugar.
Evening	:	Rice-Dal, Chapati-Vegetables.
(ii) By People	:	Puffed rice, Black gram, 'Halva', Sweets.
Off. to People (Prasadam)	:	Puffed rice, Kasaar.
Fairs and Festivals	:	Dussehra, Shivratri.
Observations		
(i) Courtyard Plants (Protected as well as for Landscaping	:	*Acacia nilotica, Acacia catechu, Araucaria columnaris, Calotropis procera, Cannabis sativa, Canna indica, Catharanthus roseus, Citrus aurantifolia, Citrus limon, Cynodon dactylon, Dalbergia sissoo, Datura innoxia, Eucalyptus tereticornis, Ficus religiosa, Hibiscus rosa-sinensis, Hibiscus syriacus, Jasminum sambac, Mirabilis jalapa, Ocimum sanctum, Pedilanthus tithymaloides, Psidium guajava, Rosa indica, Tabernaemontana coronaria, Tagetes erecta, Tectona grandis, Thevetia neriifolia, Thuja orientalis, Zizyphus mauritiana.*
(ii) Predominant Plants of the Courtyard	:	*Catharanthus roseus.*
(iii) Maintenance	:	Satisfactory.
(iv) Ethnobotanical		
Sacred	:	Leaves of *Cannabis sativa,* flowers of *Datura innoxia* and leaves and flowers of *Calotropis procera* offered to appease Lord Shiva. Stem of *Calotropis procera* used in purification ceremonies. *Ficus religiosa* and *Ocimum sanctum* worshipped as sacred plants. *Cynodon dactylon* employed in various religious ceremonies.
Medicinal	:	Bark decoction of *Acacia nilotica* used to cure diabetes and its branches as tooth brush to cure toothache. Decoction of roots of *Calotropis procera* prescribed for skin problems. Poultice of leaves of *Cannabis sativa* applied to cure body pains.

Edible : Fruits of *Psidium guajava* and *Zizyphus mauritiana* edible whereas that of *Citrus aurantifolia* and *Citrus limon* pickled.

Ornamental : *Acacia nilotica, Araucaria columnaris, Calotropis procera, Canna indica, Catharanthus roseus, Hibiscus rosa-sinensis, Hibiscus syriacus, Jasminum sambac, Mirabilis jalapa, Ocimum sanctum, Pedilanthus tithymaloides, Rosa indica, Tabernaemontana coronaria, Tagetes erecta, Tectona grandis, Thevetia neriifolia* and *Thuja orientalis* grown as ornamentals

Other Uses : 'Katha' (extracted from *Acacia catechu*) used for betel chewing. Pods of *Acacia nilotica* and *Acacia catechu* relished by sheep and goats. Wood of *Dalbergia sissoo* and *Eucalyptus tereticornis* used for furniture and building works.

Remarks : No sacrifices permitted in the sanctum sanctorum.

9. Lohara Temple (Pl. 23I)

Main Deities : Nag Devta, Bajrangbali, Baba Balak Nath, Jatheri Mata (a group of 4 temples).

Tot. Ar. of Tem. Courtyard : About 4000 sq m.

Location

(i) *Place*

Village : Lohara

Block : Amb

Tehsil : Amb

District : Una

(ii) *Approach* : 12 km from Amb.

(iii) *Altitude* : 700 m.

Management : In the name of Deity.

Historical Background : The temple was built in 1992 by Shri Jai Kishan Sharma.

Temple Architect : Sikhara style.

Offerings to God

(i) *By Priest* : Milk Bhog to Nag Devta, and Chapati, Clarified butter, Sugar to other deities. Besides, 'Halva' offered especially to Bajrangbali on Tuesday and 'Rot' to Baba Balak Nath on Sunday.

(ii) *By People* : Puffed rice, 'Patasha', Sweets, Fruits, Flowers.

Off. to People (Prasadam) : Sugar candy.

Fairs and Festivals : Nag Panchami, Navratra.

Observations

(i) *Courtyard Plants (Protected as well as for Landscaping* : *Achras zapota, Aegle marmelos, Alstonia scholaris, Amaryllis reginae, Asparagus adscendens, Asparagus racemosus, Bauhinia vahlii, Boerhaavia diffusa, Bougainvillea spectabilis, Bryophyllum calycinum, Capsicum annuum, Canna indica, Cassia fistula, Catharanthus roseus, Cedrela toona, Cestrum nocturnum, Chamaerops humilis, Chrysanthemum indicum, Citrus aurantifolia, Citrus reticulata, Clerodendrum splendens, Colocasia esculenta, Cycas revoluta, Cynodon dactylon, Dahlia rosea, Duranta repens, Euphorbia royleana, Ficus elastica, Grewia oppositifolia, Hibiscus rosa–sinensis, Jasminum primulinum, Jasminum sambac, Jatropha panduraefolia, Justicia gandarussa, Kochia trichophylla, Lagerstroemia indica, Lycopersicum esculentum, Mangifera indica, Melia azedarach, Mentha piperita, Mirabilis jalapa, Morus alba, Musa paradisiaca, Nerium odorum, Ocimum sanctum, Pedilanthus tithymaloides, Pinus roxburghii, Polyalthia longifolia, Populus nigra, Psidium guajava, Punica granatum, Quisqualis indica, Rosa indica, Sansevieria trifasciata, Syzygium cumini, Tabernaemontana coronaria, Tagetes erecta, Thuja orientalis, Zea mays, Zinnia elegans.*

(ii) *Predominant Plants of the Courtyard* : *Populus nigra, Mangifera indica, Chamaerops humilis.*

(iii) *Maintenance* : Excellent.

(iv) *Ethnobotanical*

Sacred : *Ocimum sanctum* considered sacred and worshipped on all religious ceremonies. *Cynodon dactylon* employed in religious worshipping.

Medicinal : Juice of roots of *Asparagus adscendens* prescribed to cure liver disorders. For easy delivery, roots of *Boerhaavia diffusa* kept over the navel of the pregnant lady for some time. 1-3g of its powdered leaves alongwith rock salt prescribed for curing stomachache. Also, the leaf paste applied to cure bee-sting. Raw roots of *Boerhaavia diffusa* given to livestock to increase appetite. Seed powder of *Cassia fistula* considered good as blood purifier. Decoction of its seeds given to cattle for constipation. 3-5 powdered seeds of *Syzygium cumini* taken with water in the morning everyday to cure diabetes. The milky latex of *Euphorbia royleana* applied to cure cuts and wounds.

Edible : Leaves and underground corms of *Colocasia esculenta* cooked as vegetable. As a flavorant, leaves of *Mentha piperita* used in chutney preparations. Green and dry fruits of *Capsicum annuum* used as spice in vegetables and pickles. Fruits of *Achras zapota, Aegle marmelos, Citrus reticulata, Mangifera indica, Morus alba, Musa paradisiaca, Psidium guajava, Punica granatum, Syzygium cumini* and seeds of *Pinus roxburghii* edible whereas fruits of *Citrus aurantifolia* pickled.

Ornamental : *Achras zapota, Alstonia scholaris, Amaryllis reginae, Asparagus adscendens, Asparagus racemosus, Bauhinia vahlii, Bougainvillea spectabilis, Bryophyllum calycinum, Canna indica, Cassia fistula, Catharanthus roseus, Cestrum nocturnum, Chamaerops humilis, Chrysanthemum indicum, Clerodendrum splendens, Cycas revoluta, Dahlia rosea, Euphorbia royleana, Ficus elastica, Hibiscus rosa-sinensis, Jasminum primulinum, Jasminum sambac, Jatropha panduraefolia, Justicia gandarussa, Kochia trichophylla, Lagerstroemia indica, Mirabilis jalapa, Nerium odorum, Ocimum sanctum, Pedilanthus tithymaloides, Polyalthia longifolia, Populus nigra, Quisqualis indica, Rosa indica, Sansevieria trifasciata, Tabernaemontana coronaria, Tagetes erecta, Thuja orientalis* and *Zinnia elegans* grown as ornamental plants whereas *Duranta repens* considered good for hedges.

Other Uses : Leaves of *Bauhinia vahlii* employed for making meal plates. Leaves of *Morus alba, Populus nigra* and young branches of *Grewia oppositifolia* used as fodder. Wood of *Melia azedarach* used for building works and agricultural implements, and that of *Cedrela toona* excellent for construction and furniture. *Euphorbia royleana* considered good for fencing purposes.

Remarks : No sacrifices permitted.

10. Mata Chintpurni Temple (Chhin Mastika Dham) (Pl. 23J)

Main Deity : Chintpurni (also called Chhinmastika).

Tot. Ar. of Tem. Courtyard : About 1500 sq m.

Location

(i) *Place*

Village : Chhaproh

Block : Amb

Tehsil	:	Amb
District	:	Una
(ii) Approach	:	20 km from Amb; 30 km from Jwalaji.
(iii) Altitude	:	940 m.
Management	:	Temple trust (Managed by Government since 1987).
Historical Background	:	The temple belongs to 15th-16th century A.D.
Temple Architect	:	Sikhara style.
Offerings to God		
(i) By Priest		
Morning	:	Black gram.
Noon	:	'Kheer', Rice-Dal, 'Puri', 'Halva'.
Evening	:	'Alu-Puri-Halva'.
(ii) By People	:	'Halva', Coconut, Puffed rice, 'Lachidana'.
Off. to People (Prasadam)	:	'Halva', Puffed rice, 'Lachidana'.
Fairs and Festivals	:	Fairs held during Navratra thrice a year ('Chaitra', 'Ashwin', 'Shravan'), Sawan Ashtmi (July-August).
Observations		
(i) Courtyard Plants (Protected as well as for Landscaping	:	*Ficus benghalensis.*
(ii) Maintenance	:	Excellent.
(iii) Ethnobotanical		
Sacred and Associated Myth	:	People tie sacred threads to the sacred tree of *Ficus benghalensis* for the fulfilment of their wishes.
Remarks	:	Sacrifices not permitted. Purification ceremonies held everyday in the morning.

11. Rama Temple (Pl. 24A)

Main Deity	:	Ram-Sita-Lakshman.
Other Deity	:	Radha-Krishana.
Tot. Ar. of Tem. Courtyard	:	About 1500 sq m.
Location		
(i) Place		
Village	:	Andora
Block	:	Amb
Tehsil	:	Amb
District	:	Una

(ii) Approach	:	About 3 km from Amb.
(iii) Altitude	:	700 m.
Management	:	In the name of deity.
Historical Background	:	The temple is more than 100 years old and built by locals.
Temple Architect	:	Sikhara style.
Offerings to God		
(i) By Priest	:	'Kasaar', Puffed rice, Fruits.
(ii) By People	:	'Kasaar', Puffed rice, Fruits, Seasonal crops.
Off. to People (Prasadam)	:	'Kasaar', 'Babru', Fruits.
Fairs and Festivals	:	Ram Navmi, Navratra, Janamashtmi.
Observations		
(i) Courtyard Plants (Protected as well as for Landscaping	:	*Catharanthus roseus, Ficus religiosa, Hibiscus rosa-sinensis, Jasminum sambac, Mangifera indica, Ocimum sanctum, Tagetes erecta.*
(ii) Predominant Plants of the Courtyard	:	*Ocimum sanctum.*
(iii) Maintenance	:	Satisfactory.
(iv) Ethnobotanical		
Sacred	:	*Ocimum sanctum* and *Ficus religiosa* worshipped for all religious activities.
Edible	:	Fruits of *Mangifera indica* edible.
Ornamental	:	*Catharanthus roseus, Hibiscus rosa–sinensis, Jasminum sambac* and *Tagetes erecta* grown as ornamentals.
Remarks	:	Sacrifices not permitted.

12. Sadashiv Temple (Sadashiv Dhiyunsar Mahadev Temple) (Pl. 24B)

Main Deity	:	Shivlingam.
Other Deities	:	Ganesha, Hanuman, Parvati, Durga.
Tot. Ar. of Tem. Courtyard	:	About 4500 sq m.
Location		
(i) Place		
Village	:	Bahi
Block	:	Bangana
Tehsil	:	Bangana
District	:	Una

(ii) Approach	:	About 26 km from Amb.
(iii) Altitude	:	870 m.
Management	:	Registered temple committee.
Historical Background	:	Long ago, Dhoumya Rishi (Purohit of Pandavas) meditated on Dhyunsar Hill. After his long worship, Lord Shiva was pleased and gave him a boon that the people visiting this place will get their wishes fulfilled. This place since then is worshipped as Sadashiv Dhyunsar Mahadev Temple. Later, the temple was renovated in 1948 by Shri Onkar Nand Giri (Session Judge).
Temple Architect	:	Sikhara style.
Offerings to God		
(i) By Priest	:	Chapati, Rice, Dal, Vegetables, 'Halva', 'Kheer'.
(ii) By People	:	'Halva', Flowers, Fruits, 'Bhang', 'Bil-Patri', 'Dhatura', Grains as well as flour of wheat, maize and rice.
Off. to People (Prasadam)	:	Sweets, 'Kasaar', Flowers, Fruits.
Fairs and Festivals	:	Shivratri, Janamashtmi, Local Fair (held on every Monday of Shravan Month).
Observations		
(i) Courtyard Plants (Protected as well as for Landscaping	:	*Aegle marmelos, Boerhaavia diffusa, Cannabis sativa, Capsicum annuum, Catharanthus roseus, Cynodon dactylon, Datura innoxia, Eucalyptus tereticornis, Ficus benghalensis, Ficus religiosa, Hibiscus rosa-sinensis, Jasminum sambac, Malvaviscus conzatii, Mangifera indica, Mentha piperita, Nerium odorum, Pedilanthus tithymaloides, Pinus roxburghii, Rosa indica, Salvia coccinia, Sesamum orientale, Tabernaemontana coronaria, Tagetes erecta, Tectona grandis, Thevetia neriifolia, Thuja orientalis.*
(ii) Predominant Plants of the Courtyard	:	*Datura innoxia.*
(iii) Maintenance	:	Excellent.
(iv) Ethnobotanical		
Sacred	:	Leaves of *Aegle marmelos, Cannabis sativa* and *Datura innoxia* offered to appease Lord Shiva. Seeds of *Sesamum orientale* used in purification ceremonies. *Ficus benghalensis* and *Ficus religiosa* worshipped as sacred plants. *Cynodon dactylon* employed in religious worshipping.

Medicinal : 3-5g powdered roots of *Boerhaavia diffusa* given with water daily in the morning on empty stomach to cure gall bladder stone. Seed oil of *Sesamum orientale* used for body massage to cure paralysis.

Edible : Fruits of *Aegle marmelos, Mangifera indica* and seeds of *Pinus roxburghii* edible. Unripe fruits of *Mangifera indica* pickled. Oil extracted from the seeds of *Sesamum orientale* used for cooking whereas roughage ('Khal') used as cattle feed. Leaves of *Mentha piperita* and fruits of *Capsicum annuum* used for flavouring various dishes.

Ornamental : *Catharanthus roseus, Eucalyptus tereticornis, Hibiscus rosa-sinensis, Jasminum sambac, Malvaviscus conzatii, Nerium odorum, Pedilanthus tithymaloides, Rosa indica, Salvia coccinia, Tabernaemontana coronaria, Tagetes erecta, Tectona grandis, Thevetia neriifolia* and *Thuja orientalis* grown as ornamentals.

Other Uses : Cones of *Pinus roxburghii* used for decoration purposes. Wood of *Eucalyptus tereticornis* good for construction works and furniture.

Status of Courtyard Plants in Surrounding Areas : *Tectona grandis* occurs infrequently in the surrounding areas.

Remarks : Food preparations without garlic and onion offered to the deity.

13. Santoshi Mata Temple (Pl. 24C, D)

Main Deity : Santoshi Mata.

Other Deity : Shiva.

Tot. Ar. of Tem. Courtyard : About 3000 sq m.

Location

(i) Place

Village : Khad

Block : Una

Tehsil : Una

District : Una

(ii) Approach : About 5 km from Gagret.

(iii) Altitude : 440 m.

Management : In the name of deity.

Historical Background : The temple was built in 1977 by locals.

Temple Architect : Sikhara style.

Offerings to God	
(i) By Priest	: Coconut, Puffed rice, 'Patasha', Jaggery, Gram, Flowers, Fruits.
(ii) By People	: Jaggery, Gram, Fruits, Sweets. 'Ak', 'Bhang', 'Bil-patri', Dhatura offered to appease Lord Shiva.
Off. to People (Prasadam)	: Jaggery, Gram, Fruits, Sweets.
Fairs and Festivals	: Navratra.
Observations	
(i) Courtyard Plants (Protected as well as for Landscaping	: *Acalypha wilkesiana, Aloe vera, Araucaria columnaris, Bougainvillea spectabilis, Bryophyllum calycinum, Celosia argentea* var. *cristata, Dahlia rosea, Datura stramonium, Diospyros Montana* var. *cordifolia, Ficus religiosa, Hibiscus rosa–sinensis, Melia azadirachta, Mimosa pudica, Mirabilis jalapa, Ocimum sanctum, Pelargonium hortorum, Rosa indica, Sansevieria trifasciata, Tabernaemontana coronaria, Tagetes erecta, Thevetia neriifolia, Thuja orientalis, Zinnia elegans.*
(ii) Predominant Plants of the Courtyard	: *Tagetes erecta.*
(iii) Maintenance	: Excellent.
(iv) Ethnobotanical	
Sacred and Associated Myth	: Flowers of *Thevetia neriifolia* and *Datura stramonium* offered especially to appease Lord Shiva. *Ficus religiosa* and *Ocimum sanctum* worshipped as sacred plants. People tie scarf and bengals to the tree of *Ficus religiosa* in the courtyard for fulfilment of their wishes.
Medicinal	: Decoction of leaves of *Melia azadirachta* considered good for skin disorders and that of *Ocimum sanctum* for fever and cough. Poultice of leaves of *Bryophyllum calycinum* prescribed to cure furuncles.
Edible	: Leaves of *Aloe vera* cooked as vegetable.
Ornamental	: *Acalypha wilkesiana, Araucaria columnaris, Bougainvillea spectabilis, Bryophyllum calycinum, Celosia argentea* var. *cristata, Dahlia rosea, Hibiscus rosa–sinensis, Mimosa pudica, Mirabilis jalapa, Pelargonium hortorum, Rosa indica, Sansevieria trifasciata, Tabernaemontana coronaria, Tagetes erecta, Thevetia neriifolia, Thuja orientalis* and *Zinnia elegans* grown as ornamentals.
Remarks	: No sacrifices allowed.

14. Saptdevi Sidh Peeth Ashram (Saptdevi Temple) (Pl. 24E)

Main Deities : 7 deities (Vaishno Devi, Jwalaji, Naina Devi, Chamunda Devi, Bajreshwari Devi, Kali, Chintpurni).

Other Deities : Shiva, Hanuman, Bhairav.

Tot. Ar. of Tem. Courtyard : About 3000 sq m.

Location

(i) Place

Village : Kinnu

Block : Amb

Tehsil : Amb

District : Una

(ii) Approach : 13 km from Amb; 5 km from Chintpurni.

(iii) Altitude : 720 m.

Management : In the name of deity.

Historical Background : The original temple though very old was renovated in 1983 by Mahant Yogi Suraj Nath.

Temple Architect : Sikhara style.

Offerings to God

(i) By Priest : Milk-Rot.

(ii) By People : 'Halva', 'Khir', 'Lachidana'.

Off. to People (Prasadam) : 'Lachidana'.

Fairs and Festivals : Local fair on 20th May (celebrated as 'Murti Visthapan Diwas').

Observations

(i) Courtyard Plants (Protected as well as for Landscaping : *Aegle marmelos, Aloe vera, Callistemon lanceolatus, Calotropis procera, Catharanthus roseus, Cedrela toona, Citrus jambhiri, Cynodon dactylon, Eucalyptus tereticornis, Ficus benghalensis, Ficus religiosa, Hibiscus rosa-sinensis, Jasminum sambac, Leucaena leucocephala, Mangifera indica, Mentha Piperita, Morus alba, Ocimum sanctum, Polyalthia longifolia, Prunus persica, Psidium guajava, Pyrus communis, Rosa indica, Santalum album, Tabernaemontana coronaria, Tamarindus indica, Thuja orientalis.*

(ii) Predominant Plants of the Courtyard : *Psidium guajava, Polyalthia longifolia.*

(iii) Maintenance : Excellent.

(iv) Ethnobotanical

Sacred and Associated Myth	:	Stem of *Calotropis procera* used in purification ceremonies. *Cynodon dactylon* employed in religious ceremonies as sacred plant. *Aegle marmelos, Ficus benghalensis, Ficus religiosa* and *Ocimum sanctum* considered sacred for religious worshipping. Newly married brides tie bengals and scarf to the branches of *Tabernaemontana coronaria* for their long, prosperous married life.
Medicinal	:	Decoction of roots and leaves of *Calotropis procera* used for asthma. Poultice of its roots applied to cure skin disorders and flowers chewed to remove drinking habit. Latex considered good for toothache and piles. Paste of sandalwood (*Santalum album*) applied to cure skin disorders. Sandalwood oil prescribed for urinary troubles.
Edible	:	Fruits of *Aegle marmelos, Citrus jambhiri, Mangifera indica, Prunus persica, Psidium guajava* and *Pyrus communis* edible. Leaves of *Aloe vera* cooked as vegetable. Raw pods of *Tamarindus indica* and leaves of *Mentha Piperita* used for making chutney.
Ornamental	:	*Callistemon lanceolatus, Catharanthus roseus, Eucalyptus tereticornis, Hibiscus rosa-sinensis, Jasminum sambac, Ocimum sanctum, Polyalthia longifolia Rosa indica, Santalum album, Tabernaemontana coronaria* and *Thuja orientalis* grown as ornamentals.
Other Uses	:	Branches of *Leucaena leucocephala* and leaves of *Morus alba* considered excellent fodder. Wood of *Cedrela toona* and *Eucalyptus tereticornis* employed for construction and furniture.
Status of Courtyard Plants in Surrounding Areas	:	*Santalum album* does not occur in the surrounding areas whereas *Aegle marmelos, Calotropis procera* and *Ficus benghalensis* occur infrequently.
Remarks	:	Sacrifices in any form not permitted.

15. Shitla Mata Temple (Pl. 24F)

Main Deity	:	Shitla Mata.
Other Deities	:	Kali, Bhairav, Hanuman, Shiva, Saraswati.
Tot. Ar. of Tem. Courtyard	:	About 1500 sq m.

Location	
(i) Place	
Village	: Baret
Block	: Amb
Tehsil	: Amb
District	: Una
ii) Approach	: 5 km from Chintpurni.
iii) Altitude	: 850 m.
Management	: Village temple committee.
Historical Background	: This place though worshipped for the last 400 years but the temple was built by locals about 50 years back.
Temple architect	: Sikhara style.
Offerings to God	
(i) By Priest	
Noon	: 'Kheer', Fruits, Sweets.
Evening	: Puffed rice, 'Lachidana'.
(ii) By People	: Puffed rice, 'Lachidana', Fruits, Sweets. 'Rot' and Butter milk especially offered to the deity. 'Bhang', 'Bil-patri' and Dhatura offered to appease Lord Shiva.
Off. to People (Prasadam)	: Puffed rice, 'Lachidana'.
Fairs and Festivals	: Navratra.
Observations	
(i) Courtyard Plants (Protected as well as for Landscaping	: *Citrus reticulata, Ficus benghalensis, Ficus religiosa, Ocimum sanctum, Thuja orientalis.*
(ii) Predominant Plants of the Courtyard	: *Ficus benghalensis.*
(iii) Maintenance	: Excellent.
(iv) Ethnobotanical	
Sacred	: *Ficus benghalensis, Ficus religiosa* and *Ocimum sanctum* considered sacred and worshipped by the locals.
Edible	: Fruits of *Citrus reticulata* edible.
Ornamental	: *Thuja orientalis* grown as an ornamental.
Remarks	: Sacrifices not permitted. Purification ceremonies seldom held.

16. Shri Hanuman Temple (Pl. 24G)

Main Deity	:	Hanuman.
Other Deity	:	Shiva.
Tot. Ar. of Tem. Courtyard	:	About 8000 sq m.
Location		
(i) Place		
Village	:	Thatthal
Block	:	Amb
Tehsil	:	Amb
District	:	Una
(ii) Approach	:	8 km from Amb.
(iii) Altitude	:	450 m.
Management	:	Local temple committee.
Historical Background	:	Built by locals about 40 years back.
Temple Architect	:	Sikhara style.
Offerings to God		
(i) By Priest		
Morning	:	Milk.
Noon	:	Rice-Dal, Chapati-Vegetables.
(ii) By People	:	Sweets, Fruits, 'Rot', 'Halva'. 'Bhang', 'Bil-patri' and Dhatura offered to appease Lord Shiva.
Off. to People (Prasadam)	:	Puffed rice, 'Rot', 'Halva'.
Fairs and Festivals	:	Ram-Navami, Shivratri, Sthapna Divas (13th May), Dussehra, Janamashtmi.
Observations		
(i) Courtyard Plants (Protected as well as for Landscaping	:	*Araucaria columnaris, Boerhaavia diffusa, Cannabis sativa, Canna indica, Capsicum annuum, Catharanthus roseus, Celosia argentea* var. *cristata, Chamaerops humilis, Citrus aurantifolia, Coleus blumei, Colocasia esculenta, Cynodon dactylon, Ficus elastica, Hibiscus rosa-sinensis, Malvaviscus conzatii, Mangifera indica, Ocimum sanctum, Psidium guajava, Rosa indica, Saccharum officinarum, Sansevieria trifasciata, Schefflera arboricola, Tabernaemontana coronaria, Tagetes erecta.*
(ii) Predominant Plants of the Courtyard	:	*Saccharum officinarum, Tagetes erecta.*
(iii) Maintenance	:	Satisfactory.

(iv) Ethnobotanical

Sacred	:	Leaves of *Cannabis sativa* offered to appease Lord Shiva. *Cynodon dactylon* and *Saccharum officinarum* considered sacred and employed in various religious ceremonies. Leaves of *Saccharum officinarum* used for making 'havan-vedika' at the time of purification ceremonies. *Ocimum sanctum* worshipped as sacred plant.
Medicinal	:	Tea brewed with leaves of *Ocimum sanctum* checks cold and fever. For improving eyesight, its leaves (7 in number) powdered with black pepper (5 in number) taken daily for a month. Fresh juice of leaves of *Tagetes erecta* checks earache and heels cuts. Crushed roots of *Boerhaavia diffusa* rubbed to cure backache and facilitates delivery if kept on navel of expecting women, and also given to cattle to cure fever. Its leaf paste also applied to cure insect bites. Juice of *Saccharum officinarum* considered effective against jaundice.
Edible	:	Leaves of *Colocasia esculenta* used in preparation of 'Sag' and for making 'Patoru'(snack). Fruits of *Mangifera indica* and *Psidium guajava* edible whereas that of *Citrus aurantifolia* pickled, and of *Capsicum annuum* used in the preparation of various dishes.
Ornamental	:	*Araucaria columnaris, Canna indica, Catharanthus roseus, Celosia argentea* var. *cristata, Chamaerops humilis, Coleus blumei, Colocasia esculenta, Ficus elastica, Hibiscus rosa-sinensis, Malvaviscus conzatii, Rosa indica, Sansevieria trifasciata, Schefflera arboricola, Tabernaemontana coronaria* and *Tagetes erecta* grown as ornamentals.
Other Uses	:	Wood of *Mangifera indica* used for construction works.
Remarks	:	Sacrifices not permitted.

Temple Plants of Himachal Pradesh

Acacia catechu **(Linn.f.) Willd. (Pl. 25A)**

Syn: *A. catechuoides* (Roxb.) Benth.; *A. polyacantha* Willd.; *A. sundra* (Roxb.) Bedd.; *A. wallichiana* DC; *Mimosa catechu* Linn.f.; *M. catechuoides* Roxb.

Family	:	Fabaceae.
Vernacular Name	:	Khair.
Eng., H. and Sans. Names	:	*Eng.*: Black Catechu, Catechu, Catechu Nigrum, Cutch, Pegu Catechu. *H.*: Katha, Khair, Khair-babul. *Sans.*: Dantadhavana, Jivamangalaya, Khadira, Vakrakantaka.
Distribution	:	Himalaya (Punjab to Sikkim); ascending to 1,500 m.
Habitat Ecology	:	Planted along hilly slopes, forests.
Reproductive Cycle	:	May-July.
Material Examined	:	EBH-TP-275.
Systematic Account	:	A moderate-sized tree with dark coloured, rough bark and 2-pinnate leaves. Leaflets 30-50 pairs, often ciliate; spines hooked in pairs. Flowers in peduncled axillary spikes. Calyx hairy outside. Corolla 2-3 times as long as the calyx. Pods stalked, shining and beaked at the apex.

Acacia dealbata **Link. (Pl. 25B)**

Syn: *A. affinis* Sweet; *A. decurrens* var. *dealbata* (Link.) Maiden; *A. derwentii* Siebert and Voss; *A. puberula* Dehnh.; *Racosperma dealbatum* (Link) Pedley.

Family	:	Fabaceae.
Vernacular Name	:	Pahadi Babul.
Eng. Name	:	*Eng.*: Silver Wattle.
Distribution	:	Shimla, Naini Tal and Almora Hills: 2,000-2,500 m.
Habitat Ecology	:	Commonly cultivated.
Reproductive Cycle	:	Winter, early spring.
Material Examined	:	EBH-TP-100.
Systematic Account	:	An evergreen tree upto 15 m tall and having silvery grey foliage. Leaves bipinnate, grey hoary, 8-13 cm long; leaflets small, linear. Flower heads small, yellow, in axillary and terminal panicles.

Acacia nilotica (L.) Willd. *ex* Delile subsp. *indica* (Benth.) Brenan (Pl. 25C, D)

Syn: *A. arabica* (Lam.) Willd. var. *indica* Benth.; *A. arabica* sensu Baker; *A. nilotica* (Linn.) Delile; *A. nilotica* var. *indica* (Benth.) A.F.Hill; *A. vera* Willd.; *Mimosa arabica* Lam.; *M. nilotica* Linn.

Family	:	Fabaceae.
Vernacular Names	:	Babul, Kikar.
Eng., H. and Sans. Names	:	*Eng.*: Indian Gum Arabic Tree. *H.*: Babul, Kikar. *Sans.*: Babula, Barbara, Sukshmapatra, Yugmakanta.
Distribution	:	Throughout India, ascending to 1,000 m.
Habitat Ecology	:	Drier regions.
Reproductive Cycle	:	July-September.
Material Examined	:	EBH-TP-701.
Systematic Account	:	A medium-sized, almost evergreen tree with a spreading crown, feathery foliage and zig-zag, pubescent branches. Leaves bipinnate with spinescent stipules. Leaflets 10-25 pairs, linear-oblong. Flowers golden-yellow, fragrant, crowded in axillary clusters of 2-5 heads. Pods stalked, grey-tomentose, flat.

Acalypha wilkesiana M. Arg. (Pl. 26A)

Syn: *A. amentacea* subsp. *wilkesiana* (Mull. Arg.) Fosberg; *A. circinata* A. Gray *ex* Seem.; *A. godseffiana* Mast.; *A. hamiltoniana* Briant; *A. illustris* Pax and K. Hoffm.; *A. macrophylla* Veitch; *A. tricolor* Seem.; *Ricinocarpus wilkesianus* (Mull. Arg.) Kuntze.

Family	:	Euphorbiaceae.
Vernacular Name	:	Satrangi.
Eng. Names	:	*Eng.*: Beefsteak Plant, Copper Leaf, Jacob's Coat, Match me if you can.
Distribution	:	A native of New Hebrides.
Habitat Ecology	:	Widely grown as a garden ornamental of cool climates.
Reproductive Cycle	:	Throughout the year.
Material Examined	:	EBH-TP-714.
Systematic Account	:	A handsome evergreen shrub possessing brownish-green leaves splashed with rose-pink, crimson, pink or brown and with toothed edges. Flowers reddish, in 15-20 cm slender spikes.

Acer oblongum **Wall. (Pl. 26B)**

Syn: *A. buzimpala* Hamilt.; *A. laurifolium* D. Don.

Family	:	Aceraceae.
Vernacular Names	:	Padang, Panjole.
Eng. Name	:	*Eng.*: Himalayan Maple.
Distribution	:	S.E. Asia: 1,000-2,100 m.
Habitat Ecology	:	Forests; sometimes planted in gardens as an ornamental tree.
Reproductive Cycle	:	February-April.
Material Examined	:	EBH-TP-1100.
Systematic Account	:	Evergreen trees. Leaves long-pointed, entire, glaucous beneath. Young shoots hairless. Flowers greenish-white, in a dense terminal branched hairy cluster. Fruits hairless, winged.

Achras zapota **Linn. (Pl. 26C)**

Syn: *A. sapota* Linn.; *A. dactylina* Lundell; *A. latiloba* Lundell; *A. mammosa* Linn.; *Calospermum mammosum* (Linn.) Pierre; *Manilkara zapota* (Linn.) Royen; *Sapota zapotilla* (Jacq.) Coville *ex* Safford; *Vitellaria mammosa* (Linn.) Radlk.

Family	:	Sapotaceae.
Vernacular Name	:	Chiku.
Eng. and H. Names	:	*Eng.*: Bully Tree, Sapodilla Plum, Sapota. *H.*: Chiku.
Distribution	:	Throughout the plains of India.
Habitat Ecology	:	Cultivated for its fruit.
Reproductive Cycle	:	December-January.
Material Examined	:	EBH-TP-703.
Systematic Account	:	Evergreen small trees with pubescent branches and 8-10 cm long, coriaceous leaves. Flowers axillary, solitary, white. Fruits a globose berry. Seeds black, embedded in a light-brown sweet pulp of delicious flavour.

Aegle marmelos **(Linn.) Correa (Pl. 27A)**

Syn: *Belou marmelos* (Linn.) Lyons; *Bilacus marmelos* (Linn.) Kuntze.; *Crataeva marmelos* Linn.; *C. religiosa* Ainslie; *Feronia pellucida* Roth.

Family	:	Rutaceae.
Vernacular Names	:	Bael, Bil.

Eng., H. and Sans. Names	: *Eng.*: Bael Fruit Tree, Bengal Quince, Holy Fruit, Indian Quince, Stone Apple.
	H.: Bael, Bel, Bilwa, Si-phal, Siriphal.
	Sans.: Balva, Bilva, Bilvaphalam, Malura, Shiva Druma, Sriphal, Vilva.
Distribution	: Throughout India ascending to 1,000 m; (wild or cultivated).
Habitat Ecology	: Planted. Wastelands near houses.
Reproductive Cycle	: April-July.
Material Examined	: EBH-TP-253.
Systematic Account	: A medium-sized deciduous tree armed with sharp axillary thorns and 3-foliate leaves. Flowers greenish-white, sweet-scented and in short axillary panicles. Fruits globose, rind woody. Seeds numerous, compressed and embedded in orange-coloured sweet pulp.

Aesculus indica **Colebr. (Pl. 27B)**

Family	: Hippocastanaceae.
Vernacular Names	: Akhnor, Goon, Kanor.
Eng. and H. Names	: *Eng.*: Indian Horse-chestnut.
	H.: Bankhor, Gugu, Kanor, Pankar.
Distribution	: W. Himalaya: 1,350-3,550 m.
Habitat Ecology	: Forest. Shady ravines.
Reproductive Cycle	: May-September.
Material Examined	: EBH-TP-266.
Systematic Account	: A large deciduous tree with glabrous terete branches and bark peeling off in long vertical strips. Leaves opposite, exstipulate; leaflets usually 7, unequal, distinctly petiolated. Flowers irregular, numerous, small and in pubescent cymes arranged in terminal panicles. Capsules ovoid, reddish brown, rough. Seeds dark brown, shining.

Albizia lebbeck **(Linn.) Benth. (Pl. 27C)**

Syn: *A. latifolia* Boivin; *Acacia lebbek* (Linn.) Willd.; *A. macrophylla* Bunge; *A. sirisa* Ham.; *A. speciosa* (Jacq.) Willd.; *Mimosa lebbeck* Linn.; *M. sirissa* Roxb.; *M. speciosa* Jacq.

Family	: Fabaceae.
Vernacular Names	: Sirin, Siris.

Eng., H. and Sans. Names	:	*Eng.*: East India Walnut, Lebbeck, Parrot Tree, Siris Tree, Sizzling Tree. *H.*: Garso, Kalshish, Kalsir, Lasrin, Mathirshi, Shiris, Shirish, Sirai, Sirar, Siras, Sirin, Siris, Sirsa, Tantia. *Sans.*: Barhapushpa, Bhandi, Bhandika, Kapitaka, Krishnasiris, Plavaga, Sirisha, Vishaghati, Vishahanta.
Distribution	:	Throughout India; ascending to 1,650 m.
Habitat Ecology	:	Frequently planted.
Reproductive Cycle	:	April-August.
Material Examined	:	EBH-TP-231.
Systematic Account	:	A tall deciduous tree having abruptly 2-pinnate leaves and a large gland near the base of the main petiole. Leaflets 8-18, rigidly subcoriaceous. Flowers in globose heads. Calyx funnel-shaped. Corolla greenish-yellow, twice the calyx. Stamens much longer than the corolla. Pods strap-shaped, 6-10 seeded.

Albizia stipulata (Roxb.) Boiv. (Pl. 27D)

Syn: *A. chinensis* (Osbeck) Merr.; *Acacia marginata* Ham.; *A. smithiana* Wall.; *A. stipulata* DC.; *Arthrosprion stipulatum* Hassk.; *Mimosa chinensis* Osbeck; *M. smithiana* Roxb.; *M. stipulacea* Roxb.

Family	:	Fabaceae.
Vernacular Name	:	Oyee.
Hindi Names	:	*H.*: Kanujera, Pattia, Samsundra, Siran.
Distribution	:	Throughout India; 300-1,300 m.
Habitat Ecology	:	Frequent along roadsides. Cultivated areas.
Reproductive Cycle	:	September-October.
Material Examined	:	EBH-TP-1000.
Systematic Account	:	Large-sized deciduous trees. Leaves abruptly 2-pinnate with many glands on the rachis. Leaflets 40-80; stipules broad, cordate. Flowers in small panicled heads; peduncles pubescent, solitary or in fascicles of 2-4. Calyx and corolla silky pubescent. Stamens tinged with red. Pods glabrous. Seeds 8-12, compressed, dark brown, smooth.

Alnus nitida (Spach) Endl. (Pl. 28A)

Syn: *Clethropsis nitida* Spach.

Family	:	Betulaceae.
Vernacular Name	:	Kosh.

Distribution : Kashmir–W. Nepal: 300-1,500 m.

Habitat Ecology : Riversides; common.

Reproductive Cycle : September-October.

Material Examined : EBH-TP-402.

Systematic Account : Trees with dark brown shining bark. Twigs usually hairless. Leaves usually toothed. Female cones mostly solitary, axillary. Nutlets with a narrow leathery margin.

Aloe vera **(Linn.) Burm. f. (Pl. 28B)**

Syn: *A. barbadensis* Miller; *A. chinensis* Steud. *ex* Baker; *A. elongata* Murray; *A. indica* Royle; *A. maculata* Forssk.; *A. variegata* Forssk.; *A. vulgaris* (Bauhin) Lam.

Family : Liliaceae.

Vernacular Names : Aloe, Quar, Quaria, Quar Patthal.

Eng., H. and Sans. Names : *Eng.*: Aloe, Barbados Aloe, Indian Aloe.

H.: Ghigvar, Ghi-kavar, Ghi-kumari, Gvarapatha, Kumari, Kuvarapatha.

Sans.: Ajara, Amara, Ambudhisrava, Bahupatri, Brahmagni, Ghirta-kumari, Kanya, Sukantaka, Tarani, Vipulasrava.

Distribution : Mediterranean-Planted in Indian gardens.

Habitat Ecology : Usually along field borders and slopes.

Reproductive Cycle : March-April.

Material Examined : EBH-TP-277.

Systematic Account : Perennial herbs with a short, thick stem and sessile, glaucous-green, lanceolate, erect-spreading, spiny-toothed, succulent leaves crowded in a basal rosette. Scape longer than the leaves. Racemes long, dense; bracts membranous. Flowers pendulous imbricated, yellow; anthers somewhat exserted.

Alpinia nutans **Rosc. (Pl. 28C)**

Syn: *A. speciosa* K. Schum.; *Zerumbet speciosum* Wendl.

Family : Zingiberaceae.

Vernacular Name : Kanchak.

Eng. Name : *Eng.*: The Light Galangal.

Distribution : Native of north-eastern India.

Habitat Ecology : Cultivated in gardens in many places in India.

Reproductive Cycle	:	May-August.
Material Examined	:	EBH-TP-524.
Systematic Account	:	Large-sized herbs possessing rhizomatous rootstock. Leaves oblong-lanceolate, finely pubescent beneath. Flowers in slightly compound cernuous panicles. Capsules red, globose.

Alstonia scholaris (Linn.) R. Br. (Pl. 28D)

Syn: *Echites pala* Ham. *ex* Spreng.; *E. scholaris* Linn.

Family	:	Apocynaceae.
Vernacular Name	:	Shaitan ka Jhad.
Eng., H. and Sans. Names	:	*Eng.*: Dita Bark Tree, Devil's Tree.
		H.: Chatiun, Saitankajhad, Satiun, Satni, Satwin.
		Sans.: Ayugmaparna, Brihattvaka, Devavriksha, Sarada, Vinada, Vishalatvaka, Yugmaparna.
Distribution	:	Throughout India.
Habitat Ecology	:	Often cultivated.
Reproductive Cycle	:	December-March.
Material Examined	:	EBH-TP-204.
Systematic Account	:	Trees with tall stem, bitter milky juice and branches spreading in tiers of whorls. Leaves coriaceous and in whorls of 5-10. Flowers greenish-white, in compact many-flowered pedunculate pubescent cymes. Follicles slender, numerous, pendulous.

Althaea rosea (Linn.) Cav. (Pl. 29A)

Syn: *Alcea rosea* Linn.

Family	:	Malvaceae.
Vernacular Name	:	Hollyhock.
Eng., H. and Sans. Names	:	*Eng.*: Althaea Rose, Hock Herb, Hollyhock, Round Dock.
		H.: Gul-e-khera, Gulkhairu, Gulkhirri.
		Sans.: Khatami.
Distribution	:	A native of Orient, China; planted in Indian gardens.
Habitat Ecology	:	Cultivated for its beautiful flowers; sometimes found as an escape.
Reproductive Cycle	:	August-September.
Material Examined	:	EBH-TP-300.

Systematic Account : Biennial herbs. Stem stout, hispid with fascicled branched hairs. Leaves long-petioled, acutely 5-7 lobed. Flowers short-pedicelled and in long terminal racemes. Involucre large, cup-shaped. Calyx 5-cleft, downy. Petals broad. Staminal tube short.

Amaryllis reginae **Linn. (Pl. 29B)**

Syn: *A. albertii* Lem.; *A. spectabilis* Lodd.; *Hippeastrum fulgidum* Herb.; *H. pulverulentum* Herb.; *H. reginae* (Linn.) Herb.

Family : Amaryllidaceae.

Vernacular Name : Lily.

Eng. Names : *Eng.*: Amaryllis, Mexican Lily.

Distribution : Native of Brazil and tropical America.

Habitat Ecology : Frequently cultivated for its attractive flowers.

Reproductive Cycle : Spring or early Summer.

Material Examined : EBH-TP-135.

Systematic Account : A bulbous plant upto 60 cm. Leaves basal, strap-shaped, bright green and about 4 cm wide. Flowers on smooth, naked stems; grouped, several together.

Antirrhinum majus **Linn. (Pl. 29C)**

Family : Scrophulariaceae.

Vernacular Name : Kutta Phool.

Eng. Names : *Eng.*: Dog Flower, Dragon's Mouth, Lion's Mouth, Snap Dragon.

Distribution : Native of Europe and Mediterranean region; planted in many parts of the world.

Habitat Ecology : Cultivated in various parts of India.

Reproductive Cycle : June-October.

Material Examined : EBH-TP-800.

Systematic Account : An erect bushy plant upto 90 cm tall. Flowers purplish red or white in colour and in terminal racemes.

Araucaria columnaris **(Forster) Hooker (Pl. 29D)**

Syn: *A. cookii* R. Br. *ex* Lindl.; *A. excelsa* (Lamb.) R. Br.; *A. intermedia* R. Br. *ex* Vieill.; *Columbea excelsa* (Lamb.) Spreng.; *Eutacta excelsa* (Lamb.) Link; *E. humilis* Carriere.

Family : Araucariaceae.

Vernacular Name : Aurocaria.

Eng. Name	:	*Eng.*: Fountainhead Tree.
Distribution	:	Native of New Caledonia; introduced into Indian gardens.
Habitat Ecology	:	A decorative plant in tubs, gardens, corners, and rockeries.
Reproductive Cycle	:	September-October.
Material Examined	:	EBH-TP-519.
Systematic Account	:	A slender columnar tree with older branches attached horizontally to the main trunk. Adult leaves densely imbricated, short and ovate-obtuse.

Aristolochia elegans Mast. (Pl. 30A)

Syn: *A. littoralis* Parodi.

Family	:	Aristolochiaceae.
Vernacular Name	:	Battak Phul.
Eng. Name	:	*Eng.*: Calico-flower.
Distribution	:	A native of Brazil. Widely distributed over the temperate and warmer regions of the world.
Habitat Ecology	:	An ornamental plant found in Indian gardens.
Reproductive Cycle	:	Blooms throughout the summer.
Material Examined	:	EBH-TP-274.
Systematic Account	:	Slender twining climbers having reniform-cordate leaves with rounded basal lobes. Flowers solitary, long-stalked with a short, yellow-green tube and a kidney-shaped limb, white with purple veining outside and purple-brown inside.

Artemisia moorcroftiana Wall. *ex* DC. (Pl. 30B)

Syn: *A. wallichiana* Besser.

Family	:	Asteraceae.
Vernacular Name	:	Sharkhamba.
Distribution	:	Pakistan–E. Nepal: 2,800-5,500 m.
Habitat Ecology	:	Open slopes.
Reproductive Cycle	:	August-October.
Material Examined	:	EBH-TP-1217.
Systematic Account	:	Tall herbs upto 60 cm. Leaves usually pale hairy beneath. Flower-heads often reddish, stalkless, drooping and usually in short dense lateral clusters. Involucral bracts papery, with cob-web hairs.

Artocarpus heterophyllus Lam. (Pl. 30C)

Syn: *A. integrifolia* Linn.; *Polyphema jaca* Lour.; *Sitodium cauliflorum* Gaertn.

Family	:	Moraceae.
Vernacular Name	:	Kattal.
Eng., H. and Sans. Names	:	*Eng.*: Jaca Tree, Jack-fruit Tree, Jack Wood.
		H.: Chakki, Kanthal, Kathal, Katol, Panas, Panasa.
		Sans.: Panasa, Panasah.
Distribution	:	Cultivated throughout the hotter parts of India.
Habitat Ecology	:	Cultivated.
Reproductive Cycle	:	Flowers during the winter (December-March) and fruits during the rainy season (July-August).
Material Examined	:	EBH-TP-502.
Systematic Account	:	A large evergreen tree. Leaves thickly coriaceous; petiole slender; stipules glabrous. Male receptacles without bracteoles; sepals 2. Fruiting receptacles attaining enormous size, tubercled.

Artocarpus lakoocha Roxb. (Pl. 30D)

Syn: *A. lacucha* Buch.-Ham.; *A. mollis* Wall.

Family	:	Moraceae.
Vernacular Name	:	Dheu.
Eng., H. and Sans. Names	:	*Eng.*: Barhal, Dahve, Lakuch, Monkey Jack.
		H.: Barhal.
		Sans.: Lakudi.
Distribution	:	Throughout the sub-tropical and tropical Himalaya upto 1,800 m.
Habitat Ecology	:	Occurs in evergreen, semi-evergreen and moist deciduous forests.
Reproductive Cycle	:	February-September.
Material Examined	:	EBH-TP-206.
Systematic Account	:	A tree upto 20 m with pinnatifid leaves and flowers aggregated in solitary axillary unisexual heads. Female heads green. Fruits round-lobulate, fleshy. Seeds roundish, creamy.

Asclepias curassavica **Linn. (Pl. 31A)**

Family	:	Asclepiadaceae.
Vernacular Names	:	Ghamera, Parvini.
Eng., H. and Sans. Names	:	*Eng.*: Blood Flower, Curassavian Swallow-wort, West Indian Ipecacuanha, Wild Ipecacuanha. *H.*: Kakatundi. *Sans.*: Kakatundi.
Distribution	:	Throughout India; naturalized throughout Himalaya to 1,500 m.
Habitat Ecology	:	Wasteplaces. Roadsides. Forest edges. Fields.
Reproductive Cycle	:	Flowers throughout the year.
Material Examined	:	EBH-TP-261.
Systematic Account	:	Erect undershrubs. Leaves opposite, thin, membranous, narrowed at both ends. Flowers in axillary umbels. Corolla orange-red, lobes reflexed; column stipitate. Coronal-scales 5, adnate to the stipe of the staminal column. Follicles solitary, tapering at both ends. Seeds ovoid, dark-brown.

Asclepias physocarpa **(E. Mey.) Schlecht. (Pl. 31B)**

Syn: *Gomphocarpus physocarpus* E. Mey.

Family	:	Asclepiadaceae.
Vernacular Name	:	Gol-gappa Phool.
Eng. Name	:	*Eng.*: Balloon Plant.
Distribution	:	Native of S. Africa.
Habitat Ecology	:	Planted as an ornamental.
Reproductive Cycle	:	April-July.
Material Examined	:	EBH-TP-272.
Systematic Account	:	Slender shrubs upto 2 m high and with white hairs. Leaves lanceolate, about 10 cm long. Flowers white, in axillary inflorescence. Corolla deeply lobed and often reflexed. Fruits round, inflated, yellowish green, with stiff hairs.

Asparagus adscendens **Roxb. (Pl. 31C)**

Syn: *A. satawur* James A. Murray; *A. volubilis* Wall.; *Asparagopsis adscendens* (Roxb.) Kunth.

Family	:	Liliaceae.
Vernacular Names	:	Sahastrapaud, Sanspaud, Satavari.

Hindi Names : *H.*: Hazarmuli, Satavar, Sufedmusli.

Distribution : N.W. Himalaya upto 1,600 m.

Habitat Ecology : Along edges of cultivation. Damp places.

Reproductive Cycle : March-June.

Material Examined : EBH-TP-722.

Systematic Account : A much-branched prickly shrub with white tuberous roots and tall, grooved stem. Spines stout, straight. Cladodes in tufts of 6-20, thread-like, curved. Flowers in many-flowered raceme; bracts minute. Berries 1-seeded.

Asparagus densiflorus (Kunth) Jessop (Pl. 31D)

Syn: *A. sprengeri* Regel; *Asparagopsis densiflorus* Kunth; *Protasparagus densiflorus* (Kunth) Oberm.

Family : Liliaceae.

Vernacular Name : Asparagus.

Eng. Name : *Eng.*: Asparagus.

Distribution : Native of South Africa.

Habitat Ecology : Cultivated as an ornamental plant in pots of loam based compost.

Reproductive Cycle : May-June.

Material Examined : EBH-TP-516.

Systematic Account : A tuberous-rooted plant with densely erect, plume-like branch and needle-like, rich green foliage. Flowers small, pinkish, fragrant.

Asparagus racemosus Willd. (Pl. 31E)

Syn: *A. acerosus* Roxb.; *A. jacquemontii* Baker; *A. racemosus* var. *javanicus* (Kunth) Baker; *A. saundersiae* Baker; *A. zeylanicus* (Baker) Hook.f.; *Asparagopsis abyssinica* Kunth; *A. acerosa* Kunth; *A. floribunda* Kunth; *A. javanica* Kunth; *Protasparagus racemosus* (Willd.) Oberm.

Family : Liliaceae.

Vernacular Name : Satavar.

Eng., H. and Sans. Names : *Eng.*: Indian Asparagus, Wild Asparagus.

H.: Chatwal, Satawar, Satawari, Satmuli, Shakakul.

Sans.: Atirasa, Bahusutaa, Indivari, Satamuli, Satavari, Svadurasa, Svetmuli.

Distribution	: Tropical and subtropical parts of India; ascending in Himalaya upto 1,500 m.
Habitat Ecology	: Found in the mixed forest.
Reproductive Cycle	: April-August.
Material Examined	: EBH-TP-721.
Systematic Account	: Tall climbing, much-branched, spiny shrubs with stems possessing strong, straight or recurved spines. Rootstock short; bearing numerous succulent tuberous roots. Cladodes borne in tufts of 2-6, curved. Leaves reduced to spines. Flowers white, fragrant, small, crowded in simple and branched racemes. Berries globose, red when ripe.

Bauhinia vahlii **Wight and Arn. (Pl. 32A, B)**

Syn: *B. racemosa* Vahl.; *Phanera vahlii* Benth.

Family	: Fabaceae.
Vernacular Names	: Malgen, Tor.
Eng. and H. Names	: *Eng.*: Camel's Foot Climber. *H.*: Jallur, Mahul, Malghan, Maljan, Maljhan, Malo, Malu, Maulein, Maurain.
Distribution	: Kashmir-Sikkim: upto1,500 m.
Habitat Ecology	: Forests. Rocky slopes.
Reproductive Cycle	: April-June.
Material Examined	: EBH-TP-513.
Systematic Account	: An enormous climber with widespreading stem, densely pubescent branchlets, circinate tendrils and rigidly subcoriaceous leaves. Racemes long-peduncled, terminal, densely tomentose, subcorymbose. Pods sublignose, rusty-velvety, 8-12 seeded.

Bauhinia variegata **Linn. (Pl. 32C, D)**

Syn: *B. candida* Roxb.; *B. chinensis* (DC.) Vogel; *B. decora* Uribe; *B. purpurea* Wall.; *Phanera variegata* (Linn.) Benth.

Family	: Fabaceae.
Vernacular Names	: Kachnar, Karale, Karayalae.
Eng., H. and Sans. Names	: *Eng.*: Budhist Bauhinia, Mountain Ebony, Orchid Tree. *H.*: Barial, Gurial, Gwiar, Kachnar, Kaniar, Khwairaal, Kural, Padrian.

	Sans.: Ashmantaka, Asphota, Chamarika, Gandari, Girija, Kanakaprabha, Kanchana, Kundali, Mahapushpa.
Distribution	: Common in India, Myanmar, China upto 1,800 m.
Habitat Ecology	: Forests, planted in villages, commonly grown as an agroforestry sp. in border areas of agricultural fields.
Reproductive Cycle	: March-August.
Material Examined	: EBH-TP-271.
Systematic Account	: A medium-sized deciduous tree. Leaves subcoriaceous, deeply cordate. Flowers large, fragrant, appearing when the tree is leafless in short axillary or terminal, few-flowered racemes. Bracts and bracteoles minute, deltoid. Petals clawed, much exserted. Ovary long-stalked. Pods hard, decurved, 10-15 seeded.

Beloperone guttata Brand. (Pl. 32E)

Syn: *Justicia brandegeeana* Wasshausen and L.B. Sm.

Family	: Acanthaceae.
Vernacular Name	: Shrimp.
Eng. Names	: *Eng.*: False Hop, Shrimp Bush, Shrimp Plant.
Distribution	: Native of Mexico; commonly cultivated in all parts of the world.
Habitat Ecology	: Grows in well-drained soil in partial shade.
Reproductive Cycle	: Flowers almost throughout the year.
Material Examined	: EBH-TP-512.
Systematic Account	: Dwarf shrubs with opposite, oblong–lanceolate leaves and white, tongue-shaped flowers in spikes which protrude from overlapping bracts. Corolla about 3.0 cm long, lower lip bilobed.

Berberis lycium Royle (Pl. 33A)

Family	: Berberidaceae.
Vernacular Names	: Kashmal, Sutram.
Hindi Names	: *H.*: Chitra, Kushmul.
Distribution	: Pakistan–E. Nepal: 1,500-3,000 m.
Habitat Ecology	: Shrubberies; common in cultivated areas of W. Himalaya.
Reproductive Cycle	: April-June.

Material Examined	:	EBH-TP-307.
Systematic Account	:	An evergreen shrub upto 4 m high. Twigs pale yellowish. Stem spines 6-20 mm. Leaves coriaceous, dull green above, pale and glaucous beneath. Flowers dull yellow, borne in axillary clusters. Fruits are ovoid and black in colour.

Bergenia ciliata (Haw.) Sternb. forma *ligulata* Yeo (Pl. 33B)

Syn: *B. ligulata* (Wall.) Engl.; *Saxifraga ligulata* Wall.

Family	:	Saxifragaceae.
Vernacular Name	:	Saplotri.
Eng. and Sans. Names	:	*Eng.*: Elephants-eared Saxifrage, Elephant's Ears. *Sans.*: Pashanbhed.
Distribution	:	Temperate regions: 1,800-3,000 m.
Habitat Ecology	:	Moist rocky slopes; can be grown in any soil, shade or sun, but is frost prone.
Reproductive Cycle	:	March-July.
Material Examined	:	EBH-TP-600.
Systematic Account	:	Perennial herbs possessing thick, creeping rootstock; large, thick leaves having fringed margins with short stiff hairs and flowers in spreading cymose panicles, terminating in a flexible scape. Calyx glabrous. Capsule 2-beaked.

Betula utilis D. Don (Pl. 33C)

Syn: *B. bhojpattra* Wall.; *B. castanae* Herb.; *B. jacquemontii* Spach.

Family	:	Betulaceae.
Vernacular Name	:	Bhojpattra.
Eng., H. and Sans. Names	:	*Eng.*: Himalayan Silver Birch, Indian Birch Tree, Indian Paper Birch. *H.*: Bhujpatar, Bhujpattra. *Sans.*: Bahulavalkalah, Bhurjapatra.
Distribution	:	Throughout the main Himalayan range from Bhutan westwards, ascending to 4,200 m; extends to W. Tibet and China.
Habitat Ecology	:	Drier mountainous zones.
Reproductive Cycle	:	June-September.
Material Examined	:	EBH-TP-308.

Systematic Account : A small deciduous tree or shrub having white bark peeling off in large papery layers. Shoots and young leaves pubescent. Leaves irregularly serrate, sticky when young with yellow resinous scales. Bracts of male spikes stipitate. Female spikes solitary, erect; bracts pubescent. Fruits narrower than the bracts.

Bischofia javanica **Blume (Pl. 33D)**

Syn: *B. oblongifolia* Dcne.; *B. roeperiana* Dcne.; *B. trifoliata* Hook.; *Andrachne apetala* Roxb.; *A. trifoliata* Roxb.; *Microelus roeperianus* Wight and Arn.; *Stylodiscus trifoliatus* Bennet.

Family : Euphorbiaceae.

Vernacular Name : Bhillar.

Eng. and H. Names : *Eng.*: Cedar, Java, Vinegar Wood.

H.: Bhillar, Irum, Kain, Paniala, Pankain.

Distribution : Sub-Himalayan forests and outer hills; ascending to 1,500 m.

Habitat Ecology : Shady ravines and on the banks of streams.

Reproductive Cycle : March-April.

Material Examined : EBH-TP-222.

Systematic Account : A large evergreen tree. Leaves trifoliate; leaflets crenate, glabrous. Flowers minute, greenish yellow, dioecious, in axillary or lateral paniculate racemes. Male flowers crowded. Female ones not crowded; ovary exserted, 3-4 celled. Fruits globose, smooth. Seeds 3-4 with shining.

Boerhaavia diffusa **Linn. (Pl. 34A)**

Syn: *B. glabrata* Blume; *B. repens* Linn.

Family : Nyctaginaceae.

Vernacular Names : Itsit, Ootshut.

Eng., H. and Sans. Names : *Eng.*: Pigweed.

H.: Punarnava, Sant, Santhi, Survari.

Sans.: Punarnava, Raktakanda, Raktapunarnava.

Distribution : Subtropical regions, below 1,000 m.

Habitat Ecology : Frequently met with in wastelands and roadsides.

Reproductive Cycle : Major part of the year.

Material Examined : EBH-TP-706.

Systematic Account : Diffuse herbs with stout tap root and leaves in unequal pairs. Flowers minute, in bracteate umbels of 4-10, arranged in long-stalked, axillary and terminal panicles. Perianth 10-lobed. Fruits 5-ribbed, viscidly glandular on the ribs.

Bombax ceiba **Linn. (Pl. 34B)**

Syn: *B. heptaphylla* Cav.; *B. heptaphyllum* Roxb.; *B. malabaricum* DC.; *Gossampinus rubra* Ham.; *Salmalia malabarica* (DC.) Schott and Endl.

Family : Bombacaceae.

Vernacular Name : Simbal.

Eng., H. and Sans. Names : *Eng.*: Cotton Tree, Silk Cotton-tree.

H.: Kantisembal, Pagun, Semal, Simal, Somr.

Sans.: Apurari, Mocha, Raktapushpa, Salmali.

Distribution : Throughout India; commonly in hotter forests.

Habitat Ecology : Common in lower valleys and sub-Himalayan regions.

Reproductive Cycle : February–May.

Material Examined : EBH-TP-510.

Systematic Account : A large tree having stout conical prickles and spreading branches. Leaves large; leaflets 3-7, entire. Stipules small, caducous. Flowers fascicled near the ends of the branches, appearing before the new leaves. Calyx cup-shaped. Corolla bright red. Style longer than the stamens. Capsule 5-valved.

Bougainvillea glabra **Choisy (Pl. 34C)**

Family : Nyctaginaceae.

Vernacular Name : Boganvillea.

Eng. and H. Names : *Eng.*: Paper Flower.

H.: Bag Bahar.

Distribution : Originated in Brazil.

Habitat Ecology : Widely planted in gardens.

Reproductive Cycle : Flowering in late summer and early autumn.

Material Examined : EBH-TP-270.

Systematic Account : A vigorous spiny climber having deciduous leaves and small flowers sitting in centre of 3 large bracts, grouped in axillary and terminal panicles. Bracts pale rosy-mauve changing to red tint when old.

Bougainvillea spectabilis **Willd. (Pl. 35A)**

Family	:	Nyctaginaceae.
Vernacular Names	:	Boganvillea, Sada Suhagan.
Eng. and H. Names	:	*Eng.*: Garden Glory. *H.*: Bogenvillea.
Distribution	:	Native of tropical and subtropical S. America.
Habitat Ecology	:	Generally planted as an ornamental.
Reproductive Cycle	:	Almost throughout the year.
Material Examined	:	EBH-TP-212.
Systematic Account	:	Large scrambling, thorny shrubs. Leaves alternate. Perianth cylindric. Flowers in threes and each adnate to a large purple bract.

Brassica oleracea **var.** ***capitata*** **Linn. (Pl. 35B)**

Syn: *B. oleracea* var. *acephala* DC.

Family	:	Brassicaceae.
Vernacular Name	:	Band Gobi.
Eng. and H. Names	:	*Eng.*: The Cabbage. *H.*: Band-gobi, Patagobhi.
Distribution	:	Native of some of the coastal regions of Europe; thrives best in temperate climates.
Habitat Ecology	:	Cultivated on the hills and plains especially during the cold season.
Reproductive Cycle	:	Summer.
Material Examined	:	EBH-TP-526.
Systematic Account	:	Biennials upto 0.5 m tall with glabrous stem. Leaves basal and lower cauline; all fleshy, glabrous. Flowers large, 20-40 in racemes. Fruits irregularly spreading or ascending linear pods. Seeds globose, pendulous.

Broussonetia papyrifera **(Linn.) L'Herit.** ***ex*** **Vent. (Pl. 35C)**

Syn: *Morus papyrifera* Linn.;
***Papyrius papyriferus* (Linn.) Kuntze.**

Family	:	Moraceae.
Vernacular Names	:	Jangli Toot, Karyun, Vilayati Toot.
Eng. and H. Names	:	*Eng.*: Paper-mulberry or Tapa-cloth. *H.*: Jangli-toot.

Distribution : Punjab-Arunanchal Pradesh, Meghalaya; upto 1,000m.

Habitat Ecology : Hilly slopes. Forests.

Reproductive Cycle : April-May.

Material Examined : EBH-TP-715.

Systematic Account : A small tree having sub-tomentose branchlets and sinuately 2-3 lobed, toothed leaves. Male spikes cylindric, pubescent. Female heads longer peduncled, tomentose. Fruits fleshy, red.

Bryophyllum calycinum **Salisb. (Pl. 36A)**

Syn: *B. pinnatum* (Lam.) Kurz; *Cotyledon pinnata* Lam.; *Kalanchoe pinnata* (Lam.) Pers.

Family : Crassulaceae.

Vernacular Name : Pattharchat.

Eng., H. and Sans. Names : *Eng.*: Air Plant Kalanchoe.

H.: Patharchur, Patharkunchi Zakhm-haiyat.

Sans.: Pashanabheda.

Distribution : Throughout India, ascending to 1000 m; an introduced plant.

Habitat Ecology : Excellent potted plant for houses and green houses.

Reproductive Cycle : May-June.

Material Examined : EBH-TP-917.

Systematic Account : Perennial, glabrous herbs with erect, hollow stem upto 1.2 m. Leaves petiolate, opposite, fleshy. Flowers pendulous, in a large, terminal panicle. Calyx inflated, 4-toothed. Corolla twice as long as the calyx. Stamens 8. Fruits enclosed within the persistent calyx and corolla.

Buxus sempervirens **Linn. (Pl. 36B)**

Syn: *B. emarginata* Herb.; *B. longifolia* Jacq.; *B. wallichiana* Baillon.

Family : Buxaceae.

Vernacular Names : Samsaj, Shamshad.

Eng. and H. Names : *Eng.*: Boxwood Tree, Dudgeon, Evergreen Box.

H.: Chikri, Papri.

Distribution : Temperate Himalaya: 1,800-2,700 m.

Habitat Ecology : Shady ravines.

Reproductive Cycle : April-November.

Material Examined : EBH-TP-409.

Systematic Account : A much branched shrub or small tree having nearly sessile, opposite leaves. Strongly scented flowers in small, axillary heads or spikes; the terminal flowers usually female, the rest male. Capsule ovoid, 3-horned.

Calendula officinalis **Linn. (Pl. 36C)**

Family : Asteraceae.

Vernacular Name : Zergul.

Eng. and H. Names : *Eng.*: African Marigold, Calendula, Common Marigold, Garden Marigold, Pot Marigold.

H.: Gul-e-asharfi, Sadberg.

Distribution : Mediterranean; introduced in India.

Habitat Ecology : Cultivated in gardens; preferring sunny situation and well-drained rich soil.

Reproductive Cycle : February-Early December.

Material Examined : EBH-TP-133.

Systematic Account : Erect, hispidly pubescent herbaceous annuals. Stem corymbosely branched above. Lower leaves spathulate, entire; upper lanceolate. Heads terminal; involucral bracts incurved. Ligules many, bright yellow, 3-toothed. Achenes boat-shaped.

Callistemon lanceolatus **DC. (Pl. 36D)**

Syn: *C. citrinus* (Curtis) Stapf; *Metrosideros citrina* Curtis.

Family : Myrtaceae.

Vernacular Name : Bottle Brush.

Eng. Name : *Eng.*: Bottle Brush.

Distribution : Native of Australia; widely planted in the subtropics.

Habitat Ecology : Commonly cultivated in gardens as an ornamental plant.

Reproductive Cycle : May-September.

Material Examined : EBH-TP-213.

Systematic Account : Evergreen shrubs or small trees upto 3 m tall with pendulous branches and simple, leathery, lanceolate leaves having a faint lemon odour. Flowers resembling bottle brushes, in 10 cm spikes with a tuft of leaves at the top of each stem, crimson with long, brilliant red stamens. Pods filled with small fine seeds.

Callistephus chinensis (Linn.) Nees (Pl. 37A)

Syn: *C. hortensis* Cass.; *Aster chinensis* Linn.

Family	:	Asteraceae.
Vernacular Name	:	Aster.
Eng. Name	:	*Eng.*: China Aster.
Distribution	:	A native of China and Japan; grown in Indian flower gardens.
Habitat Ecology	:	Most popular garden annuals grown throughout tropical and subtropical countries.
Reproductive Cycle	:	August-October.
Material Examined	:	EBH-TP-214.
Systematic Account	:	Annuals. Stem hispid, branched. Leaves coarsely toothed, stalked. Flower-heads solitary at ends of branches.

Calotropis procera R. Br. (Pl. 37B)

Syn: *C. hamiltonii* Wall.; *C. heterophylla* Wall.; *C. wallichii* Wight; *Asclepias procera* Willd.

Family	:	Asclepiadaceae.
Vernacular Name	:	Ak.
H. and Sans. Names	:	*H.*: Ag, Ak, Akada, Madar, Safed-ak.
		Sans.: Alarka, Arka, Dirghpuspa, Vasuka.
Distribution	:	Throughout India, ascending to 1,350 m.
Habitat Ecology	:	Wastelands, river-beds, roadsides, forest clearings; common.
Reproductive Cycle	:	February-May.
Material Examined	:	EBH-TP-215.
Systematic Account	:	An erect shrub with milky latex and young parts clothed with white cottony tomentum. Leaves subsessile, broadly ovate, with a short abrupt acumination. Flowers in umbellate cymes which are long-peduncled. Corolla lobes usually erect. Coronal-scales acute. Follicles subglobose; seeds broadly ovate, flattened.

Cannabis sativa Linn. (Pl. 37C)

Syn: *C. indica* Lamk.

Family	:	Cannabinaceae.
Vernacular Name	:	Bhang.

Eng., H. and Sans. Names : *Eng.*: Hemp, Indian Hemp.

H.: Bhang, Charas, Ganja, Ganje-ka-per, Gur, Kinnab, Phulganja, Sabzi, Siddhi.

Sans.: Bhanga, Chapola, Ganja, Ganjika, Hursini, Indrasana, Jaya, Ujaya, Vijaya, Vajradru-vrikshaha, Vrijpatta.

Distribution : Throughout India.

Habitat Ecology : Common on wastelands.

Reproductive Cycle : June-September.

Material Examined : EBH-TP-403.

Systematic Account : An aromatic scarcely branched herb with palmate, long-stalked, gland dotted, 5-7 foliate, coarsely toothed leaves. Flowers unisexual; male and female flowers in axillary clusters on different plants. Achenes flattened, glandular hairy, enclosed in the persistent perianth.

***Canna indica* Linn. (Pl. 38A)**

Syn: *C. bifida* Roem. and Schult.; *C. chinensis* Willd.; *C. coccinea* var. *bicolor* Kraenzl.; *C. edulis* Ker Gawl.; *C. floribunda* Bouche; *C. indica* var. *orientalis* Rosc.; *C. occidentalis* Roscoe; *C. orientalis* Rosc.; *C. variabilis* Willd.; *Cannacorus indicus* (Linn.) Medik.; *Xyphostylis lutea* (Mill.) Raf.

Family : Cannaceae.

Vernacular Names : Kali, Sudershan.

Eng., H. and Sans. Names : *Eng.*: Indian Bead, Indian Reed, Indian Shot.

H.: Keli, Sabbajaya, Sarvajjya.

Sans.: Devakili, Sarvajaya, Silarumba.

Distribution : Native of C. and S. America; naturalized in Asia and West Indies.

Habitat Ecology : Often cultivated in gardens as an ornamental.

Reproductive Cycle : Usually in the rainy season (June-September).

Material Examined : EBH-TP-1101.

Systematic Account : Perennial herbs with tuberous rootstock having many fibers and 0.9-1.2 m tall stem. Raceme with the pedicel 30 cm or more, erect. Flowers rather distant. Corolla-segments oblanceolate, acuminate, greenish or coloured. Staminal segments longer than the corolla. Capsule with several, globose, black seeds.

Capsicum annuum **Linn. (Pl. 38B)**

Syn: *C. frutescens* sensu Cl.

Family	:	Solanaceae.
Vernacular Name	:	Mirch.
Eng. and H. Names	:	*Eng.*: Chilli, Red Chillies.
		H.: Gachmarich, Lalmirich.
Distribution	:	Extensively cultivated throughout India.
Habitat Ecology	:	Cultivated; grows well in moist soil.
Reproductive Cycle	:	August-September.
Material Examined	:	EBH-TP-269.
Systematic Account	:	Erect herbs or under shrubs upto 1 m height. Leaves glabrous. Flowers in axillary fascicles. Calyx campanulate, much shorter in fruits. Fruits elongated, about 5.0-15.0 cm long, red, orange or yellow when ripe.

Carica papaya **Linn. (Pl. 38C)**

Syn: *Papaya carica* Gaertner.

Family	:	Caricaceae.
Vernacular Name	:	Papita.
Eng., H. and Sans. Names	:	*Eng.*: The Papaw or Papaya Tree.
		H.: Andakharbuja, Papaya or Papiya Amba, Papita, Pepiya, Popaiya.
		Sans.: Chirbhita, Erandachirbhita, Nalikadala.
Distribution	:	Cultivated mostly throughout India.
Habitat Ecology	:	Cultivated; often met as an escape.
Reproductive Cycle	:	Greater part of the year.
Material Examined	:	EBH-TP-814.
Systematic Account	:	Soft-wooded, fast-growing, small trees with large palmatifid leaves forming a round tuft at the top of the stem on long hollow petioles. Flowers on axillary panicles, fragrant, generally dioecious. Male flowers in long drooping panicles. Female flowers in short clusters. Fruits succulent, indehiscent, 1-celled. Seeds numerous, black and covered with a loose, hyaline skin or arillus.

Carissa carandas Linn. (Pl. 38D)

Syn: *C. congesta* Wight.

Family	:	Apocynaceae.
Vernacular Name	:	Kalmi Garna.
Eng., H. and Sans. Names	:	*Eng.*: Christ's Thorn, Karaunda.
		H.: Garinda, Kantakregi, Karaunda, Karondi.
		Sans.: Avighna, Kanachuka, Karamarda, Karamardaka.
Distribution	:	Indigenous to India; dry forests throughout the country.
Habitat Ecology	:	Frequently cultivated in gardens.
Reproductive Cycle	:	April-June.
Material Examined	:	EBH-TP-720.
Systematic Account	:	Evergreen shrubs or small crooked trees upto 3 m tall with dichotomous branches armed with paired axillary thorns. Leaves short-petioled, leathery, shining on both surfaces. Flowers in lax terminal cymes. Berries red turning dark purple when ripe.

Carissa opaca Stapf. *ex* Haines (Pl. 39A)

Syn: *C. spinarum* Hook. f.

Family	:	Apocynaceae.
Vernacular Name	:	Garna.
Hindi Name:		
		H.: Karaunda.
Distribution	:	Throughout the hotter parts of India.
Habitat Ecology	:	Wastelands. Open slopes. Drier areas.
Reproductive Cycle	:	May-June.
Material Examined	:	EBH-TP-802.
Systematic Account	:	Much branched thorny shrubs upto 3.6 m tall. Flowers sweet-scented, white, borne in corymbose cymes. Thorns 3.8 cm long, sharp. Berries dull to dark purple.

Carissa spinarum Linn. (Pl. 39B)

Syn: *C. diffusa* Roxb.

Family	:	Apocynaceae.
Vernacular Name	:	Garuna.
H. and Sans. Names	:	*H.*: Karaunda, Karonda.
		Sans.: Karamadika, Karamardika.

Distribution	:	Throughout India.
Habitat Ecology	:	Very common in plains; also in the scrub and chil forests.
Reproductive Cycle	:	November-February.
Material Examined	:	EBH-TP-900.
Systematic Account	:	Diffuse, evergreen bushy shrubs or rarely a tree upto 3.6 m tall with rigid branches and zig-zag branchlets armed with long thorns. Leaves short-petioled, coriaceous. Flowers fragrant, white, sometimes tinged with pink, in axillary or terminal corymbose cymes. Berries shining black or dark purple when ripe.

Caryota urens Linn. (Pl. 39C)

Family	:	Arecaceae.
Vernacular Name	:	Palm.
Eng., H. and Sans. Names	:	*Eng.*: Bastard Sago Palm, Fish-tail Palm, Indian Sago-palm, Jaggery-palm, Kitul-palm, Toddy-palm, Wine–palm. *H.*: Bankhajur, Mari. *Sans.*: Dhoajavriksha, Dirgha, Mada, Madyadru, Meda, Rajju, Vitanaka.
Distribution	:	Occurs in moist, forest regions of the W. and E. coast, Bihar, Orissa and N. E. India upto 2000 m altitude; cultivated elsewhere in our country.
Habitat Ecology	:	Commonly cultivated in gardens.
Reproductive Cycle	:	May-August.
Material Examined	:	EBH-TP-724.
Systematic Account	:	A lofty, handsome palm upto 22 m tall with a cylindrical trunk and crown of 6-10 large bipinnate leaves. Pinnae borne on 18-25 secondary rachises, each bearing 36 pairs of leaflets and a terminal leaflet. Spathe boat-shaped. Spadix very long with upto 120 secondary branches on thick, cylindrical primary axils hanging down. Flowers numerous, borne in groups of 3. Drupes 1-2 seeded, red when ripe.

Cassia fistula Linn. (Pl. 40A)

Syn: *C. excelsa* Kunth; *C. rhombifolia* Roxb.; *Bactyrilobium fistula* Willd.; *Cathartocarpus fistula* Pers.

Family	:	Fabaceae.
Vernacular Names	:	Aliya, Amaltash, Kaniar.

Eng., H. and Sans. Names	:	*Eng.*: Cassia, Indian Laburnum, Purging Cassia. *H.*: Amaltas, Girmalah, Kilvali, Kirala, Sinar. *Sans.*: Aragbadha, Rajataru, Suvarnaka.
Distribution	:	C. and E. Himalaya to Sri Lanka; ascending to 1,400 m in Kumaon, wild and often planted.
Habitat Ecology	:	Forests. Shrubberies. Open hill sides.
Reproductive Cycle	:	March-July.
Material Examined	:	EBH-TP-217.
Systematic Account	:	Erect trees with 23-40 cm long leaves having pubescent main rachis. Stipules minute. Leaflets 4-8 pairs, pale green, strongly veined. Flowers in lax racemes. Petals bright yellow, shortly clawed. Stamens all antheriferous. Pods are cylindric, nearly straight, smooth and brown-black in colour.

Cassia glauca **Lam. (Pl. 40B)**

Syn: *C. arborescens* Vahl; *C. fastigiata* Vahl; *C. surattensis* Burm.; *Senna arborescens* Roxb.; *S. sulfurea* (Collad.) H.S.Irwin and Barneby

Family	:	Fabaceae.
Vernacular Name	:	Cassia.
Distribution	:	Throughout India, cultivated.
Habitat Ecology	:	Generally grown in gardens.
Reproductive Cycle	:	April-August.
Material Examined	:	EBH-TP-218.
Systematic Account	:	A small tree with numerous spreading branches. Leaflets 4-6 pairs, subcoriaceous. Stipules caducous. Flowers in axillary corymbose racemes. Petals shortly clawed, yellow. Stamens 10, subequal. Pods flat, strap-shaped; each having 20-30, oblong, compressed, dark brown, seeds.

Casuarina equisetifolia **Linn. (Pl. 40C, D)**

Syn: *C. litorea* Linn.; *C. muricata* Roxb.

Family	:	Casuarinaceae.
Vernacular Name	:	Sut Saru.
Eng. and H. Names	:	*Eng.*: Australian Pine, Beefwood Tree, Casuarina, Coast She-oak. *H.*: Janglijhau, Janglisaru, Vilayatijhau, Vilayatisaru.

Distribution : Native of Australia, Malaysia and the pacific islands; grows profusely in the sand hills of coastal regions of tropical countries.

Habitat Ecology : Good tree in parks and large gardens. But due to the brittle nature of its branches it is not advisable to plant them along the roads.

Reproductive Cycle : February-May.

Material Examined : EBH-TP-1120.

Systematic Account : Tall, evergreen trees with straight-stem usually upto 25 m. Branches bear clusters of long, slender, green needle-like cladodes. Bark brittle, peeling off in vertical strips. Leaves whorled, reduced to minute scales. Flowers dioecious; male flowers in slender terminal spikes, female ones in cone-like heads. Fruiting heads echinate; seeds light, membranous.

Catharanthus roseus **(Linn.) G. Don (Pl. 41A)**

Syn: *Ammocallis rosea* (Linn.) Small; *Lochnera rosea* (Linn.) Reichb. *ex* Spach; *Vinca rosea* Linn.

Family : Apocynaceae.

Vernacular Name : Sadabahar.

Eng., H. and Sans. Names : *Eng.*: Madagascar Periwinkle, Old Maid, Red Periwinkle.

H.: Sadabahar, Sada Suhagan, Sada Suhagi.

Sans.: Nityakalyani, Rasna, Sadampuspa.

Distribution : Planted in gardens; occasionally naturalized to an altitude of 1,300 m.

Habitat Ecology : Commonly cultivated as an ornamental plant.

Reproductive Cycle : During the rainy season.

Material Examined : EBH-TP-265.

Systematic Account : An erect, much-branched herb upto 90 cm. Leaves oblong-elliptic, glossy. Flowers fragrant, white to pinkish-purple in terminal or axillary cymose clusters. Follicles many-seeded.

Cedrela toona **Roxb. *ex* Rottl. (Pl. 41B, C)**

Syn: *C. australis* Muell.; *C. febrifuga* Blume; *C. hexandra* Wall.; *C. serrata* Royle; *Toona ciliata* Roem.

Family : Meliaceae.

Vernacular Name : Tuni.

Eng., H. and Sans. Names : *Eng.*: Happy Tree, Indian Mahogany, Moulmein Cedar, Sandal Neem, Singapore Cedar, Toon.

H.: Lim, Lud, Mahalimbu, Mahanim, Tun, Tuna, Tuni, Tunkajhar.

Sans.: Apina, Kachhaka, Kanta, Mahanim, Nandaka, Tooni, Tunna, Tunnaka, Tunika.

Distribution : Throughout India, ascending to 1,000 m.

Habitat Ecology : Wastelands. Roadsides. Open slopes.

Reproductive Cycle : May-June.

Material Examined : EBH-TP-1131.

Systematic Account : Tall trees upto 18 m; branches numerous, forming a shady head. Leaves pinnate; leaflets 4-15 pairs. Flowers white, honey-scented, in large drooping panicles nearly as long as the leaves. Calyx-segments ciliate. Petals broadly elliptic, ciliate. Ovary hairy. Capsules oblong.

Cedrus deodara (Roxb.) Loudon (Pl. 41D)

Syn: *C. libani* Barrel. var. *deodara* Hook. f.; *Abies deodara* Lindl.; *Pinus deodara* Roxb.

Family : Pinaceae.

Vernacular Names : Deaar, Deodar, Devdar.

Eng., H. and Sans. Names : *Eng.*: Deodar, Himalayan Cedar.

H.: Deodar, Deyar, Kilan, Kilan-ka-per.

Sans.: Devadaru, Indravriksha, Svehavriksha.

Distribution : N.W. Himalaya: 1,800-3,000 m.

Habitat Ecology : Forming forests.

Reproductive Cycle : June-July.

Material Examined : EBH-TP-413.

Systematic Account : A large evergreen tree with branches usually drooping and dimorphic shoots. Long shoots with the needles solitary and spirally arranged; dwarf ones arranged in dense whorls. Leaves needle-like, triquetrous, sharp-pointed. Cones erect, woody. Seeds winged.

Celosia argentea var. *cristata* Haines (Pl. 41E)

Syn: *C. cristata* Linn.

Family : Amaranthaceae.

Vernacular Name : Cockscomb.

Eng., H. and Sans. Names : *Eng.*: Cock's Comb.

H.: Kokan, Lalmurghka, Morashikha, Pilamurghka.

Sans.: Barhichuda, Kekishikha, Mayurashikha, Shikha, Sushikha.

Distribution : Native of tropical Asia. Throughout India; gardens and as an escape.

Habitat Ecology : Suitable for growing along the garden path or in beds of mixed and annual borders or in pots.

Reproductive Cycle : May-October.

Material Examined : EBH-TP-801.

Systematic Account : An erect herb having falcate stipules, long, silky-feathery flower-spikes and flower-heads resembling a rooster's comb. The flower-heads may be about 15-23 cm across.

***Celtis australis* Linn. (Pl. 42A)**

Syn: *C. acata* Ham.; *C. caucasica* Willd.

Family : Ulmaceae.

Vernacular Names : Khadak, Khidak

Eng. Names : *Eng.*: The European Nettle-tree, The Honey-berry Tree.

Distribution : Temperate Himalaya: 1,350-2,650 m.

Habitat Ecology : Drier slopes. Screes. Edges of fields. Roadsides.

Reproductive Cycle : April-October.

Material Examined : EBH-TP-1109.

Systematic Account : A large deciduous tree with pale ashy bark having numerous lenticels and pubescent young shoots. Stipules subulate. Male flowers racemed on short leafless axillary branchlets. Sepals oblong, margins woolly. Female or bisexual flowers rather larger than the male. Ovary woolly. Fruits a globose drupe.

***Centaurea cyanus* Linn. (Pl. 42B)**

Syn: *C. lanata* Roxb.; *Leucacantha cyanus* (Linn.) Nieuwl. and Lunell.

Family : Asteraceae.

Vernacular Name : Cornflower.

Eng. Names : *Eng.*: Cornflower, Bachelor's Button, Blaver, Blue-bonnet, Blue Poppy, Corn-binks, Corn Centaury, Cornflower, Hurt-sickle, Ragged-robin.

Distribution : This hardy annual came from the British Isles; occurring in N.W. India.

Habitat Ecology : Found in cultivated places; popular for growing in beds, cut-flowers and to some extent as pot plants.

Reproductive Cycle : June-October.

Material Examined : EBH-TP-101.

Systematic Account : Erect, slender, cottony herbs with simple or only sparsely branched stems, soft, undivided, entire leaves and flower heads at the ends of the branches. Involucral bracts cottony; ray-flowers large spreading. Achenes grey silky longer than the middle row of pappus-hairs.

Centella asiatica **(Linn.) Urban (Pl. 42C)**

Syn: *Hydrocotyle asiatica* Linn.; *H. erecta* Linn.f.

Family : Apiaceae.

Vernacular Names : Brahmi, Minaki.

Eng., H. and Sans. Names : *Eng.*: Indian Pennywort.

H.: Brahma-manduki, Brihmi, Khulakhudi.

Sans.: Bhekaparni, Divya, Mandukaparni.

Distribution : Throughout India, upto 2,000 m.

Habitat Ecology : Prominent along water channels in fields.

Reproductive Cycle : May-November.

Material Examined : EBH-TP-245

Systematic Account : Prostrate herbs rooting at nodes. Stems slender. Leaves arise as a cluster from the rootstock. Petioles 7-15 cm long. Flowers pink, 3-4 in a fascicled umbel. Fruits 4 mm long, ovoid.

Cestrum diurnum **Linn. (Pl. 43A)**

Family : Solanaceae.

Vernacular Name : Jasmine.

Eng. Name : *Eng.*: Day Jasmine.

Distribution : Cultivated in gardens; wild in W. Bengal and other eastern states of India.

Habitat Ecology : Cultivated as an ornamental plant; grows well in moist soil.

Reproductive Cycle : Flowers appear in summer and continue till the end of the rainy season.

Material Examined : EBH-TP-812.

Systematic Account : A bushy shrub with erect, spreading branches and inflorescence on long axillary peduncle bearing short clusters of mildly sweet-smelling flowers. Corolla tubular, white. Fruits showy, blue and in clusters.

***Cestrum nocturnum* Linn. (Pl. 43B)**

Syn: *C. parqui* auct. non L'Her.

Family : Solanaceae.

Vernacular Name : Rat-ki-Rani.

Eng. and H. Names : *Eng.*: Lady of Night, Night Blooming Jasmine.
H.: Rat-ki-rani.

Distribution : Throughout India; cultivated upto 1,500 m.

Habitat Ecology : Wastelands; occasionally cultivated in border areas of agricultural fields or near houses.

Reproductive Cycle : July-October.

Material Examined : EBH-TP-221.

Systematic Account : Evergreen shrubs upto 3.6 m high. Leaves alternate, 8-13 cm long. Flowers greenish-yellow, fragrant, narrowly tubular, in axillary branched clusters, only opening at night.

***Chamaerops humilis* Linn. (Pl. 43C)**

Syn: *Phoenix humilis* (Linn.) Cav.

Family : Arecaceae.

Vernacular Names : Fan Palm, Khajur Palm.

Eng. Names : *Eng.*: European Fan Palm, Dwarf Fan Palm, Hair Palm.

Distribution : Indigenous to Europe and N. Africa.

Habitat Ecology : Dry scrub and rocky or sandy slopes; commonly cultivated as an ornamental plants.

Reproductive Cycle : February-March.

Material Examined : EBH-TP-318.

Systematic Account : Evergreen shrubs or small trees. Leaves fan-shaped, divided almost to the base into long narrow segments. Flowers borne in dense, erect panicles (spadix) in the axils of lower leaves. Drupes 1-3 cm long, dark yellow or reddish when mature.

Chlorophytum capense **Kuntze (Pl. 43D)**

Family	:	Liliaceae.
Vernacular Name	:	Spider Plant.
Eng. Names	:	*Eng.*: Ribbon Plant, Spider Lily, Spider Plant.
Distribution	:	Native of S. Africa; cultivated in different parts of the world.
Habitat Ecology	:	Grown for its attractive leaves in rosette in pots, hanging baskets and in ground.
Reproductive Cycle	:	July-September.
Material Examined	:	EBH-TP-508.
Systematic Account	:	Evergreen perennials with arching leaves and small, starry-white flowers appearing in loose clusters on stems upto 1.8 m long.

Chrysanthemum indicum **Linn. (Pl. 44A)**

Family	:	Asteraceae.
Vernacular Name	:	Guldaudi.
Eng., H. and Sans. Names	:	*Eng.*: Indian Chrysanthemum. *H.*: Guldaudi. *Sans.*: Bahupatrika, Bhringavallabha, Charukesara, Gandhadya, Karnika, Kumari, Ramataruni, Saha, Sevanti, Shetapatri, Taruni.
Distribution	:	Native of China and Japan; extensively cultivated in Indian gardens.
Habitat Ecology	:	Grown everywhere.
Reproductive Cycle	:	April-October.
Material Examined	:	EBH-TP-606.
Systematic Account	:	Perennial herbs upto 50 cm high. Stems sulcate. Leaves flaccid, pinnately parted. Outer involucral bracts scarious, except the herbaceous mid-nerve. Rays yellow, shorter than diameter of the disk.

Chrysanthemum leucanthemum **Linn. (Pl. 44B)**

Syn: *C. leucanthemum* var. *boecheri* Boivin; *C. leucanthemum* var. *pinnatifidum* Lecoq and Lamotte; *Leucanthemum leucanthemum* (Linn.) Rydb.; *L. vulgare* Lam.; *L. vulgare* var. *pinnatifidum* (Lecoq and Lamotte) Moldenke.

Family	:	Asteraceae.
Vernacular Name	:	Daisy.

Eng. Name	:	*Eng.*: Ox-eye Daisy.
Distribution	:	Native of Europe (including U.K.), N. America.
Habitat Ecology	:	Weed of old pastures.
Reproductive Cycle	:	May-October.
Material Examined	:	EBH-TP-114.
Systematic Account	:	Perennials. Leaves pinnatifid at base; lobes stem-clasping. Flower-heads terminal; composed of tubular disc florets and strap-shaped ray florets. Heads white; disc yellow.

Cinnamomum camphora **(Linn.) Nees and Eberm. (Pl. 44C)**

Syn: *Laurus camphora* Linn.

Family	:	Lauraceae.
Vernacular Name	:	Kapur.
Eng., H. and Sans. Names	:	*Eng.*: Camphor Tree. *H.*: Kapur, Karpur. *Sans.*: Chandra.
Distribution	:	Native of China, Japan, and Formosa; introduced into, and cultivated in many other countries including India.
Habitat Ecology	:	Cultivated as an ornamental plant or as a source of camphor.
Reproductive Cycle	:	April-August.
Material Examined	:	EBH-TP-719.
Systematic Account	:	Evergreen trees having aromatic bark which is rough from outside and smooth inside. Leaves coriaceous, aromatic, 3-veined. Flowers unisexual; female flowers usually larger than the males. Fruits globose, aromatic, and seated on enlarged perianth.

Citrus aurantifolia **(Christin) Sw. (Pl. 44D)**

Syn: *C. medica* Linn. var. *acida* Watt.;
***Limonia aurantifolia* Christm.**

Family	:	Rutaceae.
Vernacular Name	:	Nimbu.
Eng., H. and Sans. Names	:	*Eng.*: Acid or Sour Lime, Lime. *H.*: Kagji Nimbu. *Sans.*: Jambira.

Distribution : Native of India; found growing wild in the warm valleys of the Himalaya and cultivated in the plains and upto 1,500 m.

Habitat Ecology : Commonly cultivated.

Reproductive Cycle : March-June.

Material Examined : EBH-TP-210.

Systematic Account : A much branched, thorny, small bushy tree with greenish-grey bark and leaves with narrowly winged petioles. Flowers unisexual, white and 5-10 in a raceme. Fruits globular, often mamillate at the apex; rind thin, tightly attached. Pulp usually acidic with pleasant smell.

Citrus jambhiri **Lush. (Pl. 45A)**

Family : Rutaceae.

Vernacular Names : Gamiri, Jhamirdi.

Eng. and H. Names : *Eng.*: Rough Lemon.

H.: Jamberi.

Distribution : Hot valleys of Himalaya.

Habitat Ecology : Oftenly planted near border of the fields.

Reproductive Cycle : April-June.

Material Examined : EBH-TP-329.

Systematic Account : A straggling bush or small tree with thorny branches. Leaves ovate; petioles marginate or winged. Fruits ovoid-yellow with thick rind.

Citrus limon **(Linn.) Burm. f. (Pl. 45B)**

Syn: *C. limonum* Risso.; *C. medica* Linn. var. *limon* Linn.; *C. medica* var. *lomonum* Linn. (Risso) Hook. f.

Family : Rutaceae.

Vernacular Names : Galgal, Khatta.

Eng., H. and Sans. Names : *Eng.*: Lemon.

H.: Baranibu, Jambira, Paharikagazi, Paharinimbu.

Sans.: Nimbuphala.

Distribution : Cultivated in the orchards in Uttar Pradesh, Maharashtra, Tamil Nadu and Karnataka; wild in the north-west regions of India, ascending to 1,300 m.

Habitat Ecology : Cultivated.

Reproductive Cycle : April-June.

Material Examined : EBH-TP-901.

Systematic Account : A small-sized tree upto 4 m tall with thorny branches. Leaves ovate; petiole marginate or winged. Fruits are usually with a nipple-shaped extremity, bright yellow in colour and rind thick.

Citrus maxima **(Burm. f.) Merrill (Pl. 45C)**

Syn: *C. grandis* (Linn.) Osbeck; *C. decumana* Linn.; *Aurantium maximum* Burm.

Family : Rutaceae.

Vernacular Name : Chakotra.

Eng., H. and Sans. Names : *Eng.*: Forbidden-fruit, Paradise-fruit, Shaddock.

H.: Batavinebu, Chakotra, Mahanibu, Sadaphal.

Sans.: Madhukarkati.

Distribution : Native of Malaysia and Polynesia; introduced into India and Sri Lanka from Java.

Habitat Ecology : Grown mostly as a home-yard fruit.

Reproductive Cycle : November-March.

Material Examined : EBH-TP-902.

Systematic Account : A bushy tree about 5 m high. Leaves large with broadly winged petioles. Flowers white. Fruits globose, yellow, sometimes crimson when ripe; rind thick; pulp white or red.

Citrus reticulata **Blanco (Pl. 45D)**

Syn: *C. aurantium* var. *aurantium* Proper; *C. chrysocarpa* Lushington; *C. nobilis* Andrews.

Family : Rutaceae.

Vernacular Name : Santra.

Eng., H. and Sans. Names : *Eng.*: Loose-skinned Orange, Sweet Orange.

H.: Amritphal, Narangi, Sangtara, Santara.

Sans.: Nagaranga.

Distribution : Widely cultivated in subtropical regions of India.

Habitat Ecology : Cultivated; grown in orchards.

Reproductive Cycle : August-September.

Material Examined : EBH-TP-223.

Systematic Account : An evergreen, moderate-sized bushy tree with slender leaves and white, fragrant flowers. Fruits flattened on the top with thin, loose rind easily separating from the segments; bright orange when fully ripe and rich in sweet to sub-acidic juicy pulp.

Clematis grata Wall. (Pl. 46A)

Family : Ranunculaceae.

Vernacular Name : Chabadu.

Hindi Names : *H.*: Biliri, Ghantiali.

Distribution : Afghanistan–C. Nepal and China: 600-2,400 m.

Habitat Ecology : Shrubberies. Cultivated areas.

Reproductive Cycle : July-September.

Material Examined : EBH-TP-303.

Systematic Account : Climbers with deeply furrowed stems, pinnate leaves with usually 5 leaflets which are hairy beneath. Flowers in clusters, small, cream-coloured, fragrant. Petals woolly-haired beneath; stamen filaments hairless. Achenes hairy.

Clerodendrum inerme (Linn.) Gaertn. (Pl. 46B)

Syn: *Volkameria inermis* Linn.;
***V. myrtifolia* Portensch. *ex* Moldenke.**

Family : Verbenaceae.

Vernacular Name : Batraj.

Eng., H. and Sans. Names : *Eng.*: Garden Quinine;

H.: Batraj, Binjoam, Lanjai, Sangan-kuppi.

Sans.: Kundali, Vanajai.

Distribution : Indigenous to the sea coast of India extending to Polynesia.

Habitat Ecology : Swampy areas.

Reproductive Cycle : March-June.

Material Examined : EBH-TP-246.

Systematic Account : Much-branched, evergreen straggling shrubs with slender branches. Leaves opposite, base acute. Flowers borne in axillary pedunculate 3-9-flowered cymes. Corolla white, tube slender, lobes subequal. Drupes pyriform.

Clerodendrum philippinum Schauer (Pl. 46C)

Syn: *C. chinense* (Osb.) Mabberley; *C. fragrans* (Vent.) Willd.; *C. japonicum* (Thunb.) Sweet; *Volkameria japonica* Jacq.

Family : Verbenaceae.

Vernacular Names : Gandharaj, Moti-Mungra, Shiv-Darshan.

H. Names : *H.*: Madan Mohan.

Distribution : Subtropical Himalaya; grown as a medicinal plant in Himalaya upto 2,000 m.

Habitat Ecology : Moist semi-shady places.

Reproductive Cycle : August-October.

Material Examined : EBH-TP-131.

Systematic Account : An erect shrub upto 2.0 m high with quadrangular stem which is tomentose towards the tip. Leaves opposite, broadly ovate, with stiff hairs above, pubescent mainly on the veins below, toothed. Flowers double, fragrant, arranged on terminal compact sessile panicles.

Clerodendrum splendens G. Don. (Pl. 47A)

Family : Verbenaceae.

Vernacular Name : Lanjai.

Distribution : Native of tropical Africa.

Habitat Ecology : Cultivated; grows well in sun and semishade.

Reproductive Cycle : Flowers during winter months.

Material Examined : EBH-TP-710.

Systematic Account : A vigorous evergreen climber having slender, woody, slightly angled stem and simple, opposite, dark-green, glossy, 10-16 cm long leaves. Flowers in large terminal or axillary corymbose clusters. Corolla tube slender.

Clerodendrum thomsonae Balf. (Pl. 47B)

Family : Verbenaceae.

Vernacular Name : Vilayati Phool.

Eng. Name : *Eng.*: Bleeding Heart.

Distribution : A native of W. Africa.

Habitat Ecology : Grows well in pots and in ground on suitable support in partial shade.

Reproductive Cycle : June-September.

Material Examined : EBH-TP-509.

Systematic Account : Climbing shrubs with quadrangular and pubescent branches. Leaves opposite, entire, 5-9 cm long, 4-6 cm wide. Flowers very showy, in axillary cyme. Sepals inflated, membranous. Corolla tube widened at the throat.

Coix lachryma-jobi **Linn.**

Syn: *Coix agrestis* Lour.; *C. lachrymal* Linn. (Pl. 47C)

Family : Poaceae.

Vernacular Name : Suche Moti.

Eng., H. and Sans. Names : *Eng.*: Christ's Tears, Job's tears.

H.: Baru, Dabhir, Ganduta, Gurlu, Julgurgur, Kaiya, Sankru.

Sans.: Gavedi, Gavedhuka, Gojivha, Gundraguttha, Jargadi, Kshudra, Kunta.

Distribution : Widely distributed throughout tropical Asia; wild and cultivated in moist habitats throughout India upto 2,000m.

Habitat Ecology : Cultivated in the open sunny border.

Reproductive Cycle : Autumn.

Material Examined : EBH-TP-225.

Systematic Account : A stout much branched grass with culms 0.9-1.5 m tall, rooting at the lower nodes. Internodes smooth, polished. Leaves 10-50 cm long and 1.5-5 cm wide, margins spinulosely serrate; midrib stout. Flowers borne in long racemes, nodding from long peduncles.

Coleus blumei **Benth.**

Syn: *C. scutellarioides* (Linn.) Benth.; *Plectranthus blumei* (Benth.) Launert; *P. scutellarioides* (Linn.) R. Br. (Pl. 47D)

Family : Lamiaceae.

Vernacular Name : Coleus.

Eng. Names : *Eng.*: Coleus, Painted Nettle, Painter's Delight.

Distribution : Native of Java, tropical Asia and Africa.

Habitat Ecology : Widely grown in gardens for their attractive appearance and showy foliage.

Reproductive Cycle : September-October.

Material Examined : EBH-TP-527.

Systematic Account : A soft, branched perennial herb or sub-shrub having thin, ovate leaves. Flowers white and purple.

Colocasia esculenta **(Linn.) Schott. (Pl. 48A)**

Syn: *C. antiquorum* Schott.; *Alocasia dussii* Dammer; *A. illustris* W. Bull; *Aron colocasium* (Linn.) St.-Lag.; *Arum esculentum* Linn.; *Caladium colocasia* (Linn.) W.Wight; *Leucocasia esculenta* (Linn.) Nakai; *Steudnera virosa* (Roxb.) Prain.

Family : Araceae.

Vernacular Names : Kachalu, Patedu.

Eng., H. and Sans. Names : *Eng.*: Coco, Cocoyam, Dasheen, Eddo, Egyptian Arum, Elephant's Ear, Taro.

H.: Arvi, Ashukachu, Auri, Gagli, Ghuiya, Gorikachu, Kachalu, Kachu.

Sans.: Kachchi, Kachu, Kachwi.

Distribution : Wild and cultivated throughout the hotter parts of India and Sri Lanka.

Habitat Ecology : Widely grown in the tropics as a source of food in moist localities and also as an ornamental plant.

Reproductive Cycle : June-October.

Material Examined : EBH-TP-276.

Systematic Account : Tuberous herbs without above-ground stem and heart-shaped, peltate leaves. Flowers in spadix in a 30-40 cm long yellow spathe which is peduncled.

Convolvulus arvensis **Linn. (Pl. 48B)**

Syn: *C. chinensis* Ker.; *C. divaricatus* Wall.; *C. malcolmi* Roxb.; *Strophocaulos arvensis* (Linn.) Small.

Family : Convolvulaceae.

Vernacular Name : Rajbel.

Eng., H. and Sans. Names : *Eng.*: Deer's Foot Bind-weed.

H.: Harinpadi, Hiranpaddi.

Sans.: Bhadrabala, Rajbala.

Distribution : Kashmir to the Deccan, ascending to 3,000 m in the Himalaya.

Habitat Ecology : Wastelands. Dry slopes. Agricultural fields.

Reproductive Cycle : December-March.

Material Examined	:	EBH-TP-1201.
Systematic Account	:	Twining herbs having lower lobed leaves and axillary, solitary, slender peduncles. Sepals elliptic. Corolla funnel-shaped. Stamens unequal. Capsule globose, glabrous.

Cordia dichotoma Forst. (Pl. 48C)

Syn: *C. indica* Lamk.; *C. latifolia* Roxb.; *C. lowriana* Brandis; *C. myxa* Thwaites; *C. obliqua* Willd.

Family	:	Cordiaceae.
Vernacular Name	:	Lasura.
Eng., H. and Sans. Names	:	*Eng.*: Sebastan Plum.
		H.: Chhotalasura, Lasura.
		Sans.: Kshudrashleshmataka, Madhubhutadruma.
Distribution	:	Throughout India, upto 650 m.
Habitat Ecology	:	Common in moist forests.
Reproductive Cycle	:	April-July.
Material Examined	:	EBH-TP-247.
Systematic Account	:	A medium-sized tree with usually crooked trunk. Leaves ovate, elliptic to suborbicular. Flowers white. Fruits yellow berry.

Cosmos bipinnatus Cav. (Pl. 49A)

Family	:	Asteraceae.
Vernacular Name	:	Cosmos.
Eng. Names	:	*Eng.*: Cosmea, Cosmos, Mexican Aster.
Distribution	:	Native of Mexico.
Habitat Ecology	:	Commonly cultivated as an ornamental plant in gardens and houses; often an escape seen along fences, wasteplaces, barren and forest lands.
Reproductive Cycle	:	July-October.
Material Examined	:	EBH-TP-1207.
Systematic Account	:	Well-branched tall plants with deeply cut feathery foliage and large flowers. Leaves cauline and petiolate. Flowers borne on long wiry stems. Achenes glabrous, beaked.

Cosmos sulphureus **Cav. (Pl. 49B)**

Family	: Asteraceae.
Vernacular Name	: Cosmos.
Eng. Name	: *Eng.*: Yellow Cosmos.
Distribution	: Native of Mexico.
Habitat Ecology	: Grown in beds for garden decoration or for cut-flower production.
Reproductive Cycle	: March-November.
Material Examined	: EBH-TP-1121.
Systematic Account	: A medium-tall plant upto 80 cm. Leaves 2-3 pinnately cut into mucronate lobes. Flowers single, semidouble and double. Achenes hispid, beaked.

Cupressus sempervirens **Linn. (Pl. 49C)**

Syn: *C. doniana* Hort.; *C. elongata* Salisb.; *C. fastigiata* DC.; *C. pyramidalis* Targ.; *C. roylei* Carr.; *C. sempervirens* var. *indica* Parlat.

Family	: Cupressaceae.
Vernacular Names	: Saru, Shupa.
Eng., H. and Sans. Names	: *Eng.*: Mediterranean Cypress, The Cypress. *H.*: Sara, Saras, Saru. *Sans.*: Surahva.
Distribution	: Native of Turkey; cultivated in the Mediterranean region. Planted in gardens in Afghanistan and N.-W. India.
Habitat Ecology	: It is commonly grown in tropical and subtropical regions in parks and house gardens as an outdoor ornamental plant.
Reproductive Cycle	: April-May.
Material Examined	: EBH-TP-304.
Systematic Account	: An evergreen tall tree with a narrow pyramidal crown, stem fluted, bark with shallow vertical fissures, branches ascending with the leaves 4-angled. Leaves ovate-oblong with closely adpressed tips. Flowers monoecious. Male flowers numerous, small; anthers about 4. Female flower a globose cone; scales of cone 3-14, usually with a short horn or boss. Seeds ovoid or oblong, angled and narrowly winged.

Cupressus torulosa D. Don (Pl. 49D)

Family : Cupressaceae.

Vernacular Name : Saru.

Eng. Name : *Eng.*: Himalayan Cypress.

Distribution : Outer range of Himalaya from Chamba to Nepal: 2000-3000 m.

Habitat Ecology : On dry slopes. Also cultivated in gardens.

Reproductive Cycle : February-April.

Material Examined : EBH-TP-267.

Systematic Account : A large tree; branches spreading, whorled with drooping branchlets forming a broad pyramidal crown; bark peeling off in long thin strips; young foliage blue-green, becoming much darker with age. Leaves opposite, scale-like, triangular, closely imbricate. Cones terminal on branchlets, male and female on the same tree. Fruits are globose head, consisting of woody, peltate scales tightly closed along their edges at first, opening when mature. Seeds are numerous, small and winged.

Cycas revoluta Thunb. (Pl. 50A)

Family : Cycadaceae.

Vernacular Names : Cycas, Nargig.

Eng. Names : *Eng.*: China Fern Palm, China Sago Palm, Japan Fern Palm, Japan Sago Palm.

Distribution : Commonly cultivated in Indian gardens.

Habitat Ecology : Cultivated as an ornamental plant.

Reproductive Cycle : February-May.

Material Examined : EBH-TP-415.

Systematic Account : Short-trunked stout trees with deep green stiff foliage and trunk densely clothed with the old leaf-bases. Leaves pinnately divided; petiole quadrangular; leaflets margin revolute. Carpophylls laciniate nearly to the midrib, stalk longer than blade with 4-6 ovules. Immature seeds densely tomentose.

Cynodon dactylon (Linn.) Pers. (Pl. 50B)

Syn.: *C. stellatus* Willd.; *Digitaria dactylon* Scop.; *Panicum dactylon* Linn.; *Paspalum dactylon* DC.

Family : Poaceae.

Vernacular Names : Doob, Doorv, Drub.

Eng., H. and Sans. Names	:	*Eng.*: Bermuda Grass, Couch Grass, Creeping Panic Grass, Devil's Grass, Dog's Tooth Grass, Doob grass, Doorwa. *H.*: Dhoboghas, Dub, Dubra, Durrba, Hariali, Kabbar, Kalighas, Khabbal, Romghas. *Sans.*: Amari, Amrita, Ananta, Bahuvirya, Bhargavi, Dhurta, Dhurva, Durva, Gauri, Guna, Harasalika, Harita, Jaya, Mangala, Nanda, Shambhavi, Shashpa, Shatmula, Shitala, Shiveshta, Shyama, Vamini, Vijaya.
Distribution	:	Throughout India, Myanmar and Sri Lanka, ascending to 1,700 m in the Himalaya.
Habitat Ecology	:	Edges of fields. Open slopes.
Reproductive Cycle	:	August-November.
Material Examined	:	EBH-TP-319.
Systematic Account	:	A perennial grass having widely creeping stem forming matted tufts with ascending branches, soft, smooth, narrowly linear leaves and hairy ligules. Spikes 2-6, radiating from the top of a slender peduncle. Rachis very slender. Spikelets minute. Involucral glumes keeled. Grains laterally compressed.

***Cyperus flabelliformis* Rottb. (Pl. 50C)**

Syn.: *C. alternifolius* subsp. *flabelliformis* Kuk.; *C. flagellatus* Hochst.; *C. gradatus* Forssk.; *C. involucratus* Rottb.; *C. petersianus* Boeck.; *C. proximus* Steud.

Family	:	Cyperaceae.
Vernacular Name	:	Nag Buti.
Eng. Name	:	*Eng.*: Umbrella Plant.
Distribution	:	Widely distributed over the world except in cold climate.
Habitat Ecology	:	Bogs and marshes. Also grows well in moist soil in shade or semi-shade.
Reproductive Cycle	:	July-August.
Material Examined	:	EBH-TP-113.
Systematic Account	:	A perennial plant upto 1m in height with ribbed stalks. Leaves grass-like, spreading from the upper end of the stalk. Flowers green, in small clusters arising at the centre of the crown of leaves.

Dahlia rosea Cav. (Pl. 51A)

Syn.: *D. pinnata* Cav.

Family : Asteraceae.

Vernacular Name : Dalia.

Eng. Name : *Eng.*: Dahlia.

Distribution : Native of Mexico and C. America; extensively cultivated all over the world including India. Also occurs on wastelands as an escape.

Habitat Ecology : Thrives well in the plains; double flowered dahlias very common in the hills.

Reproductive Cycle : July-September.

Material Examined : EBH-TP-105.

Systematic Account : Tuberous-rooted perennial herbs having petiolate, exstipulate, pinnate leaves; leaflets oval having rounded tips. Inflorescence a large, showy heterogamous capitulum borne singly and protected by uniseriate or multiseriate involucre of green bracts.

Dalbergia sissoo Roxb. (Pl. 51B)

Syn: *D. pendula* Tenore.

Family : Fabaceae.

Vernacular Names : Shisham, Tali.

Eng., H. and Sans. Names : *Eng.*: The Sissoo.

H.: Shisham, Sisam, Sissai, Sissu, Sisu.

Sans.: Aguru, Shingshupa, Sinsapa.

Distribution : Distributed in the plains throughout India, ascending to 1,500 m in the C. Himalaya.

Habitat Ecology : On grassy slopes.

Reproductive Cycle : May-August.

Material Examined : EBH-TP-228.

Systematic Account : Erect trees with pinnate leaves, zig-zag leaf rachis, terete petioles, caducous stipules and flowers in short axillary panicles having racemoso-corymbose branches. Pods strap-shaped.

Datura fastuosa Linn. (Pl. 51C)

Syn: *D. hummatu* Bernh.

Family : Solanaceae.

Vernacular Names : Datura, Kala-Datura.

Eng., H. and Sans. Names : *Eng.*: Black Datura, Purple Datura.

H.: Kala Dhatura.

Sans.: Bhranta, Durta, Kalahemaka, Krishnadhattura, Rajadhattura, Shivapriya, Shyama, Unmatta.

Distribution : Throughout India in gardens and on waste grounds.

Habitat Ecology : Wastelands; also cultivated.

Reproductive Cycle : June-September.

Material Examined : EBH-TP-907.

Systematic Account : Annuals having zig-zag, succulent, softly pubescent, stout stem and entire or deeply toothed leaves. Flowers often double, solitary; pedicels short, stout. Calyx finely pubescent. Corolla purple, limb shortly 5-6 toothed. Capsules nodding on curved pedicels, equally spinous on all sides.

***Datura innoxia* Mill. (Pl. 52A)**

Syn: *D. metel* sensu Cl.

Family : Solanaceae.

Vernacular Name : Dhatura.

Sanskrit Name:

Sans.: Dhustura.

Distribution : N. W. Himalaya.

Habitat Ecology : Wastelands near habitations.

Reproductive Cycle : June-August.

Material Examined : EBH-TP-711.

Systematic Account : Coarse bushy annuals upto 1.2 m with dark green, ovate leaves, white and fragrant flowers and dense pubescent, 10-toothed corolla. Fruits nodding, opening at the top.

***Datura stramonium* Linn. (Pl. 52B)**

Syn: *D. ferox* Nees; *D. wallichii* Dunal;
***Stramonium vulgatum* Gaertn.**

Family : Solanaceae.

Vernacular Name : Dhatura.

Eng., H. and Sans. Names : *Eng.*: Apple of Peru, Devil's Apple, Devil's Trumphet, Mad Apple, Stink Weed, Stramonium or Thorn Apple.

H.: Dhatura.

Sans.: Devika, Dhattura, Dhuttura, Kantaphala, Shivapriya, Shivashekhara, Unmatta, Unmattaka.

Distribution	: Kashmir-Sikkim (Temperate Himalaya): 200-2,200 m.
Habitat Ecology	: Wastegrounds. Cultivated areas.
Reproductive Cycle	: June-October.
Material Examined	: EBH-TP-1106.
Systematic Account	: An erect, coarse annual upto 1m tall with spreading branches and irregularly toothed leaves. Flowers white, axillary. Capsule usually spinous on all sides. Seeds reniform.

Daucus carota **Linn. (Pl. 52C)**

Family	: Apiaceae.
Vernacular Name	: Gajar.
Eng., H. and Sans. Names	: *Eng.*: Bee's Nest, Bird's Nest, Carrot, Dauke. *H.*: Gager, Gajar. *Sans.*: Gajida, Garjara, Granthimula, Grinjana, Naranga, Pindamula, Pitamulaka, Shikha-mulam, Svadumula.
Distribution	: Kashmir and the Western Himalaya: 1,700-3,000 m; cultivated elsewhere in India.
Habitat Ecology	: Cultivated.
Reproductive Cycle	: Summer.
Material Examined	: EBH-TP-1203.
Systematic Account	: Biennials upto 1.5 m high, branching from the base and having 2-3-pinnatisect leaves. Umbel with very numerous rays. Bracts of the involucre 3-fid or pinnate. Petals radiating. Central flowers sterile, purple. Fruit 4 mm long, 3 mm broad, including the prickles; prickles setaceous.

Debregeasia hypoleuca **Wedd. (Pl. 52D)**

Syn: *D. bicolor* Wedd.; *Boehmeria hypoleuca* Hochst.; *B. salicifolia* D. Don; *Missiessya hypoleuca* Wedd.; *Morocarpus salicifolius* Blume; *Urtica bicolor* Roxb.

Family	: Urticaceae.
Vernacular Name	: Siaru.
Hindi Names	: *H.*: Seharu, Siaru.
Distribution	: Western Temperate Himalaya: 1,000-2,400 m.
Habitat Ecology	: Forests, Shrubberies; common.

Reproductive Cycle	:	May-August.
Material Examined	:	EBH-TP-1001.
Systematic Account	:	Large shrubs. Branches and leaves beneath clothed with snow-white wool. Leaves serrulate, scabrid above. Flower-heads conglobate, sessile. Stamens exserted. Fruits amber-coloured.

Deeringia amaranthoides **(Lamk.) Merr. (Pl. 53A)**

Syn: *D. baccata* Moq.; *D. celosioides* R.Br.; *D. indica* Retz.; *Achyranthes amaranthoides* Lamk.; *Celosia baccata* Retz.

Family	:	Amaranthaceae.
Vernacular Name	:	Dumbar.
H. Name	:	*H.*: Latman.
Distribution	:	Subtropical Himalaya. Shimla-Bhutan, ascending to 1,650 m.
Habitat Ecology	:	Shrubberies, cultivated areas; common.
Reproductive Cycle	:	May-November.
Material Examined	:	EBH-TP-506.
Systematic Account	:	A large scrambling shrub with long, arching branches, long-pointed, alternate leaves and shortly stalked, pale yellow-green flowers in racemes forming terminal panicles. Perianth 5-parted; segments concave. Stamens united at the base. Berries globose, scarlet.

Delphinium ajacis **Linn. (Pl. 53B)**

Syn: *D. ambiguum* Linn.; *Consolida ajacis* (Linn.) Schur; *C. ambigua* (Linn.) P.W. Ball and Heywood.

Family	:	Ranunculaceae.
Vernacular Name	:	Rocket Phool.
Eng. Names	:	*Eng.*: Dolphin Plant, Rocket Larkspur.
Distribution	:	Native of Europe and Australia.
Habitat Ecology	:	Commonly cultivated in gardens for its beautiful flowers.
Reproductive Cycle	:	May-September.
Material Examined	:	EBH-TP-913.
Systematic Account	:	Tall annuals with much-branched palmatisect leaves and spurred flowers. Stem hairy, branched. Leaves exstipulate; basal lobed, upper pinnatisect and finely

divided into long narrow segments. Inflorescence typical raceme or a spike with large numbers of blue, purple, lilac, salmon rose or white flowers arising in single or double.

Dendrophthoe falcata (Linn. f.) Ettingsh. (Pl. 53C)

Syn: *Loranthus amplexifolius* Desr.; *L. bicolor* Roxb.; *L. falcatus* Linn. f.; *L. longiflorus* Desr.

Family : Loranthaceae.

Vernacular Name : Banda.

Eng., H. and Sans. Names : *Eng.*: Mistletoe.

H.: Banda.

Sans.: Vanda, Vrikshabhaksha, Vriksharuha, Vrksadani.

Distribution : Widely distributed throughout tropical Asia.

Habitat Ecology : Throughout India as parasite on trees.

Reproductive Cycle : November-March.

Material Examined : EBH-TP-1107.

Systematic Account : A large, bushy parasitic plant with smooth grey bark, subopposite, coriaceous leaves, and flowers in short axillary racemes. Calyx tube cylindrical. Corolla split at the back, tube curved, orange-red to scarlet. Berries fleshy, bright pink when mature.

Dianthus barbatus Linn. (Pl. 54A)

Syn: *Dianthus compactus* Kit.

Family : Caryophyllaceae.

Vernacular Name : Dianthus.

Eng. Name : *Eng.*: Sweet William.

Distribution : Native of Europe, China and Korea.

Habitat Ecology : Commonly cultivated in gardens as an ornamental plant.

Reproductive Cycle : May-August.

Material Examined : EBH-TP-525.

Systematic Account : Herbaceous biennial plant upto 30-75 cm tall. Leaves opposite, lanceolate, elliptic or oblong, glabrous, sessile to short-petiolate. Flowers produced in a dense cluster at the top of the stem. Petals 5, clawed, fringed. Stamens 10. Capsule many seeded.

Diospyros lotus **Linn. (Pl. 54B)**

Family	:	Ebenaceae.
Vernacular Name	:	Barain.
Eng. and H. Names	:	*Eng.*: Dateplum, Persimmon. *H.*: Amlok.
Distribution	:	N. W. Himalaya particularly in the forests of Jammu and Kashmir.
Habitat Ecology	:	Cultivated.
Reproductive Cycle	:	April-August.
Material Examined	:	EBH-TP-229.
Systematic Account	:	Deciduous trees with rough bark and alternate, acuminate leaves. Male flowers subsessile, in clusters of 2-3. Calyx lobes ciliate. Corolla tubular, lobed; lobes ciliate. Stamens 16. Female flowers nearly sessile, solitary. Staminodes 8, hairy. Fruits globose or ovoid, purple in colour.

Diospyros montana **Roxb. var.** ***cordifolia*** **(Roxb.) Hiern. (Pl. 54C)**

Syn: *D. cordifolia* Roxb.; *D. Montana* auct. non Roxb.

Family	:	Ebenaceae.
Vernacular Name	:	Had kinnu.
Eng., H. and Sans. Names	:	*Eng.*: Mottled Ebony. *H.*: Bistendu, Dasaundu, Lohari, Tendu. *Sans.*: Tumala.
Distribution	:	Throughout India, Sri Lanka.
Habitat Ecology	:	Shrubberies. Cultivated areas.
Reproductive Cycle	:	March-May.
Material Examined	:	EBH-TP-712.
Systematic Account	:	A small tree with smooth, grey bark peeling in old trees. Branches spinescent. Leaves alternate, usually rounded at base; pubescent when young. Male flowers borne in pubescent cymes. Female flowers white, solitary; peduncles nodding. Berries reddish brown when ripe.

Dodonaea viscosa **(Linn.) Jacq. (Pl. 54D)**

Syn: *D. angustifolia* Linn.; *D. burmanniana* DC.; *D. dioica* Roxb.; *D. heterophylla* Hortul.; *D. microcarpa* DC.; *D. pallida* Miq.; *D. pentandra* Griff.; *D. wightiana* Blume; *Ptelea viscosa* Linn.

Family	:	Sapindaceae.
Vernacular Names	:	Ban Mendi, Mehndu, Mendru.

Eng., H. and Sans. Names : *Eng.*: Hop Seed Bush.

H.: Aliar, Sanatha, Sinatha, Sonalta, Walaytinahndi.

Sans.: Aliar, Sanatta.

Distribution : Throughout most of India upto 1,500 m.

Habitat Ecology : Occurring naturally or commonly cultivated as an ornamental shrub and for rehabilitating degraded sites.

Reproductive Cycle : October-March.

Material Examined : EBH-TP-602.

Systematic Account : Evergreen shrubs or a small tree upto 3 m tall with light green foliage. Young shoots and leaves sticky due to a bright yellow resin. Leaves subsessile, simple, alternate, oblanceolate. Flowers greenish-yellow; borne in a few-flowered axillary cymes. Capsule membranous, compressed with black seeds.

Dolichos lablab **Linn. (Pl. 55A)**

Syn: *D. benghalensis* Jacq.; *D. curtisii* G. Don; *D. cuspidatus* Grah.; *D. festivus* Wall.; *D. lignosus* Linn.; *D. purpureus* Linn.; *Lablab cultratus* DC.; *L. lignosus* Grah.; *L. vulgaris* Savi.

Family : Fabaceae.

Vernacular Name : Sem.

Eng., H. and Sans. Names : *Eng.*: Egyptian Kidney Been, Hyacinth Bean.

H.: Bhatavasu, Bhetarasu, Borboti, Lobia, Makhansim, Sim.

Sans.: Nespava, Nishpava, Rajashimbi, Shimbi, vallaka.

Distribution : Throughout India upto 2,000 m.

Habitat Ecology : Cultivated everywhere for its edible young pods.

Reproductive Cycle : November-December.

Material Examined : EBH-TP-254.

Systematic Account : A twining, perennial herb with slightly downy stems and leaves upto 30 cm long. Leaflets ovate, base cuneate or deltoid. Flowers racemed. Pods glabrous, 3-5 seeded, tipped with the persistent style-base.

Duranta repens **Linn. (Pl. 55B)**

Syn: *D. erecta* Linn.; *D. erecta* var. *alba* (Masters) Caro; *D. plumieri* Jacq.

Family : Verbenaceae.

Vernacular Name : Neel Kanthi.

Eng., H. and Sans. Names	: *Eng.*: Golden Dew Drop, Pigeon Berry, Sky Flower. *H.*: Daranta. *Sans.*: Vriti kanta.
Distribution	: Native of C. America.
Habitat Ecology	: Very common as a hedge-plant: 200-1,600 m.
Reproductive Cycle	: May-August.
Material Examined	: EBH-TP-203.
Systematic Account	: Spiny shrubs with lavender-blue flowers borne in terminal branched clusters and opposite, broadly elliptic leaves. Spines stout. Flowers with corolla-tube about twice as long as calyx and with rounded spreading lobes. Ripe fruits globular.

Ehretia acuminata R. Br. (Pl. 55C)

Syn: *E. ovalifolia* Hassk.; *E. pyrifolia* Don; *E. serrata* Roxb.

Family	: Boraginaceae.
Vernacular Name	: Punna.
Hindi Names	: *H.*: Arjun, Koda, Lahichan, Punjlawai, Punyan.
Distribution	: Subtropical Himalaya and adjacent plains, ascending to 1,700 m.
Habitat Ecology	: Dry slopes. Roadsides.
Reproductive Cycle	: May-July.
Material Examined	: EBH-TP-232.
Systematic Account	: Tall trees attaining upto 13 m. Leaves acuminate, serrated. Panicles crowded with flowers. Calyx-lobes obtuse. Corolla-tube very short; lobes reflexed. Anthers exserted. Drupes short-ellipsoid.

Elaeocarpus ganitrus Roxb. (Pl. 56A)

Syn: *E. sphaericus* (Gaertn.) K. Schum;
***Ganitrus sphaericus* Gaertn.**

Family	: Elaeocarpaceae.
Vernacular Name	: Rudraksh.
Eng., H. and Sans. Names	: *Eng.*: Utrasum Bead Tree. *H.*: Rudraksha. *Sans.*: Amara, Bhutanashana, Harksha, Nilakan-thaksha, Pavana, Pushpachanamara, Rudraksha, Sharvaksha, Shivaksha, Shivapriya, Trinameru.

Distribution	:	N. E. region, Bengal, Bihar, M.P. and Maharashtra.
Habitat Ecology	:	Cultivated.
Reproductive Cycle	:	April-June.
Material Examined	:	EBH-TP-511.
Systematic Account	:	Tall evergreen trees attaining 20 m. Leaves membranous. Flowers in drooping racemes. Fruits globose, bluish-purple, 1-seeded; stone tubercled.

Emblica officinalis Gaertn. (Pl. 56B)

Syn: *Cicca emblica* Kurz; *Dichelastina nodicaulis* Hance; *Phyllanthus emblica* Linn.; *P. glomeratus* Roxb.; *P. taxifolius* Don.

Family	:	Euphorbiaceae.
Vernacular Name	:	Amla.
Eng., H. and Sans. Names	:	*Eng.*: Emblic Myrobalan, Indean Goose Berry;
		H.: Amalaci, Amla, Amlaki, Anola, Anuli, Anvula, Anvurah, Anwerd, Aonla, Aungra, Aunra, Aura, Daula.
		Sans.: Amala, Amalaki, Amulki, Dhatri, Kamu, Umrita.
Distribution	:	Wild or planted throughout tropical and subtropical India.
Habitat Ecology	:	Borders of agricultural fields. Forests.
Reproductive Cycle	:	April-September.
Material Examined	:	EBH-TP-505.
Systematic Account	:	A moderate-sized deciduous tree with crooked trunk, spreading branches, greenish-grey bark peeling off in conchoidal flakes, branchlets glabrous or finely pubescent and subsessile, distichous leaves. Flowers in axillary fascicles on the leaf-bearing branchlets. Male flowers numerous on short slender pedicels. Female flowers few, subsessile. Fruits fleshy, obscurely 6-lobed.

Erigeron alpinus Linn. (Pl. 56C)

Syn: *E. acre* Clarke.

Family	:	Asteraceae.
Vernacular Name	:	Manjari.
Distribution	:	Throughout temperate and alpine Western Himalaya.
Habitat Ecology	:	Wastelands, forests; common.
Reproductive Cycle	:	May-October.
Material Examined	:	EBH-TP-305.

Systematic Account : Erect biennial or perennial, hispidly hairy or glabrate herbs. Radicle leaves petioled. Stem leaves usually sessile. Heads solitary, in terminal corymbose panicles. Involucral bracts pubescent; ligules rarely exceeding the pappus. Achenes sub-silky.

Eriobotrya japonica **(Thunb.) Lindl. (Pl. 56D)**

Syn: *Crataegus bibas* Lour.; *Mespilus japonica* Banks; *Photinia japonica* (Thunb.) Franch. and Sav.

Family : Rosaceae.

Vernacular Names : Lokat, Lugat, Lukat.

Eng. and H. Names : *Eng.*: Chinese Medlar, Japan Medlar, Japan Quince, Loquat.

H.: Logat.

Distribution : Native of China and Japan, cultivated in N. India.

Habitat Ecology : Borders of agricultural fields. Roadsides.

Reproductive Cycle : July-September.

Material Examined : EBH-TP-233.

Systematic Account : Small-sized evergreen trees with densely woolly inflorescence and narrowly oblanceolate, thick coriaceous leaves. Petiole stout woolly. Stipules sub-persistent. Flowers in terminal panicles. Petals broadly ovate. Fruits succulent, 2-5 seeded.

Eucalyptus tereticornis **Sm. (Pl. 57A)**

Syn: *E. insignis* Naudin; *E. populifolia* Desf.; *E. umbellata* Domin.

Family : Myrtaceae.

Vernacular Name : Safeda.

Eng. and H. Names : *Eng.*: Blue Gum, Eucalyptus.

H.: Safeda.

Distribution : Indigenous to E. Australia; introduced in India upto 1,200 m.

Habitat Ecology : Commonly planted along roadsides.

Reproductive Cycle : March-April.

Material Examined : EBH-TP-813.

Systematic Account : A large tree with erect trunk and ash-coloured stem. Leaves long, usually falcate. Flowers in axillary umbels. Calyx tube turbinate on a short stout pedicel. Operculum conical, exceeding the calyx-tube.

Euphorbia milii Des Moul. (Pl. 57B)

Syn: *E. breonii* Nois.; *E. milii* var. *breonii* (Nois.) Ursch and Leandri; *E. splendens* Bojer *ex* Hook.

Family : Euphorbiaceae.

Vernacular Name : Kantiali.

Eng. Names : *Eng.*: Crown of Thorns, Christ Thorn.

Distribution : Native of Madagascar.

Habitat Ecology : Cultivated as an ornamental plant.

Reproductive Cycle : March-April and throughout the summer.

Material Examined : EBH-TP-255.

Systematic Account : A succulent shrub with branched stems carrying tapering spines. Leaves few, near ends of the branches, alternate, thin, bright green. Flowers bright scarlet in groups on sticky, branched, long-stemmed inflorescence.

Euphorbia pulcherrima Willd. *ex* Klotzsch. (Pl. 57C)

Syn: *E. coccinea* Willd. *ex* Boiss.; *E. diversifolia* Willd. *ex* Boiss.; *E. erythrophylla* Bertol.; *E. fastuosa* Sessé and Moc.; *E. poinsettiana* Buist *ex* Giah.; *Poinsettia pulcherrima* (Willd. *ex* Klotzsch) Grah.

Family : Euphorbiaceae.

Vernacular Name : Ponsetia.

Eng. Name : *Eng.*: Poinsettia.

Distribution : Native of C. America; commonly cultivated in the tropics.

Habitat Ecology : A common hedge plant in the Himalaya.

Reproductive Cycle : November-January.

Material Examined : EBH-TP-1108.

Systematic Account : A soft-wooded spineless deciduous shrub upto 4 m having large vermilion bracts surrounding the yellow flower-heads; 1 cm across. Bracts leaf-like. Leaves alternate, entire. Leaf-stalks red.

Euphorbia royleana Boiss. (Pl. 57D)

Syn: *E. pentagona* Royle.

Family : Euphorbiaceae.

Vernacular Name : Chhoon.

Hindi Names : *H.*: Senhur, Shakarpitan, Thor.

Distribution	:	Himalaya, upto 1,800 m; Indus-Kumaon, Salt Range.
Habitat Ecology	:	Rocky slopes. Dry river valleys. Cultivated areas.
Reproductive Cycle	:	June-July.
Material Examined	:	EBH-TP-320.
Systematic Account	:	Shrubs. Branches whorled, 5-7-angled. Stipular spines in pair. Leaves spathulate, fleshy, sessile. Involucres yellow, in compact sessile 3-flowered cymes arising from the sinus between the nodes. Capsule 3-lobed, pedunculate.

Ficus benghalensis Linn. (Pl. 58A)

Syn: *F. indica* Linn.; *Urostigma bengalense* Gasp.

Family	:	Moraceae.
Vernacular Names	:	Bad, Bargad, Bat Vriksha.
Eng., H. and Sans. Names	:	*Eng.*: The Banyan Tree. *H.*: Bar, Bargad, Bargat, Ber, Bor. *Sans.*: Avaroha, Bahupada, Mahachhaya, Vata, Vrikshanath.
Distribution	:	Throughout India upto 1,400 m.
Habitat Ecology	:	Commonly planted as a shade tree.
Reproductive Cycle	:	February-June.
Material Examined	:	EBH-TP-235.
Systematic Account	:	Large-sized trees attaining 30 m in height; sending down many aerial roots from the branches. Leaves coriaceous, ovate, glabrescent above, minutely pubescent beneath with distinct reticulations. Receptacles sessile in pairs, axillary, globose, red when ripe, with 3 broad coriaceous basal bracts. Male flowers numerous near the mouth of the receptacles. Stamen 1. Female flowers with shorter perianth and elongated style.

Ficus carica Linn. (Pl. 58B)

Family	:	Moraceae.
Vernacular Name	:	Anjir.
Eng., H. and Sans. Names	:	*Eng.*: Fig Tree. *H.*: Anjir. *Sans.*: Anjira, Manjula, Kakodumbar.

Distribution : N.W. India. Deccan.

Habitat Ecology : Cultivated.

Reproductive Cycle : March-August.

Material Examined : EBH-TP-306.

Systematic Account : A shrub or small deciduous dioecious tree. Leaves more or less deeply lobed. Some trees have elongated receptacles with male and gall flowers while others have globose receptacles containing only female flowers. Receptacles of both sexes lengthen out into a stalk, at the base of which are the membranous bracts.

Ficus elastica **Roxb. (Pl. 58C)**

Family : Moraceae.

Vernacular Name : Rubber plant.

Eng. Name : *Eng.*: Indian Rubber Bush, Rubber Bush, Rubber Fig, Rubber plant.

Distribution : Parks and gardens throughout India.

Habitat Ecology : Commonly cultivated as a fine plant for rooms and for sub-tropical gardening.

Reproductive Cycle : April-July.

Material Examined : EBH-TP-405.

Systematic Account : Broad leaved, small evergreen tree with reddish brown bark. Leaves alternate, coriaceous, shining, dark green.

Ficus glomerata **Roxb. (Pl. 58D)**

Syn: *F. chittagonga* Miq.; *F. goolereea* Roxb.; *F. mollis* Miq.; *F. racemosa* Wall.; *Covellia glomerata* Miq.; *C. mollis* Miq.

Family : Moraceae.

Vernacular Names : Gular, Rumbal, Udumber, Umbarnae, Umro.

Eng., H. and Sans. Names : *Eng.*: Cluster Fig.

H.: Dimeri, Gular, Lelka, Paroa, Tue, Umar, Umrai.

Sans.: Brahmavriksha, Haritaksha, Udumbara.

Distribution : Outer Himalaya, plains and low hills of India, from Rajpootana and the Salt Range to the Khasia mountains., Myanmar, Deccan Peninsula and Sri Lanka.

Habitat Ecology : Moist localities and bordering streams.

Reproductive Cycle : April-July.

Material Examined : EBH-TP-236.

Systematic Account : Trees. Leaves glabrous on both surfaces, 3-nerved. Receptacles shortly pedunculate, red when ripe, with depressed umbilicus; male, female and gall flowers together in one receptacle. Male flowers near the mouth of the receptacle; gall and female flowers intermixed. Achenes granulate.

Ficus palmata Forsk. (Pl. 59A)

Syn: *F. caricoides* Roxb.; *F. pseudo-sycamorus* Dcne.; *F. virgata* Roxb.

Family : Moraceae.

Vernacular Names : Dhuda, Phakura, Khasra, Phedu, Phegura.

Hindi Names : *H.*: Anjiri, Bedu, Beru, Gular, Khabara.

Distribution : N.W. India, ascending to 2,300 m in the Himalaya.

Habitat Ecology : Rocky slopes. Along agricultural fields.

Reproductive Cycle : April-September.

Material Examined : EBH-TP-504.

Systematic Account : A deciduous shrub or small tree. Leaves often deeply lobed, scabrous above, 3-nerved. Stipules ovate. Figs axillary, contracted towards the peduncle, purple when ripe. Male flowers numerous in the upper part of the receptacles.

Ficus religiosa Linn. (Pl. 59B)

Syn: *F. affinior* Griff.; *Urostigma affine* Miq.; *U. religiosum* Gasp.

Family : Moraceae.

Vernacular Names : Peepal, Pipal.

Eng., H. and Sans. Names : *Eng.*: The Peepul Tree.

H.: Pipal.

Sans.: Asvattha, Aswaththamu, Pavitraka, Vriksharaja.

Distribution : Wild in the Sub–Himalayan Forests, Bengal and in C. India; planted elsewhere.

Habitat Ecology : Usually planted alongwith *Ficus benghalensis.*

Reproductive Cycle : March-October.

Material Examined : EBH-TP-514.

Systematic Account : A large glabrous tree. Leaves coriaceous, ovate-rotund and the apex produced into a linear-lanceolate tail, reticulations fine. Receptacles axillary, depressed-globose, dark purple when ripe; basal bracts broad. Male flowers few, sessile, only near the mouth of some receptacles. Stamen 1. Gall and female flowers sessile or pedicellate; sepals 5, style short, lateral.

Ficus roxburghii Wall. *ex* Miq. (Pl. 59C)

Syn: *F. auriculata* Lour.; *F. macrophylla* Roxb.; *F. regia* Miq.; *F. scleroptera* Griff.; *Covellia macrophylla* Miq.

Family : Moraceae.

Vernacular Names : Chimbal, Timbal, Traimbal.

Hindi Names : *H.*: Timal, Trimmal.

Distribution : Outer Himalaya, Kashmir-Bhutan; ascending to 1,600 m.

Habitat Ecology : Wastelands. Rocky slopes. Borders of agricultural fields.

Reproductive Cycle : April-August.

Material Examined : EBH-TP-1104.

Systematic Account : A spreading tree having stalked, broadly ovate leaves. Figs stalked, clustered on short thick, leafless branches, red or dull purple when ripe.

Ficus rumphii Blume (Pl. 59D)

Syn: *F. cordifolia* Roxb.; *Urostigma rumphii* Miq.; *U. cordifolium* Miq.

Family : Moraceae.

Vernacular Name : Badaar.

Hindi Names : *H.*: Gagjaira, Gajiun, Gajna, Kabar, Pakar, Pilkhan, Pipal, Pipul.

Distribution : Sub-Himalayan tract and outer hills, ascending to 1,650 m.

Habitat Ecology : Shrubberies. Open slopes.

Reproductive Cycle : May-June.

Material Examined : EBH-TP-237.

Systematic Account : A large glabrous deciduous tree having subcoriaceous, broadly ovate, glabrous leaves. Stipules ovate-lanceolate. Receptacles sessile, in pairs, axillary or from old scars, black when ripe, basal bracts small.

Ficus virens Ait. (Pl. 60A)

Syn: *F. infectoria* sensu Roxb.

Family	: Moraceae.
Vernacular Name	: Papletha.
H. and Sans. Names	: *H.*: Kahimal, Pilkha, Pilkhan.
	Sans.: Plaksha.
Distribution	: Throughout India extending to outer Himalaya upto 1,900 m; also cultivated.
Habitat Ecology	: Rocky slopes.
Reproductive Cycle	: March-August.
Material Examined	: EBH-TP-1110.
Systematic Account	: Deciduous trees. Leaves beaked, long, wavy-edged; leaf buds enclosed in 1.5 cm long scales. Figs 5-7 mm diameter, globose, in axillary pairs.

Fraxinus xanthoxyloides (Wall. *ex* G. Don) DC. (Pl. 60B)

Syn: *F. moorcroftiana* Brand.; *Ornus xanthoxyloides* G. Don.

Family	: Oleaceae.
Vernacular Name	: Sanjal.
Eng. Name	: *Eng.*: Achee Tree Ash.
Distribution	: Temperate Himalaya (Kashmir–Kumaon): 1,000-2,700 m.
Habitat Ecology	: Open slopes. Drier areas.
Reproductive Cycle	: July-September.
Material Examined	: EBH-TP-1204.
Systematic Account	: Deciduous trees possessing greyish-pale bark with reticulate cracks, stiff branches and pinnate leaves with winged midrib. Leaflets 5-11, scattered, saw-toothed. Flowers in dense heads, in the axils of fallen leaves. Petals absent. Fruits a winged nut.

Gaillardia pulchella Foug. (Pl. 60C)

Family	: Asteraceae.
Vernacular Name	: Aster.
Eng. Name	: *Eng.*: Gaillardia.
Distribution	: An annual from America.

Habitat Ecology : Cultivated; best for growing in the beds, in rockeries and also in pots.

Reproductive Cycle : July-October.

Material Examined : EBH-TP-905.

Systematic Account : Erect, branched, softly hairy annuals upto 60 cm high. Leaves oblong, nearly sessile. Flower-heads 5 cm across or more. Ray florets yellow with crimson base.

Gomphrena globosa **Linn. (Pl. 61A)**

Family : Amaranthaceae.

Vernacular Name : Phulnu.

Eng. Names : *Eng.*: Batchelor's Button, Globe Amaranth.

Distribution : Native of America.

Habitat Ecology : Cultivated for its bright coloured flower-heads; and frequently met with as an escape.

Reproductive Cycle : Flowers during rainy and cold seasons in India.

Material Examined : EBH-TP-1122.

Systematic Account : Erect, hairy, dichotomously branched annuals. Leaves shortly petioled. Flower heads large globose with 2 leafy bracts; bracteoles pinkish purple. Sepals densely woolly.

Grevillea robusta **A. Cunn. *ex* R. Br. (Pl. 61B)**

Family : Proteaceae.

Vernacular Names : Silver Fir, Silver Oak.

Eng. Names : *Eng.*: Australian Silk Oak, Silver Oak.

Distribution : Indigenous to Queensland and New South Wales; often planted along roadsides in Himalaya.

Habitat Ecology : Planted in gardens and avenues as a shade tree.

Reproductive Cycle : March-July.

Material Examined : EBH-TP-238.

Systematic Account : Large trees having rusty-tomentose young shoots compound pinnate type, 20-30 cm long, spirally arranged leaves and orange coloured flowers in one-sided fascicled racemes on short leafless branchlets. Perianth segments recurved to one side. Follicles oblique coriaceous, 1 or 2 seeded.

***Grewia oppositifolia* Roxb. (Pl. 61C)**

Syn: *G. optiva* Drumm.

Family	: Tiliaceae.
Vernacular Names	: Beul, Tamman.
Hindi Names	: *H.*: Bahul, Bhengal, Bhimal, Biul, Biung.
Distribution	: N.W. Himalaya (Indus to Nepal), ascending to 1,800 m.
Habitat Ecology	: Edges of agricultural fields, wasteplaces; also frequently cultivated.
Reproductive Cycle	: April-June.
Material Examined	: EBH-TP-239.
Systematic Account	: Trees with ash-coloured bark and broadly ovate-lanceolate, 3-nerved leaves. Stipules caducous. Cymes umbellate, opposite the leaves. Flower-buds pilose, yellowish. Sepals 3-nerved. Petals half the length of the sepals. Torus hairy. Drupes glabrescent, blackish, 1-4 lobed.

***Hamelia patens* Jacq. (Pl. 62A)**

Syn: *H. erecta* Jacq.; *H. patens* var. *axillaroides* Wernham; *Duhamelia patens* (Jacq.) Pers.

Family	: Rubiaceae.
Vernacular Name	: Hamelia.
Distribution	: Indigenous to tropical and subtropical America.
Habitat Ecology	: Widely grown as an ornamental plant in gardens.
Reproductive Cycle	: August-September.
Material Examined	: EBH-TP-240.
Systematic Account	: Evergreen shrubs with dense foliage of greenish-bronze leaves in whorls of 3. Flowers almost scarlet, in a terminal pedunculate cymose umbel. Berries blood-red.

***Hedychium coronarium* Koenig. (Pl. 62B)**

Syn: *H. chrysoleucum* Hook.; *H. flavum* Roxb.; *H. maximum* Roscoe; *Amomum filiforme* Hunter *ex* Ridl.; *Gandasulium lingulatum* (Hassk.) Kuntze; *Kaempferia hedychium* Lam.

Family	: Zingiberaceae.
Vernacular Name	: Vantosh.
Eng. Names	: *Eng.*: Common Ginger Lily, Garland Flower.

Distribution	:	Moist tropical and subtropical regions of Asia from India eastwards to Malaysia.
Habitat Ecology	:	Commonly grown in gardens as an ornamental.
Reproductive Cycle	:	May-July.
Material Examined	:	EBH-TP-404.
Systematic Account	:	Tall perennial herbs upto 2 m having stout, fleshy, jointed rootstock bearing leafy shoots. Leaves distichous with long clasping sheaths. Ligule membranous. Flowers fragrant, borne in dense oblong spikes; bracts closely imbricate. Capsules yellow inside, many seeded with a crimson aril.

Helianthus annuus **Linn. (Pl. 62C)**

Syn: *H. annuus* var. *texanus* (Heiser) Shinners; *H. aridus* Rydb.; *H. lenticularis* Dougl. *ex* Lindl.

Family	:	Asteraceae.
Vernacular Name	:	Surajmukhi.
Eng., H. and Sans. Names	:	*Eng.*: Lady Eleven o'Clock, Marigold of Peru, Sunflower. *H.*: Hurhuja, Surajmukhi. *Sans.*: Adityabhakta, Suria-mukhi, Suryavarta, Suvarchala.
Distribution	:	Native of W. United States.
Habitat Ecology	:	Commonly cultivated as an ornamental plant throughout India. Grows best in rich, deeply cultivated soil.
Reproductive Cycle	:	July-September.
Material Examined	:	EBH-TP-918.
Systematic Account	:	Pubescent annuals attaining 2 m height. Leaf blades broadly ovate. Flower-heads large, flat, circular; ray florets yellow; disc florets brownish-purple. Achenes cylindrical, obovoid-compressed.

Hemerocallis fulva **(Linn.) Linn. (Pl. 63A)**

Syn: *H. lilioasphodelus* var. *fulva* Linn.

Family	:	Liliaceae.
Vernacular Name	:	Lily.
Eng. Names	:	*Eng.*: Orange Day-lily, Tawny Day-lily.
Distribution	:	Native of China.

Habitat Ecology : Cultivated and naturalized in the Himalaya: 2,400-3,500 m.

Reproductive Cycle : May-June.

Material Examined : EBH-TP-1112.

Systematic Account : Rhizomatous perennials having keeled leaves and large funnel-shaped flowers with out-curved petals borne on stems which are shortly branched above. Inner petals distinctly broader than the outer. Filaments of stamens yellow.

Hibiscus mutabilis Linn. (Pl. 63B)

Syn: *H. aestuans* Rottler.

Family : Malvaceae.

Vernacular Name : Local Gulab.

Eng., H. and Sans. Names : *Eng.*: Changeable Rose, Confederate Rose, Cotton Rose.

H.: Guliajaib, Shalapara, Sthalkamal.

Sans.: Padmacharini, Sthalpadma.

Distribution : Native of China.

Habitat Ecology : Cultivated in gardens; do best in warm humid regions.

Reproductive Cycle : September-December.

Material Examined : EBH-TP-906.

Systematic Account : A deciduous shrub of 2.5 m in height. Leaves 3-5 lobed, hairy on both surfaces. Flowers axillary, pedunculate, red. Involucral bracts 7-10, shorter than the velvety-pubescent calyx-lobes. Staminal tube shorter than the corolla. Capsules globose, flattened, hairy with reniform, hispid seeds.

Hibiscus rosa-sinensis Linn. (Pl. 63C)

Family : Malvaceae.

Vernacular Name : Chini Gulab.

Eng., H. and Sans. Names : *Eng.*: Chinese Hibiscus, Scarlet Rose-mallow, Shoe Flower.

H.: Jasum, Jasun, Jasut.

Sans.: Arkapriya, Aruna, Harivallabha, Japa-pushpam, Java, Joba, Odhrapushpa, Pratiko, Rudra Pushpam.

Distribution : Native of China.

Habitat Ecology : Cultivated throughout India in gardens for its handsome flowers.

Reproductive Cycle : February-October.

Material Examined : EBH-TP-242.

Systematic Account : Shrubs having woody, branched stem and ovate, acuminate leaves. Pedicels axillary, solitary and very long. Calyx with very minute stellate hairs, lanceolate. Corolla tubular below, red. Staminal tube exserted far beyond the petals. Capsules many seeded.

Hibiscus syriacus **Linn. (Pl. 63D)**

Syn: *H. rhombifolius* Cav.

Family : Malvaceae.

Vernacular Name : Brahm phool.

Eng. Name : *Eng.*: Rose of Sharon.

Distribution : Native of Syria.

Habitat Ecology : Cultivated in gardens throughout India; prefers subtropical and mild temperate climates.

Reproductive Cycle : July-September.

Material Examined : EBH-TP-805.

Systematic Account : A deciduous, much-branched erect shrub with 3-lobed, coarsely dentate, short-petioled leaves. Stipules minute. Flowers solitary, on short peduncles. Bracteoles linear, shorter than calyx.

Holoptelea integrifolia **(Roxb.) Planch. (Pl. 64A)**

Syn: *Ulmus integrifolia* Roxb.

Family : Ulmaceae.

Vernacular Names : Papri, Pardesi.

Eng., H. and Sans. Names : *Eng.*: Entire-leaved Elm, Indian Elm.

H.: Banchilla, Begana, Chilbil, Dhamna, Kandru, Kanju, Kumba, Kunja, Papri.

Sans.: Chirabilva.

Distribution : Throughout the greater part of India upto 700 m.

Habitat Ecology : Hot drier regions.

Reproductive Cycle : February-May.

Material Examined : EBH-TP-243.

Systematic Account : A large deciduous tree with bark exfoliating in corky scales. Flowers greenish yellow, polygamous, in short racemes or fascicles. Fruits a samara with membranous wing and flat seed.

Humulus lupulus **Linn. (Pl. 64B)**

Family : Cannabinaceae.

Vernacular Name : Hops.

Eng. Name : *Eng.*: The Hop.

Distribution : Native of Europe.

Habitat Ecology : Cultivated on commercial scale in N. W. Himalaya.

Reproductive Cycle : June-July.

Material Examined : EBH-TP-1205.

Systematic Account : A scabrid twining, perennial herb possessing stout, branched rootstock and stems with reversed bristles. Bracts and bracteoles covered with resinous glands. Flowers dioecious, males panicled; female heads yellow, styles purple.

Hydrangea macrophylla **(Thunb.) DC. (Pl. 64C)**

Syn: *H. hortensis* Sm.

Family : Hydrangeaceae.

Vernacular Name : Hydrangea.

Eng. and H. Names : *Eng.*: Common Hydrangea.
H.: Hydrangea.

Distribution : A native of Japan.

Habitat Ecology : Grown in gardens of hills and temperate areas for their handsome foliage and large showy clusters of flowers.

Reproductive Cycle : June-July.

Material Examined : EBH-TP-106.

Systematic Account : A bushy shrub with brownish branches and oblong-ovate to elliptic-lanceolate leaves. Inflorescence ball-shaped; the flowers range from white through pink to violet. Heads globular (mainly of sterile flowers) or 'lace-caps' with the outer sterile flowers and the inner fertile ones.

Hymenocallis littoralis **(Jacq.) Salisb. (Pl. 64D)**

Family : Amaryllidaceae.

Vernacular Name : Safed Lily.

Eng. Name : *Eng.*: Spider Lily.

Distribution : A native of Southern America.

Habitat Ecology : Grown on the edges of gardens or as potted plant.

Reproductive Cycle : July-August.

Material Examined : EBH-TP-227.

Systematic Account : A bulbous perennial herb with strap-shaped erect leaves. Flowers white, fragrant, borne on a flattened scape. Petals long, narrow, reflexed. Stamens united to corona.

Hyoscyamus niger **Linn. (Pl. 64E)**

Family : Solanaceae.

Vernacular Names : Bajra Bhang, Dhandru.

Eng., H. and Sans. Names : *Eng.*: Henbane.

H.: Khurasani-ajvayan, Khurasani Jamani, Khurasani Yamani.

Sans.: Madakarini, Mani, Parasikaya.

Distribution : Temperate W. Himalaya: 1,600-3,000 m (Kashmir-Garwhal).

Habitat Ecology : Wastelands. Rubbish heaps. Forest clearings.

Reproductive Cycle : April-September.

Material Examined : EBH-TP-1206.

Systematic Account : An erect viscid herb with sessile leaves. Flowers nearly sessile, lower ones in the forks of the branches, upper solitary in the axils of leaf-like bracts. Corolla purple-veined. Stamens protruding. Capsule enclosed in the globose tube of the enlarged calyx.

Hypericum cernuum **Roxb. (Pl. 65A)**

Syn: *H. govanianum* Wall.; *H. oblongifolium* Choisy; *H. speciosum* Wall.

Family : Hypericaceae.

Vernacular Name : Pitrika.

Eng. Name : *Eng.*: St. John's Wort.

Distribution : W. Temperate Himalaya (Kumaon to Sirmour): 1,700-2,350 m.

Habitat Ecology : Common on rocks, ravines, open slopes, shrubberies, banks, steep rocky slopes, etc.

Reproductive Cycle : April-October.

Material Examined : EBH-TP-107.

Systematic Account : Glabrous shrubs upto 180 cm. Branches cylindrical. Leaves minutely dotted. Cymes terminal, 3-5 flowered; flowers yellow. Capsule conical.

Impatiens balsamina Linn. (Pl. 65B)

Family : Balsaminaceae.

Vernacular Name : Balsam.

Eng., H. and Sans. Names : *Eng.*: The Garden Balsam.

H.: Gul-mendi.

Sans.: Dushpatrijati.

Distribution : Throughout India and Sri Lanka, upto 1,500 m. China. Malaya.

Habitat Ecology : Abundant in fields, wastelands; also cultivated.

Reproductive Cycle : July-September.

Material Examined : EBH-TP-1113.

Systematic Account : Erect annuals. Leaves deeply serrate. Petiole glandular. Flowers rose-coloured, showy; pedicels 1-3, axillary. Sepals minute. Wings broad. Lip boat-shaped. Spur incurved. Capsules tomentose. Seeds reticulate.

Ipomoea cairica (Linn.) Sweet (Pl. 65C)

Syn: *I. palmata* Forsk.; *I. pulchella* Roth.; *Convolvulus cairicus* Linn.

Family : Convolvulaceae.

Vernacular Name : Bahubel.

Eng. Name : *Eng.*: Railway Creeper.

Distribution : Deccan Peninsula, Nepal, H.P.

Habitat Ecology : Roadsides. Forests edges. Hedges. Wasteplaces.

Reproductive Cycle : June-November.

Material Examined : EBH-TP-806.

Systematic Account : Perennial climbers with a tubercled stem and deeply palmately 5-7-lobed leaves. Flowers borne on pedicels 1.2-3 cm long. Corolla purplish, funnel-shaped. Capsule ovoid, 4-seeded.

Ipomoea purpurea (Linn.) Roth. (Pl. 65D)

Syn: *I. hirsutula* Jacq. f.; *I. purpurea* var. *diversifolia* (Lindl.) O'Donell; *Pharbitis purpurea* (Linn.) Voigt; *Convolvulus purpureus* Linn.

Family : Convolvulaceae.

Vernacular Name : Gonani.

Eng. Name	:	*Eng.*: Morning Glory.
Distribution	:	India; ascending to 1,000-2,350 m in Khasia; W. Deccan Peninsula.
Habitat Ecology	:	Cultivated areas. Wastelands.
Reproductive Cycle	:	April-October.
Material Examined	:	EBH-TP-116.
Systematic Account	:	Twining herbs with ovate-cordate leaves and large pink flowers in small, stalked clusters. Corolla funnel-shaped. Sepals hirsuate near the base. Seeds glabrous.

Ipomoea quamoclit Linn. (Pl. 66A)

Syn: *Convolvulus pennatus* Lamk.; *C. quamoclit* Spreng.; *Quamoclit pinnata* (Desr.) Bojer; *Q. vulgaris* Chois.

Family	:	Convolvulaceae.
Vernacular Name	:	Nagar Bel.
Eng., H. and Sans. Names	:	*Eng.*: Cypress Vine, Indian Pink, Red Jasmine. *H.*: Kamalata. *Sans.*: Kamalata, Turulata.
Distribution	:	Naturalised in the Himalaya upto 1,200 m.
Habitat Ecology	:	Common in cultivated areas as a climber on shrubs and herbaceous plants.
Reproductive Cycle	:	August-October.
Material Examined	:	EBH-TP-241.
Systematic Account	:	A slender glabrous twinner with leaves pinnately divided into numerous distant segments. Cymes axillary, few-flowered. Flowers erect. Sepals apiculate. Corolla deep red. Capsule ovoid.

Iris germanica Linn. (Pl. 66B)

Family	:	Iridaceae.
Vernacular Name	:	Iris.
Eng. Names	:	*Eng.*: Blue Iris, Flag Iris, German Iris.
Distribution	:	Native of S. Europe.
Habitat Ecology	:	Widely cultivated as an ornamental plant; grow well in exposed soil with plenty of lime or old mortar rubble.
Reproductive Cycle	:	June-July.
Material Examined	:	EBH-TP-109.

Systematic Account : Rhizomatous perennial herbs having a beard at the top of the three lower petals and sword-like leaves. Flowers purple and lilac, fragrant, borne on 61-90 cm long stems.

Ixia chinensis Linn. (Pl. 66C)

Syn: *Belamcanda chinensis* (Linn.) DC; *B. punctata* Moench; *Gemmingia chinensis* (Linn.) Kuntze; *Iris domestica* Goldblatt and Mabb.

Family : Iridaceae.

Vernacular Name : Lily.

Eng. Names : *Eng.*: Blackberry Lily, Leopard Flower, Leopard Lily.

Distribution : Native of Southeastern Asia.

Habitat Ecology : Cultivated in gardens as an ornamental plant.

Reproductive Cycle : July-October.

Material Examined : EBH-TP-919.

Systematic Account : Perennial herb growing upto 1.5m tall with flat, sword-shaped leaves. Flowers yellow to orange in colour with darker spots on the petals, produced in loose terminal branching cymes. Fruits a membranous obovoid Capsule.

Jacaranda acutifolia Humb. and Bonpl. (Pl. 67A)

Syn: *J. mimosifolia* D. Don; *J. ovalifolia* R. Br.

Family : Bignoniaceae.

Vernacular Name : Jackranda.

Eng. Name : *Eng.*: Jacaranda.

Distribution : Outer foothills of the W. Himalaya.

Habitat Ecology : Commonly planted along roadsides and railwaytracks.

Reproductive Cycle : March-June.

Material Examined : EBH-TP-200.

Systematic Account : A medium-sized, deciduous tree attaining 8-10 m in height. Leaves opposite, bipinnate. Flowers bluish-purple, in loose pyramidal panicles. Fruits a flat woody capsule.

Jasminum grandiflorum Linn. (Pl. 67B)

Syn: *J. aureum* Don.; *J. officinale* Linn. var. *grandiflorum* (Linn.) Stokes.

Family : Oleaceae.

Vernacular Names : Chameli, Malti.

Eng., H. and Sans. Names : *Eng.*: Catalonian Jasmine, Common White Garden Jasmine, Spanish Jasmine, White Jasmine.

H.: Chameli, Jati.

Sans.: Bahugandha, Gajavhaya, Jati, Punyagandha.

Distribution : N. W. Himalaya (600-2,500 m); wild. Often cultivated in Indian gardens.

Habitat Ecology : Shrubberies. Ravines.

Reproductive Cycle : April-September.

Material Examined : EBH-TP-400.

Systematic Account : A climbing shrub with subpendent angular branches and imparipinnate leaves. Leaflets 7-11. Flowers white, in lax axillary and terminal cymes. Calyx lobes subulate. Carpels 2.

Jasminum humile **Linn. (Pl. 67C)**

Syn: *J. bignoniaceum* Wall.; *J. chrysanthemum* Roxb.; *J. inodorum* Jacquem; *J. revolutum* Sims.; *J. wallichianum* Lindl.

Family : Oleaceae.

Vernacular Names : Chameli, Samsaun, Sunajui.

Eng., H. and Sans. Names : *Eng.*: Golden Jasmine, Italian Jasmine, Nepal Jasmine, Yellow Jasmine.

H.: Malto, Peeli Chameli, Pitmalti.

Sans.: Gandadhya, Haima, Hemapushpi, Hemapushpika, Pitika, Svarnajuthika.

Distribution : Subtropical Himalaya: 1,500-3,000 m.

Habitat Ecology : Shrubberies, forests; common.

Reproductive Cycle : May-September.

Material Examined : EBH-TP-112.

Systematic Account : Erect shrubs with numerous green angular branches. Leaves pinnate; leaflets elliptic-acute at both ends. Flowers solitary or in dense cymes at the ends of the twigs, yellow; pedicels pubescent. Fruits a juicy black berry.

Jasminum primulinum **Hemsl. *ex* Baker (Pl. 67D)**

Syn: *J. mesnyi* Hance.

Family : Oleaceae.

Vernacular Name : Chameli.

Eng. Names : *Eng.*: Jasmine, Primrose Jasmine.

Distribution : Native of Yunnan (China), cultivated throughout the tropical and subtropical parts of the world.

Habitat Ecology : Grown as an ornamental plant; thrives on poor soil and under adverse conditions.

Reproductive Cycle : March-May.

Material Examined : EBH-TP-111.

Systematic Account : An evergreen twiggy shrub having 4-angled, stiff branches. Leaves opposite, 3-foliolate; leaflets almost sessile, glabrous, thick in texture. Flowers solitary, axillary, primrose-yellow in colour. Calyx lobes lanceolate. Corolla lobes obovate.

Jasminum sambac **(Linn.) Ait. (Pl. 68A)**

Syn: *J. fragrans* Salisb.; *J. pubescens* Wall.; *J. quinqueflorum* Heyne; *J. undulatum* Willd.; *J. zambac* Roxb.; *Mogorium sambac* Lamk.; *M. undulatum* Lamk.; *Nyctanthes sambac* Linn.

Family : Oleaceae.

Vernacular Name : Chameli.

Eng., H. and Sans. Names : *Eng.*: Arabian Jasmine, Lily Jasmine, Sambac Jasmine, Tuscan Jasmine.

H.: Banmallika, Bel, Chamba, Mogra, Motia, Mugra.

Sans.: Ananga, Ashtapadi, Asphota, Devalata, Gandharaja, Gauri, Pramodini, Saptala, Vanachandrika, Varshiki, Vitapriya.

Distribution : Cultivated throughout India and tropics of both hemispheres.

Habitat Ecology : Shrubberies; commonly cultivated.

Reproductive Cycle : Summer–rainy seasons.

Material Examined : EBH-TP-807.

Systematic Account : A suberect shrub with pubescent branchlets and opposite, membranous leaves. Flowers white, very fragrant, usually in 3-flowered terminal cymes. Bracts linear-subulate. Corolla-tube 1.3 cm long; lobes as long as the tube. Ripe carpels 1-2, subglobose, black.

Jatropha panduraefolia **Andr. (Pl. 68B)**

Syn: *J. integerrima* Jacq.; *J. hastata* Jacq.

Family : Euphorbiaceae.

Vernacular Name : Dudhal.

Distribution	: Native of W. Indies; widely distributed throughout warm regions.
Habitat Ecology	: Cultivated in gardens.
Reproductive Cycle	: Summer.
Material Examined	: EBH-TP-723.
Systematic Account	: Flowering shrubs upto 3 m tall. Leaves simple, fiddle-shaped. Flowers in axillary or terminal cyme, unisexual, red. Stamens yellow.

Juglans regia **Linn. (Pl. 68C)**

Syn: *J. regia* var. *Kumaonica* DC.; *J. arguta* Wall.

Family	: Juglandaceae.
Vernacular Names	: Akhrot, Khor.
Eng., H. and Sans. Names	: *Eng.*: Common Walnut Tree, European Walnut, Himalayan Walnut, Walnut Tree. *H.*: Akhrot, Akrot. *Sans.*: Akhoda, Akhota, Akschoda, Akshota.
Distribution	: Temperate Himalaya and W. Tibet: 1,500-3,000 m (wild and cultivated). Khasia Hills (cultivated).
Habitat Ecology	: Forests; widely cultivated in West Himalaya.
Reproductive Cycle	: April-October.
Material Examined	: EBH-TP-110.
Systematic Account	: A large deciduous tree possessing aromatic leaves and bark and imparipinnate leaves. Leaflets ovate, usually entire, sessile. Male spikes lateral on the shoots of the previous year; bracts stipitate. Stamens apiculate. Female flowers 1-3, sessile, in a short terminal spike. Calyx densely tomentose. Petals linear-lanceolate. Nuts distinctly 2-valved, thick-shelled.

Juniperus chinensis **Linn. (Pl. 69A)**

Family	: Cupressaceae.
Vernacular Name	: Juniperus.
Eng. Name	: *Eng.*: Chinese Juniper.
Distribution	: Native of China, Mangolia, Manchuria, Korea and Japan.
Habitat Ecology	: Cultivated, particularly in its shrub and dwarf forms.
Reproductive Cycle	: March-April.

Material Examined	: EBH-TP-521.
Systematic Account	: Dioecious plants upto 20 m tall with broad crown, thin branchlets and with 2 leaf forms; scale leaves tightly appressed, arranged in 4 rows; needle-like leaves in whorls of 3 or in deccusate pairs, apex prickly. Male flowers yellow. Fruits globose, brown, 2-3 seeded.

Juniperus macropoda Boiss. (Pl. 69B)

Syn: *J. excelsa* Brand.; *J. gossainthaneana* Loddig.; *J. polycarpos* K. Koch; *Sabina seravschanica* (Kom.) Nevski.

Family	: Cupressaceae.
Vernacular Names	: Devidiar, Shur.
Eng. Names	: *Eng.*: The Himalayan Pencil Cedar, Indian Juniper.
Distribution	: Afghanistan-C. Nepal and N. Temperate Zone: 1,800-3,600 m.
Habitat Ecology	: Stony slopes, drier areas.
Reproductive Cycle	: September-March.
Material Examined	: EBH-TP-1202.
Systematic Account	: Trees possessing stout trunk with irregular branches and reddish brown, vertically fissured bark. Leaves exhibit mostly scaly leaves on the upper branches whereas lower leaves adpressed to stem. Flowers monoecious; the male cones arise at the tips of the branchlets, the female ones spreading as short lateral branchlets. Fruits globose, blue black, resinous with 2-5 seeds.

Justicia gendarussa Linn. f. (Pl. 69C)

Syn: *Gendarussa vulgaris* Nees.

Family	: Acanthaceae.
Vernacular Name	: Kali Basuti.
H. and Sans. Names	: *H.*: Nilinargandi, Udisanbhalu. *Sans.*: Bhutakeshi, Gandharasa, Indrani, Krishnanirgundi, Nirgundi.
Distribution	: Cultivated throughout India; often as an escape.
Habitat Ecology	: Grown in gardens and naturalized.
Reproductive Cycle	: January-March.

Material Examined : EBH-TP-903.

Systematic Account : An undershrub with erect stem and short-petioled leaves. Spikes terminal often forming a panicle; flowers clustered. Bracts linear. Corolla white with purple spots. Capsule 4-seeded.

Kochia trichophylla Voss. (Pl. 70A)

Family : Chenopodiaceae.

Vernacular Name : Kochia.

Eng. Names : *Eng.*: Fire Bush, Kochia, Summer Cypress.

Distribution : S. France; eastwards to Japan.

Habitat Ecology : Usually grown in garden beds and along the pathways for indoor and outdoor decoration.

Reproductive Cycle : Flowering after the rainy season.

Material Examined : EBH-TP-220.

Systematic Account : Erect, much-branched annuals attaining a height upto one metre. Leaves narrow, needle-like which turn purple when old. Flowers red.

Lagerstroemia indica Linn. (Pl. 70B)

Syn: *L. elegans* Wall.; *Velaga globosa* Gaertn.

Family : Lythraceae.

Vernacular Name : Har-Shingar.

Eng. and H. Names : *Eng.*: China Privet, Crape, Crape Myrtle, Indian Lilac, Myrtle.

H.: Phurush, Saoni, Telinga-china.

Distribution : Throughout India; usually cultivated in gardens. Originally from China.

Habitat Ecology : Planted.

Reproductive Cycle : June-October.

Material Examined : EBH-TP-810.

Systematic Account : A leaf-shedding tree possessing slenderly pedicelled flowers occurring solitary or in peduncled cymes at the end of the branchlets. Calyx deeply 6-lobed. Petals crumpled and curled. Outer 6 stamens longer and larger than the numerous inner ones. Capsules globose.

Lannea coromandelica (Houtt.) Merrill (Pl. 70C)

Syn: *L. grandis* (Dennst.) Engler; *Dialium coromandelicum* Houtt.; *Haberlia grandis* Dennst.; *Odina pinnata* Rottl.; *O. wodier* Roxb.

Family	:	Anacardiaceae.
Vernacular Name	:	Kaimbal.
H. and Sans. Names	:	*H.*: Jhingan, Jingan, Kaimil, Mohi, Mohin, Mohni, Moyen, Thingan. *Sans.*: Jingini.
Distribution	:	Throughout India to an altitude of 1,500 m.
Habitat Ecology	:	Common in dry locales of mixed deciduous forests of the hills and plains; often planted as an avenue tree.
Reproductive Cycle	:	February-June.
Material Examined	:	EBH-TP-224.
Systematic Account	:	Deciduous trees with a spreading crown, stout branches, bark exfoliating in irregular, rounded plates and imparipinnate leaves crowded at the ends of branches. Leaflets 7-9, oblong-ovate. Flowers small, unisexual; male fascicled in slender compound racemes and female in simple pubescent racemes. Fruits reniform, 1-seeded.

Lawsonia alba Lam. (Pl. 70D)

Syn: *L. inermis* Roxb.; *L. purpurea* Lamk.; *L. spinosa* Linn.

Family	:	Lythraceae.
Vernacular Name	:	Mendi.
Eng., H. and Sans. Names	:	*Eng.*: Camphire, Cypress Shrub, Egyptian Privet, Henna Plant. *H.*: Hena, Mehndi, Mhindi. *Sans.*: Dvivranta, Kuravaka, Medika, Mendhi, Mendika, Nakharanjani, Ragagarbha, Ranjaka, Sakachara, Timira.
Distribution	:	Throughout India, very common, cultivated and naturalized.
Habitat Ecology	:	Widely cultivated in tropical countries as a dye plant.
Reproductive Cycle	:	April-July.
Material Examined	:	EBH-TP-503.
Systematic Account	:	A glabrous much branched erect shrub with lateral branches often ending in a spinous point and shortly

petioled leaves narrowed at the base. Flowers numerous, fragrant, in large terminal pyramidal panicled cymes. Petals as broad as long. Stamens 8, inserted in pairs on the calyx-tube. Capsules supported by the persistent calyx and tipped with the style.

Lespedeza sericea (Thunb.) Miq. (Pl. 70E)

Syn: *L. cuneata* (Dum.-Cours) G. Don; *L. juncea* (Linn.f.) Pers. var. *sericea* (Miq.) Forbes and Hemsl.; *Hedysarum sericeum* Thunb.

Family : Fabaceae.

Vernacular Name : Karkura.

Distribution : Throughout Himalaya: 1,500-2,500 m.

Habitat Ecology : Cultivated.

Reproductive Cycle : July-August.

Material Examined : EBH-TP-1114.

Systematic Account : Erect shrubs. Stems densely pubescent with nearly sessile, overlapping leaves. Leaflets wedge-shaped; lower surface densely white-silky. Flowers in small, axillary clusters. Calyx white-silky. Keel slightly curved. Pods thinly silky, hardly longer than the calyx.

Leucaena leucocephala (Lam.) de Wit (Pl. 70F)

Syn: *L. glauca* Benth.; *Acacia biceps* Willd.; *A. frondosa* Willd.; *A. glauca* (Linn.) Moench; *Mimosa glauca* Linn.; *M. leucocephala* Lam.

Family : Fabaceae.

Vernacular Names : Alseenia, Lasini, Lasunia.

Eng. and H. Names : *Eng.*: Lead Tree, Leucaena, White Popinac.

H.: Kadam, Kubabul, Toira.

Distribution : Throughout India.

Habitat Ecology : Hedges, along boundaries of fields; also grown in gardens and plantations.

Reproductive Cycle : September-November.

Material Examined : EBH-TP-1115.

Systematic Account : Unarmed shrubs or a small tree upto 9 m having bipinnate, channelled leaves ending in a weak spine. Pinnae 3-8 pairs with pubescent rachises. Leaflets 10-15 pairs, membranous. Flowers borne in dense globose heads. Pods straight, flat, obliquely triangular at apex, 15-25 seeded.

Ligustrum lucidum Aiton (Pl. 71A)

Syn: *L. nepalense* var. *glabrum* Hook. f.; *L. spicatum* Hort.

Family	:	Oleaceae.
Vernacular Name	:	Pataal Ped.
Eng. Names	:	*Eng.*: Chinese Privet, Glossy Privet.
Distribution	:	Found in Khasi hills upto 1,500 m; often cultivated in gardens.
Habitat Ecology	:	Cultivated.
Reproductive Cycle	:	July-November.
Material Examined	:	EBH-TP-407.
Systematic Account	:	A large evergreen shrub possessing grey bark and coriaceous glossy leaves. Flowers in panicles, creamy white, fragrant. Fruits oblong, blue black when ripe.

Litchi chinensis Sonn. (Pl. 71B)

Syn: *Nephelium litchi* Camb.

Family	:	Sapindaceae.
Vernacular Name	:	Litchi.
Eng. and H. Names	:	*Eng.*: Litchi. *H.*: Litchi.
Distribution	:	Native of S. China; widely cultivated in India.
Habitat Ecology	:	Widely grown in the tropics.
Reproductive Cycle	:	April-June.
Material Examined	:	EBH-TP-226.
Systematic Account	:	An evergreen shrub or small tree with exstipulate leaves and greenish flowers in terminal panicles. Calyx cup-shaped. Petals 0. Stamens 6-10, inserted inside a fleshy, glabrous disk. Fruits red when ripe, containing one large obovoid brown seed enclosed in a fleshy edible aril. Pericarp dry, thin, brittle.

Litsea polyantha Juss. (Pl. 71C)

Syn: *L. monopetala* (Roxb.) Pers.; *Tetranthera alnoides* Miquel; *T. hexantha* Sieb.; *T. macrophylla* Wall.; *T. monopetala* Roxb.; *Tomex pubescens* Herb.

Family	:	Lauraceae.
Vernacular Name	:	Sangran.
Eng., H. and Sans. Names	:	*Eng.*: Common Grey Mango Laurel.

		H.: Gwa, Kakuri, Kari, Meda, Papria, Randkari, Sangran.
		Sans.: Gajapippali.
Distribution	:	Sub-Himalayan tract from the Ravi eastwards, ascending to 1,500 m.
Habitat Ecology	:	Preferring moist localities.
Reproductive Cycle	:	March-May.
Material Examined	:	EBH-TP-809.
Systematic Account	:	Trees. Bark pale brown, corky. Young shoots rusty pubescent. Leaves with pale yellowish brown pubescence beneath. Flowers pale greenish yellow, in rounded umbellate heads. Fruits ovoid, black, seated on the persistent base of perianth.

Litsea sebifera Pers. (Pl. 71D)

Syn: *L. chinensis* Lam.; *Sebifera glutinosa* Lour.; *Tetranthera apetala* Roxb.; *T. laurifolia* Roxb.

Family	:	Lauraceae.
Vernacular Name	:	Kala Rajaan.
Eng., H. and Sans. Names	:	*Eng.*: Common Tallow Laurel.
		H.: Garbijaur, Maidalakri, Medh, Menda, Singrauf.
		Sans.: Adhavara, Drava, Jivani, Manichhidra, Meda, Medovati, Rasa, Sadhavi, Shalyada, Shreshtha, Snigdha.
Distribution	:	Throughout the hotter parts of India.
Habitat Ecology	:	Wastelands. Roadsides. Dry slopes.
Reproductive Cycle	:	May-July.
Material Examined	:	EBH-TP-248.
Systematic Account	:	A medium-sized tree having pubescent branchlets, inflorescence and underside of leaves. Flowers yellowish, in umbellate heads arranged in lax pedunculate axillary corymbs. Stamens upto 20 or more. Filaments clothed with long soft hairs. Fruits globose, black, supported by the thick pedicel.

Lonicera japonica Thunb. (Pl. 72A)

Syn: *L. macrantha* DC.; *Caprifolium japonicum* Don; *C. macranthum* Don.

Family	:	Caprifoliaceae.
Vernacular Name	:	Lushai.

Eng. Names	:	*Eng.*: Honeysuckle, Japanese Honeysuckle.
Distribution	:	C. Nepal-Bhutan, China and S.E. Asia: 1,500-2,400 m.
Habitat Ecology	:	Shrubberies, planted.
Reproductive Cycle	:	May-August.
Material Examined	:	EBH-TP-104.
Systematic Account	:	Semi-evergreen shrubby climbers having coriaceous leaves which are villous beneath. Peduncles in subterminal panicles. Calyx-tube more or less hairy. Corolla-lobes fading to yellow. Berries black.

Lycopersicum esculentum **Mill. (Pl. 72B)**

Syn: *L. cerasiforme* Dunal.; *Solanum humboldtii* Willd.; *S. lycopersicum* Linn.; *S. pseudolycopersicum* Jacq.

Family	:	Solanaceae.
Vernacular Name	:	Tamatar.
Eng. and H. Names	:	*Eng.*: Love-apple, Tomato.
		H.: Gur-begun, Tamatar, Tamati, Timoti.
Distribution	:	Cultivated in India.
Habitat Ecology	:	Cultivated; often an escape on wastelands and along borders of agricultural fields.
Reproductive Cycle	:	May-June.
Material Examined	:	EBH-TP-249.
Systematic Account	:	A trailing herb with pinnate leaves and yellow flowers in cymes. Berries globose, red. Seeds compressed.

Machilus gamblei **King *ex* Hook. f. (Pl. 72C)**

Family	:	Lauraceae.
Vernacular Name	:	Khadamb.
Distribution	:	Outer Himalaya; the sub-Himalayan tract from Chamba eastwards and in Assam upto 1,500 m.
Habitat Ecology	:	Moist and swampy places along streams and ravines.
Reproductive Cycle	:	March-June.
Material Examined	:	EBH-TP-1116.
Systematic Account	:	A moderate-sized tree having rough bark and thinly coriaceous leaves. Flowers in panicles, yellowish green; perianth lobes villous within. Fruits globose, black.

Magnolia grandiflora **Linn. (Pl. 73A)**

Syn: *M. angustifolia* Millais; *M. elliptica* Link; *M. foetida* (Linn.) Sarg.; *M. lanceolata* Link; *M. rotundifolia* Millais.

Family	:	Magnoliaceae.
Vernacular Names	:	Champa, Champakali.
Eng. and H. Names	:	*Eng.*: Bull Bay, Southern Magnolia. *H.*: Andachampa, Him Champa.
Distribution	:	Indigenous to Southeastern United States (N. Carolina, Florida and Texas).
Habitat Ecology	:	Planted in gardens.
Reproductive Cycle	:	May-August.
Material Examined	:	EBH-TP-1002.
Systematic Account	:	Semideciduous pyramidal trees with young branches covered in brown hairs and simple leathery leaves. Flowers creamy white, fragrant and measure upto 20 cm across when fully open. Fruits brown colored and measure upto 10 cm long.

Mallotus philippinensis **(Lam.) Muell.-Arg. (Pl. 73B)**

Syn: *Croton cascarilloides* Rauesch. *ex* Steud; *C. coccineus* Vahl.; *C. distans* Wall.; *C. montanus* Willd.; *C. philippensis* Lamk.; *C. punctatus* Retz.; *Rottlera affinis* Hassk.; *R. aurantiaca* Hook. and Arn.; *R. montana* Wall.; *R. tinctoria* Roxb.

Family	:	Euphorbiaceae.
Vernacular Names	:	Kaamal, Kaambal.
Eng., H. and Sans. Names	:	*Eng.*: Kamala Tree, Monkey Face Tree. *H.*: Kamala, Kambhal, Kambila, Kamela, Kamila, Kamud, Rauni, Rohni, Ruin, Rulu. *Sans.*: Bahupushpa, Kampilla, Kapila, Karkatha, Lohitanga, Madhuka, Nadivasa, Pikaksha, Punnaga, Raktachurnaka, Raktaphala, Ranjaka, Rechanaka, Rochana.
Distribution	:	Throughout Tropical India, Sri Lanka, Myanmar.
Habitat Ecology	:	Often gregarious; moist forests.
Reproductive Cycle	:	April-October.
Material Examined	:	EBH-TP-702.
Systematic Account	:	A small much-branched rusty-pubescent tree. Leaves with numerous orbicular red glands beneath,

reticulately veined. Flowers dioecious; males clustered in erect terminal spikes, the females in short spikes. Capsules 3-lobed with globose black seeds.

Malva sylvestris **Linn. (Pl. 73C)**

Syn: *M. mauritiana* Linn.; *M. sylvestris* var. *mauritiana* (Linn.) Boiss.

Family : Malvaceae.

Vernacular Name : Gulkhair.

Eng. and H. Names : *Eng.*: Blue Mallow, Bread-and-cheese, Common Mallow, Cheese Cake, Country Mallow, Marsh Mallow, Pick Cheese.

H.: Gulkhair, Kunzi, Vilayati-kangai.

Distribution : Western Temperate Himalaya, from Punjab to Kumaon, upto 2,400 m.

Habitat Ecology : Wastelands. Drier regions. Periphery of the forests.

Reproductive Cycle : May-October.

Material Examined : EBH-TP-1117.

Systematic Account : Tall, erect, glabrous herbs with strong, branched stem. Flowers large, stalked. Bracteoles entire. Corolla purple, much longer than the calyx. Seeds many, reniform.

Malva verticillata **Linn. (Pl. 73D)**

Syn: *M. alchemillaefolia* Wall.; *M. neilgherrensis* Wight.

Family : Malvaceae.

Vernacular Name : Pinzi.

Eng. Name : *Eng.*: Cluster Mallow.

Distribution : Throughout Himalayas upto 4,000 m.

Habitat Ecology : Cultivated areas. Wastelands.

Reproductive Cycle : May-September.

Material Examined : EBH-TP-1118.

Systematic Account : An erect, branched, pubescent herb upto 120 cm high with flowers in dense axillary clusters. Bracteoles narrowly lanceolate, acute. Petals pale pink, nearly twice as long as the calyx.

Malvaviscus conzatii **Green. (Pl. 74A)**

Family : Malvaceae.

Vernacular Name : Gudhal.

Distribution : Cultivated in tropical gardens.

Habitat Ecology : Hedges; near houses.

Reproductive Cycle : Major part of the year.

Material Examined : EBH-TP-250.

Systematic Account : Shrubs upto 2.5 m tall. Leaves alternate, simple, broadly ovate, coarsely-toothed with upper surface rough. Flowers usually solitary, axillary, scarlet.

Mangifera indica **Linn. (Pl. 74B)**

Syn: *M. domestica* Gaertn.

Family : Anacardiaceae.

Vernacular Name : Amb.

Eng., H. and Sans. Names : *Eng.*: Mango Tree.

H.: Am, Amb.

Sans.: Amra, Chutu, Madha-dut.

Distribution : Tropical Himalaya.

Habitat Ecology : Planted along roadsides, field borders, local pathways.

Reproductive Cycle : March-July.

Material Examined : EBH-TP-251.

Systematic Account : An evergreen tree having coriaceous leaves at the ends of the branches. Flowers monoecious, in many-flowered pubescent panicles. Drupes large, fleshy, obliquely pyriform.

Melia azadirachta **Linn. (Pl. 74C)**

Syn: *M. indica* Brandis; *M. parviflora* Moon.; *Antelaea azadirachta* (Linn.) Adelb.; *Azadirachta indica* A. Juss.

Family : Meliaceae.

Vernacular Names : Neem, Nim.

Eng., H. and Sans. Names : *Eng.*: Indian Lilac, Margosa Tree, Neem Tree.

H.: Bal-nimb, Nim, Nimb, Ninb.

Sans.: Arishta, Nimba, Nimba-vrikshaha.

Distribution : Common, cultivated, throughout the greater part of India and Myanmar.

Habitat Ecology : Throughout the drier parts.

Reproductive Cycle : February-April.

Material Examined : EBH-TP-522.

Systematic Account : Trees with a straight trunk, imparipinnate leaves, subopposite leaflets and flowers in axillary panicles. Staminal tube a little shorter than the petals. Ovary 3-celled. Fruits a 1-seeded drupe having endocarp woody.

Melia azedarach Linn. (Pl. 74D)

Syn: *M. bukayun* Royle; *M. sempervirens* Sw.

Family : Meliaceae.

Vernacular Names : Drek, Kachen.

Eng., H. and Sans. Names : *Eng.*: Barbados Lilac, Bead Tree, Indian Lilac, Persian Lilac.

H.: Bakain, Bakarja, Bakayan, Betain, Deikna, Drek, Maha-ninb.

Sans.: Mahanimba, Himadruma, Parvata-nimba-vrikshaha.

Distribution : Throughout India: 600-1,000 m.

Habitat Ecology : Planted in villages, along roadsides.

Reproductive Cycle : May-October.

Material Examined : EBH-TP-252.

Systematic Account : Trees with short trunk and broad crown. Leaves bipinnate; ultimate leaflets 3-12. Flowers lilac, honey-scented, in long-peduncled, axillary panicles. Drupes subglobose, 4-seeded.

Mentha longifolia (Linn.) Huds. (Pl. 74E)

Family : Lamiaceae.

Vernacular Names : Jungli Pudina, Madeni.

Eng., H. and Sans. Names : *Eng.*: Horse-mint.

H.: Jungli Pudina, Podina.

Sans.: Ajirnahara, Pudina, Rochani, Shakashobana, Sugandhipatra, Vantihara, Vyanjana.

Distribution : Western Himalaya extending upto Central Nepal. Tibet, N. W. Asia, Europe.

Habitat Ecology : Common along streamsides, waysides, bogs, irrigation channels, borders of paddy fields.

Reproductive Cycle : November-February.

Material Examined : EBH-TP-605.

Systematic Account : Perennial sweet smelling herbs with branched underground suckers, erect stem upto 1 m, and saw-toothed, densely hoary leaves which are pale beneath. Flowers in whorls forming slender spikes borne at the ends of the branches.

Mentha piperita **Linn. (Pl. 75A)**

Family : Lamiaceae.

Vernacular Name : Pudina.

Eng. and H. Names : *Eng.*: Brandy Mint, Peppermint.

H.: Gamathi Pudina, Paparaminta.

Distribution : Cultivated in Indian gardens.

Habitat Ecology : Commonly near water channels, moist places.

Reproductive Cycle : July-September.

Material Examined : EBH-TP-707.

Systematic Account : An erect, fragrant, perennial herb having coarsely serrated petioled leaves and whorls in terminal cylindric spikes which are interrupted below. Bracts minute. Pedicels and flowers very sparingly hispid.

Michelia champaca **Linn. (Pl. 75B)**

Syn: *M. aurantiaca* Wall.; *M. doldsopa* Ham.; *M. rheedii* Wight.; *M. rufinervis* DC.; *Champaca velutina* Kuntze.

Family : Magnoliaceae.

Vernacular Name : Champa.

Eng., H. and Sans. Names : *Eng.*: Golden Champa, Yellow Champa.

H.: Champa, Champac, Champaca.

Sans.: Anjana, Champaka, Champeya, Deepapushpa, Hemapushpa, Kanchana, Kusuma, Peetapushpa, Surabhi, Svarnapushpa.

Distribution : Wild in E. sub-Himalayan tract and lower hills upto 1,000 m; cultivated in various parts of India and Myanmar.

Habitat Ecology : Commonly cultivated as a garden ornamental.

Reproductive Cycle : April-September.

Material Examined : EBH-TP-523.

Systematic Account : Evergreen trees with a straight trunk and ascending branches forming a close head. Flowers fragrant, axillary, solitary, each enclosed in bud by a coriaceous deciduous bract. Capsules dark brown, opening on the back by two valves.

Mimosa pudica Linn. (Pl. 75C)

Syn: *M. hispidula* Kunth.

Family	: Fabaceae.
Vernacular Name	: Chhui-mui.
Eng., H. and Sans. Names	: *Eng.*: The Sensitive Plant, Touch-me-not. *H.*: Chui-mui, Lajalu, Lajjavati. *Sans.*: Lajjalu, Varahakranta.
Distribution	: Native of tropical America; naturalized throughout tropical and subtropical parts of India.
Habitat Ecology	: Open wastelands.
Reproductive Cycle	: Flowers and fruits throughout the year.
Material Examined	: EBH-TP-908.
Systematic Account	: A diffuse undershrub having stems and branches clothed with long bristles from bulbous bases and sensitive, digitate leaves. Pinnae 1-2 pairs; leaflets 12-20 pairs, subcoriaceous. Flowers pink, in small peduncled heads, 1-2 from each axil, all down the branches. Stamens 4, exserted. Pods with densely prickly sutures.

Mimusops elengi Linn. (Pl. 75D, E)

Syn: *M. lucida* Poir.; *M. parvifolia* R.Br.; *Imbricaria perroudii* Montrouz.

Family	: Sapotaceae.
Vernacular Name	: Molsari.
H. and Sans. Names	: *H.*: Bakul, Maulsarau, Maulser, Mulsari. *Sans.*: Bakul, Vakula.
Distribution	: Deccan and Malay Peninsulas; frequently cultivated in N. India.
Habitat Ecology	: Often planted in temple courtyards of H.P.
Reproductive Cycle	: March-May.
Material Examined	: EBH-TP-201.
Systematic Account	: Evergreen trees with a compact leafy head, and short erect trunk, elliptic leaves and white, fragrant flowers

which are solitary or in fascicles of 2-6. Corolla-lobes 16-20, white. Staminodes 8, alternate with the stamens. Berries 1-seeded, yellow when ripe.

Mirabilis jalapa Linn. (Pl. 76A)

Syn: *M. jalapa* subsp. *lindheimeri* Standl.; *M. lindheimeri* (Standl.) Shinners.

Family	:	Nyctaginaceae.
Vernacular Name	:	Gulal.
Eng., H. and Sans. Names	:	*Eng.*: Four O'Clock Flower, Marvel of Peru.
		H.: Gulabash, Gulabbas.
		Sans.: Krishnakeli, Sandhyakali.
Distribution	:	Throughout India; often planted.
Habitat Ecology	:	Moist localities. Wasteplaces near habitations.
Reproductive Cycle	:	June-September.
Material Examined	:	EBH-TP-604.
Systematic Account	:	Leafy herbs with large perennial tuberous roots, fleshy stems, cordate leaves, long tubular fragrant flowers and elliptical, leathery, black fruits.

Momordica charantia Linn. (Pl. 76B)

Syn: *M. humilis* Wall.; *M. muricata* DC; *M. senegalensis* Lamk.

Family	:	Cucurbitaceae.
Vernacular Name	:	Karela.
Eng., H. and Sans. Names	:	*Eng.*: Bitter Gourd, Carilla Fruit.
		H.: Karela, Kareli, Karola.
		Sans.: Ambuvallika, Chiripatra, Kandura, Kantaphalla, Karaka, Kara-valli-lata, Karavella, Pitapushpa, Sukanda, Sushavi, Visakantaki.
Distribution	:	Throughout India.
Habitat Ecology	:	Cultivated over much of the tropics for their edible fruits.
Reproductive Cycle	:	July-October.
Material Examined	:	EBH-TP-716.
Systematic Account	:	Annuals having much branched, angled and grooved stem with simple tendrils and leaves deeply divided into 5-7 lobes. Flowers monoecious. Male peduncle furnished with a large bract. Corolla lemon-yellow. Female peduncle slender. Staminodes 3, glanduliform. Fruits many-ribbed, covered with triangular tubercles.

Morus alba **Linn. (Pl. 76C)**

Syn: *M. serrata* Wall.; *M. tatarica* Linn.

Family	:	Moraceae.
Vernacular Name	:	Toot.
Eng., H. and Sans. Names	:	*Eng.*: Mulberry, White Mulberry.
		H.: Chinni, Chun, Tul, Tulklu, Tut.
		Sans.: Tula.
Distribution	:	Cultivated in Panjab, N. W. Himalaya and Western Tibet, ascending to 3,300 m.
Habitat Ecology	:	Common along agricultural fields.
Reproductive Cycle	:	March-May.
Material Examined	:	EBH-TP-717.
Systematic Account	:	Deciduous trees with brown bark, lobed leaves and greenish flowers. Male spikes puberulous, flowers minutely pedicellate. Female spikes ovoid, pedunculate. Mature fruits black.

Morus serrata **Roxb. (Pl. 76D)**

Syn: *M. alba* var. *serrata* Bureau; *M. pabularia* Dcne.;
***M. vicorum* Jacquem.**

Family	:	Moraceae.
Vernacular Names	:	Cheun, Kroon.
Eng. and H. Names	:	*Eng.*: Himalayan Mulberry.
		H.: Himu, Kimu.
Distribution	:	Throughout Himalaya: 1,200-2,700 m.
Habitat Ecology	:	Forests, shrubberies; also cultivated in drier areas.
Reproductive Cycle	:	April-June.
Material Examined	:	EBH-TP-603.
Systematic Account	:	A large deciduous tree upto 15 m high. Leaves serrated. Male and female flowers spikes on different trees. Male spikes fall off after flowering. Style arms united near the base. Fruits 1.5-2 cm long, purple when ripe.

Murraya koenigii **(Linn.) Spreng. (Pl. 76E)**

Syn: *Bergera koenigii* Linn.

Family	:	Rutaceae.
Vernacular Names	:	Gandla, Kari patta.

Eng., H. and Sans. Names : *Eng.*: Curry-leaf Tree.

H.: Bursunga, Gandhela, Harri, Katnim.

Sans.: Krishna-nimba, Nimba-patram, Surabhi, Surabhi-nimba.

Distribution : Along the foot of Himalaya, ascending to 1,650 m.

Habitat Ecology : Open rocky slopes.

Reproductive Cycle : March-June.

Material Examined : EBH-TP-230.

Systematic Account : Strong smelling, small-sized trees with dark grey bark, imparipinnate leaves, terete petiole, 11-25 leaflets and white flowers in much-branched terminal peduncled corymbose cymes. Petals gland-dotted. Filaments dilated at the base. Ovary 2-celled. Fruits black, rugose.

Murraya paniculata (Linn.) Jack. (Pl. 77A)

Syn: *M. exotica* Linn.; *Chalcas exotica* (Linn.) Millsp.; *C. paniculata* Linn.

Family : Rutaceae.

Vernacular Name : Kamini.

Eng. and H. Names : *Eng.*: Orange Jessamine, The China Box Tree.

H.: Bibsar, Juti, Kamini, Marchula.

Distribution : Throughout the plains of India.

Habitat Ecology : Shady ravine slopes; also cultivated.

Reproductive Cycle : March-September.

Material Examined : EBH-TP-234.

Systematic Account : Evergreen shrubs or small trees possessing smooth, yellowish white bark and slender branches. Leaves imparipinnate; leaflets 3-9, usually acuminate with a notched tip. Flowers white, fragrant, solitary or in terminal and axillary corymbs. Berries, red or deep orange when ripe.

Musa paradisiaca Linn. (Pl. 77B)

Family : Musaceae.

Vernacular Name : Kela.

Eng., H. and Sans. Names : *Eng.*: Banana.

H.: Kela.

Sans.: Kadali, Rambha.

Distribution : Indigenous to tropical Asia; cultivated throughout India.

Habitat Ecology : Universally cultivated in the tropics for food.

Reproductive Cycle : Throughout the year.

Material Examined : EBH-TP-136

Systematic Account : A monoecious perennial herb with rhizome and false aerial stem upto 10 m tall. Leaves oblong. Inflorescence drooping and appear at the top of the aerial stem.

Narcissus tazetta **Linn. (Pl. 77C)**

Syn: *N. linnaeanus* Rouy; *N. linnaeanus* subsp. *tazetta* Rouy; *Hermione tazetta* (Linn.) Haw.; *Jonquilla tazetta* (Linn.) Raf.; *Pancratium tazetta* (Linn.) Sesse and Moc.

Family : Amaryllidaceae.

Vernacular Name : Nargis.

Eng. Name : *Eng.*: Narcissus.

Distribution : Widely distributed in Europe and Japan.

Habitat Ecology : Commonly grown in garden as an ornamental plant.

Reproductive Cycle : January-February.

Material Examined : EBH-TP-406.

Systematic Account : Herbs possessing copiously truncated bulbs with brown membranous coats and bluntly keeled leaves at the back. Flowers 4 cm across, white and yellow in colour, fragrant.

Nerium odorum **Soland. (Pl. 78A)**

Syn: *N. indicum* Mill.; *N. odoratum* Lamk.; *N. oleander* Linn.

Family : Apocynaceae.

Vernacular Names : Kaner, Gandheera, Lal Kaner.

Eng., H. and Sans. Names : *Eng.*: Indian Oleander, Sweet-scented Oleander.

H.: Kanel, Kaner, Karber, Kuruvira.

Sans.: Asvamaraca, Chandata, Hayamaraca, Karavira, Pratihasa, Sataprasa.

Distribution : Western Himalaya (Nepal westwards, ascending to 2,000 m). Extensively cultivated throughout the greater part of India as well as in China and Japan.

Habitat Ecology : Frequently grown in gardens.

Reproductive Cycle : April-November.

Material Examined : EBH-TP-102.

Systematic Account : A glabrous evergreen milky shrub with coriaceous leaves and sweet-scented flowers borne in terminal cymes. Corolla lobes rounded. Filaments hairy. Follicles cylindrical, green.

Nicandra physaloides (Linn.) Gaertn. (Pl. 78B)

Syn: *Atropa physaloides* Linn.; *Physalis daturaefolia* Lamk.

Family : Solanaceae.

Vernacular Name : Lohia.

Eng. Name : *Eng.*: Apple of Peru.

Distribution : Subtemperate Himalaya (1,000-2,000 m), W. Deccan Peninsula.

Habitat Ecology : Cultivated areas. Wastelands.

Reproductive Cycle : May-October.

Material Examined : EBH-TP-115.

Systematic Account : Herbaceous glabrous annuals with ovate-lanceolate, toothed, petioled leaves, blue campanulate corolla, globose berries and many, compressed, subdiscoid seeds.

Nyctanthes arbor-tristis Linn. (Pl. 78C)

Syn: *Parilium arbor-tristis* Gaertn.; *Scabrita scabra* Linn.; *S. triflora* Linn.

Family : Oleaceae.

Vernacular Name : Har-Shingar.

Eng., H. and Sans. Names : *Eng.*: Coral Jasmine, Indian Mourner, Night-flowering Jasmine, Sorrowful Tree.

H.: Har, Harsinghar, Nibari, Saherwa, Seoli, Siharu.

Sans.: Parijataka, Rajanikasa, Sephalica.

Distribution : Throughout sub-Himalayan tract ascending to 1,500 m.

Habitat Ecology : Cultivated; often in gardens as an ornamental plant.

Reproductive Cycle : July-October.

Material Examined : EBH-TP-718.

Systematic Account : Large shrubs or small trees with stiff whitish hairs, sharply quadrangular young branches, and sweet-scented, sessile flowers in pedunculate bracteate fascicles of 3-5. Peduncles 4-angled. Corolla tube orange-coloured. Capsules compressed with 2-celled.

Ocimum basilicum Linn. (Pl. 78D)

Syn: *O. album* Linn.; *O. americanum* Jacq.; *O. caryophyllatum* Roxb.; *O. ciliatum* Hornem.; *O. hispidum* Lamk.; *O. menthaefolium* Benth.; *O. minimum* Burm.; *O. pilosum* Willd.; *Plectranthus barrelieri* Spreng.

Family	:	Lamiaceae.
Vernacular Name	:	Tulsi.
Eng., H. and Sans. Names	:	*Eng.*: Common Basil, Common Sweet Basil, Sweet Basil.
		H.: Babui-tulsi, Babul, Kali-tulsi, Sabzah.
		Sans.: Ajagandhika, Asurasa, Karahi, Munjariki, Surabhi, Tulasidvesha, Varvara.
Distribution	:	Cultivated throughout the greater part of India, Sri Lanka and Myanmar.
Habitat Ecology	:	Commonly cultivated, rarely an escape in moist slopes.
Reproductive Cycle	:	June-September.
Material Examined	:	EBH-TP-256.
Systematic Account	:	Hispidly pubescent branching herbs. Flowers in densely racemose whorls. Fruiting calyx shortly pedicelled. Stamens slightly exserted. Nutlets ellipsoid, black and pitted.

Ocimum gratissimum Linn. (Pl. 78E)

Syn: *O. viride* Willd.

Family	:	Lamiaceae.
Vernacular Name	:	Jangli Tulsi.
Eng., H. and Sans. Names	:	*Eng.:* Large Basil, Lemon Basil, Shrubby Basil, Tea Bush.
		H.: Bantulsi, Malatulsi, Ramtulsi.
		Sans.: Ajaka, Doshakleshi, Nidralu, Ramatulsi, Shophahari, Sugandhi, Sumukha, Vriddhatulasi.
Distribution	:	Throughout India.
Habitat Ecology	:	Roadsides and wasteplaces; often cultivated.
Reproductive Cycle	:	September-October.
Material Examined	:	EBH-TP-257.
Systematic Account	:	Strongly scented, much-branched tall perennial shrubs having subquadrangular stem and branches, gland-dotted leaves pubescent on both surfaces and pale greenish-yellow flowers borne in simple or branched racemes. Bracts sessile. Nutlets subglobose.

Ocimum kilimandscharicum Gurke (Pl. 79A)

Family	: Lamiaceae.
Vernacular Name	: Tulsi.
Eng., H. and Sans. Names	: *Eng.*: Camphor Basil, Hoary Basil. *H.*: Kapurtulsi. *Sans.*: Karpuratulasi.
Distribution	: Native of East Africa; introduced and cultivated in India.
Habitat Ecology	: Drier areas; also cultivated.
Reproductive Cycle	: August-September.
Material Examined	: EBH-TP-408.
Systematic Account	: Aromatic undershrubs with pubescent branchlets and opposite leaves pubescent on both surfaces. Flowers in 4-6 whorls on long villose racemes. Nutlets oblong, black to brown.

Ocimum sanctum Linn. (Pl. 79B)

Syn: *O. inodorum* Burm; *O. tenuiflorum* Linn.

Family	: Lamiaceae.
Vernacular Names	: Ram Tulsi, Tulsi.
Eng., H. and Sans. Names	: *Eng.*: The Sacred Basil. *H.*: Baranda, Kala-tulsi, Tulsi, Varanda. *Sans.*: Ajaka, Manjarika, Parnasa, Sorasaw, Tulashi Tulasi.
Distribution	: Throughout India, ascending Himalaya to 2,000 m.
Habitat Ecology	: Plant held sacred by Hindus, and grown in pots on pedestals in homes and temples throughout India.
Reproductive Cycle	: July-October.
Material Examined	: EBH-TP-258.
Systematic Account	: Softly hairy, erect herbaceous plants having purplish stems and branches, leaves pubescent on both sides, and flowers in racemose whorls. Corolla small, hardly longer than the calyx. Stamens exserted. Nutlets broadly ellipsoid, pale red-brown in colour.

Oenothera drummondii Hook f. (Pl. 79C)

Syn: *Raimannia drummondii* (Hook.) Rose *ex* Sprague and Riley.

Family	: Onagraceae.
Vernacular Name	: Salsa.

Eng. Names : *Eng.*: Evening Primrose, Sun Drops.

Distribution : A hardy annual from Texas.

Habitat Ecology : Generally found as an escape.

Reproductive Cycle : August-September.

Material Examined : EBH-TP-301.

Systematic Account : Annuals with soft-hairy leaves and bright yellow flowers, each measuring about 5 cm or more in diameter and opening in the evening. Fruits narrow, pointed.

***Olea glandulifera* Wall. (Pl. 79D)**

Family : Oleaceae.

Vernacular Name : Jhadenu.

Distribution : N. W. Himalaya upto 1,800 m.

Habitat Ecology : River banks. Shady ravines.

Reproductive Cycle : March-May.

Material Examined : EBH-TP-1119.

Systematic Account : A medium-sized glabrous tree possessing lanceolate leaves and white flowers in terminal and axillary panicles. Calyx teeth ciliate. Corolla-tube short. Anthers large, dehiscing laterally. Drupes black when ripe.

***Papaver rhoeas* L. var. *latifolia* Prain (Pl. 79E)**

Syn: *P. hookeri* Baker.

Family : Papaveraceae.

Vernacular Name : Poppy.

Eng. Name : *Eng.*: Shirley Poppy.

Distribution : Native of Great Britain and adjacent areas of Europe.

Habitat Ecology : Ideal for cultivation in beds and mixed borders for garden decoration and cut flower production.

Reproductive Cycle : Summer.

Material Examined : EBH-TP-108.

Systematic Account : Branched annuals attaining 60 cm in height. Leaves pinnately divided and flowers on long peduncles clothed with spreading hairs. Capsules sub-globose, smooth. Seeds dark brown.

Passiflora caerulea **Linn. (Pl. 80A)**

Family	:	Passifloraceae.
Vernacular Name	:	Ghari-Phool.
Eng. Name	:	*Eng.*: Passion Flower.
Distribution	:	Native of Central and Western S. America.
Habitat Ecology	:	Hardy on walls.
Reproductive Cycle	:	June-September.
Material Examined	:	EBH-TP-117.
Systematic Account	:	An evergreen climber with leaves divided into 5 lanceolate segments. Flowers solitary, axillary; corona 5 cm across, outer filaments blue at the tip, white in the middle and purple at the base.

Pedilanthus tithymaloides **(L.) Poit. (Pl. 80B)**

Syn: *Euphorbia tithymaloides* Linn.;
***Tithymalus tithymaloides* (Linn.) Croizat.**

Family	:	Euphorbiaceae.
Vernacular Name	:	Nagan buti.
Eng. and Sans. Names	:	*Eng.*: Devil's Backbone, Jacob's Ladder, Jewbush, Redbird Flower, Slipper Plant. *Sans.*: Vilayatisij.
Distribution	:	Native of Mexico and S. America.
Habitat Ecology	:	An ornamental plant grown in hedges and as a pot plant; occasionally met as an escape in Indian gardens.
Reproductive Cycle	:	Flowers mainly in summer, but intermittently throughout the year.
Material Examined	:	EBH-TP-815.
Systematic Account	:	Erect or scandant subshrubs with zig-zag subsucculent branchlets, spiral, distichous, deciduous leaves and stalked, terminal, cyathia subtended by petaloid, caducous bracts. Involucre scarlet red, slipper-shaped.

Pelargonium hortorum **Bailey (Pl. 80C)**

Family	:	Geraniaceae.
Vernacular Name	:	Geranium.
Eng. Name	:	*Eng.*: Zonal Geranium.
Distribution	:	Grown in gardens on hills.

Habitat Ecology : Excellent for growing indoors; require full sunlight, plenty of water, cool temperature and well-drained soil.

Reproductive Cycle : Summer season.

Material Examined : EBH-TP-118.

Systematic Account : Plants with succulent stems upto 50 cm high. Leaves 5-10 cm across, variegated. Flowers large, semi-double, pink.

Petunia hybrida **Vilm. (Pl. 80D)**

Syn: *P. atkinsiana* D. Don *ex* Loud.

Family : Solanaceae.

Vernacular Name : Petunia.

Eng. Name : *Eng.*: Petunia.

Distribution : Indigenous to S. America.

Habitat Ecology : Showy plants, effective when grown in pots and also in summer flower-beds outdoors.

Reproductive Cycle : Summer.

Material Examined : EBH-TP-119.

Systematic Account : Annuals having a trailing habit and rather small leaves. Flowers violet or white, showy, on solitary pedicels.

Phaseolus vulgaris **Linn. (Pl. 81A)**

Syn: *P. communis* Pritzel; *P. compessus* DC.; *P. esculentus* Salisb.; *P. nanus* Kudr.

Family : Fabaceae.

Vernacular Name : Makkan Bean.

Eng. and H. Names : *Eng.*: Dwarf Bean, French Bean, Haricot Bean, Kidney Bean.

H.: Bakla, Frash Bean, Lobia.

Distribution : Extensively cultivated in tropical and temperate regions.

Habitat Ecology : Cultivated.

Reproductive Cycle : June-September.

Material Examined : EBH-TP-1102.

Systematic Account : A subglabrous annual with 3-foliolate leaves and suberect or twining stems. Racemes lax, few-flowered; pedicels longer than the calyx. Bracteoles persistent. Pods linear recurved, 4-6-seeded.

Phoenix sylvestris (L.) Roxb. (Pl. 81B)

Syn: *P. humilis* Royle; *Elate sylvestris* Linn.; *Elate versicolor* Salisb.

Family	:	Arecaceae.
Vernacular Name	:	Khajur.
Eng., H. and Sans. Names	:	*Eng.*: Date-sugar Palm, Wild Date Palm.
		H.: Kejur, Khaji, Khajur, Salma, Sendhi, Thakil, Thalma.
		Sans.: Kharjura, Khurjjuri, Madhukshir.
Distribution	:	Common throughout India; wild or more often cultivated.
Habitat Ecology	:	Dry areas, common.
Reproductive Cycle	:	May-June.
Material Examined	:	EBH-TP-725
Systematic Account	:	Palms reaching upto 15 m and having solitary, stout trunk clothed with persistent bases of petioles. Crown hemispherical. Leaves pinnate; petiole spinous. Pinnules numerous, densely fascicled. Male flowers scented. Female flowers distant, roundish. Fruits scattered on long pendulous spikes.

Phytolacca acinosa Roxb. (Pl. 81C)

Syn: *P. latbenia* (Moq.) Walter; *Pircunia latbenia* Moq.

Family	:	Phytolaccaceae.
Vernacular Name	:	Zorvo.
Eng. and H. Names	:	*Eng.*: Indian Poke, Sweet Belladonna.
		H.: Matazor.
Distribution	:	Temperate Himalaya: 1,700-3,000 m.
Habitat Ecology	:	Forests. Shrubberies. Cultivated areas.
Reproductive Cycle	:	June-September.
Material Examined	:	EBH-TP-302.
Systematic Account	:	Herbs with robust succulent stems, large, lanceolate leaves and greenish-white flowers borne opposite the leaves in clusters. Fruits with dark purple carpels.

Pinus gerardiana Wall. *ex* D. Don (Pl. 81D)

Syn: *P. gerardii* Forbes.

Family	:	Pinaceae.
Vernacular Name	:	Chilgoza.

Eng., H. and Sans. Names	:	*Eng.*: The Neosia or Edible Pine.
		H.: Gunober, Newr, Rhi, SEEDS=Chilgoza, Neoza.
		Sans.: Bhutamari, Chida, Mangalya, Tara, Taruni.
Distribution	:	N.W. Himalaya: 1,800-3,000 m.
Habitat Ecology	:	Open forests.
Reproductive Cycle	:	June-July.
Material Examined	:	EBH-TP-309.
Systematic Account	:	Trees with a stout trunk, roundish head, grey bark peeling off in large thin flakes and leaves in bundles of 3. Sheath of bud-scales deciduous. Cones peduncled; scales woody with a stout recurved beak. Seeds possess a short caducous wing.

Pinus roxburghii Sarg. (Pl. 82A)

Syn: *P. longifolia* Roxb. *ex* Lamb.

Family	:	Pinaceae.
Vernacular Names	:	Chil, Chir.
Eng., H. and Sans. Names	:	*Eng.*: Chir Pine, Three-leaved Pine, Long-leaved Pine.
		H.: Chil, Chir, Dhupasarala, Salla, Saral.
		Sans.: Bhadradaru, Dhupavrikshaka, Manojna, Marichapatraka, Pitadaru, Putikashtha, Sarala, Snigdhadarusanjnaka, Surabhidaruka.
Distribution	:	Outer Himalayan ranges from Indus to Bhutan: 500-2,500 m.
Habitat Ecology	:	Forming extensive forests.
Reproductive Cycle	:	March-May.
Material Examined	:	EBH-TP-259.
Systematic Account	:	Trees having tall straight trunk; thick, furrowed bark; leaves in bundles of 3 each surrounded at the base by a persistent sheath of bud-scales. Male catkins short, cylindric. Female cones on short stiff stalks, solitary or 2-5 together. Seeds with a thin membranous wing and 10-14 cotyledons.

Pinus wallichiana A. B. Jackson (Pl. 82B)

Syn: *P. excelsa* Wall.; *P. griffithii* M'Clelland; *P. pendula* Griff.

Family	:	Pinaceae.
Vernacular Name	:	Kail.

Eng. and H. Names : *Eng.*: Five-leaved Pine, Indian Blue Pine.

H.: Kail.

Distribution : Himalaya: 1,800-4,300 m.

Habitat Ecology : Open hill sides and abandoned cultivated tracts.

Reproductive Cycle : April-June.

Material Examined : EBH-TP-130.

Systematic Account : A tall tree with blue or grey-green foliage and leaves in clusters of 5. Male catkins short. Female cones at the ends of the shoots. Ripe cones 15-30 cm long, cylindric, scales slightly woody with obtuse tips. Seeds winged. Cotyledons 8-12.

Pistacia integerrima J. L. Stewart *ex* Brand. (Pl. 82C)

Syn: *Rhus integerrima* Wall.; *R. kakrasingee* Royle.

Family : Anacardiaceae.

Vernacular Names : Kakad, Kakadan, Kakadeyin, Kakderan.

Eng., H. and Sans. Names : *Eng.*: Crab's Claw.

H.: Kakra.

Sans.: Chakrangi, Ghosha, Karkatakhya, Karkati, Kulingi, Vakra.

Distribution : N.W. Himalaya: 400-2,400 m.

Habitat Ecology : Dry slopes. Field borders.

Reproductive Cycle : March-May.

Material Examined : EBH-TP-1105.

Systematic Account : Glabrous trees having terete petiole and coriaceous leaflets. Panicles lateral; male panicles compact, female ones lax. Drupes broader than long, grey.

Pisum sativum Linn. (Pl. 82D)

Syn: *P. arvense* Linn.; *P. vulgare* Jundz.; *Lathyrus oleraceus* Lam.

Family : Fabaceae.

Vernacular Name : Mattar.

Eng., H. and Sans. Names : *Eng.*: Pea.

H.: Batra, Matar.

Sans.: Satila.

Distribution : Native of S. Europe and W. Asia.

Habitat Ecology	:	Commonly cultivated in drier Tibetan borderland areas in the Himalaya.
Reproductive Cycle	:	June-July.
Material Examined	:	EBH-TP-1210.
Systematic Account	:	Climbing annuals with pinnate leaves ending in a branched tendril. Leaflets 2-8; stipules oval-oblong, toothed. Flowers axillary, borne on a stalk shorter than the subtending leaf. Pods 4-6 cm by 1-1.7 cm broad.

Platanus orientalis **Linn. (Pl. 82E)**

Syn: *P. vulgaris* Spach.

Family	:	Platanaceae.
Vernacular Name	:	Chinar.
Eng. and H. Names	:	*Eng.*: Oriental Plane Tree, Plane Tree.
		H.: Chinar.
Distribution	:	N. W. Himalaya: 1,500-2,800 m.
Habitat Ecology	:	Commonly cultivated as an ornamental tree for shade in parks and roadsides.
Reproductive Cycle	:	April-May.
Material Examined	:	EBH-TP-401.
Systematic Account	:	Deciduous trees having spreading branches and 5-7 lobed leaves. Stipules caducous. Flowers monoecious, densely crowded in unisexual globose heads on a slender peduncle. Fruits a 1-seeded achene.

Plumeria acutifolia **Poiret (Pl. 82F)**

Syn: *P. acuminata* Roxb.; *P. rubra* Linn.

Family	:	Apocynaceae.
Vernacular Name	:	Champa.
Eng., H. and Sans. Names	:	*Eng.*: Frangipani, Jasmine Tree, Pagoda Tree, Temple Tree.
		H.: Chameli, Champa, Goburchamp, Golainchi, Gulachin.
		Sans.: Devaganagalu, Gosampige, Kshira champa.
Distribution	:	Cultivated and naturalized in many parts of India.
Habitat Ecology	:	Extensively planted throughout the tropics.
Reproductive Cycle	:	June-September.

Material Examined	:	EBH-TP-501.
Systematic Account	:	A small deciduous tree with thick branches, spirally arranged leaves, copious milky juice, and fragrant flowers in terminal panicles. Stamens inserted near the base of the corolla-tube. Follicles cylindric.

Polianthes tuberosa **Linn. (Pl. 83A)**

Syn: *P. gracilis* Link; *P. tuberosa* var. *gracilis* (Link) Baker; *P. tuberosa* forma *plena* Moldenke; *Agave tuberosa* (Linn.) Thiede and Eggli; *Crinum angustifolium* Houtt.; *Tuberosa amica* Medik.

Family	:	Amaryllidaceae.
Vernacular Name	:	Rajnigandha.
Eng., H. and Sans. Names	:	*Eng.*: Tuberose.
		H.: Gulcheri, Gulshabbo.
		Sans.: Rajanigandha.
Distribution	:	Cultivated in gardens throughout India.
Habitat Ecology	:	Warm humid climate.
Reproductive Cycle	:	Throughout greater part of the year.
Material Examined	:	EBH-TP-410.
Systematic Account	:	An erect 60-120 cm high herb with stout bulbs, and sessile, funnel-shaped flowers. Filaments attached on upper part of corolla. Florets borne on long terminal racemes blooming from base upward.

Polyalthia longifolia **(Sonn.) Thwaites (Pl. 83B)**

Syn: *Guatteria longifolia* (Sonn.) Wallich; *Unona longifolia* (Sonn.) Dunal; *Uvaria longifolia* Sonn.

Family	:	Annonaceae.
Vernacular Name	:	Ashoka.
Eng., H. and Sans. Names	:	*Eng.*: Cemetery Tree, Mast Tree.
		H.: Asok, Asoka, Debdari, Devadaru, Devidaru.
		Sans.: Asoka, Asopalava, Devadaru, Kasthadaru, Putranjiva, Ulkatah.
Distribution	:	Cultivated throughout India.
Habitat Ecology	:	Planted in gardens and along avenues.
Reproductive Cycle	:	February-May.
Material Examined	:	EBH-TP-520.

Systematic Account	:	Evergreen tall columnar trees with a dense crown, straight trunk, greyish-brown bark and glossy green leaves. Flowers yellowish-green, in axillary fascicles or umbels. Fruits a cluster of 1-seeded carpels, purple when ripe.

Polygonum alpinum All. (Pl. 83C)

Syn: *P. acidum* Pall.; *P. divaricatum* Vill.; *P. polymorphum* Ledeb.; *P. sibiricum* Linn.

Family	:	Polygonaceae.
Vernacular Name	:	Choda.
Distribution	:	N.W. Himalaya: 2,750-4,000 m.
Habitat Ecology	:	Rocky slopes, meadows, edges of cultivation.
Reproductive Cycle	:	June-September.
Material Examined	:	EBH-TP-1209.
Systematic Account	:	Undershrubs upto 1.5 m high with softly pubescent young parts, shortly petioled leaves, and flowers in terminal, long panicles. Nuts acutely trigonus, black.

Pongamia pinnata (Linn.) Pierre (Pl. 83D)

Syn: *P. glabra* Vent.; *P. mitis* (Linn.) Kurz; *P. xerocarpa* Hassk.; *Cytisus pinnatus* Linn.; *Derris indica* (Lam.) Benn.; *Millettia novo-guineensis* Kaneh. and Hatus.; *Pterocarpus flavus* Lour.; *Robinia mitis* Linn.

Family	:	Fabaceae.
Vernacular Name	:	Kanji.
Eng., H. and Sans. Names	:	*Eng.*: Indian Beech, Karanj, Pongam Oil Tree. *H.*: Kanji, Karanj, Karanjaka, Kiramal, Papar. *Sans.*: Angaravalli, Badhaphala, Chirabilva, Gaura, Karanja, Vishari, Vrittaparna.
Distribution	:	Throughout most of India to an elevation of 1,200 m.
Habitat Ecology	:	Planted as an avenue tree along roadsides as well as in gardens.
Reproductive Cycle	:	March-June.
Material Examined	:	EBH-TP-914.
Systematic Account	:	Trees upto 18 m tall. Leaves imparipinnate; leaflets opposite, shortly acuminate. Flowers in fascicles of 2-4 in axillary racemes. Calyx truncate. Corolla much exserted. Pods indehiscent, woody, curved at apex.

Populus ciliata Wall. (Pl. 83E)

Syn: *P. pyriformis* Royle; *P. rotundifolia* Griff.

Family	:	Salicaceae.
Vernacular Names	:	Poplar, Safeda.
Eng. and H. Names	:	*Eng.*: Himalayan Poplar.
		H.: Paharipipal, Sharphara, Tilaunja.
Distribution	:	N.W. Himalaya: 1,500-3,000 m.
Habitat Ecology	:	Streamsides, drier areas in coniferous forests.
Reproductive Cycle	:	April-July.
Material Examined	:	EBH-TP-1208.
Systematic Account	:	Lofty trees having catkins compact in flower. Male flowers possess oblanceolate bracts fringed with long hairs and many stamens. Female flowers have 3-4 very large stigmas. Capsules 3-4 valved.

Populus nigra Linn. (Pl. 83F)

Syn: *P. dilatata* Ait.; *P. italica* (Du Roi) Moench; *P. nigra* var. *italica* Du Roi.

Family	:	Salicaceae.
Vernacular Names	:	Poplar, Popular.
Eng. Names	:	*Eng.*: Black Poplar, Lombardy Poplar.
Distribution	:	N.W. Himalaya.
Habitat Ecology	:	Generally planted along roadsides.
Reproductive Cycle	:	April-July.
Material Examined	:	EBH-TP-1111.
Systematic Account	:	A tall umbrageous tree having tough wood thick, blackish, spongy bark and smooth branches. Leaves deltoid, deep green, very smooth, pointed, serrated. Catkins long, loose and pendulous.

Prinsepia utilis Royle (Pl. 84A)

Family	:	Rosaceae.
Vernacular Names	:	Bekhal, Bhikhad, Bhikwa.
Hindi Names	:	*H.*: Bekhra, Bhekal, Cherara, Dhatila, Jhatela, Karanga.
Distribution	:	Temperate Himalaya: 1,200-2,700 m.
Habitat Ecology	:	Forests. Shrubberies. Hedges.

Reproductive Cycle : February-March.

Material Examined : EBH-TP-121.

Systematic Account : Spinescent deciduous shrubs with stout, axillary, leaf-bearing thorns and coriaceous leaves. Flowers in short axillary racemes. Calyx cup-shaped. Petals shortly clawed. Stamens numerous. Carpel 1, sessile. Fruits fleshy, purple drupe.

Prunus amygdalus Batsch. (Pl. 84B)

Syn: *Amygdalus communis* Linn.; *A. dulcis* Mill.; *Druparia amygdalus* Clairv.

Family : Rosaceae.

Vernacular Name : Badam.

Eng., H. and Sans. Names : *Eng.*: Almond.

H.: Badam.

Sans.: Badama, Badamitte.

Distribution : Cultivated in cooler parts of India.

Habitat Ecology : Planted.

Material Examined : EBH-TP-1123.

Reproductive Cycle : May-September.

Systematic Account : Medium-sized trees. Leaves serrulate appearing after the flowers. Stipules fimbricate. Flowers peduncled. Drupes velvety; stone compressed.

Prunus armeniaca Linn. (Pl. 84C)

Syn: *P. americana* Marsh.; *Armeniaca vulgaris* Lam.

Family : Rosaceae.

Vernacular Name : Khumani.

Eng. and H. Names : *Eng.*: Apricot, Common Apricot, Mishmus, Moon of the Faithful.

H.: Chilu, Chuari, Chulu, Khubani, Sardalu.

Distribution : N.W. India.

Habitat Ecology : Cultivated.

Reproductive Cycle : May-August.

Material Examined : EBH-TP-601.

Systematic Account : Deciduous trees having glandular petiole, lanceolate stipules and flowers appearing before the leaves.

Stamens inserted at the mouth of the calyx-tube. Drupes downy or glabrous; stones smooth with a thickened grooved margin.

Prunus domestica Linn. (Pl. 84D)

Syn: *P. communis* Huds.; *P. domestica* var. *damascena* Ser.; *Druparia prunus* Clairv.

Family : Rosaceae.

Vernacular Name : Plum.

Eng. and H. Names : *Eng.*: Plum.
H.: Alubukhara, Alucha.

Distribution : Throughout sub-Himalayan region.

Habitat Ecology : Cultivated for their fruits.

Reproductive Cycle : February-March.

Material Examined : EBH-TP-1124.

Systematic Account : A small tree with serrated leaves and small, white flowers appearing before the appearance of leaves. Fruits yellow or red when ripe, edible.

Prunus persica (Linn.) Batsch (Pl. 84E)

Syn: *Amygdalus collinus* Wall.; *A. persica* Linn.; *Persica saligna* Royle; *P. vulgaris* Miller.

Family : Rosaceae.

Vernacular Name : Aru.

Eng. and H. Names : *Eng.*: Peach Tree.
H.: Aru, Shaftalu.

Distribution : Cooler parts of India, upto 3,350 m in the N.W. Himalaya.

Habitat Ecology : Commonly planted near habitations.

Reproductive Cycle : February-May.

Material Examined : EBH-TP-1125.

Systematic Account : Small trees having fimbriate stipules and flowers usually appearing before the leaves. Stamens inserted at the mouth of the calyx-tube. Drupes succulent. Pericarp indehiscent.

Prunus puddum (Roxb. *ex* Ser.) Brandis (Pl. 85A)

Syn: *P. cerasoides* D. Don; *P. sylvatica* Roxb.; *Cerasus phoshia* Ham.; *C. puddum* Wall.

Family : Rosaceae.

Vernacular Name : Pajja.

Eng., H. and Sans. Names : *Eng.*: Himalayan Wild Cherry.
H.: Paddam, Padma-kashtha, Paya.
Sans.: Padmaka, Padmaksh.

Distribution : Temperate Himalaya: 1,200-2,500 m.

Habitat Ecology : Forests; often planted by wayside resting places.

Reproductive Cycle : November-May.

Material Examined : EBH-TP-122.

Systematic Account : A middle-sized tree. Leaves sharply serrate with glandular petiole. Flowers appearing before the leaves in umbellate fascicles. Drupes obtuse at both ends; stone bony and furrowed.

Psidium guajava Linn. (Pl. 85B)

Family : Myrtaceae.

Vernacular Name : Amrood.

Eng., H. and Sans. Names : *Eng.*: Guava.
H.: Am, Amrud, Amrut.
Sans.: Amruta-phalam, Bahu-bija-phalam.

Distribution : Naturalised throughout India.

Habitat Ecology : Cultivated for their edible fruits.

Reproductive Cycle : July-September.

Material Examined : EBH-TP-260.

Systematic Account : A small tree. Leaves on very short petioles, acuminate. Peduncles axillary, 1-3 flowered. Flowers white. Calyx-tube adnate to the ovary. Fruits a globose or pyriform berry.

Pterospermum acerifolium Willd. (Pl. 85C)

Family : Sterculiaceae.

Vernacular Names : Kanchan Champa, Letter Chamba.

H. and Sans. Names : *H.*: Kanakchampa, Kaniar, Kathachampa, Muchkund.
Sans.: Karnikara, Mushukunda, Padapotpala, Parivyadha.

Distribution : Sub-Himalayan tract and outer Himalaya upto 1,200m.

Habitat Ecology : Commonly planted in gardens and avenues.

Reproductive Cycle : April-June.

Material Examined : EBH-TP-207.

Systematic Account : Tall evergreen trees possessing ash-coloured bark; large leaves and fragrant, white, axillary flowers. Sepals densely tomentose outside. Petals somewhat obliquely cuneate. Staminodes club-shaped. Capsules pentagonal. Seeds with large brown membranous wings.

***Punica granatum* Linn. (Pl. 85D)**

Syn: *P. nana* Linn.

Family : Punicaceae.

Vernacular Names : Anar, Dadu.

Eng., H. and Sans. Names : *Eng.*: Pomegranate.

H.: Anar, Dhalim, Dharimb.

Sans.: Bijapura, Dadima, Dantabijaka, Suphala.

Distribution : Cultivated throughout India; wild in the N. W. regions.

Habitat Ecology : Stony slopes. Roadsides. Edges of agricultural fields.

Reproductive Cycle : May-September.

Material Examined : EBH-TP-515.

Systematic Account : Large deciduous shrubs or small trees armed with axillary or terminal thorns and opposite, glabrous leaves. Flowers mostly solitary, terminating short shoots. Calyx-tube coriaceous, adnate to the ovary below. Petals inserted between the calyx-lobes. Stamens numerous, inserted on the calyx below the petals at various levels. Fruits globose, containing red juice round the seeds.

***Putranjiva roxburghii* Wall. (Pl. 86A)**

Syn: *P. amblyocarpa* Mull. Arg.; *Cyclostemon racemosus* Zipp. *ex* Span.; *Drypetes roxburghii* (Wall.) Hurus.; *Nageia putranjiva* Roxb.

Family : Euphorbiaceae.

Vernacular Names : Patajan, Patranjia.

H. and Sans. Names : *H.*: Jiaputa, Juti, Patigia, Patji, Putajan, Putijia, Putrajiva, Putranjiva.

Sans.: Putrajiva, Putranjiva, Putranjivah, Sutrajiva.

Distribution	:	Throughout tropical India, wild and cultivated.
Habitat Ecology	:	Shaded valleys or in forested wetlands and in evergreen to moist deciduous forests; often planted.
Reproductive Cycle	:	March-April.
Material Examined	:	EBH-TP-262.
Systematic Account	:	Evergreen trees with drooping branches; corky bark; coriaceous leaves; small stipules and small, dioecious flowers. Male flowers yellowish, in dense rounded axillary clusters. Female flowers green, axillary, solitary or 2-3 together. Drupes pointed at the apex and narrowed at base; stone very hard.

***Pyrostegia venusta* (Ker.-Gawl.) Miers (Pl. 86B)**

Syn: *P. ignea* (Vell.) Presl.; *Bignonia ignea* Vell.; *B. venusta* Ker.-Gawl.

Family	:	Bignoniaceae.
Vernacular Name	:	Kamuki.
Eng. Name	:	*Eng.*: Golden Shower.
Distribution	:	Native of Brazil.
Habitat Ecology	:	Warm gardens everywhere.
Reproductive Cycle	:	January to April.
Material Examined	:	EBH-TP-1126.
Systematic Account	:	An evergreen climber with brilliant orange colored flowers in the form of long tubes in hanging clusters. Stamens well exserted. Leaves borne opposite in twos and threes, sharp-pointed and 7-10 cm long.

***Pyrus communis* Linn. (Pl. 86C)**

Syn: *P. pyrifolia* Nakai var. *Culta* Nakai.

Family	:	Rosaceae.
Vernacular Name	:	Nashpati.
Eng., H. and Sans. Names	:	*Eng.*: The Common Pear, Pear. *H.*: Nashpati. *Sans.*: Amritaphala.
Distribution	:	N.W. Himalaya.
Habitat Ecology	:	Cultivated for their fruits.
Reproductive Cycle	:	June-August.

Material Examined : EBH-TP-126.

Systematic Account : Trees or shrubs. Leaves obtusely serrate. Corymbs terminating short shoots. Calyx-lobes persistent. Petals white. Fruits crowned with the persistent calyx.

Pyrus malus Linn. (Pl. 86D)

Syn: *Malus communis* Desf.; *M. domestica* Borkh.

Family : Rosaceae.

Vernacular Names : Seb, Seu.

Eng., H. and Sans. Names : *Eng.*: Apple.

H.: Seb, Seo, Sev.

Sans.: Seba, Seva.

Distribution : Cultivated in the cooler parts of India.

Habitat Ecology : Planted for their fruits.

Reproductive Cycle : May-October.

Material Examined : EBH-TP-310.

Systematic Account : A tree upto 10 m tall. Leaves ovate, obtusely serrate. Flowers white, tinged with red. Calyx lobes persistent. Petals with a tomentose claw. Fruits globose, intruded at both ends.

Pyrus pashia Buch.-Ham. *ex* D. Don (Pl. 87A)

Syn: *P. nepalensis* Herb.; *P. variolosa* Wall.; *P. verruculosa* Bertol.

Family : Rosaceae.

Vernacular Names : Kainth, Shegal.

Hindi Names : *H.*: Mehal, Mol.

Distribution : Temperate Himalaya (Kashmir-Bhutan): 800-2,700 m. Khasia mountains, 1,800 m.

Habitat Ecology : Shrubberies, cultivated areas.

Reproductive Cycle : March-May.

Material Examined : EBH-TP-411.

Systematic Account : Deciduous trees having woolly young parts. Flowers in simple corymbs, white, tinged with pink. Calyx tomentose. Stamens about 30. Style base woolly. Fruits yellow-brown, warted, rough with small, white spots.

Quercus dilatata **Lindl.** ***ex*** **A. DC. (Pl. 87B)**

Syn: *Q. dealbata* Royle; *Q. dilatata* Royle;
***Q. floribunda* Lindl. *ex* A. Camus.**

Family	:	Fagaceae.
Vernacular Name	:	Moru.
Eng. and H. Names	:	*Eng.*: Green Oak, Moru Oak. *H.*: Moru.
Distribution	:	N.W. Himalaya: 2,000-2,700 m.
Habitat Ecology	:	Often mixed with other oaks in damp cool situations.
Reproductive Cycle	:	April-May.
Material Examined	:	EBH-TP-124.
Systematic Account	:	A large evergreen tree. Bark peeling off in longitudinal scales. Leaves glabrous, entire or with sharp spinescent teeth. Male flowers distant from each other. Female spikes short. Acorns usually solitary; cup covering about half the nut which is ovoid, pointed, smooth, brown.

Quercus incana **Roxb. (Pl. 87C)**

Syn: *Q. dealbata* Wall.; *Q. lanata* Don var. *incana* Wenzig;
***Q. lanata* Smith.**

Family	:	Fagaceae.
Vernacular Name	:	Ban.
Eng. and H. Names	:	*Eng.*: Grey Oak, Himalayan White Oak, Kumaon Oak. *H.*: Ban, Banj.
Distribution	:	Temperate Himalaya: 1,200-2,400 m.
Habitat Ecology	:	Common in forests; often associated with *Rhododendron arboreum*.
Reproductive Cycle	:	April-October.
Material Examined	:	EBH-TP-125.
Systematic Account	:	Evergreen trees possessing dark grey bark and hoary or woolly young shoots and petioles. Leaves cuspidate-serrate, coriaceous, glabrous above and densely white-tomentose beneath. Male spikes softly hairy. Female flowers solitary or clustered in the leaf-axils on short stout peduncles. Acorns half-enclosed by the campanulate cup when ripe.

Quisqualis indica **Linn. (Pl. 87D)**

Syn: *Q. densiflora* Wall. *ex* Miq.; *Q. ebracteata* Beauv.; *Q. glabra* Burm.; *Q. loureiri* G. Don; *Q. pubescens* Burm.; *Q. sinensis* Lindl.; *Q. villosa* Roxb.

Family : Combretaceae.

Vernacular Name : Rangoon Bel.

Eng. and H. Names : *Eng.*: Chinese Honeysuckle, Rangoon Creeper.
H.: Rangoon-ki-bel.

Distribution : Cultivated throughout India.

Habitat Ecology : Commonly planted in gardens for its showy flowers.

Reproductive Cycle : May-September.

Material Examined : EBH-TP-910.

Systematic Account : Woody climbing shrubs. Flowers numerous, fragrant, borne in pendulous axillary racemes. Fruits dry, coriaceous, acutely 5-angled or 5-winged, 1-seeded.

Ribes orientale **Desf. (Pl. 88A)**

Family : Saxifragaceae.

Vernacular Name : Iguta.

Distribution : Inner arid Himalaya: 2,100-4,200 m.

Habitat Ecology : Open slopes. Rocks. Moist places.

Reproductive Cycle : July-October.

Material Examined : EBH-TP-1211.

Systematic Account : Unarmed, sticky-glandular, minutely hairy small shrubs. Flowers unisexual, in erect racemes with ciliate deciduous bracts. Male racemes with many and female with few flowers. Berries red or yellow when ripe, with scattered glandular hairs.

Ricinus communis **Linn. (Pl. 88B)**

Syn: *R. inermis* Jacq.; *R. lividus* Jacq.; *R. speciosus* Burm.; *R. spectabilis* Blume; *R. viridis* Willd.; *Croton spinosus* Linn.

Family : Euphorbiaceae.

Vernacular Name : Arind.

Eng., H. and Sans. Names : *Eng.*: Caster-oil Plant.
H.: Arand, Arend, Erand, Rand.
Sans.: Eranda, Ruvuka, Vatari.

Distribution : Throughout India, naturalized near habitations.

Habitat Ecology : Common in cultivated areas and wastelands.

Reproductive Cycle : March-May.

Material Examined : EBH-TP-1127.

Systematic Account : An evergreen bush or small tree possessing palmately lobed leaves and monoecious, rather large flowers in terminal subpaniculate, racemes; male flowers crowded in upper portion of the inflorescence, the females below. Fruits a prickly capsule of three 2-valved cocci. Seeds oblong, smooth, mottled.

Robinia pseudoacacia **Linn. (Pl. 88C)**

Syn: *R. pringlei* Rose; *R. pseudoacacia* var. *rectissima* Raber.

Family : Fabaceae.

Vernacular Names : Robinia, Shulu.

Eng. Names : *Eng.*: Black-locust, False Acacia, Robinia.

Distribution : W. Himalaya upto 2,400 m.

Habitat Ecology : Commonly planted by roadsides.

Reproductive Cycle : April-May.

Material Examined : EBH-TP-317.

Systematic Account : Small deciduous trees with fissured bark, pinnate leaves, and pendulous clusters of white sweet-scented flowers. Leaflets 7-21, oval; spines present on branches. Pods 5-10 cm long, flat, hairless.

Rosa alba **Linn. (Pl. 88D)**

Syn: *R. glandulifera* Roxb.

Family : Rosaceae.

Vernacular Name : Gulab.

Eng., H. and Sans. Names : *Eng.*: Common White Dog Rose, White Cottage Rose.

H.: Gulab.

Sans.: Bahupatrika, Bhrigavallabha, Gandhadhya, Shatapatri, Shivaballabha.

Distribution : Throughout India.

Habitat Ecology : Cultivated as garden ornamental for their beautiful flowers.

Reproductive Cycle : June-August.

Material Examined	:	EBH-TP-123.
Systematic Account	:	A tall spreading shrub possessing stems with scattered hooked prickles and broadly ovate leaflets. Flowers more or less double, white, fragrant with glandular-hispid pedicels. Receptacle usually smooth. Fruits ovate, scarlet.

Rosa banksiae R. Br. (Pl. 89A)

Syn: *R. banksiae* var. *alboplena* Rehder; *R. inermis* Roxb.

Family	:	Rosaceae.
Vernacular Name	:	Gulabri.
Eng. Name	:	*Eng.*: Banksian Rose.
Distribution	:	Indigenous to China and Japan; cultivated in India.
Habitat Ecology	:	Cultivated.
Reproductive Cycle	:	May-June.
Material Examined	:	EBH-TP-127.
Systematic Account	:	Unarmed evergreen climbing shrubs with 3-5 leaflets. Flowers on smooth pedicels in many-flowered umbels, white or yellow, slightly fragrant.

Rosa foetida Herrm. (Pl. 89B)

Syn: *R. chlorophylla* Ehrh.; *R. eglanteria* Linn.; *R. lutea* Mill.

Family	:	Rosaceae.
Vernacular Name	:	Laybala.
Eng. Names	:	*Eng.*: Austrian Briar, Yellow Persian Rose.
Distribution	:	W. Himalaya: 2,100-3,600 m.
Habitat Ecology	:	Forests. Villages. Commonly planted in border areas of agricultural fields.
Reproductive Cycle	:	June-September.
Material Examined	:	EBH-TP-1212.
Systematic Account	:	An upright prickly shrub with yellow, sweet–scented flowers either solitary or 2-3 together, compound leaves with 2-4 pairs of deeply serrate leaflets which are glandular pubescent on the lower side. Carpels enclosed in a fleshy receptacle. Fruits flask-shaped, red with persistent sepals.

Rosa indica **Linn. (Pl. 89C)**

Syn: *R. chinensis* Jacq.; *R. fragrans* Redoute; *R. sinica* Linn.

Family : Rosaceae.

Vernacular Name : Gulab.

Eng. and H. Names : *Eng.*: Rose, China Rose, Monthly Rose.
H.: Gulab.

Distribution : Indigenous to China.

Habitat Ecology : Extensively cultivated in Indian gardens for their beautiful flowers.

Reproductive Cycle : June-September.

Material Examined : EBH-TP-263.

Systematic Account : An evergreen glabrous shrub. Stipules adnate nearly to the apex. Flowers large on long slender pedicels; sepals reflexed.

Rosa jacquemontii **Crep. *ex* Hook. f. (Pl. 89D)**

Family : Rosaceae.

Vernacular Name : Gulab.

Distribution : Kashmir-Kumaon: 1,500-4,100 m.

Habitat Ecology : Commonly grown as an ornamental plant.

Reproductive Cycle : June-September.

Material Examined : EBH-TP-312.

Systematic Account : Prickly shrubs with white, sweet-scented flowers either solitary or in dense clusters. Leaflets coarsely toothed. Sepals lanceolate. Petals obcordate. Fruit flask-shaped.

Rosa moschata **Herrm. (Pl. 89E)**

Syn: *R. arborea* Pers.; *R. brunonii* Lindl.; *R. glandulifera* Roxb.; *R. lindleyi* Herb.; *R. pubescens* Roxb.

Family : Rosaceae.

Vernacular Name : Yal.

Eng., H. and Sans. Names : *Eng.*: Himalayan Musk Rose.
H.: Karer, Kuji, Kunja, Kwiala.
Sans.: Sewati.

Distribution : Temperate C. and W. Himalaya: 1,200-2,500 m.

Habitat Ecology : Forests. Shrubberies. Open places and frequently on other shrubs and trees.

Reproductive Cycle : April-October.

Material Examined : EBH-TP-313.

Systematic Account : A tall climber with recurved prickles. Leaflets puberulous beneath. Flowers in compound corymbs, white. Calyx-lobes lanceolate, entire or pinnatifid. Petals orbicular-obovate. Fruits small, subglobose.

Rosa multiflora Thunb. (Pl. 90A)

Family : Rosaceae.

Vernacular Name : Gulabari.

Eng. Names : *Eng.*: Baby Rose, Japanese Rose.

Distribution : Indigenous to Japan and China; cultivated in India.

Habitat Ecology : Often planted for ornamental purposes in hills and mountains.

Reproductive Cycle : June-July.

Material Examined : EBH-TP-128.

Systematic Account : Deciduous shrubs with long, recurving or climbing branches. Leaflets usually 9; stipules and bracts pectinate. Flowers in many-flowered pyramidal corymbs, rosy; petioles and inflorescence with long soft hairs. Fruits small, globular.

Rosa webbiana Wall. *ex* Royle (Pl. 90B)

Syn: *R. pimpinellifolia* Herb.; *R. unguicularis* Bertol.

Family : Rosaceae.

Vernacular Names : Shaybala, Socialic.

Distribution : Drier Himalaya (Kashmir-Kumaon):1,500-4,100 m.

Habitat Ecology : Rocky and drier areas, villages.

Reproductive Cycle : June-September.

Material Examined : EBH-TP-1213.

Systematic Account : Prickly shrubs. Twigs glabrate. Leaves pinnate; leaflets 5-9, serrate. Flowers large-sized, usually solitary. Prickles straight. Petals obcordate. Fruits red when ripe.

Roylea cinerea (D. Don) Baill. (Pl. 90C)

Syn: *R. calycina* (Roxb.) Briq.; *R. elegans* Wall.; *Ballota cinerea* D. Don; *Phlomis calycina* Roxb.

Family : Lamiaceae.

Vernacular Names : Kadve, Karkoi.

H. Name	:	*H.*: Patkarru.
Distribution	:	Subtropical W. Himalaya: 600-2,400 m.
Habitat Ecology	:	Frequent on open, rocky slopes.
Reproductive Cycle	:	March-November.
Material Examined	:	EBH-TP-517.
Systematic Account	:	Shrubs having coarsely toothed, shortly-stalked leaves. Flowers white, tinged with pink. Nutlets smooth.

Ruellia tuberosa **Linn. (Pl. 90D)**

Family	:	Acanthaceae.
Vernacular Name	:	Kapila.
Eng. Name	:	*Eng.*: Menow-weed.
Distribution	:	W. Indies and S.W. United States.
Habitat Ecology	:	Cultivated; a light rich soil.
Reproductive Cycle	:	March-September.
Material Examined	:	EBH-TP-911.
Systematic Account	:	Erect herbs with hispid tender parts, purplish nodes, and blue to violet flowers in axillary dichasial cymes. Seeds orbicular, hairy.

Russelia equisetiformis **Cham. and Schl. (Pl. 90E)**

Syn: *R. floribunda* Woodrow; *R. juncea* Zucc.

Family	:	Scrophulariaceae.
Vernacular Name	:	Coral.
Eng. Names	:	*Eng.*: Coral Plant, Fountain Bush, Fountain Plant.
Distribution	:	Native of Mexico.
Habitat Ecology	:	Grown in gardens; in all types of soils but flowers best in porous soil.
Reproductive Cycle	:	Summer.
Material Examined	:	EBH-TP-264.
Systematic Account	:	Much-branched small shrubs with drooping rush-like stems and very small, elliptic leaves. Flowers brilliant scarlet in clusters of 1-3, narrowly tubular with 5 segments.

***Saccharum officinarum* Linn. (Pl. 90F)**

Syn: *S. sinense* Roxb.

Family : Poaceae.

Vernacular Name : Ganna.

Eng., H. and Sans. Names : *Eng.*: Noble Cane, Sugar-cane.
H.: Ganna, Ikh, Kumad, Nai Shakar, Rikhu, Uk, Ukh.
Sans.: Ikshu, Kanguruku, Pundra, Rusala.

Distribution : Inhabiting tropical and subtropical regions.

Habitat Ecology : Cultivated in the hotter parts of India.

Reproductive Cycle : January-April.

Material Examined : EBH-TP-709.

Systematic Account : Tall perennial herbs. Stem leafy, many-noded. Leaves overlapping, grassy. Flowers in dense woolly plumes.

***Salix alba* Linn. (Pl. 91A)**

Syn: *S. alba* subsp. *caerulea* (Sm.) Rech.f.; *S. vitellina* Linn.

Family : Salicaceae.

Vernacular Name : Benusa.

Eng. Names : *Eng.*: Huntingdon Willow, White Willow.

Distribution : N. W. Himalaya, W. Tibet.

Habitat Ecology : Extensively planted along irrigation channels.

Reproductive Cycle : March-May.

Material Examined : EBH-TP-316.

Systematic Account : Trees having silky-hairy young twigs, lanceolate deciduous stipules and catkins on leafy peduncles. Female catkins a little longer than the male. Capsule glabrous, subsessile.

***Salix babylonica* Linn. (Pl. 91B)**

Family : Salicaceae.

Vernacular Name : Majnu.

Eng. and H. Names : *Eng.*: The Weeping Willow.
H.: Majnun.

Distribution : Plains of India; Himalaya, ascending to 3,000 m.

Habitat Ecology : Often planted in gardens.

Reproductive Cycle : February-May.

Material Examined : EBH-TP-315.

Systematic Account : A medium-sized tree with pendulous branchlets and linear-lanceolate leaves. Catkins on leafy peduncles. Female catkins as long as the male. Bracts small, pale. Capsules sessile, glabrous.

Salix fragilis **Linn. (Pl. 91C)**

Family : Salicaceae.

Vernacular Name : Beli.

Eng. Name : *Eng.*: White Willow.

Distribution : N. W. Asia; cultivated in H.P. and W. Tibet. 2,400-4,000 m.

Habitat Ecology : Extensively cultivated around habitations.

Reproductive Cycle : April-June.

Material Examined : EBH-TP-1214.

Systematic Account : Trees with fragile branches, lanceolate, glandular-serrate, short-stalked, silky leaves and 1-sexed catkins. Fruits a globose catkin.

Salix oxycarpa **Anders. (Pl. 91D)**

Syn: *S. zygostemon* Boiss.

Family : Salicaceae.

Vernacular Name : Changma.

Distribution : W. Himalaya: 2,000-3,500 m.

Habitat Ecology : Shady slopes.

Reproductive Cycle : June-September.

Material Examined : EBH-TP-1215.

Systematic Account : Small trees. Leaves lanceolate, serrate. Flowers appearing with or a little before the leaves. Male catkins nearly erect, densely hairy; bracts black. Female catkins drooping; bracts brown. Capsules glabrous, shortly-stalked.

Salix tetrasperma **Roxb. (Pl. 91E)**

Syn: *S. disperma* Don; *S. horsfieldiana* Miq.; *S. nilagirica* Miq.

Family : Salicaceae.

Vernacular Names : Besu, Biyunsh.

Eng., H. and Sans. Names : *Eng.*: Indian Willow.

H.: Bains, Baishi, Bed, Bet, Bhainsh, Laila, Bent.

Sans.: Burum, Varuna.

Distribution : Throughout tropical and subtropical India, ascending Himalaya to 2,000 m.

Habitat Ecology : Open slopes. Roadsides.

Reproductive Cycle : April-July.

Material Examined : EBH-TP-1128.

Systematic Account : A medium-sized tree with silky young shoots, deciduous stipules, and flowers appearing after the leaves. Male catkins sweet-scented. Capsules glabrous, on slender distinct stipes.

Salvia coccinia **Juss. (Pl. 91F)**

Family : Lamiaceae.

Vernacular Name : Phultulsi.

Distribution : Native of C. and S. America.

Habitat Ecology : Dry slopes; also cultivated as an ornamental plant.

Reproductive Cycle : July-August.

Material Examined : EBH-TP-808.

Systematic Account : Herbs with erect, quadrangular stem. Leaves simple, petiolate. Flowers in a twiggy raceme; calyx lips half as long as tube; corolla deep scarlet, lower lip twice as long as upper.

Salvia splendens **Sellow *ex* Roemer and J.A. Schultes (Pl. 92A)**

Family : Lamiaceae.

Vernacular Name : Salvia.

Eng. Name : *Eng.*: Scarlet Sage.

Distribution : Native of Brazil.

Habitat Ecology : Grown as an ornamental plant.

Reproductive Cycle : Summer onwards.

Material Examined : EBH-TP-1129.

Systematic Account : Sub-shrubs with quadrangular branched stem. Leaves opposite, deccusate, petiolate. Tubular scarlet flowers borne in long terminal spikes. Calyx bell-shaped with 3 broad-ovate teeth. Corolla tube exserted, lower lip short.

Sambucus nigra **Linn. (Pl. 92B)**

Family	:	Sambucaceae.
Vernacular Name	:	Salja.
Eng. Names	:	*Eng.*: Bour Tree, Box Tree, Common Elder, Elder Berry.
Distribution	:	Native of Europe and Asia Minor.
Habitat Ecology	:	Woods. Hedges. Wastegrounds; often cultivated in most temperate regions.
Reproductive Cycle	:	May-September.
Material Examined	:	EBH-TP-811.
Systematic Account	:	Shrubs or small trees with branches as well as annual shoots possessing a thick white pith. Flowers yellowish, fragrant, arranged in flat terminal inflorescences. Fruits clustered in drooping heads, shining, blackish violet, globose, three-seeded.

Sansevieria trifasciata **Prain (Pl. 92C)**

Syn: *S. jacquinii* N.E.Br.; *S. laurentii* De Wild.; *S. trifasciata* var. *laurentii* (De Wild.) N.E.Br.; *S. zeylanica* var. *laurentii* (De Wild.) L.H.Bailey.

Family	:	Liliaceae.
Vernacular Name	:	Snake Plant.
Eng. Names	:	*Eng.*: Mother-in-law's Tongue, Snake Plant.
Distribution	:	Native of W. tropical Africa.
Habitat Ecology	:	Cultivated; also naturalized in some areas from sea-level to 800 m.
Reproductive Cycle	:	June-July.
Material Examined	:	EBH-TP-219.
Systematic Account	:	Evergreen perennials. Leaves stiff, straight, erect, sword-shaped, tapering to a sharp point, cross banded on both sides with shades of dark and yellow.

Santalum album **Linn. (Pl. 92D)**

Syn: *S. myrtifolium* Roxb.; *S. verum* Linn.; *Sandalum album* Rumph.; *Sirium myrtifolium* Roxb.

Family	:	Santalaceae.
Vernacular Name	:	Chandan.
Eng., H. and Sans. Names	:	*Eng.*: Sandal Tree, Sandal Wood, White Sandal Wood, Yellow Sandal Wood.

		H.: Chandal, Chandan, Chandoie, Safed Chandan, Sandal.
		Sans.: Bhadrasri, Chandana, Gandhasra, Hari chandana, Krishna Chandanam, Malayaja, Pitachandana, Srikhanda.
Distribution	:	Deccan Peninsula, ascending to 1,000 m; cultivated elsewhere.
Habitat Ecology	:	Widely scattered in dry deciduous forests.
Reproductive Cycle	:	February-May.
Material Examined	:	EBH-TP-708.
Systematic Account	:	Small evergreen trees having brownish purple, inodourous flowers in terminal and axillary paniculate cymes which are shorter than the leaves. Perianth campanulate. Stamens exserted, alternating with 4 rounded obtuse scales. Drupes purple-black; endocarp ribbed.

Sapindus mukorossi **Gaertn. (Pl. 93A)**

Syn: *S. abruptus* Lour.

Family	:	Sapindaceae.
Vernacular Name	:	Ritha.
Eng., H. and Sans. Names	:	*Eng.*: Soap-nut Tree of North India.
		H.: Aritha, Dodan, Kanmar, Ritha.
		Sans.: Phenila, Urista.
Distribution	:	Cultivated throughout N.W. India, ascending to 1,650 m.
Habitat Ecology	:	Edges of fields. Roadsides.
Reproductive Cycle	:	April-September.
Material Examined	:	EBH-TP-1003.
Systematic Account	:	Deciduous trees. Bark rough. Leaves pinnate; leaflets 5-8 pairs, shortly petioled. Flowers subsessile, bisexual, in large terminal pubescent pyramidal panicles. Fruits wrinkled before falling. Seeds solitary, globose, black.

Sapium insigne **(Royle) Benth. (Pl. 93B)**

Syn: *Carumbium insigne* (Royle) Kurz.; *Falconeria insignis* Royle; *F. wallichiana* Royle.

Family	:	Euphorbiaceae.
Vernacular Name	:	Balodar.

Hindi Names	: *H.*: Khina, Khindra.
Distribution	: Subtropical Himalaya: 500-1,800 m.
Habitat Ecology	: Open slopes. Drier areas.
Reproductive Cycle	: May-August.
Material Examined	: EBH-TP-912.
Systematic Account	: A robust deciduous tree with branchlets leafy at the tips. Leaves crenate-serrate; petiole glandular below the tip. Spikes solitary terminal. Fruits 1-3 celled, obscurely lobed.

Sapium sebiferum **(Linn.) Roxb. (Pl. 93C)**

Syn: *Carumbium sebiferum* Kurz; *Croton sebiferus* Linn; *Excoecaria sebifera* Muell.; *Stillingfleetia sebifera* Bojer; *Stillingia sebifera* Michaux; *S. sinensis* Baill.

Family	: Euphorbiaceae.
Vernacular Names	: Makhan, Pahadi-Taali.
Eng., H. and Sans. Names	: *Eng.*: Chinese Tallow Tree. *H.*: Pahari-shisham, Pippalyang, Tarcharbi, Vilayati-shisham. *Sans.*: Toyapippali.
Distribution	: A native of China. Cultivated in various parts of India; and elsewhere in warm countries.
Habitat Ecology	: Forests. Shrubberies.
Reproductive Cycle	: June-August.
Material Examined	: EBH-TP-202.
Systematic Account	: A glabrous tree possessing rhomboid leaves. Flowers in terminal racemes. Ovary glabrous, narrowed into a style with 3 long spreading stigmas.

Schefflera arboricola **(Hayata) Merr. (Pl. 93D)**

Family	: Araliaceae.
Vernacular Name	: Umbrella plant.
Eng. Name	: *Eng.*: Dwarf Umbrella Tree.
Distribution	: Native of China.
Habitat Ecology	: Cultivated as an ornamental plant.
Reproductive Cycle	: August-September.
Material Examined	: EBH-TP-244.

Systematic Account : An evergreen shrub upto 3-4 m tall. Leaves palmate, glossy green; leaflets 7-9, obovate. Flowers in a panicle of small umbels.

Sempervivum arboreum **Linn. (Pl. 93E)**

Syn: *Aeonium arboreum* (Linn.) Webb. and Berthel.

Family : Crassulaceae.

Vernacular Name : Kanchan.

Distribution : Origin from Portugal, Morocco and eastwards to Crete.

Habitat Ecology : Cultivated; requires warm temperate, well lighted conditions.

Reproductive Cycle : January–March.

Material Examined : EBH-TP-1130.

Systematic Account : Perennials upto 90 cm tall. Rosette of 5-7.5 cm leaves borne at the end of the stout light brown branches. Flowers bright yellow, forming a dense ovoid head about 10 cm long.

Sesamum orientale **Linn. (Pl. 94A)**

Syn: *S. indicum* Linn.; *S. luteum* Retz.; *S. occidentale* Heer and Regel.

Family : Pedaliaceae.

Vernacular Name : Til.

Eng., H. and Sans. Names : *Eng.*: Gingelly, Sesame.

H.: Til.

Sans.: Snehaphala, Tila, Tilah, Tilaha.

Distribution : Throughout the plains of India upto 1,200 m.

Habitat Ecology : Cultivated; often naturalized along roadsides, abandoned fields and waste areas in drier localities.

Reproductive Cycle : August-September.

Material Examined : EBH-TP-705.

Systematic Account : Strong-smelling annuals reaching upto 180 cm. Leaves simple or pinnately compound. Flowers borne in racemes in leaf axils. Fruits capsular, slightly compressed, deeply 4-grooved.

Shorea robusta Gaertn. f. (Pl. 94B)

Syn: *Vatica robusta* Steud.

Family	:	Dipterocarpaceae.
Vernacular Name	:	Sal.
Eng., H. and Sans. Names	:	*Eng.*: The Sal Tree.
		H.: Damar, Dhuna, Ral, Sakher, Sakhu, Sakhua, Sakoh, Sal, Sala, Salwa.
		Sans.: Asvakarna, Guggilam, Koushi-kaha, Rala, Sala.
Distribution	:	Tropical Himalaya, and along its base from Assam to the Sutlej. Eastern districts of C. India, Western Bengal Hills.
Habitat Ecology	:	Dominant over much of the sub-Himalayan areas.
Reproductive Cycle	:	March-June.
Material Examined	:	EBH-TP-518.
Systematic Account	:	A deciduous tree. Leaves thinly coriaceous. Flowers in terminal axillary racemose panicles. Stamens upto 50, shorter than the petals. Ovary 3-celled. Fruits indehiscent and pubescent.

Solanum tuberosum Linn. (Pl. 94C)

Family	:	Solanaceae.
Vernacular Name	:	Alu.
Eng., H. and Sans. Names	:	*Eng.*: Potato.
		H.: Alu.
		Sans.: Alukam.
Distribution	:	Cultivated in different parts of India excluding the drier regions.
Habitat Ecology	:	An important crop in the Himalaya.
Reproductive Cycle	:	May-September.
Material Examined	:	EBH-TP-1216.
Systematic Account	:	Bushy herbs having underground stem tuber, winged aerial stem, odd-pinnate leaves with a large terminal leaflet, and flowers borne in cymose panicles. Corolla rotate. Berries globose. Seeds oval.

Syzygium cumini (Linn.) Skeels (Pl. 94D)

Syn: *S. jambolanum* DC.; *Calyptranthes capitellata* Ham.; *C. jambolana* Willd.; *Eugenia caryophyllifolia* Lam.; *E. frondosa* Wall.; *E. jambolana* Lam.; *E. moorei* F. Muell.; *Myrtus cumini* Linn.

Family	:	Myrtaceae.
Vernacular Name	:	Jamun.
Eng., H. and Sans. Names	:	*Eng.*: Black Plum, Java Plum.
		H.: Jam, Jaman, Jamni Phalani, Jamun, Paiman, Phalanda, Phalinda, Pharenda, Phaunda.
		Sans.: Jambavam, Jambu, Jambula, Rajphala.
Distribution	:	Distributed throughout India (wild or cultivated) in both peninsulas and in Sri Lanka; from the plains to 1,500 m.
Habitat Ecology	:	Planted near water sources in fields.
Reproductive Cycle	:	April-August.
Material Examined	:	EBH-TP-268.
Systematic Account	:	A large-sized tree with a thick crooked trunk, coriaceous leaves and whitish, fragrant, sessile flowers arranged mostly in trichotomous panicles. Calyx shortly turbinate. Petals united into a thin membranous calyptra. Berries often obliquely oblong.

Syzygium jambos (Linn.) Alston (Pl. 94E)

Syn: *Eugenia jambos* Linn.; *Jambosa vulgaris* DC.

Family	:	Myrtaceae.
Vernacular Name	:	Gulab Jamun.
Eng. and H. Names	:	*Eng.*: Rose-apple.
		H.: Gulabjaman.
Distribution	:	Indigenous to Sikkim terai. Cultivated in S. India and many other parts of India upto 1,300 m.
Habitat Ecology	:	Planted.
Reproductive Cycle	:	February-June.
Material Examined	:	EBH-TP-216.
Systematic Account	:	An evergreen tree upto 10 m high. Crown spreading. Leaves 10-20 cm long, oblong or elliptic-lanceolate. Flowers greenish white or white, in short terminal corymbs. Fruits globose or pear-shaped.

Tabernaemontana coronaria **(Jacq.) Willd. (Pl. 95A)**

Syn: *T. divaricata* (Linn.) R.Br.; *Ervatamia coronaria* (Jacq.) Stapf; *Nerium divaricatum* Linn.

Family	:	Apocynaceae.
Vernacular Name	:	Chandni.
Eng., H. and Sans. Names	:	*Eng.*: Cape Jasmine, Wax Flower.
		H.: Chandni, Chandui, Sugandabala, Taggai, Taggar.
		Sans.: Ashvathabheda, Gajapadapa, Kshayataru, Nandivriksha, Tagara, Vishnupriya.
Distribution	:	Common in India and Myanmar.
Habitat Ecology	:	Cultivated in gardens; often met as an escape.
Reproductive Cycle	:	May-December.
Material Examined	:	EBH-TP-500.
Systematic Account	:	Evergreen dichotomously branched shrubs with grey bark having copious lenticels and membranous leaves. Flowers white, fragrant at night, often double. Follicles ribbed, recurved, red inside, 3-6 seeded.

Tagetes erecta **Linn. (Pl. 95B)**

Family	:	Asteraceae.
Vernacular Names	:	Bowdu, Genda, Sartajad.
Eng., H. and Sans. Names	:	*Eng.*: French Marigold, Marigold.
		H.: Genda, Gultera.
		Sans.: Sthulapushpa, Zandu, Zanduga.
Distribution	:	Native of Mexico, extensively cultivated throughout tropics and subtropics; occasionally naturalized in India.
Habitat Ecology	:	Extensivly cultivated in gardens as an ornamental plant.
Reproductive Cycle	:	October-April.
Material Examined	:	EBH-TP-1200.
Systematic Account	:	Stout branching herbs attaining upto 60 cm. Leaves strong-scented, pinnately divided; segments 11-17 having serrate margins. Flower heads solitary, lemon yellow-orange. Rays many, long-clawed, involucre campanulate. Achenes 6-7 mm long; pappus scaly.

Tamarindus indica **Linn. (Pl. 95C)**

Syn: *T. occidentalis* Gaertn.; *T. officinalis* Hook.; *Tamarindus umbrosa* Salisb.

Family	: Fabaceae.
Vernacular Name	: Imli.
Eng., H. and Sans. Names	: *Eng.*: The Tamarind Tree.
	H.: Amli, Amlica, Anbli, Imli, Nuli, Teter.
	Sans.: Ambia, Amlika, Tentrani, Tintidi, Tintili, Tintiri.
Distribution	: Probably native of Asia and Africa; cultivated throughout India.
Habitat Ecology	: Cultivated.
Reproductive Cycle	: June-July.
Material Examined	: EBH-TP-704.
Systematic Account	: Large-sized trees with spreading branches, abruptly pinnate leaves and glabrous leaflets. Flowers in copious lax racemes at the end of the branchlets. Pedicels articulated at the base of the calyx. Petals yellowish with pink stripes. Pods slightly curved, subcompressed, 3-10 seeded.

Tecoma radicans **Juss. (Pl. 96A)**

Syn: *Bignonia radicans* Linn.; *Campsis radicans* (Linn.) Seem.

Family	: Bignoniaceae.
Vernacular Name	: Shehnai Phool.
Eng. Names	: *Eng.*: Trumpet Climber, Trumpet-vine.
Distribution	: Native of S.E. United States.
Habitat Ecology	: Useful climber on bare walls.
Reproductive Cycle	: August-September.
Material Examined	: EBH-TP-904.
Systematic Account	: A deciduous scandent plant with adventitious roots arising from almost every node. Leaves opposite, compound; leaflets 9-11, toothed. Flowers funnel-shaped, orange coloured; 4-12 in a cluster.

Tecoma stans **(Linn.) Juss. *ex* H.B.K. (Pl. 96B)**

Syn: *Bignonia stans* Linn.; *Stenolobium stans* Seem.

Family	: Bignoniaceae.
Vernacular Name	: Peela Phool.

Eng. Name	:	*Eng.*: Yellow Trumpet Bush.
Distribution	:	Native of tropical S. America; planted as a hedge plant in gardens throughout India upto 1,500 m.
Habitat Ecology	:	Often naturalized on open degraded sites.
Reproductive Cycle	:	Flowers and fruits throughout the year.
Material Examined	:	EBH-TP-713.
Systematic Account	:	Shrubs or small trees upto 4 m tall. Leaves opposite, imparipinnate; leaflets usually 5 or 7, margins sharply serrate. Flowers fragrant, borne in terminal panicles. Calyx green. Corolla yellow, 4-4.5 cm long, tube inflated upward. Capsules linear, compressed; with seeds numerous.

Tecomaria capensis **(Thunberg) Spach. (Pl. 96C)**

Syn: *Bignonia capensis* Thunberg; *Tecoma capensis* (Thunberg) Lindley.

Family	:	Bignoniaceae.
Vernacular Name	:	Tecomeria.
Eng. Name	:	*Eng.*: Cape-honeysuckle.
Distribution	:	A native of S. Africa.
Habitat Ecology	:	Cultivated in subtropical gardens everywhere.
Reproductive Cycle	:	May-September.
Material Examined	:	EBH-TP-607.
Systematic Account	:	Evergreen smooth rambling shrubs. Leaves pinnate; leaflets toothed at the edges. Flowers in groups at the ends of upstanding branches, funnel-shaped. Petals 5, turned-back, fiery orange-red.

Tectona grandis **Linn. f. (Pl. 96D)**

Family	:	Verbenaceae.
Vernacular Name	:	Sagwan.
Eng., H. and Sans. Names	:	*Eng.*: The Teak Tree. *H.*: Sagun, Sagwan, Sakhu, Segun. *Sans.*: Saka.
Distribution	:	W. Deccan Peninsula. Malay Peninsula.
Habitat Ecology	:	Cultivated.
Reproductive Cycle	:	August-December.
Material Examined	:	EBH-TP-211.

Systematic Account : Deciduous trees having stellately tomentose quadrangular branchlets. Lower surface of leaves clothed with dense stellate grey tomentum. Flowers in terminal branched cymose panicles with lanceolate bracts at the forks. Corolla white, 5-6 lobed. Drupes subglobose, 4-lobed.

Terminalia bellirica (Gaertn.) Roxb. (Pl. 96E)

Syn: *Myrobalanus bellirica* Gaertn.

Family : Combretaceae.

Vernacular Name : Bheda.

Eng., H. and Sans. Names : *Eng.*: Belleric Myrobalan.

H.: Bahera, Behara, Behra, Bhaira, Bhairah, Bharla, Buhura, Bulla, Sagona.

Sans.: Baheruka, Bahira, Vibhitaka, Vibhitaki, Vipitakaha.

Distribution : Common in the plains and lower hills throughout India.

Habitat Ecology : Planted near houses, temples, agricultural fields.

Reproductive Cycle : May-September.

Material Examined : EBH-TP-208.

Systematic Account : A large deciduous tree with coriaceous leaves gathered about the extremities of the branches. Flowers greenish-yellow, in axillary slender spikes; upper flowers of the spikes male and lower ones hermaphrodite. Calyx pubescent outside, inside woolly. Drupes ovoid, velvety, 5-angled when dried.

Terminalia chebula (Gaertn.) Retz. (Pl. 97A)

Syn: *T. aruta* Ham.; *T. reticulata* Roth.; *Embryogonia arborea* Teys. and Binn.; *Myrobalanus chebula* Gaertn.

Family : Combretaceae.

Vernacular Name : Harad.

Eng., H. and Sans. Names : *Eng.*: The Chebulic or Black Myrobalan.

H.: Har, Harara, Harra.

Sans.: Abhaya, Amogha, Amruta, Avyatha, Haritaki.

Distribution : Throughout India.

Habitat Ecology : Barren Lands. Open slopes. Along agricultural fields.

Reproductive Cycle : April-June.

Material Examined : EBH-TP-209.

Systematic Account : Deciduous trees having leaf-buds, branchlets and youngest leaves covered with rust-coloured shining hairs. Flowering spikes usually in short terminal panicles arising in the axils of the uppermost leaves. Drupes more or less 5-ribbed.

Thevetia neriifolia Juss *ex* Steud. (Pl. 97B)

Syn: *T. peruviana* (Pers.) K. Schum.; *Cascabela thevetia* (Linn.) Lippoid; *Cerbera thevetia* Linn.

Family : Apocynaceae.

Vernacular Names : Ganeera, Kaneer, Pila Kaner.

Eng., H. and Sans. Names : *Eng.*: Exile Oleander, Yellow Oleander.

H.: Pila Kaner, Zard kunel.

Sans.: Ashvaghna, Divyapushpa, Sidhapushpa, Vira.

Distribution : Throughout India.

Habitat Ecology : Often cultivated in gardens in the plains; frequently near temples.

Reproductive Cycle : July-September.

Material Examined : EBH-TP-1004.

Systematic Account : An evergreen milky glabrous shrub with spirally arranged leaves which are narrowed at both ends. Flowers yellow, in a few-flowered terminal cymes. Corolla tubular at the base. Corona of 5 scales arching over the stamens inserted at the base of the corolla throat. Drupes broader than long.

Thuja orientalis Linn. (Pl. 97C)

Syn: *T. acuta* Moench; *T. chengii* Borderes and Gaussen; *T. decora* Salisb.; *Biota orientalis* (Linn.) Endl.; *Platycladus orientalis* (Linn.) Franco; *P. stricta* Spach.

Family : Cupressaceae.

Vernacular Name : Morpankh.

Eng. and H. Names : *Eng.*: Chinese Arborvitae, Chinese Thuja, Oriental Arborvitae.

H.: Mayur Pankh, Morpankhi.

Distribution : Tropical parts of India.

Habitat Ecology : Grown in gardens for its dense attractive foliage and bushy habit.

Reproductive Cycle : July-October.

Material Examined : EBH-TP-129.

Systematic Account : A small-sized evergreen conical tree with spreading and ascending branches. Branchlets thin. Leaves opposite, rhombic-ovate, acute, appressed, minute, bright green.

Thuja plicata D. Don (Pl. 97D)

Syn: *T. gigantea* Nutt.

Family : Cupressaceae.

Vernacular Name : Morpankh.

Eng. Names : *Eng.*: Giant Arbor-vitae, Western Red Cedar.

Distribution : Grows from Alaska (Baranof Island) through British Columbia, Oregon and Washington to California.

Habitat Ecology : Cultivated in gardens; perform well in deep moist soil and full sun.

Reproductive Cycle : June-September.

Material Examined : EBH-TP-314.

Systematic Account : Evergreen conifers with horizontally spreading short pendulous branches forming a narrow pyramidal head and bark fissured into long ridges. Shoots rounded and clothed with decussate scale-like leaves which are pointed on the leading shoots. Leaves with whitish markings beneath. Cones urn-shaped; seeds flattened with two narrow wings.

Thunbergia erecta (Benth.) T. Anders. (Pl. 98A)

Syn: *Meyenia erecta* Benth.

Family : Acanthaceae.

Vernacular Name : Punarva.

Eng. Name : *Eng.*: Bush Clockvine.

Distribution : Native of Tropical to S. Africa.

Habitat Ecology : Cultivated; thrives well in sun and all types of soils.

Reproductive Cycle : June-July.

Material Examined : EBH-TP-909.

Systematic Account : An erect hardy bushy shrub almost upto 1.5 m. Stems square with opposite decussate leaves which are toothed in middle. Flowers axillary, dark violet, 4 cm across, throat orange.

Tithonia tagetiflora Desf. (Pl. 98B)

Family : Asteraceae.

Vernacular Name : Aster.

Distribution : Himalaya; native of Mexico.

Habitat Ecology : Grown for their showy, dahlia-like flowers; perform well on any good soil.

Reproductive Cycle : May-September.

Material Examined : EBH-TP-1103.

Systematic Account : Plants with stout stems bearing pale green, three–lobed leaves and large heads of deep orange or orange-yellow flowers, each measuring about 10 cm or more in diameter.

Tropaeolum majus Linn. (Pl. 98C)

Family : Tropaeolaceae.

Vernacular Name : Tindu Phool.

Eng. Names : *Eng.*: Garden Nastrutium, Indian Cress, Nasturtium.

Distribution : Commonly cultivated in Indian gardens.

Habitat Ecology : Hilly slopes.

Reproductive Cycle : May-November.

Material Examined : EBH-TP-120.

Systematic Account : Sprawling and climbing, tuberous rooted annual herbs. Leaves large, divided into 5-narrow, pointed lobes, borne on long petioles. Flowers orange, red or yellow in colour, with a nectar-filled spur at the base.

Ulmus villosa Brandis (Pl. 98D)

Syn: *U. laevigata* Royle.

Family : Ulmaceae.

Vernacular Names : Moldu, Mullang, Moun.

H. Name : *H.*: Bhamri.

Distribution : N. W. Himalaya: 1,800-3,200 m.

Habitat Ecology : Planted along roadsides in villages.

Reproductive Cycle : February-March.

Material Examined : EBH-TP-311.

Systematic Account : A large-sized tree with drooping branches and glabrous branchlets. Leaves ovate-oblong; secondary nerves 14-

18 pairs, each nerve terminating in a large serrature. Flowers 10-15, in dense sessile fascicles. Perianth sessile and articulate on a short hairy peduncle. Ovary and styles densely woolly with long hairs. Samara densely villous.

Viola canescens Wall. (Pl. 98E)

Syn: *V. serpens* Wall. var. *canescens* (Wall.) Hook. f. and Thoms.

Family : Violaceae.

Vernacular Name : Nani.

Distribution : Hilly districts throughout India: 1,500-2,500 m.

Habitat Ecology : Shrubberies. Shady banks.

Reproductive Cycle : April-November.

Material Examined : EBH-TP-608.

Systematic Account : Densely pubescent soft herbs. Stems producing long, leafy runners. Leaves tufted, crenate; stipules fringed. Flowers lilac. Stigma truncate, not beaked.

Viola tricolor Linn. (Pl. 99A)

Family : Violaceae.

Vernacular Name : Dog Flower.

Eng. Names : *Eng.*: Flame Flower, Heartsease, John-of-my-pink, Love-in-idleness, Pansy, Pancy Violet.

Distribution : Cultivated in India. Indigenous to Europe, N. Asia and N. America.

Habitat Ecology : Cultivated as an ornamental plant.

Reproductive Cycle : April-May.

Material Examined : EBH-TP-803.

Systematic Account : Glabrous herbs having long, branched stems with pinnately parted large stipules toward the base. Flowers large; spur twice as long as the appendages of the calyx.

Wedelia calendulacea Less. (Pl. 99B)

Syn: *W. chinensis* (Osbeck) Merr.; *Joegeria calendulacea* Spreng.; *Verbesina calendulacea* Linn.

Family : Asteraceae.

Vernacular Name : Bhanjara.

H. and Sans. Names : *H.*: Bhangra, Bhanra.

		Sans.: Bhringaraja, Devapriya, Haripriya, Kesaraja, Pitabhringi.
Distribution	:	Bengal. Assam. Silhet. Eastern and Western Peninsulas and Sri Lanka.
Habitat Ecology	:	Cultivated in Gardens.
Reproductive Cycle	:	March-April.
Material Examined	:	EBH-TP-915.
Systematic Account	:	A perennial herb with stem terete, procumbent at the base and rooting at the lower nodes. Leaves opposite, subsessile, lanceolate-oblong. Heads solitary on long slender axillary peduncles. Involucral bracts herbaceous, hairy. Ray-flowers ligulate; ligules yellow, 2-3 toothed. Pappus a toothed membranous cup. Achenes of the ray tapering, slightly pubescent.

Wisteria sinensis (Sims) Sweet (Pl. 99C)

Syn: *W. chinensis* DC.; *W. praecox* Hand.-Mazz.; *Glycine sinensis* Sims; *Millettia chinensis* Benth.; *Rehsonia sinensis* (Sims) Stritch.

Family	:	Fabaceae.
Vernacular Name	:	Wisteria.
Eng. Name	:	*Eng.*: Chinese Wisteria.
Distribution	:	A native of China; found in various parts of India.
Habitat Ecology	:	Cultivated; suitable for growing on arch, wall, arbour and trellis.
Reproductive Cycle	:	April-May.
Material Examined	:	EBH-TP-103.
Systematic Account	:	Woody climbers. Leaves odd-pinnately compound, alternate, pale green; leaflets 7-11 pairs, light coppery when young. Flowers blue-violet, showy, arranged on pendulous terminal racemes upto 30 cm long; corolla pea-shaped, 3 cm long.

Withania somnifera (Linn.) Dunal (Pl. 100A)

Syn: *Physalis arborescens* Thunb.; *P. flexuosa* L.; *P. somnifera* Link.; *P. tomentosa* Thunb.

Family	:	Solanaceae.
Vernacular Name	:	Ashwagandha.
Eng., H. and Sans. Names	:	*Eng.*: Asvagandha, Winter Cherry.
		H.: Asgand, Ashwagandha, Punir.
		Sans.: Ashvagandha, Asvagandha, Turangigandha.

Distribution	:	Throughout drier and subtropical parts of India.
Habitat Ecology	:	Wasteplaces and hedges throughout the drier regions of India; occasionally cultivated for its medicinal roots.
Reproductive Cycle	:	November-February.
Material Examined	:	EBH-TP-132.
Systematic Account	:	An erect greyish tomentose evergreen shrub upto 2 m tall with stout, fleshy, whitish-brown roots and simple, broadly ovate leaves. Flowers greenish or lurid-yellow, borne singly or in few-flowered axillary umbellate cymes. Berries orange-red when mature and enclosed in the inflated, papery calyx. Seeds yellow, discoid or reniform.

Wrightia tomentosa **(Roxb.) Roem and Schult. (Pl. 100B)**

Syn: *W. wallichii* A. DC.

Family	:	Apocynaceae.
Vernacular Name	:	Khlawa.
H. and Sans. Names	:	*H.*: Daira, Dharauli, Dudhi. *Sans.*: Kutaja, Kutajah, Kutajastri.
Distribution	:	Throughout tropical India, ascending to 1,200 m.
Habitat Ecology	:	Moist forests.
Reproductive Cycle	:	March-July.
Material Examined	:	EBH-TP-916.
Systematic Account	:	A deciduous tree with divaricate branches and corky bark yielding abundant milky sap when cut. Leaves opposite, densely tomentose. Flowers with orange scales on the corona, borne in short dense erect terminal corymbose cymes. Follicles paired, pendant, cylindrical pod covered with rough tubercles.

Zanthoxylum armatum **DC. (Pl. 100C)**

Syn: *Z. alatum* Roxb.; *Z. hostile* Wall.; *Z. planispinum* Sieb and Zucc.

Family	:	Rutaceae.
Vernacular Name	:	Tirmir.
H. and Sans. Names	:	*H.*: Darmar, Nipalidhanya, Tejbal, Tejphal, Tezmal, Timbur, Timur, Tumru, Tun. *Sans.*: Andhaka, Dviha, Gandhalu, Mahamuni, Sanuja, Saurabh, Sauraja, Sphutal, Sugandhi, Tikshnapatra, Tumbaru.

Distribution : Kashmir-S.W. China; S.E. Asia: 1,100-2,500 m.

Habitat Ecology : Shrubberies, cultivated areas.

Reproductive Cycle : April-June.

Material Examined : EBH-TP-273.

Systematic Account : Shrubs or small trees upto 6 m high. Branches armed with straight prickles. Twigs smooth with scattered pale lenticels. Bark pale brown, deeply furrowed, corky. Leaves imparipinnate; leaflets 5-11. Flowers polygamous, yellow, in dense pubescent lateral panicles. Drupes globose.

Zea mays **Linn. (Pl. 100D)**

Family : Poaceae.

Vernacular Names : Kukri, Makki.

Eng., H. and Sans. Names : *Eng.*: Corn, Indian Corn, Maize.

H.: Bara-jowar, Bhutta, Junri, Kukri, Makai, Makka.

Sans.: Makaya, Yavanala.

Distribution : Throughout India.

Habitat Ecology : Widely cultivated crop in both tropical and temperate regions.

Reproductive Cycle : August-September.

Material Examined : EBH-TP-725.

Systematic Account : A robust, monoecious annual grass developing adventitious fibrous roots from the lower nodes of stem. Leaves with a short ciliate ligule and broadly lanceolate blade. Male flowers borne in numerous long spike-like racemes comprised of 2-flowered paired awnless spikelets, forming large panicles ('tassels'). Female inflorescence axillary, comprised of 8-16 (or more) rows or spikelets on a woody axis ('cob') being enclosed in large leafy bracts or spathes with long, silky styles protruding from the summit.

Zephyranthes candida **(Lindl.) Herb. (Pl. 101A)**

Syn: *Z. nivea* (Schult. and Schult. f.) D. Dietr.; *Amaryllis candida* Lindl.; *A. nivea* Schult. and Schult. f.; *Argyropsis candida* (Lindl.) M. Roem.; *Atamosco candida* (Lindl.) Sasaki; *Plectronema candida* (Lindl.) Raf.

Family : Amaryllidaceae.

Vernacular Name : Punkesar.

Eng. Name : *Eng.*: Flower of the Western Wind.

Distribution : Argentina and Uruguay along the La Plata River.

Habitat Ecology : Cultivated as an ornamental plant.

Reproductive Cycle : September-October.

Material Examined : EBH-TP-412.

Systematic Account : A bulbous plant with dark green, fleshy leaves. Stem bears white blooms, which are crocus-like and opening outwardly to show 6 spreading or recurved petals.

***Zephyranthes grandiflora* Lindl. (Pl. 101B)**

Syn: *Z. ackermannia* (Herb.) M.Roem.; *Z. carinata* Herb.; *Z. nervosa* D. Dietr.; *Z. pallida* M. Roem.; *Z. striata* Herb.; *Amaryllis minima* Ker. Gawl.; *A. pallida* Willd. *ex* Schult. and Schult. f.

Family : Amaryllidaceae.

Vernacular Name : Lily

Eng. Names : *Eng.*: Rain Lily, Thunder Flower, Zephyr Lily.

Distribution : Guatemala, W. Indies; naturalized in many warm countries.

Habitat Ecology : Often cultivated; also found as an escape.

Reproductive Cycle : June-July.

Material Examined : EBH-TP-804.

Systematic Account : Small bulbous plants with evergreen, basal, grassy, fleshy leaves, crocus-like, funnel-shaped flowers; segments 6, pink, about 5 cm long.

***Zinnia elegans* Jacq. (Pl. 101C)**

Syn: *Z. violacea* Cav.; *Crassina elegans* (Jacq.) Kuntze.

Family : Asteraceae.

Vernacular Name : Zinnia.

Eng. Name : *Eng.*: Zinnia.

Distribution : A native of Mexico.

Habitat Ecology : These plants prefer a sunny location and a moist soil rich in organic matter.

Reproductive Cycle : Summer, Autumn.

Material Examined : EBH-TP-1132.

Systematic Account : A coarse, upright, bushy annuals upto 1m high, with large solitary daisy like flowerheads on long stems, and opposite, sandpapery, lance-shaped leaves. The ray flowers are purple, the discs yellow and black, and the entire head is about 5 cm across.

Zizyphus mauritiana Lamk. (Pl. 101D)

Syn: *Z. jujuba* (Linn.) Gaertn.

Family	:	Rhamnaceae.
Vernacular Names	:	Baer, Ber.
Eng., H. and Sans. Names	:	*Eng.*: Common Jujube, Indian Jujube, Jujube. *H.*: Baer, Ber, Beri. *Sans.*: Ajapriya, Badara, Karkandlm, Kurala, Madhuraphala.
Distribution	:	Throughout India ascending in the outer Himalaya and hills upto 1,800 m; often cultivated.
Habitat Ecology	:	Wastelands; occasionally planted along field borders and habitations.
Reproductive Cycle	:	April-September.
Material Examined	:	EBH-TP-205.
Systematic Account	:	A small glabrous deciduous tree with deeply furrowed bark and spines solitary or in pairs. Leaves with buff hairs on the under sides. Flowers small, yellowish-green. Drupes elliptic, fleshy when ripe and with a 2-celled stone.

Epilogue

Statistical analysis of plant diversity available within the boundary of the temple-courtyards of H.P. revealed altogether 305 species (edible-77 spp; fodder-32 spp; material culture-60 spp; medicinal-96 spp; ornamental-232 spp; sacred-114 spp) belonging to 223 genera under 95 families (Figures 2, 17-19; Table 3). The overall diversity includes 131 trees, 87 herbs, 64 shrubs and 23 climbers (Figure 3; Table 4) of which 12 species belong to Gymnosperms (*Araucaria columnaris, Cedrus deodara, Cupressus sempervirens, Cupressus torulosa, Cycas revoluta, Juniperus chinensis, Juniperus macropoda, Pinus gerardiana, Pinus roxburghii, Pinus wallichiana, Thuja orientalis, Thuja plicata*), 28 species to Monocots (*Aloe vera, Alpinia nutans, Amaryllis reginae, Asparagus adscendens, Asparagus densiflorus, Asparagus racemosus, Canna indica, Caryota urens, Chamaerops humilis, Chlorophytum capense, Coix lachryma-jobi, Colocasia esculenta, Cynodon dactylon, Cyperus flabelliformis, Hedychium coronarium, Hemerocallis fulva, Hymenocallis littoralis, Iris germanica, Ixia chinensis, Musa paradisiaca, Narcissus tazetta, Phoenix sylvestris, Polianthes tuberosa, Saccharum officinarum, Sansevieria trifasciata, Zea mays, Zephyranthes candida, Zephyranthes grandiflora*) and 265 species to Dicots (191 genera) with bulk share of having 86.89 per cent representation of the total diversity (Figure 2; Tables 3, 4). As is inferred from the Table 4, the plant diversity is predominantly woody with trees accounting for 42.95 per cent of the overall diversity (Figure 4). Evidently, the Dicotyledonous families are over 8 times larger than Monocotyledonous families, the genera are more than 7 times the Monocots and the species are more than 9 times the Monocots (Figure 2; Table 3). The proportion of species belonging to Monocots to Dicots is 1: 9.46, of genera 1: 7.64 and of families is 1: 8.1. However, the ratio of the total number of genera to species is low, *i.e.*, 1: 1.37 in comparison to corresponding ratio for the whole India, *i.e.*, 1: 7 (Hooker, 1872-1897). Nevertheless, Good (1964) had also observed the same that within the same floral region, flora of smaller areas and remote islands have lower genus-species ratio as ecological diversity

Table 3: Statistical Synopsis of the Temple Plants of H.P.

Category	*Dicotyledons*		*Monocotyledons*		*Gymnosperms*		*Total*
	Total	*per cent*	*Total*	*per cent*	*Total*	*per cent*	
Family	81	85.26	10	10.53	4	4.21	95
Genus	191	85.65	25	11.21	7	3.14	223
Species	265	86.89	28	9.18	12	3.93	305

Table 4: Relative Percentage of the Temple Species in Different Groups of Plant Kingdom

Growth Form	*Angiosperms*		*Gymnosperms*	*Total*	*Percentage (per cent)*
	Dicots	*Monocots*			
Trees	118	2	11	131	42.95
Herbs	64	23	-	87	28.53
Shrubs	61	2	1	64	20.98
Climbers	22	1	-	23	7.54
Total	265	28	12	305	100

tends to be smaller. Regarding phyto-geographical enumeration of the collected plant resources, the study highlights the predominance of Asian elements (Afganistan, Bangladesh, Bhutan, China, India, Japan, Myanmar, Nepal, Pakistan, Sri Lanka,

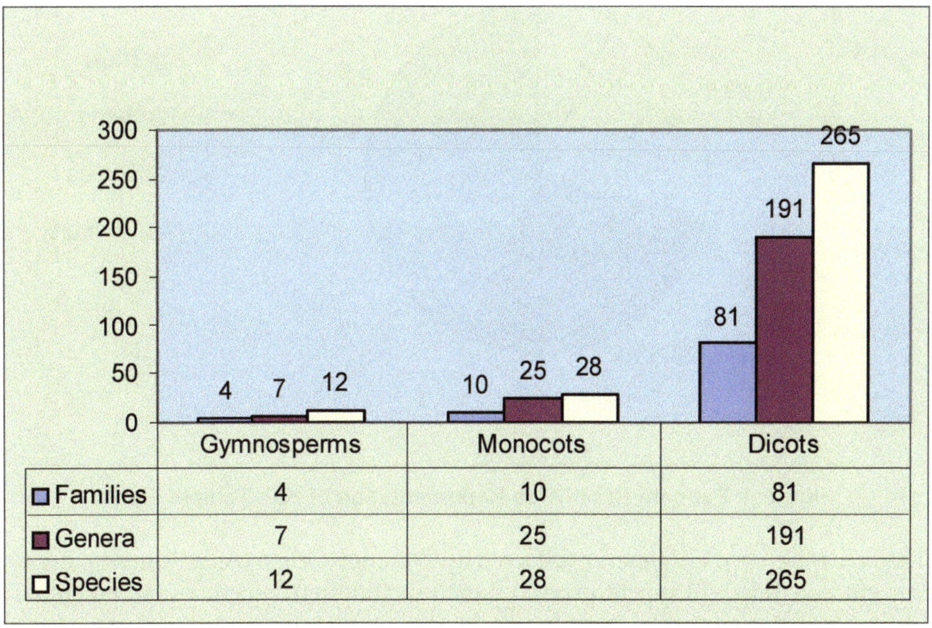

Figure 2: Families, Genera and Species of the Temple Plants of H.P.

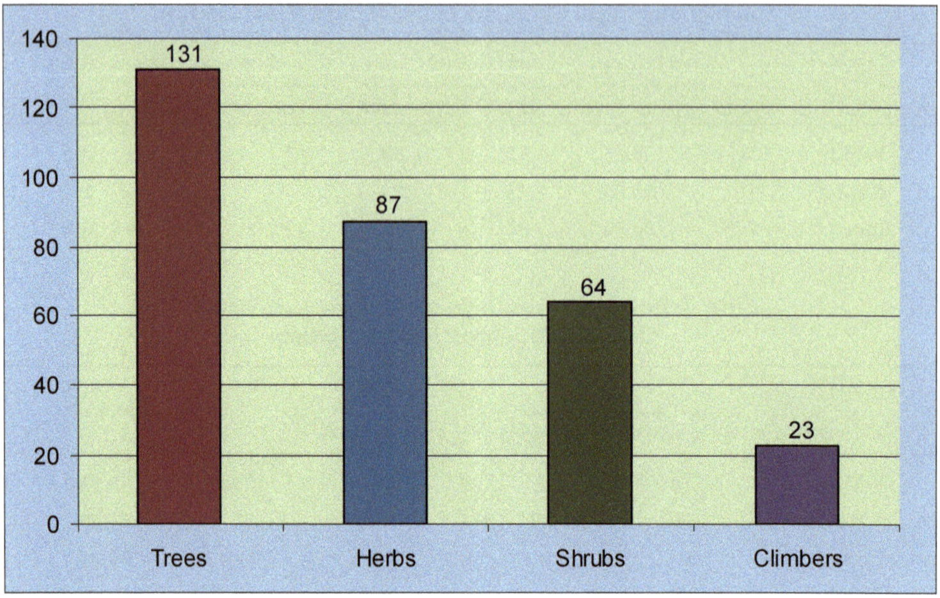

Figure 3: Histogram Reflecting Quantitative Number of Trees, Herbs, Shrubs and Climbers of H.P. Temple Plants

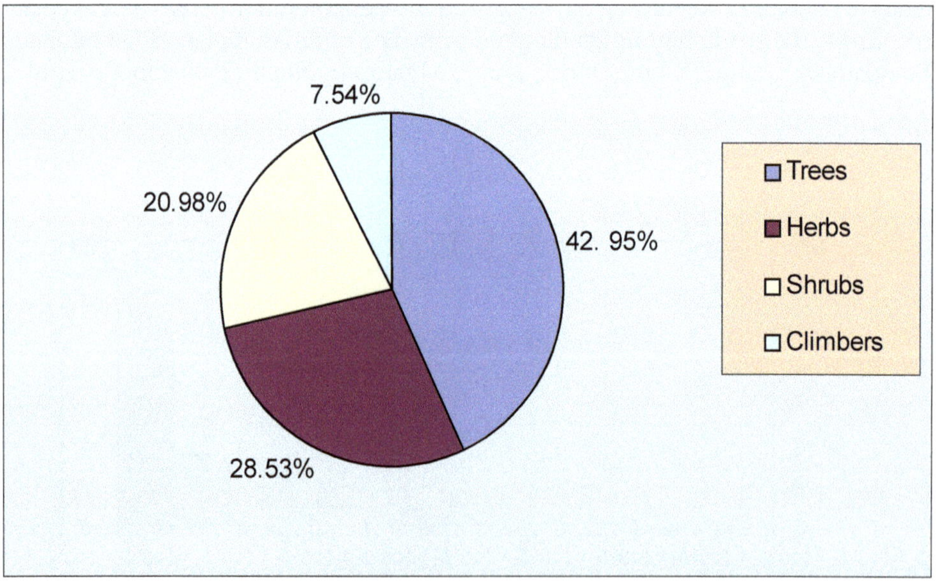

Figure 4: Per cent Habit-Wise Representation of H.P. Temple Plants

Thailand)–208 spp (148 genera), followed by Common Origin (Asia, America, Africa, Europe, Australia)–73 spp (66 genera), North and South American Continent (Brazil, Chile, Mexico, Venezuela, West Indies, United States)–70 spp (60 genera), African Continent (Algeria, Egypt, Ethiopia, Ghana, Kenya, Morocco, South Africa, Tunisia,

Zimbabwe)–46 spp (44 genera), European Continent (Austria, Belgium, Bulgaria, France, Germany, Greece, Italy, Romania, Spain, Switzerland, Ukraine)–45 spp (40 genera) and Australian Continent–32 spp (30 genera) with least representation (Figures 13, 14; Table 7). No doubt, exotics or aliens contribute to taxonomic richness of the area.

As a matter of fact, the most dominant families of flowering plants with respect to species richness pertain to Fabaceae (20 spp), Rosaceae (18 spp), Asteraceae (16 spp), Moraceae (14 spp), Euphorbiaceae (13 spp), Solanaceae (12 spp), Apocynaceae (10 spp), Lamiaceae (10 spp), Rutaceae (9 spp), Oleaceae (8 spp), Liliaceae, Malvaceae, Salicaceae (7 spp each), Amaryllidaceae, Cupressaceae, Verbenaceae (6 spp each), etc. Moreover, these 16 families account for about 55.40 per cent of the overall recorded species diversity from the study area, the predominant Fabaceae amongst Dicots contributing to 6.56 per cent, predominant Liliaceae amongst Monocots to 2.30 per cent and Cupressaceae amongst Gymnosperms to 1.97 per cent (Figures 5-9; Table 5). Amongst the predominant genera (Table 6), these are *Ficus* (9 spp), *Rosa* (8 spp), *Citrus, Prunus, Salix* (5 spp each), *Clerodendrum, Jasminum, Ocimum* (4 spp each), *Acacia, Carissa, Datura, Euphorbia, Hibiscus, Ipomoea* and *Pyrus* (3 spp each) for Dicots (Figure 10); *Asparagus* (3 spp) and *Zephyranthes* (2 spp) for Monocots (Figure 11) and *Pinus* (3 spp), *Cupressus, Juniperus* and *Thuja* (2 spp each) for Gymnosperms (Figure 12). The Dicotyledonous taxa predominate in their use. There could be no two opinions that the temple-courtyards harbour rich plant diversity and the local communities occurring in the surrounding areas have vast knowledge for sustainable utilization of the native plant wealth.

Table 5: Relative Percentage of the Pre-dominant Temple Families of H.P.

Family	*Species of Ethnic Use*	*Relative Percentage*
Fabaceae	20	6.56
Rosaceae	18	5.90
Asteraceae	16	5.25
Moraceae	14	4.59
Euphorbiaceae	13	4.26
Solanaceae	12	3.93
Apocynaceae	10	3.28
Lamiaceae	10	3.28
Rutaceae	9	2.95
Oleaceae	8	2.62
Liliaceae	7	2.30
Malvaceae	7	2.30
Salicaceae	7	2.30
Amaryllidaceae	6	1.97
Cupressaceae	6	1.97
Verbenaceae	6	1.97

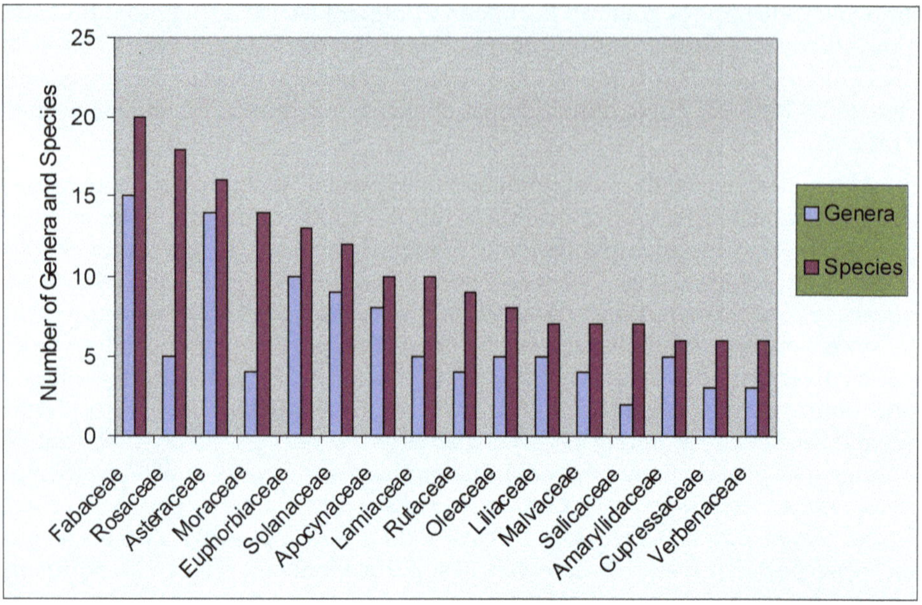

Figure 5: Pre-dominant Ethnobotanical Families of the Temple Plants

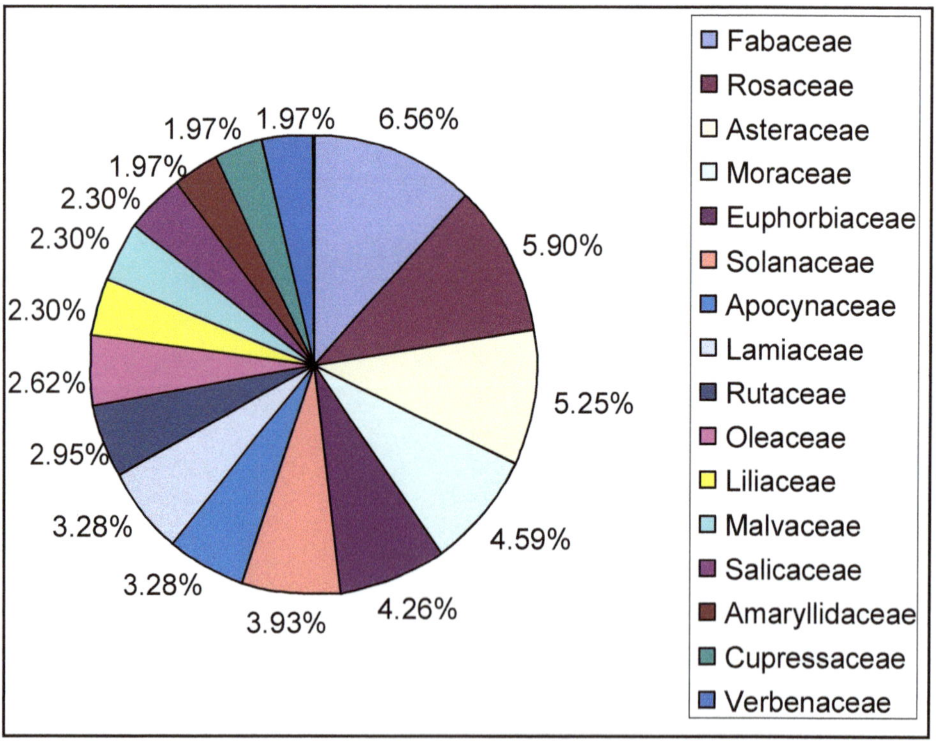

Figure 6: Relative Percentage of Taxa Belonging to Pre-dominant Ethnobotanical Families of the Temple Plants

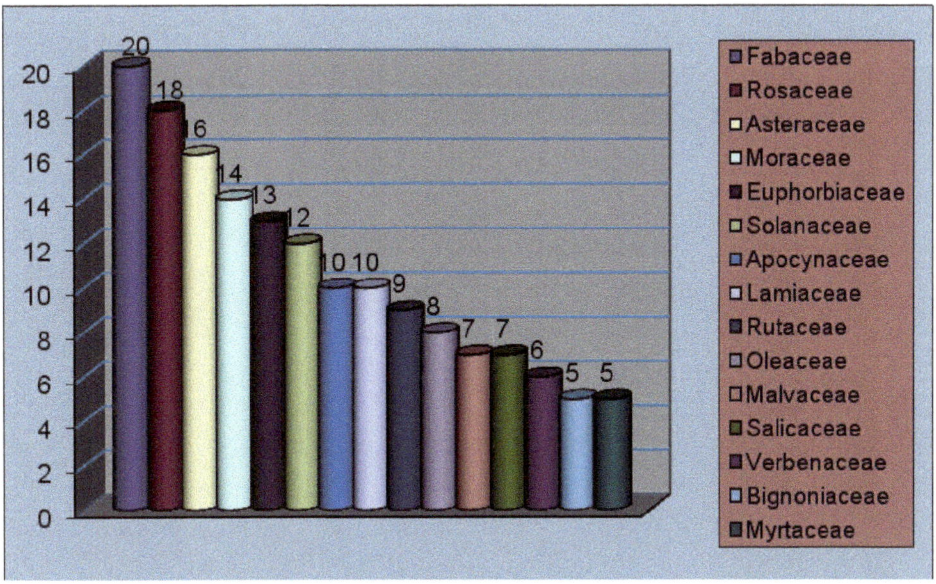

Figure 7: Pre-dominant Dicotyledonous Families of the Temple Plants

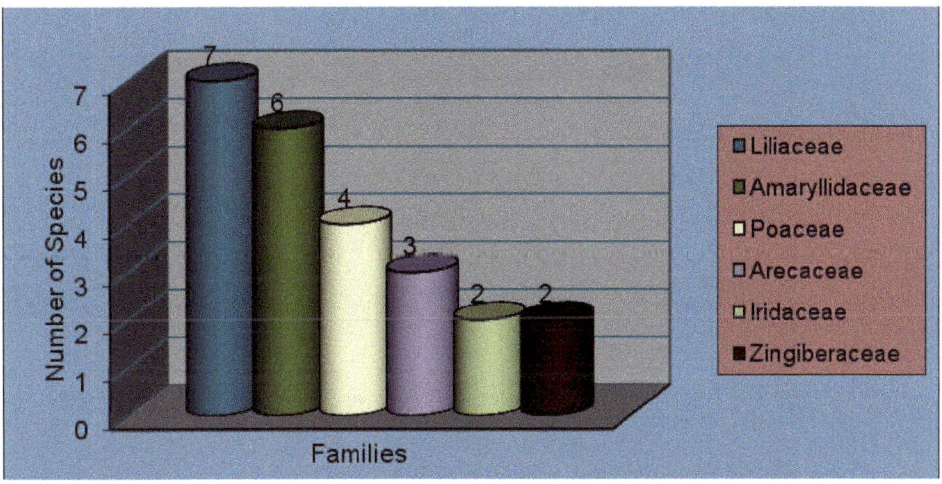

Figure 8: Pre-dominant Monocotyledonous Families of the Temple Plants

In all medical uses of 96 species covering 239 prescriptions for prevalent 71 ailments have been recorded during field studies in the temple-courtyards of H. P. (Table 8). These plant species belong to 82 genera and 43 families. Of the enlisted 96 species for medicinal uses, 56 occur in the wild state (Table 15) and the remaining 39 such as *Aloe vera, Alstonia scholaris, Bougainvillea spectabilis, Bryophyllum calycinum, Carica papaya, Catharanthus roseus, Cinnamomum camphora, Citrus aurantifolia, Citrus reticulata, Eucalyptus tereticornis, Hedychium coronarium, Jasminum humile, Juglans regia, Lagerstroemia indica, Lawsonia alba, Lonicera japonica, Lycopersicum esculentum, Melia*

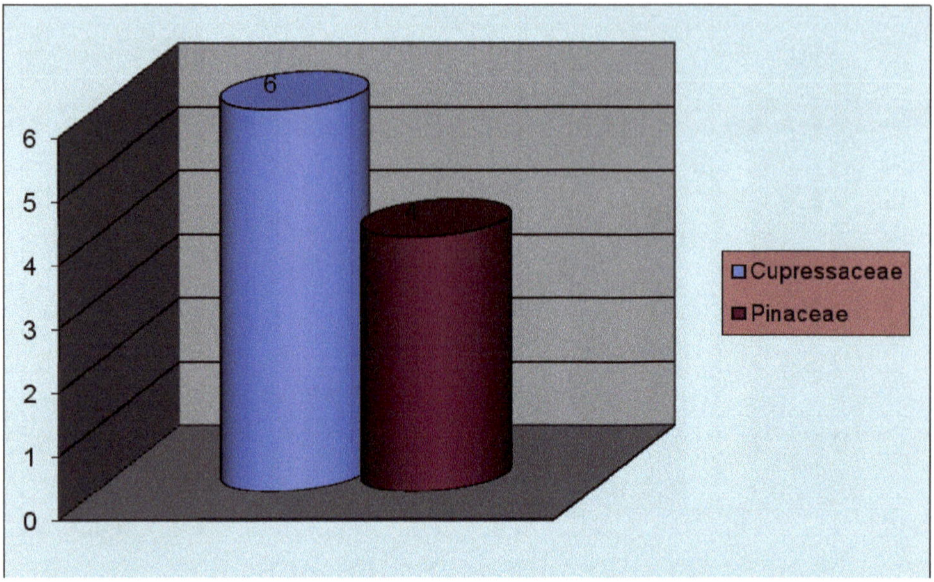

Figure 9: Pre-dominant Gymnospermous Families of the Temple Plants

Table 6: Pre-dominant Genera of the Temple Plants

Genus	*Total No. of Species*
Ficus	9
Rosa	8
Citrus	5
Prunus	5
Salix	5
Clerodendrum	4
Jasminum	4
Ocimum	4
Acacia	3
Asparagus	3
Carissa	3
Datura	3
Euphorbia	3
Hibiscus	3
Ipomoea	3
Pinus	3
Pyrus	3

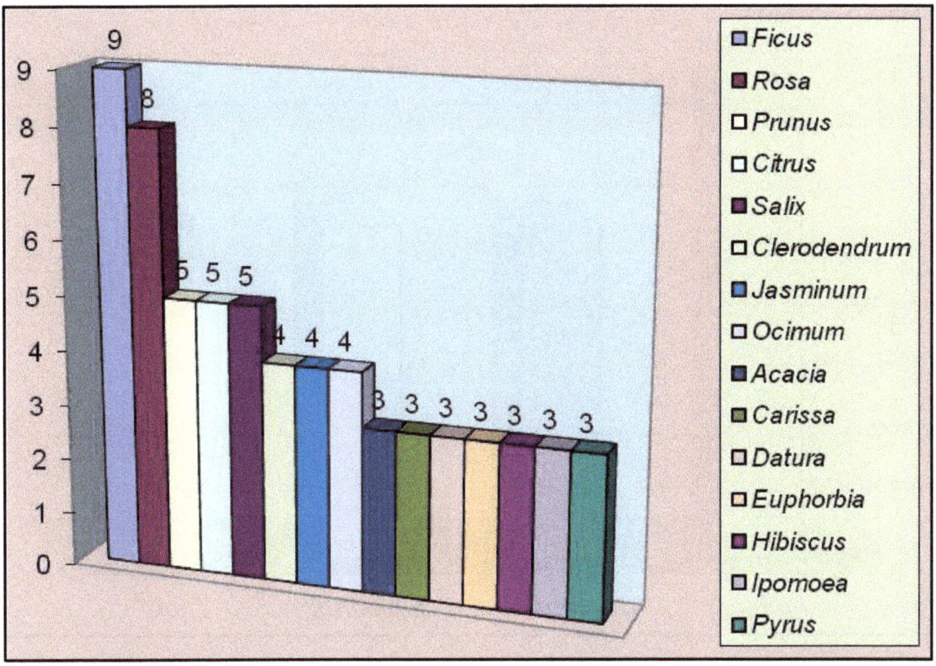

Figure 10: Pre-dominant Dicotyledonous Genera of the Temple Plants

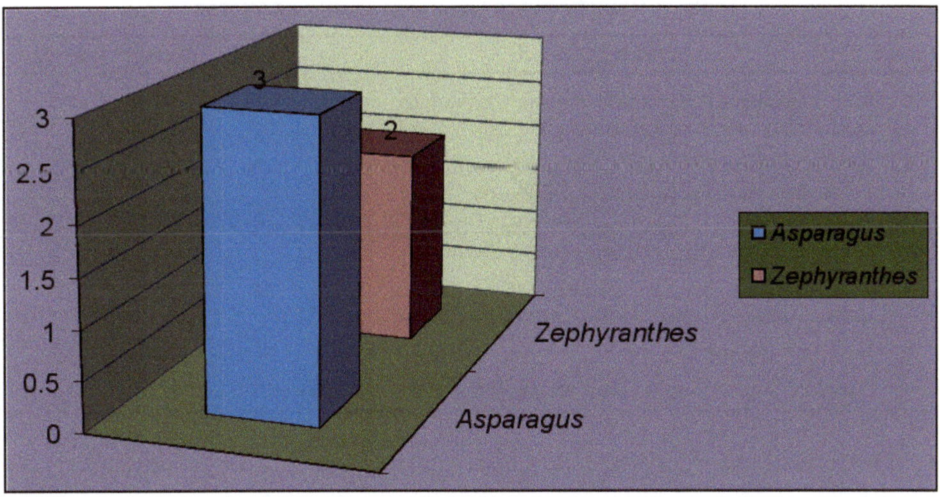

Figure 11: Pre-dominant Monocotyledonous Genera of the Temple Plants

azadirachta, Mentha piperita, Michelia champaca, Mimosa pudica, Mirabilis jalapa, Momordica charantia, Musa paradisiaca, Ocimum kilimandscharicum, Ocimum sanctum, Prunus amygdalus, Prunus persica, Psidium guajava, Punica granatum, Rosa foetida, Rosa indica, Saccharum officinarum, Santalum album, Sesamum orientale, Tabernaemontana coronaria, Tagetes erecta, Tamarindus indica and *Thevetia neriifolia* are cultivated

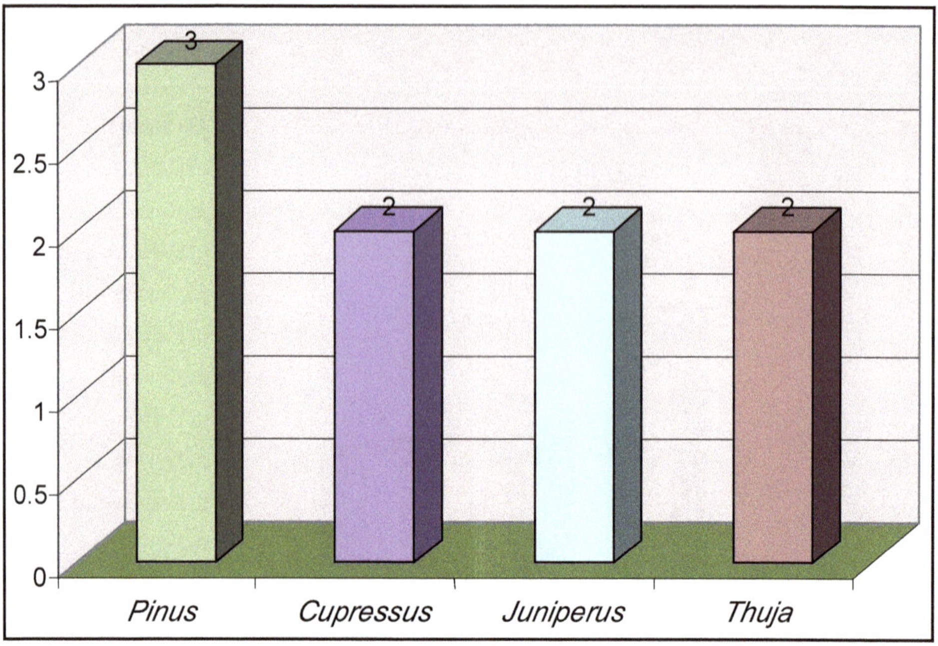

Figure 12: Pre-dominant Gymnospermous Genera of the Temple Plants

Table 7: Nativity of the Collected Plant Resources

Sl.No.	*Continent/Country*	*Species*	*Genera*	*Families*
1.	Asia (Afganistan, Bangladesh, Bhutan, China, India, Japan, Myanmar, Nepal, Pakistan, Sri Lanka, Thailand, etc.)	208	148	77
2.	North and South American Continent (Brazil, Chile, Mexico, Venezuela, West Indies, United States, etc.)	70	60	32
3.	African Continent (Algeria, Egypt, Ethiopia, Ghana, Kenya, Morocco, South Africa, Tunisia, Zimbabwe, etc.)	46	44	31
4.	Europe (Austria, Belgium, Bulgaria, France, Germany, Greece, Italy, Romania, Spain, Switzerland, Ukraine, etc.)	45	40	32
5.	Australian Continent	32	30	25
6.	Common Origin (Asia, America, Africa, Europe and Australia)	73	66	44

(Table 14) whereas *Mangifera indica* occurs in both wild as well as cultivated state. Out of all medicinal uses, leaves are used for 90 times, fruits for 43 times, seeds for 32 times, stem-bark for 20 times, roots for 19 times, latex for 15 times, flowers for 5 times, kernels for 3 times, fruit peel for 2 times, heartwood for 2 times, needles for 2 times, stem for 2 times, twigs for 2 times, whole plant for 2 times, rhizome, root bark, softwood and wood for once each. As evident from Table 10, the most prevalent ailments happen to be stomach disorders (acidity, stomachache-15 spp) and gastrointestinal disorders (dysentery, indigestion, diarrhoea-14 spp), followed by diabetes (11 spp), skin ailments (11 spp), cough (10 spp), liver complaints (hepatic complaints, jaundice-

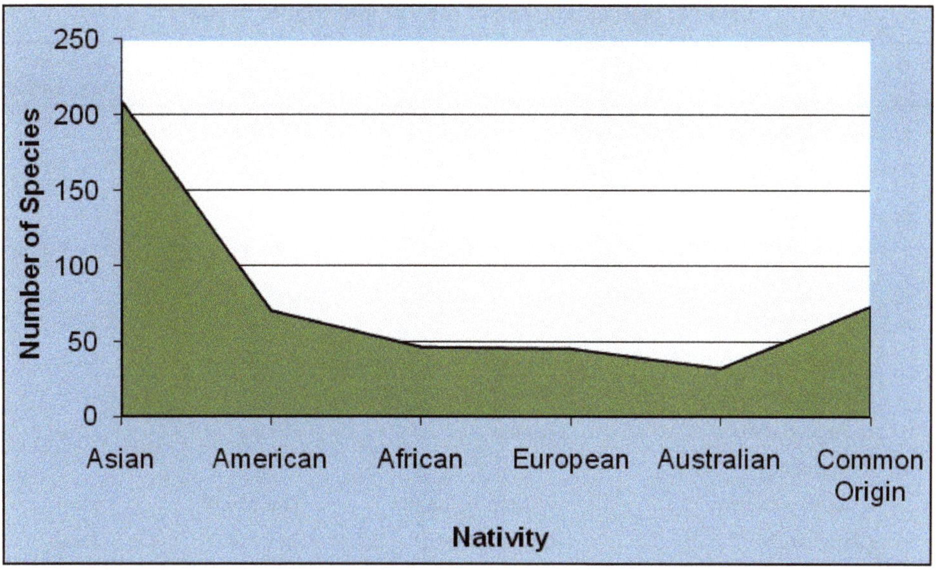

Figure 13: Pre-dominant Elements in the Temple-Courtyards of H.P.

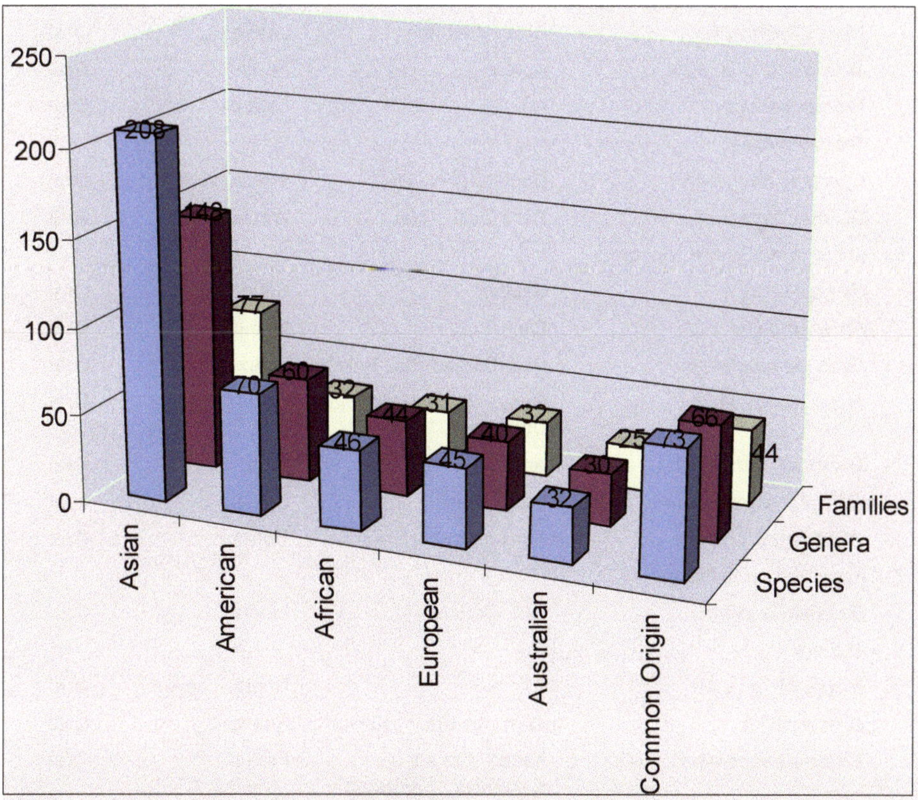

Figure 14: Histogram Showing Nativity and Respective Representation of Species, Genera and Families of the Temple Plants of H.P.

Table 8: Medicinal Uses of the Temple-Plants of H.P.

Sl.No.	*Botanical Name of Taxa*	*Local Name*	*Part/s Used*	*Mode of Administration*
1.	Accumulation of uric acid			
	Cannabis sativa	Bhang	Seeds	Oral
2.	Allergic disorders			
	Lagerstroemia indica	Har-Shingar	Leaves	Oral
3.	Anaemia			
	Carissa spinarum	Garuna	Fruits	Oral
	Lycopersicum esculentum	Tamatar	Fruits	Oral
	Punica granatum	Anar	Fruits	Oral
	Syzygium cumini	Jamun	Fruits	Oral
4.	Anthelmintic			
	Alstonia scholaris	Shaitan ka Jhad	Root-bark	Oral
	Citrus limon	Galgal, Khatta	Leaves	Oral
	Mallotus philippinensis	Kaamal, Kaambal	Fruits	Oral
	Melia azadirachta	Neem, Nim	Leaves	Oral
	Melia azedarach	Drek, Kachen	Leaves	Oral
	Momordica charantia	Karela	Fruits	Oral
	Prunus persica	Aru	Leaves	Oral
5.	Anticancerous			
	Cynodon dactylon	Doob, Doorv, Drub	Whole plant	Oral
	Ocimum sanctum	Ram Tulsi, Tulsi	Whole plant	Oral
6.	Aphrodisiac/Sexual vigour			
	Bergenia ciliata	Saplotri	Stem	Oral
	Bombax ceiba	Simbal	Roots	Oral
	Ficus benghalensis	Bad, Bargad, Bat Vriksha	Latex	Oral
	Ficus glomerata	Gular, Rumbal, Udumber, Umbarnae, Umro	Latex	Oral
	Mangifera indica	Amb	Fruit	Oral
	Ocimum gratissimum	Jangli Tulsi	Leaves	Oral
	Withania somnifera	Ashwagandha	Roots	Oral
7.	Appetizer (veterinary)			
	Boerhaavia diffusa	Itsit, Ootshut	Leaves	Oral
8.	Asthma			
	Calotropis procera	Ak	Roots, Leaves	Oral
	Cassia fistula	Aliya, Amaltash, Kaniar	Seeds	Oral
	Pistacia integerrima	Kakad, Kakadan, Kakadeyin, Kakderan	Fruits	Oral

Contd...

Table 8–*Contd...*

Sl.No.	*Botanical Name of Taxa*	*Local Name*	*Part/s Used*	*Mode of Administration*
9.	Bad smell from mouth			
	Zanthoxylum armatum	Tirmir	Leaves	Gargle
10.	Bed wetting			
	Aloe vera	Aloe, Quar, Quaria, Quar Patthal	Leaves	Oral
11.	Blood disorders			
	Bauhinia variegata	Kachnar, Karale, Karayalae	Stem-Bark	Oral
	Melia azadirachta	Neem, Nim	Leaves	Oral
12.	Blood purifier			
	Cassia fistula	Aliya, Amaltash, Kaniar	Seeds	Oral
	Mallotus philippinensis	Kaamal, Kaambal	Seeds	Oral
	Melia azadirachta	Neem, Nim	Leaves	Oral
	Murraya koenigii	Gandla, Kari patta	Leaves	Oral
	Syzygium cumini	Jamun	Stem-Bark, Fruits	Oral
13.	Body massage			
	Prunus amygdalus	Badam	Seeds	External
	Sesamum orientale	Til	Seeds	External
14.	Body swelling			
	Boerhaavia diffusa	Itsit, Ootshut	Roots	Oral
	Ricinus communis	Arind	Leaves	External
15.	Boils			
	Mirabilis jalapa	Gulal	Leaves	External
16.	Bone dislocation			
	Lawsonia alba	Mendi	Leaves	External
17.	Burning sensation (soles)			
	Lawsonia alba	Mendi	Leaves	External
18.	Cholera			
	Bischofia javanica	Bhillar	Leaves, Stem-Bark	Oral
19.	Cholesterol level			
	Emblica officinalis	Amla, Aonla	Fruits	Oral
	Juglans regia	Akhrot, Khor	Kernels	Oral
20.	Cold			
	Clerodendrum philippinum	Gandharaj, Moti-Mungra, Shiv-Darshan	Leaves	Oral
	Eucalyptus tereticornis	Safeda	Leaves	Oral
	Ocimum sanctum	Ram Tulsi, Tulsi	Leaves	Oral
	Roylea cinerea	Kadve, Karkoi	Leaves	Nasal drops

Contd...

Table 8–*Contd...*

Sl.No.	*Botanical Name of Taxa*	*Local Name*	*Part/s Used*	*Mode of Administration*
21.	Conception			
	Putranjiva roxburghii	Patajan, Patranjia	Seeds	Oral
22.	Constipation			
	Aloe vera	Aloe, Quar, Quaria, Quar Patthal	Leaves	External
	Cassia fistula	Aliya, Amaltash, Kaniar	Seeds	Oral
	Emblica officinalis	Amla, Aonla	Fruits	Oral
	Ficus palmata	Dhuda, Phakura, Khasra, Phedu, Phegura	Fruits	Oral
	Ipomoea purpurea	Gonani	Leaves	Oral
	Ricinus communis	Arind	Seeds	Oral
	Rosa indica	Gulab	Flower petals	Oral
	Terminalia bellirica	Bheda	Fruits	Oral
	Terminalia chebula	Harad	Fruits	Oral
23.	Constipation (veterinary)			
	Cassia fistula	Aliya, Amaltash, Kaniar	Seeds	Oral
24.	Cough			
	Aloe vera	Aloe, Quar, Quaria, Quar Patthal	Leaves	Oral
	Artocarpus lakoocha	Dheu	Fruits	Oral
	Calotropis procera	Ak	Flowers	Oral
	Citrus reticulata	Santra	Fruit peels	Oral
	Clerodendrum philippinum	Gandharaj, Moti-Mungra, Shiv-Darshan	Leaves	Oral
	Eucalyptus tereticornis	Safeda	Leaves	Oral
	Ocimum sanctum	Ram Tulsi, Tulsi	Leaves	Oral
	Pistacia integerrima	Kakad, Kakadan, Kakadeyin, Kakderan	Fruits	Oral
	Syzygium cumini	Jamun	Seeds	Oral
	Terminalia chebula	Harad	Fruits	Chewed
25.	Cuts			
	Alstonia scholaris	Shaitan ka Jhad	Latex	External
	Euphorbia royleana	Chhoon	Latex	External
	Ficus glomerata	Gular, Rumbal, Udumber, Umbarnae, Umro	Latex	External
	Ficus palmata	Dhuda, Phakura, Khasra, Phedu, Phegura	Latex	External
	Tagetes erecta	Bowdu, Genda, Sartajad	Leaves	External

Contd...

Table 8–*Contd...*

Sl.No.	*Botanical Name of Taxa*	*Local Name*	*Part/s Used*	*Mode of Administration*
26.	Delivery			
	Boerhaavia diffusa	Itsit, Ootshut	Roots	External
27.	Dental problems (toothache, gum-bleeding, etc.)			
	Acacia nilotica	Babul, Kikar	Twigs	Tooth brush
	Bergenia ciliata	Saplotri	Roots, Leaves	External
	Calotropis procera	Ak	Latex	Scouring teeth
	Ficus religiosa	Peepal, Pipal	Twigs	Chewed
	Hyoscyamus niger	Bajra Bhang, Dhandru	Seeds	Smoked
	Murraya koenigii	Gandla, Kari patta	Leaves	Gargle
	Psidium guajava	Amrood	Leaves	Gargle
	Thevetia neriifolia	Ganeera, Kaneer, Pila Kaner	Latex	Scouring teeth (Cotton plug)
	Zanthoxylum armatum	Tirmir	Leaves	Gargle
28.	Diabetes			
	Acacia nilotica	Babul, Kikar	Stem-Bark	Oral
	Aloe vera	Aloe, Quar, Quaria, Quar Patthal	Leaves	Oral
	Bougainvillea spectabilis	Boganvillea, Sada Suhagan	Leaves	Oral
	Catharanthus roseus	Sadabahar	Leaves, Roots	Oral
	Duranta repens	Neel Kanthi	Leaves	Oral
	Mangifera indica	Amb	Leaves	Oral
	Melia azadirachta	Neem, Nim	Leaves	Oral
	Momordica charantia	Karela	Fruits	Oral
	Murraya koenigii	Gandla, Kari patta	Leaves	Oral
	Psidium guajava	Amrood	Fruits	Oral
	Syzygium cumini	Jamun	Seeds	Oral
29.	Drinking habit			
	Calotropis procera	Ak	Flowers	Chewed
30.	Earache			
	Tagetes erecta	Bowdu, Genda, Sartajad	Leaves	Ear drop
31.	Eye disorders			
	Berberis lycium	Kashmal, Sutram	Roots	External
	Bombax ceiba	Simbal	Thorns	External
32.	Eyesight			
	Ficus glomerata	Gular, Rumbal, Udumber, Umbarnae, Umro	Fruits	Oral
	Ocimum sanctum	Ram Tulsi, Tulsi	Leaves	Oral

Contd...

Table 8–*Contd...*

Sl.No.	*Botanical Name of Taxa*	*Local Name*	*Part/s Used*	*Mode of Administration*
33.	Face pimples			
	Citrus reticulata	Santra	Fruit peel	External
	Tagetes erecta	Bowdu, Genda, Sartajad	Leaves	External
34.	Fever			
	Ficus religiosa	Peepal, Pipal	Leaves	Oral
	Hedychium coronarium	Vantosh	Rhizome	Oral
	Lagerstroemia indica	Har-Shingar	Leaves	Oral
	Michelia champaca	Champa	Stem-Bark	Oral
	Ocimum sanctum	Ram Tulsi, Tulsi	Leaves	Oral
	Tamarindus indica	Imli	Fruit	Oral
35.	Fever (veterinary)			
	Boerhaavia diffusa	Itsit, Ootshut	Roots	Oral
36.	Fracture			
	Bombax ceiba	Simbal	Stem-Bark	External
37.	Fungal infection			
	Lawsonia alba	Mendi	Leaves	External
38.	Furuncles			
	Bombax ceiba	Simbal	Stem-Bark	External
	Bryophyllum calycinum	Pattharchat	Leaves	External
	Ficus glomerata	Gular, Rumbal, Udumber, Umbarnae, Umro	Latex	External
39.	Gall bladder stone			
	Boerhaavia diffusa	Itsit, Ootshut	Roots	Oral
40.	Gastrointestinal disorders (dysentery, indigestion, diarrhoea)			
	Aegle marmelos	Bil	Fruit	Oral
	Bombax ceiba	Simbal	Stem-Bark	Oral
	Cedrela toona	Tuni	Stem-Bark	Oral
	Citrus aurantifolia	Nimbu	Seeds	Oral
	Dalbergia sissoo	Shisham, Tali	Leaves	Oral
	Ficus palmata	Dhuda, Phakura, Khasra, Phedu, Phegura	Fruits	Oral
	Lagerstroemia indica	Har-Shingar	Stem-Bark	Oral
	Lonicera japonica	Lushai	Leaves	Oral
	Mentha piperita	Pudina	Leaves	Oral
	Musa paradisiaca	Kela	Fruits	Oral
	Psidium guajava	Amrood	Leaves	Oral
	Punica granatum	Anar, Dadu	Seeds	Oral
	Pyrus pashia	Kainth, Shegal	Fruits	Oral
	Terminalia bellirica	Bheda	Fruits	Oral

Contd...

Table 8–*Contd...*

Sl.No.	*Botanical Name of Taxa*	*Local Name*	*Part/s Used*	*Mode of Administration*
41.	Hair-care (dandruff, hair colouring, hair–tonic)			
	Cinnamomum camphora	Kapur	Stem-bark	External
	Emblica officinalis	Amla, Aonla	Fruits	External
	Lawsonia alba	Mendi	Leaves	External
	Prunus amygdalus	Badam	Seeds	External
42.	Headache			
	Putranjiva roxburghii	Patajan, Patranjia	Seeds	External
	Roylea cinerea	Kadve, Karkoi	Leaves	contd.
	Nasal drops			
	Sapindus mukorossi	Ritha	Seeds	Oral
43.	Heart ailments			
	Emblica officinalis	Amla, Aonla	Fruit	Oral
	Juglans regia	Akhrot, Khor	Kernels	Oral
44.	Insect bites			
	Boerhaavia diffusa	Itsit, Ootshut	Leaves	External
	Cannabis sativa	Bhang	Leaves	External
	Catharanthus roseus	Sadabahar	Leaves	External
	Tagetes erecta	Bowdu, Genda, Sartajad	Leaves	External
45.	Internal injury			
	Bombax ceiba	Simbal	Stem-Bark	External
46.	Iron deficiency			
	Psidium guajava	Amrood	Fruits	Oral
47.	Joint pains (rheumatism)			
	Calotropis procera	Ak	Leaves	External
	Cannabis sativa	Bhang	Seeds	Oral
	Duranta repens	Neel Kanthi	Leaves	Oral
	Juglans regia	Akhrot, Khor	Kernel oil	External (massage)
	Ricinus communis	Arind	Seeds	External (massage)
	Sapium sebiferum	Makhan, Pahadi-Taali	Leaves, Stem-Bark	External
	Withania somnifera	Ashwagandha	Leaves	External
48.	Kidney stone			
	Bryophyllum calycinum	Pattharchat	Leaves	Oral
	Mimosa pudica	Chhui-mui	Roots	Oral

Contd...

Table 8–*Contd...*

Sl.No.	*Botanical Name of Taxa*	*Local Name*	*Part/s Used*	*Mode of Administration*
49.	Leucoderma			
	Calotropis procera	Ak	Latex	External
50.	Leucorrhoea			
	Aegle marmelos	Bil	Fruit	Oral
51.	Liver problems (hepatic complaints, jaundice)			
	Asparagus adscendens	Sanspaud, Satavari	Roots	Oral
	Bauhinia variegata	Kachnar, Karale, Karayalae	Stem-Bark	Oral
	Carica papaya	Papita	Fruits	Oral
	Centella asiatica	Brahmi, Minaki	Leaves	Oral
	Emblica officinalis	Amla, Aonla	Seeds	Oral
	Ficus religiosa	Peepal, Pipal	Leaves	Oral
	Lagerstroemia indica	Har-Shingar	Roots	Oral
	Litsea polyantha	Sangran	Stem-Bark	Oral
	Rosa foetida	Laybala	Flower petals	Oral
	Saccharum officinarum	Ganna	Stem juice	Oral
52.	Memory			
	Centella asiatica	Brahmi, Minaki	Leaves	Oral
	Prunus amygdalus	Badam	Seeds	Oral
53.	Menstrual problems			
	Pinus roxburghii	Chil, Chir	Needles	Oral
54.	Mental disorders			
	Cannabis sativa	Bhang	Seeds	Oral
	Datura fastuosa	Datura, Kala-Datura	Seeds	Oral
	Withania somnifera	Ashwagandha	Roots	Oral
55.	Mouth ulcers			
	Acacia catechu	Khair	Heartwood	Oral
	Jasminum humile	Chameli, Samsaun, Sunajui	Leaves	Chewed
	Jasminum sambac	Chameli	Leaves	Chewed
	Tabernaemontana coronaria	Chandni	Stem-Bark	External
56.	Nose bleeding			
	Emblica officinalis	Amla, Aonla	Fruits	External
57.	Pains and aches (body pain, backache, waist pain)			
	Aloe vera	Aloe, Quar, Quaria, Quar Patthal	Leaves	External
	Boerhaavia diffusa	Itsit, Ootshut	Leaves, Roots	External
	Cannabis sativa	Bhang	Leaves	External

Contd...

Table 8–*Contd...*

Sl.No.	Botanical Name of Taxa	Local Name	Part/s Used	Mode of Administration
58.	Paralysis			
	Sesamum orientale	Til	Seeds	External (massage)
59.	Piles			
	Berberis lycium	Kashmal, Sutram	Roots	Oral
	Bryophyllum calycinum	Pattharchat	Leaves	External
	Calotropis procera	Ak	Latex	External
	Cinnamomum camphora	Kapur	Stem-bark	Oral
	Melia azadirachta	Neem, Nim	Seeds	Oral
	Melia azedarach	Drek, Kachen	Seeds	Oral
	Sapindus mukorossi	Ritha	Fruit	Oral
60.	Pregnancy problems			
	Aegle marmelos	Bil	Fruit	Oral
61.	Prostate			
	Aloe vera	Aloe, Quar, Quaria, Quar Patthal	Leaves	Oral
62.	Skin ailments			
	Aesculus indica	Akhnor, Goon, Kanor	Fruit	External
	Albizia stipulata	Oyee	Stem-Bark	External
	Calotropis procera	Ak	Roots	External, Oral
	Carica papaya	Papita	Softwood	External
	Jasminum humile	Chameli, Samsaun, Sunajui	Leaves	External
	Juglans regia	Akhrot, Khor	Leaves	External
	Mallotus philippinensis	Kaamal, Kaambal	Seeds	External
	Melia azadirachta	Neem, Nim	Leaves	External, Oral
	Santalum album	Chandan	Stem-bark	External
	Withania somnifera	Ashwagandha	Leaves	External
	Zizyphus mauritiana	Baer, Ber	Leaves	External
63.	Spermatorrhoea			
	Citrus aurantifolia	Nimbu	Seeds	Oral
64.	Sprains			
	Lannea coromandelica	Kaimbal	Leaves	External
	Ocimum kilimandscharicum	Tulsi	Leaves	External
65.	Stomach disorders (acidity, stomachache)			
	Aegle marmelos	Bil, Bael	Fruit	Oral
	Aloe vera	Aloe, Quar, Quaria, Quar Patthal	Leaves	External, Oral

Contd...

Table 8–*Contd...*

Sl.No.	*Botanical Name of Taxa*	*Local Name*	*Part/s Used*	*Mode of Administration*
	Boerhaavia diffusa	Itsit, Ootshut	Leaves	Oral
	Centella asiatica	Brahmi, Minaki	Leaves	Oral
	Cordia dichotoma	Lasura	Fruits	Oral
	Emblica officinalis	Amla, Aonla	Fruits	Oral
	Mallotus philippinensis	Kaamal, Kaambal	Fruits	Oral
	Melia azadirachta	Neem, Nim	Leaves	Oral
	Mentha longifolia	Jangli Pudina, Madeni	Leaves	Oral
	Mentha piperita	Pudina	Leaves	Chewed
	Ricinus communis	Arind	Seeds	Oral
	Rosa indica	Gulab	Flower petals	Oral
	Syzygium cumini	Jamun	Seeds	Oral
	Terminalia bellirica	Bheda	Fruits	Oral
	Terminalia chebula	Harad	Fruits	Oral
66.	Throat infection, tonsils and respiratory problems			
	Acacia catechu	Khair	Heartwood	Oral
	Citrus limon	Galgal, Khatta	Fruit	Oral
	Pyrus pashia	Kainth, Shegal	Fruits	Oral
67.	Tuberculosis			
	Pinus roxburghii	Chil, Chir	Needles	Oral
68.	Typhoid			
	Citrus aurantifolia	Nimbu	Seeds	Oral
69.	Urinary troubles			
	Bauhinia variegata	Kachnar, Karale, Karayalae	Stem-Bark	Oral
	Boerhaavia diffusa	Itsit, Ootshut	Roots	Oral
	Santalum album	Chandan	Wood	Oral
	Tagetes erecta	Bowdu, Genda, Sartajad	Leaves	Oral
70.	Vomiting			
	Mentha piperita	Pudina	Leaves	Oral
71.	Wounds			
	Alstonia scholaris	Shaitan ka Jhad	Latex	External
	Bergenia ciliata	Saplotri	Roots, Leaves	External
	Euphorbia royleana	Chhoon	Latex	External
	Ficus glomerata	Gular, Rumbal, Udumber, Umbarnae, Umro	Latex	External
	Ficus palmata	Dhuda, Phakura, Khasra, Phedu, Phegura	Latex	External
	Melia azadirachta	Neem, Nim	Leaves	External

Table 9: Prescriptions

Method of Application	*No. of Prescriptions*	*Percentage (per cent)*
Administered orally	155	64.85
Applied externally	67	28.03
Chewed	6	2.51
Gargle	4	1.67
Nasal Drops	2	0.84
Scouring Teeth	2	0.84
Ear Drop	1	0.42
Smoked	1	0.42
Tooth brush	1	0.42

Table 10: Major Ailments in the Study Area

Name of the Ailment	*No. of Species*
Stomach disorders (acidity, stomachache)	15
Gastrointestinal disorders (dysentery, indigestion, diarrhoea)	14
Diabetes	11
Skin ailments	11
Cough	10
Liver problems (hepatic complaints, jaundice)	10
Constipation	9
Dental problems (toothache, gum-bleeding)	9
Anthelmintic	7
Aphrodisiac/Sexual vigour	7
Joint pains (rheumatism)	7
Piles	7
Fever	6
Wounds	6
Blood purifier	5
Cuts	5
Anaemia	4
Cold	4
Hair-care (dandruff, hair colouring, hair–tonic)	4
Insect bites	4
Mouth ulcers	4
Urinary troubles	4
Asthma	3
Furuncles	3

Contd...

Table 10–*Contd...*

Name of the Ailment	*No. of Species*
Headache	3
Mental disorders	3
Pains and aches (body pain, backache, Waist pain)	3
Throat infection, tonsils and respiratory problems	3
Anticancerous	2
Blood disorders	2
Body massage	2
Body swelling	2
Cholesterol level	2
Eye disorders	2
Eyesight	2
Face pimples	2
Heart ailments	2
Kidney stone	2
Memory	2
Sprains	2
Accumulation of uric acid	1
Allergic disorders	1
Appetiser (Veterinary)	1
Bad smell from mouth	1
Bed wetting	1
Boils	1
Bone dislocation	1
Burning sensation (soles)	1
Cholera	1
Conception	1
Constipation (Veterinary)	1
Delivery	1
Drinking habit	1
Earache	1
Fever (Veterinary)	1
Fractures	1
Fungal infection	1
Gall bladder stone	1
Internal injury	1
Iron deficiency	1

Contd...

Table 10–*Contd...*

Name of the Ailment	*No. of Species*
Leucoderma	1
Leucorrhoea	1
Menstrual problems	1
Nose bleeding	1
Paralysis	1
Pregnancy problems	1
Prostrate	1
Spermatorrhoea	1
Tuberculosis	1
Typhoid Fever	1
Vomiting	1

Table 11: Simultaneous Uses of a Particular Plant for the Treatment of Prevalent Diseases

Plant Species	*Ailment*
Acacia catechu	Mouth ulcers, Throat infection
Acacia nilotica	Dental problems, Diabetes
Aegle marmelos	Gastrointestinal disorders, Leucorrhoea, Pregnancy problems, Stomach disorders
Aesculus indica	Skin ailments
Albizia stipulata	Skin ailments
Aloe vera	Bed wetting, Constipation, Cough, Diabetes, Pains and aches, Prostate, Stomach disorders
Alstonia scholaris	Anthelmintic, Cuts, Wounds
Artocarpus lakoocha	Cough
Asparagus adscendens	Liver problems
Bauhinia variegata	Blood disorders, Liver problems, Urinary troubles
Berberis lycium	Eye disorders, Piles
Bergenia ciliata	Dental problems, Sexual vigour, Wounds
Bischofia javanica	Cholera
Boerhaavia diffusa	Appetiser (veterinary), Body swelling, Delivery, Fever (Veterinary), Gall bladder stone, Insect bites, Pains and aches, Stomach disorders, Urinary troubles
Bombax ceiba	Eye disorders, Fractures, Furuncles, Gastrointestinal disorders, Internal injury, Sexual vigour
Bougainvillea spectabilis	Diabetes
Bryophyllum calycinum	Furuncles, Kidney stone, Piles

Contd...

Table 11–*Contd...*

Plant Species	*Ailment*
Calotropis procera	Asthma, Cough, Dental problems, Drinking habit, Joint pains, Leucoderma, Piles, Skin ailments
Cannabis sativa	Accumulation of uric acid, Insect bites, Joint pains, Mental disorders, Pains and aches
Carica papaya	Liver problems, Skin ailments
Carissa spinarum	Anaemia
Cassia fistula	Asthma, Blood purifier, Constipation, Constipation (veterinary)
Catharanthus roseus	Diabetes, Insect bites
Cedrela toona	Gastrointestinal disorders
Centella asiatica	Liver problems, Memory, Stomach disorders
Cinnamomum camphora	Hair-care, Piles
Citrus aurantifolia	Gastrointestinal disorders, Spermatorrhoea, Typhoid
Citrus limon	Anthelmintic, Throat infection
Citrus reticulata	Cough, Face pimples
Clerodendrum philippinum	Cold, Cough
Cordia dichotoma	Stomach disorders
Cynodon dactylon	Anticancerous
Dalbergia sissoo	Gastrointestinal disorders
Datura fastuosa	Mental disorders
Duranta repens	Diabetes, Joint pains
Emblica officinalis	Cholesterol level, Constipation, Hair-care, Heart ailments, Liver problems, Nose bleeding, Stomach disorders
Eucalyptus tereticornis	Cold, Cough
Euphorbia royleana	Cuts, Wounds
Ficus benghalensis	Sexual vigour
Ficus glomerata	Cuts, Wounds, Eyesight, Furuncles, Sexual vigour
Ficus palmata	Constipation, Cuts, Wounds, Gastrointestinal disorders
Ficus religiosa	Dental problems, Fever, Liver problems
Hedychium coronarium	Fever
Hyoscyamus niger	Dental problems
Ipomoea purpurea	Constipation
Jasminum humile	Mouth ulcers, Skin ailments
Jasminum sambac	Mouth ulcers
Juglans regia	Cholesterol level, Heart ailments, Joint pains, Skin ailments
Lagerstroemia indica	Allergic disorders, Fever, Gastrointestinal disorders, Liver problems
Lannea coromandelica	Sprains

Contd...

Table 11–*Contd...*

Plant Species	*Ailment*
Lawsonia alba	Bone dislocation, Burning sensation (soles), Fungal infection, Hair-care
Lonicera japonica	Gastrointestinal disorders
Lycopersicum esculentum	Anaemia
Litsea polyantha	Liver problems
Mallotus philippinensis	Anthelmintic, Blood purifier, Skin ailments, Stomach disorders
Mangifera indica	Aphrodisiac, Diabetes
Melia azadirachta	Anthelmintic, Blood disorders, Blood purifier, Diabetes, Piles, Skin ailments, Stomach disorders, Wounds
Melia azedarach	Anthelmintic, Piles
Mentha longifolia	Stomach disorders
Mentha piperita	Gastrointestinal disorders, Stomach disorders, Vomiting
Michelia champaca	Fever
Mimosa pudica	Kidney stone
Mirabilis jalapa	Boils
Momordica charantia	Anthelmintic, Diabetes
Murraya koenigii	Blood purifier, Dental problems, Diabetes
Musa paradisiaca	Gastrointestinal disorders
Ocimum gratissimum	Aphrodisiac
Ocimum kilimandscharicum	Sprains
Ocimum sanctum	Anticancerous, Cold, Cough, Eyesight, Fever
Pinus roxburghii	Menstrual problems, Tuberculosis
Pistacia integerrima	Asthma, Cough
Prunus amygdalus	Body massage, Hair-care, Memory
Prunus persica	Anthelmintic
Psidium guajava	Dental problems, Diabetes, Gastrointestinal disorders, Iron deficiency
Punica granatum	Anaemia, Gastrointestinal disorders
Putranjiva roxburghii	Conception, Headache
Pyrus pashia	Gastrointestinal disorders, Throat infection
Ricinus communis	Body swelling, Constipation, Joint pains, Stomach disorders
Rosa foetida	Liver problems
Rosa indica	Constipation, Stomach disorders
Roylea cinerea	Cold, Headache
Saccharum officinarum	Liver problems
Santalum album	Skin ailments, Urinary troubles
Sapindus mukorossi	Headache, Piles

Contd...

Table 11–*Contd...*

Plant Species	*Ailment*
Sapium sebiferum	Joint pains
Sesamum orientale	Body massage, Paralysis
Syzygium cumini	Anaemia, Blood purifier, Cough, Diabetes, Stomach disorders
Tabernaemontana coronaria	Mouth ulcers
Tagetes erecta	Cuts, Earache, Face pimples, Insect bites, Urinary troubles
Tamarindus indica	Fever
Terminalia bellirica	Constipation, Gastrointestinal disorders, Stomach disorders
Terminalia chebula	Constipation, Cough, Stomach disorders
Thevetia neriifolia	Dental problems
Withania somnifera	Joint pains, Mental disorders, Sexual vigour, Skin ailments
Zanthoxylum armatum	Bad smell from mouth, Dental problems
Zizyphus mauritiana	Skin ailments

Table 12: Temple Plants of H.P. and the Number of Diseases Cured

Plant Species	*No. of Diseases*	*Plant Species*	*No. of Diseases*
Boerhaavia diffusa	9	*Calotropis procera*	8
Melia azadirachta	8	*Aloe vera*	7
Emblica officinalis	7	*Bombax ceiba*	6
Cannabis sativa	5	*Ficus glomerata*	5
Ocimum sanctum	5	*Syzygium cumini*	5
Tagetes erecta	5	*Aegle marmelos*	4
Cassia fistula	4	*Ficus palmata*	4
Juglans regia	4	*Lagerstroemia indica*	4
Lawsonia alba	4	*Mallotus philippinensis*	4
Psidium guajava	4	*Ricinus communis*	4
Withania somnifera	4	*Alstonia scholaris*	3
Bauhinia variegata	3	*Bergenia ciliata*	3
Bryophyllum calycinum	3	*Centella asiatica*	3
Citrus aurantifolia	3	*Ficus religiosa*	3
Mentha piperita	3	*Murraya koenigii*	3
Prunus amygdalus	3	*Terminalia bellirica*	3
Terminalia chebula	3	*Acacia catechu*	2
Acacia nilotica	2	*Berberis lycium*	2
Carica papaya	2	*Catharanthus roseus*	2
Cinnamomum camphora	2	*Citrus limon*	2

Contd...

Table 12–*Contd...*

Plant Species	*No. of Diseases*	*Plant Species*	*No. of Diseases*
Citrus reticulata	2	*Clerodendrum philippinum*	2
Duranta repens	2	*Eucalyptus tereticornis*	2
Euphorbia royleana	2	*Jasminum humile*	2
Mangifera indica	2	*Melia azedarach*	2
Momordica charantia	2	*Pinus roxburghii*	2
Pistacia integerrima	2	*Punica granatum*	2
Putranjiva roxburghii	2	*Pyrus pashia*	2
Rosa indica	2	*Roylea cinerea*	2
Santalum album	2	*Sapindus mukorossi*	2
Sesamum orientale	2	*Zanthoxylum armatum*	2
Aesculus indica	1	*Albizia stipulata*	1
Artocarpus lakoocha	1	*Asparagus adscendens*	1
Bischofia javanica	1	*Bougainvillea spectabilis*	1
Carissa spinarum	1	*Cedrela toona*	1
Cordia dichotoma	1	*Cynodon dactylon*	1
Dalbergia sissoo	1	*Datura fastuosa*	1
Ficus benghalensis	1	*Hedychium coronarium*	1
Hyoscyamus niger	1	*Ipomoea purpurea*	1
Jasminum sambac	1	*Lannea coromandelica*	1
Lonicera japonica	1	*Lycopersicum esculentum*	1
Litsea polyantha	1	*Mentha longifolia*	1
Michelia champaca	1	*Mimosa pudica*	1
Mirabilis jalapa	1	*Musa paradisiaca*	1
Ocimum gratissimum	1	*Ocimum kilimandscharicum*	1
Prunus persica	1	*Rosa foetida*	1
Saccharum officinarum	1	*Sapium sebiferum*	1
Tabernaemontana coronaria	1	*Tamarindus indica*	1
Thevetia neriifolia	1	*Zizyphus mauritiana*	1

10 spp), constipation (9 spp), dental problems (toothache, gum-bleeding-9 spp), anthelmintic (7 spp), aphrodisiac/sexual vigour (7 spp), joint pains (rheumatism-7 spp), piles (7 spp), fever (6 spp), wounds (6 spp), blood purifier (5 spp), cuts (5 spp), anaemia (4 spp), cold (4 spp), hair-care (dandruff, hair–tonic-4 spp), insect bites (4 spp), mouth ulcers (4 spp), urinary troubles (4 spp), asthma (3 spp), furuncles (3 spp), headache (3 spp), mental disorders (3 spp), pains and aches (body pain, backache, waist pain-3 spp), throat infection, tonsils and respiratory problems (3 spp), anticancerous (2 spp), blood disorders (2 spp), body massage (2 spp), body swelling (2 spp), cholesterol level (2 spp), eye disorders (2 spp), eyesight (2 spp), face

pimples (2 spp), heart ailments (2 spp), kidney stone (2 spp), memory (2 spp) and sprains (2 spp). However, only one plant species has been recommended for the remaining 31 ailments thereby indicating less prevalence of these diseases in the study area. For alleviating their problems, the simultaneous uses of a particular plant for the treatment of a number of diseases have been reported for *Boerhaavia diffusa, Calotropis procera, Melia azadirachta, Aloe vera, Emblica officinalis, Bombax ceiba, Cannabis sativa, Ficus glomerata, Ocimum sanctum, Syzygium cumini, Tagetes erecta, Aegle marmelos, Cassia fistula, Ficus palmata, Juglans regia, Lagerstroemia indica, Lawsonia alba, Mallotus philippinensis, Psidium guajava, Ricinus communis, Withania somnifera, Alstonia scholaris, Bauhinia variegata, Bergenia ciliata, Bryophyllum calycinum, Centella asiatica, Citrus aurantifolia, Ficus religiosa, Mentha piperita, Murraya koenigii, Prunus amygdalus, Terminalia bellirica, Terminalia chebula,* etc. (Tables 11 and 12). Most of these non-cultivated medicinal species can be utilized as potential germplasm of important medicinal plants of the region and hold the key for taking up further researches for new leads for systematic pharmacotherapeutic and clinical research.

Of the 43 families used, high use families are: Fabaceae (8 spp), Euphorbiaceae (7 spp), Lamiaceae (6 spp), Rutaceae (6 spp), Apocynaceae (5 spp), Moraceae (5 spp), Rosaceae (5 spp), Solanaceae (4 spp), Anacardiaceae (3 spp), Meliaceae (3 spp), Myrtaceae (3 spp), Nyctaginaceae (3 spp), Combretaceae (2 spp), Lauraceae (2 spp), Liliaceae (2 spp), Lythraceae (2 spp), Oleaceae (2 spp), Poaceae (2 spp), Verbenaceae (2 spp), etc. and the remaining families such as Apiaceae, Asclepiadaceae, Asteraceae, Berberidaceae, Bombacaceae, Cannabinaceae, Caprifoliaceae, Caricaceae, Convolvulaceae, Cordiaceae, Crassulaceae, Cucurbitaceae, Hippocastanaceae, Juglandaceae, Magnoliaceae, Musaceae, Pedaliaceae, Pinaceae, Punicaceae, Rhamnaceae, Santalaceae, Sapindaceae, Saxifragaceae and Zingiberaceae are represented by one species each (Table 13). So, Fabaceae deserve priority attention for their phytochemical confirmation, followed by Euphorbiaceae, Lamiaceae, Rutaceae, Apocynaceae, Moraceae and Rosaceae. It is also seen that Poaceae which has given cereals to the world are not as important medicinally as other families. Notably, high use genera are: *Ficus* (4 spp), *Citrus* (3 spp), *Ocimum* (3 spp), *Acacia* (2 spp), *Jasminum* (2 spp), *Melia* (2 spp), *Mentha* (2 spp), *Prunus* (2 spp), *Rosa* (2 spp), *Terminalia* (2 spp), etc. (Table 13). With regard to the mode of administration of drugs, the study shows the predominance of oral administration (155 times) having 64.85 per cent of overall representation than others (Table 9) either singly or in combination with others. Basically, religious-cultural faith, weak economy, inaccessibility and poor health service availability in the study area seem to be the cause of their dependence on these medicinal plant species. Regarding bioperspective assessment of the phytodiversity from medicinal point of view, it is seen that the plants marked with asterisk in Tables 14 and 15 have been enlisted as commercial plants for trade and drug industry and are collected regularly on small scale or often on large scale as and when there is a demand from the market (Chauhan, 1999) whereas the others in the lists are those for whom most of the remedies either less known or are hitherto unrecorded and worthy of consideration as perspective healthcare plants for the welfare of mankind. These studies have also shown that much of the valuable information has remained unverified and unutilized. The detailed phytochemical and pharmacological studies are, however, required to determine the effective constituents and characteristic biological activity of these potential medicinal plants for wider human welfare, and ultimately helping in integrating traditional with modern system of medicine. At the same time the organized collection, storage and processing of crude herbal drugs for pharmaceutical industries will generate employment to the rural populace for their upliftment.

Table 13: Pre-dominant Families of the Temple Plants of H.P. Used for Various Ailments

Family	*Species*	*Genera/Species*
Anacardiaceae	*Lannea coromandelica, Mangifera indica, Pistacia integerrima*	3/3
Apiaceae	*Centella asiatica*	1/1
Apocynaceae	*Alstonia scholaris, Carissa spinarum, Catharanthus roseus, Tabernaemontana coronaria, Thevetia neriifolia*	5/5
Asclepiadaceae	*Calotropis procera*	1/1
Asteraceae	*Tagetes erecta*	1/1
Berberidaceae	*Berberis lycium*	1/1
Bombacaceae	*Bombax ceiba*	1/1
Cannabinaceae	*Cannabis sativa*	1/1
Caprifoliaceae	*Lonicera japonica*	1/1
Caricaceae	*Carica papaya*	1/1
Combretaceae	*Terminalia bellirica, Terminalia chebula*	½
Convolvulaceae	*Ipomoea purpurea*	1/1
Cordiaceae	*Cordia dichotoma*	1/1
Crassulaceae	*Bryophyllum calycinum*	1/1
Cucurbitaceae	*Momordica charantia*	1/1
Euphorbiaceae	*Bischofia javanica, Emblica officinalis, Euphorbia royleana, Mallotus philippinensis, Putranjiva roxburghii, Ricinus communis, Sapium sebiferum*	7/7
Fabaceae	*Acacia catechu, Acacia nilotica, Albizia stipulata, Bauhinia variegata, Cassia fistula, Dalbergia sissoo, Mimosa pudica, Tamarindus indica*	7/8
Hippocastanaceae	*Aesculus indica*	1/1
Juglandaceae	*Juglans regia*	1/1
Lamiaceae	*Mentha longifolia, Mentha piperita, Ocimum gratissimum, Ocimum kilimandscharicum, Ocimum sanctum, Roylea cinerea*	3/6
Lauraceae	*Litsea polyantha, Cinnamomum camphora*	2/2
Liliaceae	*Aloe vera, Asparagus adscendens,*	2/2
Lythraceae	*Lawsonia alba, Lagerstroemia indica*	2/2
Magnoliaceae	*Michelia champaca*	1/1
Meliaceae	*Cedrela toona, Melia azadirachta, Melia azedarach*	2/3

Contd...

Table 13–*Contd...*

Family	*Species*	*Genera/Species*
Moraceae	*Artocarpus lakoocha, Ficus benghalensis, Ficus glomerata, Ficus palmata, Ficus religiosa*	2/5
Musaceae	*Musa paradisiaca*	1/1
Myrtaceae	*Eucalyptus tereticornis, Psidium guajava, Syzygium cumini*	3/3
Nyctaginaceae	*Boerhaavia diffusa, Bougainvillea spectabilis, Mirabilis jalapa*	3/3
Oleaceae	*Jasminum humile, Jasminum sambac*	1/2
Pedaliaceae	*Sesamum orientale*	1/1
Pinaceae	*Pinus roxburghii*	1/1
Poaceae	*Cynodon dactylon, Saccharum officinarum*	2/2
Punicaceae	*Punica granatum*	1/1
Rhamnaceae	*Zizyphus mauritiana*	1/1
Rosaceae	*Prunus amygdalus, Prunus persica, Pyrus pashia, Rosa foetida, Rosa indica*	3/5
Rutaceae	*Aegle marmelos, Citrus aurantifolia, Citrus limon, Citrus reticulata, Murraya koenigii, Zanthoxylum armatum*	4/6
Santalaceae	*Santalum album*	1/1
Sapindaceae	*Sapindus mukorossi*	1/1
Saxifragaceae	*Bergenia ciliata*	1/1
Solanaceae	*Datura fastuosa, Hyoscyamus niger, Lycopersicum esculentum, Withania somnifera*	4/4
Verbenaceae	*Clerodendrum philippinum, Duranta repens*	2/2
Zingiberaceae	*Hedychium coronarium*	1/1

Table 14: Cultivated Plants Having Medicinal Uses

Species	*Family*	*Edible Part/s*
Aloe vera	Liliaceae	Leaves
Alstonia scholaris	Apocynaceae	Root-bark, Latex
Bougainvillea spectabilis	Nyctaginaceae	Leaves
Bryophyllum calycinum	Crassulaceae	Leaves
**Carica papaya*	Caricaceae	Fruit, Softwood
Catharanthus roseus	Apocynaceae	Leaves, Root
**Cinnamomum camphora*	Lauraceae	Stem-bark
**Citrus aurantifolia*	Rutaceae	Seed
**Citrus reticulata*	Rutaceae	Fruit Peel
Eucalyptus tereticornis	Myrtaceae	Leaves
Hedychium coronarium	Zingiberaceae	Rhizome
**Jasminum humile*	Oleaceae	Leaves

Contd...

Table 14–*Contd...*

Species	*Family*	*Edible Part/s*
**Juglans regia*	Juglandaceae	Kernels, Leaves
Lagerstroemia indica	Lythraceae	Leaves, Stem-Bark, Root
Lawsonia alba	Lythraceae	Leaves
Lonicera japonica	Caprifoliaceae	Leaves
Lycopersicum esculentum	Solanaceae	Fruit
Mangifera indica	Anacardiaceae	Fruit, Leaves
Melia azadirachta	Meliaceae	Leaves, Seed
**Mentha piperita*	Lamiaceae	Leaves
Michelia champaca	Magnoliaceae	Stem-Bark
**Mimosa pudica*	Fabaceae	Root
**Mirabilis jalapa*	Nyctaginaceae	Leaves
Momordica charantia	Cucurbitaceae	Fruits
Musa paradisiaca	Musaceae	Fruit
Ocimum kilimandscharicum	Lamiaceae	Leaves
**Ocimum sanctum*	Lamiaceae	Whole Plant, Leaves
**Prunus amygdalus*	Rosaceae	Seed
Prunus persica	Rosaceae	Leaves
Psidium guajava	Myrtaceae	Leaves, Fruit
**Punica granatum*	Punicaceae	Fruit, Seed
Rosa foetida	Rosaceae	Flower
Rosa indica	Rosaceae	Flower
Saccharum officinarum	Poaceae	Stem Juice
Santalum album	Santalaceae	Stem-bark, Wood
**Sesamum orientale*	Pedaliaceae	Seed
Tabernaemontana coronaria	Apocynaceae	Stem-Bark
Tagetes erecta	Asteraceae	Leaves
Tamarindus indica	Fabaceae	Fruit
Thevetia neriifolia	Apocynaceae	Latex

*: Plants for Trade and Drug Industry.

The study highlights the use of 77 species belonging to 57 genera under 37 families for edible purposes (Table 16) of which 37 spp are cultivated and the remaining 39 spp grow as wild whereas *Mangifera indica* grows wild as well as cultivated (Tables 18 and 19) thereby representing 1.97 per cent of the total edible diversity reported from India (Anonymous, 1994). Of these, flower buds of 1 species, root of 1 species, kernel of 1 species, tender branches and stem of 4 species, tubers and corms of 2 species, leaves of 8 species, seeds of 8 species and fruits of 56 species (maximum

Table 15: Wild Medicinal Plants

Species	*Family*	*Edible Part/s*
**Acacia catechu*	Fabaceae	Heartwood
**Acacia nilotica*	Fabaceae	Twig, Stem-Bark
**Aegle marmelos*	Rutaceae	Fruit
**Aesculus indica*	Hippocastanaceae	Fruit
Albizia stipulata	Fabaceae	Stem-Bark
**Artocarpus lakoocha*	Moraceae	Fruit
**Asparagus adscendens*	Liliaceae	Root
**Bauhinia variegata*	Fabaceae	Stem-Bark
**Berberis lycium*	Berberidaceae	Root
**Bergenia ciliata*	Saxifragaceae	Root, Leaves, Stem
Bischofia javanica	Euphorbiaceae	Leaves, Stem-Bark
**Boerhaavia diffusa*	Nyctaginaceae	Leaves, Root
**Bombax ceiba*	Bombacaceae	Thorn, Stem-Bark, Root
**Calotropis procera*	Asclepiadaceae	Root, Leaves, Flower, Latex
**Cannabis sativa*	Cannabinaceae	Seed, Leaves
Carissa spinarum	Apocynaceae	Fruit
**Cassia fistula*	Fabaceae	Seed
Cedrela toona	Meliaceae	Stem-Bark
**Centella asiatica*	Apiaceae	Leaves
**Citrus limon*	Rutaceae	Leaves, Fruit
**Clerodendrum philippinum*	Verbenaceae	Leaves
**Cordia dichotoma*	Cordiaceae	Fruit
**Cynodon dactylon*	Poaceae	Whole Plant
Dalbergia sissoo	Fabaceae	Leaves
**Datura fastuosa*	Solanaceae	Seed
Duranta repens	Verbenaceae	Leaves
**Emblica officinalis*	Euphorbiaceae	Fruit, Seed
**Euphorbia royleana*	Euphorbiaceae	Latex
**Ficus benghalensis*	Moraceae	Latex
Ficus glomerata	Moraceae	Latex, Fruit
Ficus palmata	Moraceae	Fruit, Latex
**Ficus religiosa*	Moraceae	Twig, Leaves
**Hyoscyamus niger*	Solanaceae	Seed
Ipomoea purpurea	Convolvulaceae	Leaves
Jasminum sambac	Oleaceae	Leaves
**Lannea coromandelica*	Anacardiaceae	Leaves

Contd...

Table 15–*Contd...*

Species	*Family*	*Edible Part/s*
Litsea polyantha	Lauraceae	Stem-Bark
**Mallotus philippinensis*	Euphorbiaceae	Fruit, Seed
Mangifera indica	Anacardiaceae	Fruit, Leaves
Melia azedarach	Meliaceae	Leaves, Seed
**Mentha longifolia*	Lamiaceae	Leaves
**Murraya koenigii*	Rutaceae	Leaves
Ocimum gratissimum	Lamiaceae	Leaves
Pinus roxburghii	Pinaceae	Needle
**Pistacia integerrima*	Anacardiaceae	Fruit
**Putranjiva roxburghii*	Euphorbiaceae	Seed
Pyrus pashia	Rosaceae	Fruit
**Ricinus communis*	Euphorbiaceae	Leaves, Seed oil
Roylea cinerea	Lamiaceae	Leaves
Sapindus mukorossi	Sapindaceae	Seed, Fruit
Sapium sebiferum	Euphorbiaceae	Leaves, Stem-Bark
**Syzygium cumini*	Myrtaceae	Fruit, Stem-Bark, Seed
**Terminalia bellirica*	Combretaceae	Fruit
**Terminalia chebula*	Combretaceae	Fruit
**Withania somnifera*	Solanaceae	Leaves, Root
**Zanthoxylum armatum*	Rutaceae	Leaves
**Zizyphus mauritiana*	Rhamnaceae	Leaves

*: Plants for Trade and Drug Industry.

use) (Table 17) are consumed either in common or in time of scarcity as raw and/or cooked as well as pickled or as flavorant and chutney with maximum diversity represented by Rosaceae (11 spp), followed by Moraceae (9 spp), Rutaceae (7 spp), Fabaceae (6 spp), Apocynaceae (3 spp), Myrtaceae (3 spp), Solanaceae (3 spp), Combretaceae (2 spp), Lamiaceae (2 spp), Pinaceae (2 spp), Poaceae (2 spp), Sapotaceae (2 spp), etc. Low use families such as Anacardiaceae, Apiaceae, Araceae, Arecaceae, Berberidaceae, Bombacaceae, Boraginaceae, Brassicaceae, Caricaceae, Cordiaceae, Euphorbiaceae, Hippocastanaceae, Juglandaceae, Liliaceae, Musaceae, Nyctaginaceae, Pedaliaceae, Phytolaccaceae, Polygonaceae, Punicaceae, Rhamnaceae, Sapindaceae, Saxifragaceae, Ulmaceae and Urticaceae are represented by 1 species each (Table 16). Among the genera *Citrus* and *Prunus* with 5 spp each showed highest species diversity, followed by *Ficus* (4 spp), *Carissa* (3 spp), *Pyrus* (3 spp), *Artocarpus* (2 spp), *Morus* (2 spp), *Pinus* (2 spp), *Syzygium* (2 spp), *Terminalia* (2 spp) and the remaining genera like *Achras, Aegle, Aesculus, Aloe, Bauhinia, Berberis, Boerhaavia, Bombax, Brassica, Broussonetia, Capsicum, Carica, Cassia, Celtis, Colocasia, Cordia, Daucus, Debregeasia, Dolichos, Ehretia, Emblica, Eriobotrya, Juglans, Litchi,*

Lycopersicum, Mangifera, Mentha, Mimusops, Murraya, Musa, Ocimum, Phaseolus, Phoenix, Phytolacca, Pisum, Polygonum, Prinsepia, Psidium, Punica, Ribes, Rosa, Saccharum, Sesamum, Solanum, Tamarindus, Zea and *Zizyphus* are represented by 1 species each. It is worthy of mention that there exists no record of the use of *Berberis lycium* (fruit), *Boerhaavia diffusa* (leaves), *Broussonetia papyrifera* (fruit), *Celtis australis* (fruit), *Debregeasia hypoleuca* (fruit), *Ehretia acuminata* (fruit), *Ficus glomerata* (fruit), *Phytolacca acinosa* (leaves, tender branches), *Polygonum alpinum* (young stem) and *Rosa webbiana* (fruit) in literature so far, and these form new records of alternative food plants. Hopefully, the data generated would be of immense use to horticulturists and others in evolving improved varieties of edibles for meeting the food demand of growing population as well as to overcome the sufferings due to the deficiency of nutritional constituents.

Table 16: Families, Number of Genera and Species of the Edible Temple-Plants of H.P.

Family	*Plant Species*	*Genera/Species*
Anacardiaceae	*Mangifera indica*	1/1
Apiaceae	*Daucus carota*	1/1
Apocynaceae	*Carissa carandas, Carissa opaca, Carissa spinarum*	1/3
Araceae	*Colocasia esculenta*	1/1
Arecaceae	*Phoenix sylvestris*	1/1
Berberidaceae	*Berberis lycium*	1/1
Bombacaceae	*Bombax ceiba*	1/1
Boraginaceae	*Ehretia acuminata*	1/1
Brassicaceae	*Brassica oleracea* var. *capitata*	1/1
Caricaceae	*Carica papaya*	1/1
Combretaceae	*Terminalia bellirica, Terminalia chebula*	1/2
Cordiaceae	*Cordia dichotoma*	1/1
Euphorbiaceae	*Emblica officinalis*	1/1
Fabaceae	*Bauhinia variegata, Cassia fistula, Dolichos lablab, Phaseolus vulgaris, Pisum sativum, Tamarindus indica*	6/6
Hippocastanaceae	*Aesculus indica*	1/1
Juglandaceae	*Juglans regia*	1/1
Lamiaceae	*Mentha piperita, Ocimum sanctum*	2/2
Liliaceae	*Aloe vera*	1/1
Moraceae	*Artocarpus heterophyllus, Artocarpus lakoocha, Broussonetia papyrifera, Ficus cairica, Ficus glomerata, Ficus palmata, Ficus roxburghii, Morus alba, Morus serrata*	4/9
Musaceae	*Musa paradisiaca*	1/1
Myrtaceae	*Psidium guajava, Syzygium cumini, Syzygium jambos*	2/3
Nyctaginaceae	*Boerhaavia diffusa*	1/1
Pedaliaceae	*Sesamum orientale*	1/1

Contd...

Table 16–*Contd...*

Family	*Plant Species*	*Genera/Species*
Phytolaccaceae	*Phytolacca acinosa*	1/1
Pinaceae	*Pinus gerardiana, Pinus roxburghii*	½
Poaceae	*Sachharum officinarum, Zea mays*	2/2
Polygonaceae	*Polygonum alpinum*	1/1
Punicaceae	*Punica granatum*	1/1
Rhamnaceae	*Zizyphus mauritiana*	1/1
Rosaceae	*Eriobotrya japonica, Prinsepia utilis, Prunus amygdalus, Prunus armeniaca, Prunus domestica, Prunus persica, Prunus puddum, Pyrus communis, Pyrus malus, Pyrus pashia, Rosa webbiana*	5/11
Rutaceae	*Aegle marmelos, Citrus aurantifolia, Citrus jambhiri, Citrus limon, Citrus maxima, Citrus reticulata, Murraya koenigii*	3/7
Sapindaceae	*Litchi chinensis*	1/1
Sapotaceae	*Achras zapota, Mimusops elengi*	2/2
Saxifragaceae	*Ribes orientale*	1/1
Solanaceae	*Capsicum annuum, Lycopersicum esculentum, Solanum tuberosum*	3/3
Ulmaceae	*Celtis australis*	1/1
Urticaceae	*Debregeasia hypoleuca*	1/1

Table 17: Edible Part/s of the Temple-Plants of H.P.

Species	*Edible Part/s*
Achras zapota	Fruit
Aegle marmelos	Fruit, pulp used for cool drinks
Aesculus indica	Fruit
Aloe vera	Leaves (vegetable)
Artocarpus heterophyllus	Fruit (vegetable); also pickled
Artocarpus lakoocha	Fruits eaten; also used for making chutney, vegetable and pickling
Bauhinia variegata	Flower buds (vegetable)
**Berberis lycium*	Fruit
**Boerhaavia diffusa*	Leaves (vegetable)
Bombax ceiba	Seed
Brassica oleracea var. *capitata*	Leaves (eaten raw or as salads and with stem cooked as vegetable)
**Broussonetia papyrifera*	Fruit
Capsicum annuum	Fruit (pickled, flavorants, spice)
Carica papaya	Fruit eaten, unripe ones cooked as vegetable
Carissa carandas	Fruit

Contd...

Table 17–*Contd...*

Species	*Edible Part/s*
Carissa opaca	Fruit
Carissa spinarum	Fruit
**Cassia fistula*	Young pods (vegetable, pickled)
**Celtis australis*	Fruit
Citrus aurantifolia	Fruit (pickled, refreshing drinks)
Citrus jambhiri	Fruit
Citrus limon	Fruit (pickled, candy)
Citrus maxima	Fruit
Citrus reticulata	Fruit (juice refreshing and cooling)
Colocasia esculenta	Leaves, underground corm (vegetable)
Cordia dichotoma	Fruit (vegetable, pickled)
Daucus carota	Roots (vegetable)
**Debregeasia hypoleuca*	Fruits
Dolichos lablab	Pods (vegetable)
**Ehretia acuminata*	Fruit
Emblica officinalis	Fruit, also pickled, candy–rich source of vitamin C
Eriobotrya japonica	Fruit
Ficus carica	Fruit
**Ficus glomerata*	Fruit
Ficus palmata	Fruit
Ficus roxburghii	Fruit
Juglans regia	Kernels
Litchi chinensis	Fruit
Lycopersicum esculentum	Fruit (vegetable)
Mangifera indica	Fruit, raw fruit (pickled, chutney 'amchur' 'am-papad' candy); juice refreshing and cooling
Mentha piperita	Leaves (flavorants, chutney)
Mimusops elengi	Fruit
Morus alba	Fruit
Morus serrata	Fruit
Murraya koenigii	Leaves (flavorants)
Musa paradisiaca	Fruit; unripe fruit (vegetable)
Ocimum sanctum	Leaves (flavorants)
Phaseolus vulgaris	Pods (vegetable)
Phoenix sylvestris	Fruit
**Phytolacca acinosa*	Leaves, tender branches (vegetable)
Pinus gerardiana	Seed

Contd...

Table 17–*Contd...*

Species	*Edible Part/s*
Pinus roxburghii	Seed
Pisum sativum	Seed
**Polygonum alpinum*	Young stem
Prinsepia utilis	Fruit
Prunus amygdalus	Seed
Prunus armeniaca	Fruit
Prunus domestica	Fruit
Prunus persica	Fruit
Prunus puddum	Fruit
Psidium guajava	Fruit
Punica granatum	Fruit, dried seeds (chutney)
Pyrus communis	Fruit
Pyrus malus	Fruit
Pyrus pashia	Fruit
Ribes orientale	Fruit
**Rosa webbiana*	Fruit
Saccharum officinarum	Cane juice (cool drinks)
Sesamum orientale	Seed oil (cooking)
Solanum tuberosum	Tuber (vegetable)
Syzygium cumini	Fruit
Syzygium jambos	Fruit
Tamarindus indica	Fruit, pulp as chutney
Terminalia bellirica	Fruit
Terminalia chebula	Fruit
Zea mays	Grain
Zizyphus mauritiana	Fruit

*: New record.

There is a bewildering wealth of ornamental plants in the study area. As many as 232 species belonging to 180 genera under 84 families have been characterised (Table 20) of which 162 species are cultivated and 70 species grow as wild (Tables 21 and 22). In order of usefulness the family Asteraceae (15 spp) is at the top, followed by Fabaceae (14 spp), Euphorbiaceae (9 spp), Rosaceae (9 spp), Apocynaceae (8 spp), Lamiaceae (8 spp), Oleaceae (8 spp), Solanaceae (8 spp), Malvaceae (7 spp), Salicaceae (7 spp), Amaryllidaceae (6 spp), Cupressaceae (6 spp), Liliaceae (6 spp), Verbenaceae (6 spp), Bignoniaceae (5 spp), Acanthaceae (4 spp), Convolvulaceae (4 spp), Lauraceae (4 spp), Pinaceae (4 spp), Amaranthaceae (3 spp), Arecaceae (3 spp), Meliaceae (3 spp), Moraceae (3 spp), Myrtaceae (3 spp), Nyctaginaceae (3 spp), Rutaceae (3 spp),

Table 18: Cultivated Edible Temple-Courtyard Plants of H.P.

Family	*Species*
Anacardiaceae	*Mangifera indica*
Apiaceae	*Daucus carota*
Apocynaceae	*Carissa carandas*
Araceae	*Colocasia esculenta*
Brassicaceae	*Brassica oleracea* var. *capitata*
Caricaceae	*Carica papaya*
Fabaceae	*Dolichos lablab, Phaseolus vulgaris, Pisum sativum, Tamarindus indica*
Juglandaceae	*Juglans regia*
Lamiaceae	*Mentha piperita, Ocimum sanctum*
Liliaceae	*Aloe vera*
Moraceae	*Ficus carica*
Musaceae	*Musa paradisiaca*
Myrtaceae	*Psidium guajava, Syzygium jambos*
Pedaliaceae	*Sesamum orientale*
Poaceae	*Saccharum officinarum, Zea mays*
Punicaceae	*Punica granatum*
Rosaceae	*Eriobotrya japonica, Pyrus communis, Pyrus malus, Prunus amygdalus, Prunus armeniaca, Prunus domestica, Prunus persica*
Rutaceae	*Citrus aurantifolia, Citrus maxima, Citrus reticulata*
Sapindaceae	*Litchi chinensis*
Sapotaceae	*Achras zapota, Mimusops elengi*
Solanaceae	*Capsicum annuum, Lycopersicum esculentum, Solanum tuberosum*

Table 19: Wild Edible Temple-Courtyard Plants of H.P.

Family	*Species*
Anacardiaceae	*Mangifera indica*
Apocynaceae	*Carissa opaca, Carissa spinarum*
Arecaceae	*Phoenix sylvestris*
Berberidaceae	*Berberis lycium*
Bombacaceae	*Bombax ceiba*
Boraginaceae	*Ehretia acuminata*
Combretaceae	*Terminalia bellirica, Terminalia chebula*
Cordiaceae	*Cordia dichotoma*
Euphorbiaceae	*Emblica officinalis*
Fabaceae	*Bauhinia variegata, Cassia fistula*
Hippocastanaceae	*Aesculus indica*

Contd...

Table 19–*Contd...*

Family	*Species*
Moraceae	*Artocarpus heterophyllus, Artocarpus lakoocha, Broussonetia papyrifera, Ficus glomerata, Ficus palmata, Ficus roxburghii, Morus alba, Morus serrata*
Myrtaceae	*Syzygium cumini*
Nyctaginaceae	*Boerhaavia diffusa*
Phytolaccaceae	*Phytolacca acinosa*
Pinaceae	*Pinus gerardiana, Pinus roxburghii*
Polygonaceae	*Polygonum alpinum*
Rhamnaceae	*Zizyphus mauritiana*
Rosaceae	*Prinsepia utilis, Prunus puddum, Pyrus pashia, Rosa webbiana*
Rutaceae	*Aegle marmelos, Citrus jambhiri, Citrus limon, Murraya koenigii*
Saxifragaceae	*Ribes orientale*
Ulmaceae	*Celtis australis*
Urticaceae	*Debregeasia hypoleuca*

Ulmaceae (3 spp), etc. Likewise, high used genera commonly employed for ornamental purposes are: *Rosa* (8 spp), *Salix* (5 spp), *Clerodendrum* (4 spp), *Jasminum* (4 spp), *Datura* (3 spp), *Euphorbia* (3 spp), *Hibiscus* (3 spp), *Ipomoea* (3 spp), *Ocimum* (3 spp), *Pinus* (3 spp), *Acacia* (2 spp), *Asclepias* (2 spp), *Asparagus* (2 spp), *Bauhinia* (2 spp), *Bougainvillea* (2 spp), *Cassia* (2 spp), *Cestrum* (2 spp), *Chrysanthemum* (2 spp), *Cosmos* (2 spp), *Cupressus* (2 spp), *Diospyros* (2 spp), *Ficus* (2 spp), *Juniperus* (2 spp), *Litsea* (2 spp), *Malva* (2 spp), *Melia* (2 spp), *Murraya* (2 spp), *Populus* (2 spp), *Salvia* (2 spp), *Sapium* (2 spp), *Tecoma* (2 spp), *Thuja* (2 spp), and *Zephyranthes* (2 spp) whereas the rest of genera are represented by one species each (Table 20). By looking at the nativity of the ornamental species (Tables 37 and 38), it is apparent that most of the exotic/alien species due to their usefulness have been introduced intentionally in the temple-courtyards of H.P. over a period of time. It is seen that predominant ornamental exotics belong to families such as Asteraceae (14 spp), Amaryllidaceae, Euphorbiaceae, Fabaceae, Malvaceae, Solanaceae (6 spp each), Bignoniaceae, Rosaceae (5 spp each), Acanthaceae, Convolvulaceae, Cupressaceae, Lamiaceae, Liliaceae, Salicaceae, Verbenaceae (4 spp each), Myrtaceae, Nyctaginaceae, Oleaceae (3 spp each), Apocynaceae, Asclepiadaceae, Moraceae (2 spp each), etc. (Table 45). Surprisingly, 33 families, *viz.*, Acanthaceae, Amaryllidaceae, Araliaceae, Araucariaceae, Aristolochiaceae, Asclepiadaceae, Bignoniaceae, Buxaceae, Cannabinaceae, Cannaceae, Caprifoliaceae, Caricaceae, Caryophyllaceae, Chenopodiaceae, Convolvulaceae, Crassulaceae, Cycadaceae, Cyperaceae, Geraniaceae, Hydrangeaceae, Myrtaceae, Nyctaginaceae, Onagraceae, Papaveraceae, Passifloraceae, Platanaceae, Proteaceae, Punicaceae, Rubiaceae, Sambucaceae, Scrophulariaceae, Tropaeolaceae and Violaceae are characterised by the absence of native ornamentals in the study area whereas the representation of Aceraceae, Annonaceae, Araceae, Balsaminaceae, Betulaceae, Bombacaceae, Boraginaceae,

Casuarinaceae, Combretaceae, Dipterocarpaceae, Fagaceae, Hippocastanaceae, Hypericaceae, Meliaceae, Phytolaccaceae, Poaceae, Pinaceae, Polygonaceae, Rhamnaceae, Rutaceae, Santalaceae, Saxifragaceae, Sterculiaceae, Ulmaceae, Urticaceae and Zingiberaceae for exotics are also lacking (Table 45). It may also be noted that all these species may not occur together in a temple-courtyard and their density/dominance varies in different temple-courtyards.

Table 20: Families, Number of Genera and Species of Ornamental Plants in the Temple-Courtyards of H.P.

Family	*Plant Species*	*Genus/Species*
Acanthaceae	*Beloperone guttata, Justicia gendarussa, Ruellia tuberosa, Thunbergia erecta*	4/4
Aceraceae	*Acer oblongum*	1/1
Amaranthaceae	*Celosia argentea* var. *cristata, Deeringia amaranthoides, Gomphrena globosa*	3/3
Amaryllidaceae	*Amaryllis reginae, Hymenocallis littoralis, Narcissus tazetta, Polianthes tuberosa, Zephyranthes candida, Zephyranthes grandiflora*	5/6
Annonaceae	*Polyalthia longifolia*	1/1
Apocynaceae	*Alstonia scholaris, Carissa carandas, Catharanthus roseus, Nerium odorum, Plumeria acutifolia, Tabernaemontana coronaria, Thevetia neriifolia, Wrightia tomentosa*	8/8
Araceae	*Colocacia esculenta*	1/1
Araliaceae	*Schefflera arboricola*	1/1
Araucariaceae	*Araucaria columnaris*	1/1
Arecaceae	*Caryota urens, Chamaerops humilis, Phoenix sylvestris*	3/3
Aristolochiaceae	*Aristolochia elegans*	1/1
Asclepiadaceae	*Asclepias curassavica, Asclepias physocarpa*	1/2
Asteraceae	*Calendula officinalis, Callistephus chinensis, Centaurea cyanus, Chrysanthemum indicum, Chrysanthemum leucanthemum, Cosmos bipinnatus, Cosmos sulphureus, Dahlia rosea, Erigeron alpinus, Gaillardia pulchella, Helianthus annuus, Tagetes erecta, Tithonia tagetiflora, Wedelia calendulacea, Zinnia elegans*	13/15
Balsaminaceae	*Impatiens balsamina*	1/1
Betulaceae	*Alnus nitida, Betula utilis*	2/2
Bignoniaceae	*Jacaranda acutifolia, Pyrostegia venusta, Tecoma radicans, Tecoma stans, Tecomaria capensis*	4/5
Bombacaceae	*Bombax ceiba*	1/1
Boraginaceae	*Ehretia acuminata*	1/1
Buxaceae	*Buxus sempervirens*	1/1
Cannabinaceae	*Humulus lupulus*	1/1
Cannaceae	*Canna indica*	1/1
Caprifoliaceae	*Lonicera japonica*	1/1

Contd...

Table 20–*Contd...*

Family	*Plant Species*	*Genus/Species*
Caricaceae	*Carica papaya*	1/1
Caryophyllaceae	*Dianthus barbatus*	1/1
Casuarinaceae	*Casuarina equisetifolia*	1/1
Chenopodiaceae	*Kochia trichophylla*	1/1
Combretaceae	*Quisqualis indica*	1/1
Convolvulaceae	*Convolvulus arvensis, Ipomoea cairica, Ipomoea purpurea, Ipomoea quamoclit*	2/4
Crassulaceae	*Bryophyllum calycinum, Sempervivum arboreum*	2/2
Cupressaceae	*Cupressus sempervirens, Cupressus torulosa, Juniperus chinensis, Juniperus macropoda, Thuja orientalis, Thuja plicata*	3/6
Cycadaceae	*Cycas revoluta*	1/1
Cyperaceae	*Cyperus flabelliformis*	1/1
Dipterocarpaceae	*Shorea robusta*	1/1
Ebenaceae	*Diospyros lotus, Diospyros montana* var. *cordifolia*	1/2
Elaeocarpaceae	*Elaeocarpus ganitrus*	1/1
Euphorbiaceae	*Acalypha wilkesiana, Euphorbia milii, Euphorbia pulcherrima, Euphorbia royleana, Jatropha panduraefolia, Pedilanthus tithymaloides, Putranjiva roxburghii, Sapium insigne, Sapium sebiferum*	6/9
Fabaceae	*Acacia dealbata, Acacia nilotica, Bauhinia vahlii, Bauhinia variegata, Cassia fistula, Cassia glauca, Dalbergia sissoo, Lespedeza sericea, Leucaena leucocephala, Mimosa pudica, Pongamia pinnata, Robinia pseudoacacia, Tamarindus indica, Wistoria sinensis*	11/14
Fagaceae	*Quercus dilatata*	1/1
Geraniaceae	*Pelargonium hortorum*	1/1
Hippocastanaceae	*Aesculus indica*	1/1
Hydrangeaceae	*Hydrangea macrophyla*	1/1
Hypericaceae	*Hypericum cernuum*	1/1
Iridaceae	*Iris germanica, Ixia chinensis*	2/2
Lamiaceae	*Coleus blumei, Mentha piperita, Ocimum basilicum, Ocimum kilimandscharicum, Ocimum sanctum, Roylea cinerea, Salvia coccinia, Salvia splendens*	5/8
Lauraceae	*Cinnamomum camphora, Litsea polyantha, Litsea sebifera, Machilus gamblei*	3/4
Liliaceae	*Aloe vera, Asparagus densiflorus, Asparagus racemosus, Chlorophytum capense, Hemerocallis fulva, Sansevieria trifasciata*	5/6
Lythraceae	*Lagerstroemia indica, Lawsonia alba*	2/2
Magnoliaceae	*Magnolia grandiflora, Michelia champaca*	2/2

Contd...

Table 20–*Contd...*

Family	*Plant Species*	*Genus/Species*
Malvaceae	*Althaea rosea, Hibiscus mutabilis, Hibiscus rosa-sinensis, Hibiscus syriacus, Malva sylvestris, Malva verticillata, Malvaviscus conzatii*	4/7
Meliaceae	*Cedrela toona, Melia azadirachta, Melia azedarach*	2/3
Moraceae	*Ficus carica, Ficus elastica, Morus alba*	2/3
Myrtaceae	*Callistemon lanceolatus, Eucalyptus tereticornis, Syzygium jambos*	3/3
Nyctaginaceae	*Bougainvillea glabra, Bougainvillea spectabilis, Mirabilis jalapa*	2/3
Oleaceae	*Fraxinus xanthoxyloides, Jasminum grandiflorum, Jasminum humile, Jasminum primulinum, Jasminum sambac, Ligustrum lucidum, Nyctanthes arbor-tristis, Olea glandulifera*	5/8
Onagraceae	*Oenothera drummondii*	1/1
Papaveraceae	*Papaver rhoeas* var. *latifolia*	1/1
Passifloraceae	*Passiflora caerulea*	1/1
Phytolaccaceae	*Phytolacca acinosa*	1/1
Pinaceae	*Cedrus deodara, Pinus gerardiana, Pinus roxburghii, Pinus wallichiana*	2/4
Platanaceae	*Platanus orientalis*	1/1
Poaceae	*Coix lachryma-jobi*	1/1
Polygonaceae	*Polygonum alpinum*	1/1
Proteaceae	*Grevillea robusta*	1/1
Punicaceae	*Punica granatum*	1/1
Ranunculaceae	*Clematis grata, Delphinium ajacis*	2/2
Rhamnaceae	*Zizyphus mauritiana*	1/1
Rosaceae	*Eriobotrya japonica, Rosa alba, Rosa banksiae, Rosa foetida, Rosa indica, Rosa jacquemontii, Rosa moschata, Rosa multiflora, Rosa webbiana*	2/9
Rubiaceae	*Hamelia patens*	1/1
Rutaceae	*Murraya koenigii, Murraya paniculata, Zanthoxylum armatum*	2/3
Salicaceae	*Populus ciliata, Populus nigra, Salix alba, Salix babylonica, Salix fragilis, Salix oxycarpa, Salix tetrasperma*	2/7
Sambucaceae	*Sambucus nigra*	1/1
Santalaceae	*Santalum album*	1/1
Sapindaceae	*Dodonaea viscosa, Sapindus mukorossi*	2/2
Sapotaceae	*Achras zapota, Mimusops elengi*	2/2
Saxifragaceae	*Bergenia ciliata, Ribes orientale*	2/2
Scrophulariaceae	*Antirrhinum majus, Russelia equisetiformis*	2/2
Solanaceae	*Cestrum diurnum, Cestrum nocturnum, Datura fastuosa, Datura innoxia, Datura stramonium, Hyoscyamus niger, Nicandra physaloides, Petunia hybrida*	5/8

Contd...

Table 20–*Contd...*

Family	*Plant Species*	*Genus/Species*
Sterculiaceae	*Pterospermum acerifolium*	1/1
Tropaeolaceae	*Tropaeolum majus*	1/1
Ulmaceae	*Celtis australis, Holoptelea integrifolia, Ulmus villosa*	3/3
Urticaceae	*Debregeasia hypoleuca*	1/1
Verbenaceae	*Clerodendrum inerme, Clerodendrum philippinum, Clerodendrum splendens, Clerodendrum thomsonae, Duranta repens, Tectona grandis*	3/6
Violaceae	*Viola tricolour*	1/1
Zingiberaceae	*Alpinia nutans, Hedychium coronarium*	2/2

For raising fence, six species, *viz., Acacia catechu, Bougainvillea spectabilis, Carissa opaca, Duranta repens, Euphorbia royleana* and *Zizyphus mauritiana* are being used (Figure 16; Table 34). Due to spines, thorns and profuse branching these species are preferred to raise fence in their cropland and/or in the temple-courtyards to counter any damage to the growing vegetation by any external agents. Such a data will be of great use to policy planners in developing effective strategies for evolving a meaningful sustainable rural development plan for traditional farmers and rural communities.

Table 21: Cultivated Ornamental Plants in the Temple-Courtyards of H.P.

Plant Species	*Family*
Acacia dealbata	Fabaceae
Acalypha wilkesiana	Euphorbiaceae
Achras zapota	Sapotaceae
Aloe vera	Liliaceae
Alpinia nutans	Zingiberaceae
Alstonia scholaris	Apocynaceae
Althaea rosea	Malvaceae
Amaryllis reginae	Amaryllidaceae
Antirrhinum majus	Scrophulariaceae
Araucaria columnaris	Araucariaceae
Aristolochia elegans	Aristolochiaceae
Asclepias curassavica	Asclepiadaceae
Asclepias physocarpa	Asclepiadaceae
Asparagus densiflorus	Liliaceae
Beloperone guttata	Acanthaceae
Betula utilis	Betulaceae
Bougainvillea glabra	Nyctaginaceae

Contd...

Table 21–*Contd...*

Plant Species	*Family*
Bougainvillea spectabilis	Nyctaginaceae
Bryophyllum calycinum	Crassulaceae
Buxus sempervirens	Buxaceae
Calendula officinalis	Asteraceae
Callistemon lanceolatus	Myrtaceae
Callistephus chinensis	Asteraceae
Canna indica	Cannaceae
Carica papaya	Caricaceae
Carissa carandas	Apocynaceae
Caryota urens	Arecaceae
Cassia glauca	Fabaceae
Casuarina equisetifolia	Casuarinaceae
Catharanthus roseus	Apocynaceae
Celosia argentea var. *cristata*	Amaranthaceae
Centaurea cyanus	Asteraceae
Cestrum diurnum	Solanaceae
Cestrum nocturnum	Solanaceae
Chamaerops humilis	Arecaceae
Chlorophytum capense	Liliaceae
Chrysanthemum indicum	Asteraceae
Chrysanthemum leucanthemum	Asteraceae
Cinnamomum camphora	Lauraceae
Clerodendrum inerme	Verbenaceae
Clerodendrum splendens	Verbenaceae
Clerodendrum thomsonae	Verbenaceae
Coleus blumei	Lamiaceae
Colocacia esculenta	Araceae
Cosmos bipinnatus	Asteraceae
Cosmos sulphureus	Asteraceae
Cupressus sempervirens	Cupressaceae
Cupressus torulosa	Cupressaceae
Cycas revoluta	Cycadaceae
Cyperus flabelliformis	Cyperaceae
Dahlia rosea	Asteraceae
Delphinium ajacis	Ranunculaceae

Contd...

Table 21–*Contd...*

Plant Species	*Family*
Dianthus barbatus	Caryophyllaceae
Diospyros lotus	Ebenaceae
Duranta repens	Verbenacae
Elaeocarpus ganitrus	Elaeocarpaceae
Eriobotrya japonica	Rosaceae
Eucalyptus tereticornis	Myrtaceae
Euphorbia milii	Euphorbiaceae
Euphorbia pulcherrima	Euphorbiaceae
Ficus carica	Moraceae
Ficus elastica	Moraceae
Gaillardia pulchella	Asteraceae
Gomphrena globosa	Amaranthaceae
Grevillea robusta	Proteaceae
Hamelia patens	Rubiaceae
Hedychium coronarium	Zingiberaceae
Helianthus annuus	Asteraceae
Hemerocallis fulva	Liliaceae
Hibiscus mutabilis	Malvaceae
Hibiscus rosa-sinensis	Malvaceae
Hibiscus syriacus	Malvaceae
Humulus lupulus	Cannabinaceae
Hydrangea macrophyla	Hydrangeaceae
Hymenocallis littoralis	Amaryllidaceae
Impatiens balsamina	Balsaminaceae
Ipomoea quamoclit	Convolvulaceae
Iris germanica	Iridaceae
Ixia chinensis	Iridaceae
Jacaranda acutifolia	Bignoniaceae
Jasminum grandiflorum	Oleaceae
Jasminum humile	Oleaceae
Jasminum primulinum	Oleaceae
Jatropha panduraefolia	Euphorbiaceae
Juniperus chinensis	Cupressaceae
Justicia gendarussa	Acanthaceae
Kochia trichophylla	Chenopodiaceae

Contd...

Table 21–*Contd...*

Plant Species	*Family*
Lagerstroemia indica	Lythraceae
Lawsonia alba	Lythraceae
Lespedeza sericea	Fabaceae
Lonicera japonica	Caprifoliaceae
Magnolia grandiflora	Magnoliaceae
Malva sylvestris	Malvaceae
Malva verticillata	Malvaceae
Malvaviscus conzatii	Malvaceae
Melia azadirachta	Meliaceae
Mentha piperita	Lamiaceae
Michelia champaca	Magnoliaceae
Mimosa pudica	Fabaceae
Mimusops elengi	Sapotaceae
Mirabilis jalapa	Nyctaginaceae
Murraya paniculata	Rutaceae
Narcissus tazetta	Amaryllidaceae
Nerium odorum	Apocynaceae
Nyctanthes arbor-tristis	Oleaceae
Ocimum basilicum	Lamiaceae
Ocimum kilimandscharicum	Lamiaceae
Ocimum sanctum	Lamiaceae
Papaver rhoeas var. *latifolia*	Papaveraceae
Passiflora caerulea	Passifloraceae
Pedilanthus tithymaloides	Euphorbiaceae
Pelargonium hortorum	Geraniaceae
Petunia hybrida	Solanaceae
Platanus orientalis	Platanaceae
Plumeria acutifolia	Apocynaceae
Polianthes tuberosa	Amaryllidaceae
Polyalthia longifolia	Annonaceae
Pongamia pinnata	Fabaceae
Populus ciliata	Salicaceae
Populus nigra	Salicaceae
Pterospermum acerifolium	Sterculiaceae
Punica granatum	Punicaceae
Pyrostegia venusta	Bignoniaceae

Contd...

Table 21–*Contd...*

Plant Species	*Family*
Quisqualis indica	Combretaceae
Rosa alba	Rosaceae
Rosa banksiae	Rosaceae
Rosa foetida	Rosaceae
Rosa indica	Rosaceae
Rosa jacquemontii	Rosaceae
Rosa moschata	Rosaceae
Rosa multiflora	Rosaceae
Ruellia tuberosa	Acanthaceae
Russelia equisetiformis	Scrophulariaceae
Salix babylonica	Salicaceae
Salvia coccinia	Lamiaceae
Salvia splendens	Lamiaceae
Sambucus nigra	Sambucaceae
Sansevieria trifasciata	Liliaceae
Santalum album	Santalaceae
Sapindus mukorossi	Sapindaceae
Schefflera arboricola	Araliaceae
Sempervivum arboreum	Crassulaceae
Syzygium jambos	Myrtaceae
Tabernaemontana coronaria	Apocynaceae
Tagetes erecta	Asteraceae
Tamarindus indica	Fabaceae
Tecoma radicans	Bignoniaceae
Tecoma stans	Bignoniaceae
Tecomaria capensis	Bignoniaceae
Tectona grandis	Verbenaceae
Thevetia neriifolia	Apocynaceae
Thuja orientalis	Cupressaceae
Thuja plicata	Cupressaceae
Thunbergia erecta	Acanthaceae
Tithonia tagetiflora	Asteraceae
Tropaeolum majus	Tropaeolaceae
Viola tricolour	Violaceae
Wedelia calendulacea	Asteraceae
Wisteria sinensis	Fabaceae
Zephyranthes candida	Amaryllidaceae
Zephyranthes grandiflora	Amaryllidaceae
Zinnia elegans	Asteraceae

Table 22: Wild Ornamental Plants in the Temple-Courtyards of H.P.

Plant Species	*Family*
Acacia nilotica	Fabaceae
Acer oblongum	Aceraceae
Aesculus indica	Hippocastanaceae
Alnus nitida	Betulaceae
Asparagus racemosus	Liliaceae
Bauhinia vahlii	Fabaceae
Bauhinia variegata	Fabaceae
Bergenia ciliata	Saxifragaceae
Bombax ceiba	Bombacaceae
Cassia fistula	Fabaceae
Cedrela toona	Meliaceae
Cedrus deodara	Pinaceae
Celtis australis	Ulmaceae
Clematis grata	Ranunculaceae
Clerodendrum philippinum	Verbenaceae
Coix lachryma-jobi	Poaceae
Convolvulus arvensis	Convolvulaceae
Dalbergia sissoo	Fabaceae
Datura fastuosa	Solanaceae
Datura innoxia	Solanaceae
Datura stramonium	Solanaceae
Debregeasia hypoleuca	Urticaceae
Deeringia amaranthoides	Amaranthaceae
Diospyros montana var. *cordifolia*	Ebenaceae
Dodonaea viscosa	Sapindaceae
Ehretia acuminata	Boraginaceae
Erigeron alpinus	Asteraceae
Euphorbia royleana	Euphorbiaceae
Fraxinus xanthoxyloides	Oleaceae
Holoptelea integrifolia	Ulmaceae
Hyoscyamus niger	Solanaceae
Hypericum cernuum	Hypericaceae
Ipomoea cairica	Convolvulaceae
Ipomoea purpurea	Convolvulaceae
Jasminum sambac	Oleaceae
Juniperus macropoda	Cupressaceae
Leucaena leucocephala	Fabaceae

Contd...

Table 22–*Contd...*

Plant Species	*Family*
Ligustrum lucidum	Oleaceae
Litsea polyantha	Lauraceae
Litsea sebifera	Lauraceae
Machilus gamblei	Lauraceae
Melia azedarach	Meliaceae
Morus alba	Moraceae
Murraya koenigii	Rutaceae
Nicandra physaloides	Solanaceae
Oenothera drummondii	Onagraceae
Olea glandulifera	Oleaceae
Phoenix sylvestris	Arecaceae
Phytolacca acinosa	Phytolaccaceae
Pinus gerardiana	Pinaceae
Pinus roxburghii	Pinaceae
Pinus wallichiana	Pinaceae
Polygonum alpinum	Polygonaceae
Putranjiva roxburghii	Euphorbiaceae
Quercus dilatata	Fagaceae
Ribes orientale	Saxifragaceae
Robinia pseudoacacia	Fabaceae
Rosa webbiana	Rosaceae
Roylea cinerea	Lamiaceae
Salix alba	Salicaceae
Salix fragilis	Salicaceae
Salix oxycarpa	Salicaceae
Salix tetrasperma	Salicaceae
Sapium insigne	Euphorbiaceae
Sapium sebiferum	Euphorbiaceae
Shorea robusta	Dipterocarpaceae
Ulmus villosa	Ulmaceae
Wrightia tomentosa	Apocynaceae
Zanthoxylum armatum	Rutaceae
Zizyphus mauritiana	Rhamnaceae

Regarding the diversity in forages, altogether 32 species belonging to 28 genera under 18 families are valued (Table 23). Species richness predominantly belongs to Fabaceae (7 spp), Moraceae (4 spp), Poaceae (3 spp), Salicaceae (3 spp), Ulmaceae (2 spp), and the rest of the families, *viz.*, Apocynaceae, Betulaceae, Boraginaceae, Convolvulaceae, Euphorbiaceae, Fagaceae, Hippocastanaceae, Pedaliaceae, Rhamnaceae, Rosaceae, Sapindaceae, Tiliaceae and Urticaceae are represented by one species each (Table 23). Plant species used locally for fodder are: *Acacia catechu, Acacia nilotica, Aesculus indica, Albizia lebbeck, Albizia stipulata, Alnus nitida, Bauhinia variegata, Broussonetia papyrifera, Carissa opaca, Celtis australis, Coix lachryma-jobi, Convolvulus arvensis, Cynodon dactylon, Debregeasia hypoleuca, Dodonaea viscosa, Ehretia acuminata, Ficus palmata, Ficus roxburghii, Grewia oppositifolia, Holoptelea integrifolia, Leucaena leucocephala, Morus alba, Populus nigra, Prunus puddum, Quercus incana, Robinia pseudoacacia, Salix babylonica, Salix fragilis, Sapium sebiferum, Sesamum orientale, Zea mays* and *Zizyphus mauritiana* (Table 23). Of these, 4 taxa like *Populus nigra* (Salicaceae), *Salix babylonica* (Salicaceae), *Sesamum orientale* (Pedaliaceae) and *Zea mays* (Poaceae) are cultivated for forages (Table 24) whereas the remaining 28 plant species belonging to 25 genera under 17 families grow as wild (Table 25). Interestingly, many of these plant species are not evenly widely known and enumerated in the list of fodder plants of our country (Anon., 1994).

Table 23: Families, Number of Genera and Species of Fodder Plants in the Temple-Courtyards of H.P.

Family	*Plant Species*	*Genus/Species*
Apocynaceae	*Carissa opaca*	1/1
Betulaceae	*Alnus nitida*	1/1
Boraginaceae	*Ehretia acuminata*	1/1
Convolvulaceae	*Convolvulus arvensis*	1/1
Euphorbiaceae	*Sapium sebiferum*	1/1
Fabaceae	*Acacia catechu, Acacia nilotica, Albizia lebbeck, Albizia stipulata, Bauhinia variegata, Leucaena leucocephala, Robinia pseudoacacia*	5/7
Fagaceae	*Quercus incana*	1/1
Hippocastanaceae	*Aesculus indica*	1/1
Moraceae	*Broussonetia papyrifera, Ficus palmata, Ficus roxburghii, Morus alba*	3/4
Pedaliaceae	*Sesamum orientale*	1/1
Poaceae	*Coix lachryma-jobi, Cynodon dactylon, Zea mays*	3/3
Rhamnaceae	*Zizyphus mauritiana*	1/1
Rosaceae	*Prunus puddum*	1/1
Salicaceae	*Populus nigra, Salix babylonica, Salix fragilis*	2/3
Sapindaceae	*Dodonaea viscosa*	1/1
Tiliaceae	*Grewia oppositifolia*	1/1
Ulmaceae	*Holoptelea integrifolia, Celtis australis*	2/2
Urticaceae	*Debregeasia hypoleuca*	1/1

Table 24: Cultivated Fodder Plants in the Temple-Courtyards of H.P.

Plant Species	*Family*	*Plant Part*
Populus nigra	Salicaceae	Leaves
Salix babylonica	Salicaceae	Branches
Sesamum orientale	Pedaliaceae	Seeds
Zea mays	Poaceae	Whole Plant

Table 25: Wild Fodder Plants in the Temple-Courtyards of H.P.

Plant Species	*Family*	*Plant Part*
Acacia catechu	Fabaceae	Pods
Acacia nilotica	Fabaceae	Pods
Aesculus indica	Hippocastanaceae	Leaves
Albizia lebbeck	Fabaceae	Leaves
Albizia stipulata	Fabaceae	Leaves
Alnus nitida	Betulaceae	Leaves
Bauhinia variegata	Fabaceae	Leaves
Broussonetia papyrifera	Moraceae	Leaves
Carissa opaca	Apocynaceae	Leaves
Celtis australis	Ulmaceae	Leaves
Coix lachryma-jobi	Poaceae	Whole Plant
Convolvulus arvensis	Convolvulaceae	Whole Plant
Cynodon dactylon	Poaceae	Whole Plant
Debregeasia hypoleuca	Urticaceae	Leaves
Dodonaea viscosa	Sapindaceae	Leaves
Ehretia acuminata	Boraginaceae	Leaves
Ficus palmata	Moraceae	Leaves
Ficus roxburghii	Moraceae	Leaves
Grewia oppositifolia	Tiliaceae	Leaves
Holoptelea integrifolia	Ulmaceae	Leaves
Leucaena leucocephala	Fabaceae	Leaves
Morus alba	Moraceae	Leaves
Prunus puddum	Rosaceae	Leaves
Quercus incana	Fagaceae	Leaves
Robinia pseudoacacia	Fabaceae	Leaves
Salix fragilis	Salicaceae	Young Branches
Sapium sebiferum	Euphorbiaceae	Leaves
Zizyphus mauritiana	Rhamnaceae	Leaves

For tribals and rural communities of H. P., plants are indispensable to rituals and religion as evidenced from the occurrence of 114 plant species belonging to 80 genera under 41 families, accounting for 37.38 per cent of the overall diversity in the 205 temple-courtyards having range in size varying from 384 sq m (Murli Dhar Temple, Shiva Temple-Kullu) to 10, 000 sq m (Markanday Temple, Nav Durga Temple-Bilaspur) (Figure 15; Table 26). An analysis of data indicates that the most pre-dominantly employed families for offerings, religious ceremonies and rituals are Rosaceae (16 spp), Asteraceae (8 spp), Fabaceae (7 spp), Poaceae (7 spp), Apocynaceae (5 spp), Malvaceae (5 spp), Solanaceae (5 spp), Cupressaceae (4 spp), Moraceae (4 spp), Oleaceae (4 spp), Amaryllidaceae (3 spp), Arecaceae (3 spp), Lamiaceae (3 spp), Nyctaginaceae (3 spp), Pinaceae (3 spp), Anacardiaceae (2 spp), Bignoniaceae (2 spp), Brassicaceae (2 spp), Rutaceae (2 spp), Sapotaceae (2 spp), etc. The rest of the families namely Apiaceae, Asclepiadaceae, Cannabinaceae, Cannaceae, Cucurbitaceae, Elaeocarpaceae, Juglandaceae, Musaceae, Myrtaceae, Nymphaeaceae, Pedaliaceae, Piperaceae, Punicaceae, Theaceae, Tropaeolaceae, Verbenaceae, Violaceae and Vitaceae are represented by one species each (Table 29). High used genera associated for religious rituals are: *Rosa* (8 spp), *Jasminum* (4 spp), *Prunus* (4 spp), *Datura* (3 spp), *Ficus* (3 spp), *Ocimum* (3 spp), *Pyrus* (3 spp), *Bougainvillea, Brassica, Cosmos, Cupressus, Hibiscus, Pinus, Tecoma, Vigna* and *Zephyranthes* (2 spp each) (Table 29). For religious rituals and worshipping of the local deities, Gods and Goddesses in the temple-yards, the various plant parts employed for offerings are, *viz.*, flowers (46 spp), fruits (17 spp), leaves (16 spp), seeds (15 spp), wood (14 spp), whole plant (6 spp), kernels (3 spp), root (2 spp), inflorescence (1 sp), stem (1 sp), stigma (1 sp) and tuber (1 sp) (Table 27). Of these, the use of flowers in offerings and worshipping is more predominant as compared to others in H.P. Interestingly, for religious rituals, dried leaves of *Cupressus sempervirens, Cupressus torulosa, Juniperus macropoda, Thuja orientalis* (Cupressaceae), whole plant of *Artemisia moorcroftiana,* (Asteraceae) and roots of *Viola canescens* (Violaceae) are burnt as incense. For purification of the environment, wood of *Acacia catechu* (Fabaceae), *Aegle marmelos* (Rutaceae), *Cedrus deodara* (Pinaceae), *Ficus benghalensis, Ficus glomerata, Ficus religiosa* (Moraceae), *Mangifera indica* (Anacardiaceae), *Pinus roxburghii* (Pinaceae), *Prinsepia utilis* (Rosaceae) and seeds of *Sesamum orientale* are extensively employed for burning in 'havan' ceremonies by the deeply religious communities of the study area (Table 26) as most of such plant species contain essential oils and due to burning they emit a fragrance that brings about purification of the air all around. Generally, dried plant parts are burnt as such. In addition, offerings comprising of wine prepared from fruits (*Prunus armeniaca, Prunus persica, Pyrus malus, Vitis vinifera*) and grains (*Hordeum vulgare*) to appease the deity Chandika Devi in district Kinnaur and rosaries made up of seeds (*Prunus armeniaca*), kernels (*Juglans regia*) and grains (*Coix lachryma-jobi*) are all unique to the tribals and rural communities of the study area (Tables 26 and 27). Earlier, the use of roots of *Viola canescens* as incense, flowers of *Mimusops elengi* for decorating bed ('Shaiya') of the deity, and unique offering of rosaries made up of seeds (*Prunus armeniaca*), kernels (*Juglans regia*) and grains (*Coix lachryma-jobi*) have not been reported so far and new to the knowledge on ethnobotany.

Table 26: Plant Species Used for Offerings, Religious Ceremonies and Rituals in the Temple-Courtyards of H.P.

Species	*Family*	*Use*
Acacia catechu (L.f.) Willd.	Fabaceae	Wood used in 'havan'.
Achras zapota Linn.	Sapotaceae	Fruits used as offering to the various deities.
Aegle marmelos (L.) Correa	Rutaceae	Leaves ('bil-patri') offered to Lord Shiva and Gayatri Mata as special offering. Wood and fruits (after drying) used in 'havan'. Leaves also used for making 'havan-vedika'.
Althaea rosea (L.) Cav.	Malvaceae	Flowers used as offering to the various deities.
Anacardium occidentale Linn.+	Anacardiaceae	Cashew kernels ('kaju') used as 'bhog' to the various deities in the form of 'panchmeva' alongwith other dry fruits.
Arachis hypogaea Linn.+	Fabaceae	Seeds used as offering to the various deities.
Areca catechu Linn.+	Arecaceae	Nuts ('supari') used as offering to Goddess Kali.
Artemisia moorcroftiana Wall. *ex* DC.	Asteraceae	The plant used as incense.
**Bauhinia vahlii* Wight and Arn.	Fabaceae	Wood used for the preparation of sacred ash ('vibhuti') for distribution as 'prasadam' to the people visiting Baba Sidh Chaino Temple.
Bougainvillea glabra Choisy	Nyctaginaceae	Flowers considered sacred for offering to the various deities.
Bougainvillea spectabilis Willd.	Nyctaginaceae	Flowers used as offering to the various deities.
Brassica campestris Linn.+	Brassicaceae	Seeds and seed oil used as special offering to Lord Shani Dev.
Brassica oleracea var. *botrytis* Linn.+	Brassicaceae	Vegetable prepared from inflorescence used as 'bhog' to the various deities.
Brassica oleracea var. *capitata* Linn.	Brassicaceae	Vegetable prepared from leaves used as 'bhog' to the various deities.
Camellia sinensis (Linn.) O. Kuntze+	Theaceae	Tea ('cha') prepared from leaves used as 'bhog' to Guru Sunder Muni Nirwan.
Calotropis procera R. Br.	Asclepiadaceae	Leaves and flowers ('ak') offered to Lord Shiva and Lord Shani Dev as special offering.
Cannabis sativa Linn.	Cannabinaceae	Leaves ('bhang') offered to appease Lord Shiva. Leaves also employed to make drink ('ghota') during Shivratri.
Canna indica Linn.	Cannaceae	Flowers considered sacred for offering to the various deities.

Contd...

Table 26–*Contd...*

Species	*Family*	*Use*
Catharanthus roseus (Linn.) G. Don	Apocynaceae	Flowers used as offering to the various deities.
**Cedrus deodara* (Roxb.) Loudon	Pinaceae	Wood employed for manufacturing palanquin for deities, and articles used in various religious rites and also in purification ceremonies.
Chrysanthemum indicum Linn.	Asteraceae	Flowers considered sacred for offering to the various deities.
Cicer arietinum Linn.+	Fabaceae	Seeds ('chana') used as 'bhog' to the various deities. Boiled seeds or curry prepared from it also used as 'bhog' alongwith cooked rice to the various deities. 'dal' also used as special offering to Lord Shani Dev. Sweets ('besan' and 'ladoo') prepared from the flour of 'dal' used as special offering to Lord Vishnu and Goddess Bagula Mukhi.
Citrus reticulata Blanco	Rutaceae	Fruits used as offering to the various deities.
Clerodendrum philippinum Schauer	Verbenaceae	Flowers used as special offering to Markanday Rishi and Lord Shiva.
Cocos nucifera Linn.+	Arecaceae	Coconut ('nariyal') or kernel ('gari') used as 'bhog' to the various deities. Kernel also given as 'prasadam' to the visitors.
**Coix lachryma-jobi* Linn.	Poaceae	Necklace made of grains offered to the deity (Lord Raghunath).
Cosmos bipinnatus Cav.	Asteraceae	Flowers offered to the various deities.
Cosmos sulphureus Cav.	Asteraceae	Flowers used as offering to the various deities.
Cucurbita hispida Thunb.+	Cucurbitaceae	Fruits ('petha') used as special offering to Lord Shani Dev.
Cupressus sempervirens Linn.	Cupressaceae	Leaves used as incense.
Cupressus torulosa D. Don	Cupressaceae	Leaves used as incense in various monasteries.
Cynodon dactylon (Linn.) Pers.	Poaceae	Plant employed for performing almost all religious ceremonies and rituals.
Dahlia rosea Cav.	Asteraceae	Flowers used as offering to the various deities.
Datura fastuosa Linn.	Solanaceae	Flowers ('dhatura') used as special offering to appease Lord Shiva. Also used in religious worshippings.
Datura innoxia Mill.	Solanaceae	Flowers ('dhatura') considered sacred for offering to appease Lord Shiva. Also used in various religious ceremonies.

Contd...

Table 26–*Contd...*

Species	*Family*	*Use*
Datura stramonium Linn.	Solanaceae	Flowers ('dhatura') specially offered to appease Lord Shiva. Also used in various religious worshippings.
Daucus carota Linn.	Apiaceae	Vegetable prepared from roots used in 'bhog' to the various deities.
Elaeocarpus ganitrus Roxb.	Elaeocarpaceae	Rosaries and bracelets of nuts ('rudraksha') worn by Brahmins, 'Fakirs' and Buddhist devotees, and also employed for repeated chanting of sacred 'mantras'.
Euryale ferox Salisb+	Nymphaeaceae	Seeds (Fox nut) used as 'bhog' to the various deities and also given to the visitors as 'prasadam'.
Ficus benghalensis Linn.	Moraceae	Worshipped as sacred tree. Wood used in 'havan'.
Ficus glomerata Roxb.	Moraceae	Wood and leaves used in 'havan' and several other religious worshipping.
Ficus religiosa Linn.	Moraceae	Worshipped as sacred tree. Wood and leaves used in 'havan' for performing religious rites.
Helianthus annuus Linn.	Asteraceae	Flowers used as offering to the various deities.
Hibiscus rosa-sinensis Linn.	Malvaceae	Flowers specially offered to appease Goddess Kali and various other deities during religious worshipping.
Hibiscus syriacus Linn.	Malvaceae	Flowers considered sacred for offering to the various deities.
Hordeum vulgare Linn.+	Poaceae	Wine prepared from its grains used as 'bhog' to the various deities (especially Chandika devi) in Kinnaur. 'Sattu' made of barley flour used as offering to the various deities. 'Sow' made of barley flour used as special 'bhog' to Lord Gautam Buddha in various monasteries and also given as 'prasadam' to the visitors.
Jasminum grandiflorum Linn.	Oleaceae	Flowers considered sacred for offering to the various deities.
Jasminum humile Linn.	Oleaceae	Flowers offered to the various deities.
Jasminum primulinum Hemsl. *ex* Baker	Oleaceae	Flowers specially offered to Goddess Kali and also given to the visitors alongwith 'prasadam'.
Jasminum sambac (Linn.) Ait.	Oleaceae	Flowers considered sacred for offering to the various deities.

Contd...

Table 26–*Contd...*

Species	*Family*	*Use*
**Juglans regia* Linn.	Juglandaceae	Fruits ('akhrot') given as 'prasadam' to the people visiting the temples. Fruits also used as 'bhog' to the various deities. Necklace made of its kernels used as offering to the various deities especially Usha Devi in Kinnaur District.
Juniperus macropoda Boiss.	Cupressaceae	Leaves used as incense.
Malva sylvestris Linn.	Malvaceae	Flowers offered to the various deities.
Malvaviscus conzatii Green.	Malvaceae	Flowers considered sacred for offering to the various deities.
**Mangifera indica* Linn.	Anacardiaceae	Wood and leaves used in 'havan'. Wood also used for the preparation of sacred ash ('vibhuti') for distributing to the people visiting Baba Sidh Chaino Temple. Leaves also used in various other religious ceremonies.
**Mimusops elengi* Linn.	Sapotaceae	Flower garlands used especially for decorating bed ('shaiya') of the deity (Goddess Jwalaji).
Mirabilis jalapa Linn.	Nyctaginaceae	Flowers offered to the various deities.
Morus serrata Roxb.	Moraceae	Small piece of wood of *Morus serrata*, growing in the courtyard of Chaurasi Temple used as sacred wood for burning dead bodies.
Musa paradisiaca Linn.	Musaceae	Fruits ('kela') used as offering ('bhog') to the various deities and also given as 'prasadam' to the visitors. Fruits also used as special offering to Lord Vishnu. Leaves used for making 'havan-vedika' during 'havan'.
Narcissus tazetta Linn.	Amaryllidaceae	Flowers considered sacred for offering to the various deities.
Nerium odorum Soland.	Apocynaceae	Flowers used as special offering to Lord Shiva.
Nicandra physaloides (Linn.) Gaertn.	Solanaceae	Flowers used as offering to the various deities (especially Mahasu Devta).
Ocimum basilicum Linn.	Lamiaceae	Plant worshipped.
Ocimum kilimandscharicum Gurke	Lamiaceae	Plant worshipped.
Ocimum sanctum Linn.	Lamiaceae	Plant worshipped. Leaves added for preparing holy water ('charnamat') and given to the visitors as 'prasadam'. Leaves also offered to Lord Vishnu as special offering.

Contd...

Table 26–*Contd...*

Species	*Family*	*Use*
Oryza sativa Linn.+	Poaceae	Puffed rice ('khillan') used as 'bhog' to various deities and given to the visitors as 'prasadam'. Cooked rice or in the form of pudding ('kheer') offered to the deities as 'bhog'. 'Kaleya', *i.e.*, soaked rice mixed in ghee distributed to the visitors on 'Sakranti' in Kamru Nag Temple. Newly harvested grains offered to the various deities by locals after every harvesting season.
Phoenix dactylifera Linn.+	Arecaceae	Dried fruits used as 'bhog' to the various deities.
Pinus roxburghii Sarg.	Pinaceae	Wood used in 'havan'.
**Pinus wallichiana* A. B. Jackson	Pinaceae	Wood employed for making musical instruments and manufacturing palanquin of the deities.
Piper betle Linn.+	Piperaceae	Betel leaves used as offering to Goddess Kali.
Pisum sativum Linn.	Fabaceae	Newly harvested crop offered to the various deities by locals after every harvesting season in Lahul-Spiti and other districts. Vegetable prepared from its seeds used as 'bhog' to the various deities.
Plumeria acutifolia Poiret	Apocynaceae	Flowers used as offering to the various deities.
Prinsepia utilis Royle	Rosaceae	Wood used in 'havan'.
Prunus amygdalus Batsch.	Rosaceae	Seeds alongwith other dry fruits offered as 'bhog' to the various deities in the form of 'panchmeva' (five dried fruits). Newly harvested fruits offered to the local deities after every harvesting season.
**Prunus armeniaca* Linn.	Rosaceae	Fruits ('khumani') and seeds offered as 'bhog' to the various deities. Wine of its fruits offered to appease Goddess Chandika devi in Kinnaur. Necklace made of its seeds offered to appease Goddess Usha Devi in Kinnaur District.
Prunus persica (Linn.) Batsch	Rosaceae	Fruits offered to the various deities. Wine of its fruits used as 'bhog' to Goddess Chandika Devi in Kinnaur.
Prunus puddum (Roxb. *ex* Ser.) Brandis	Rosaceae	Wood used in religious worshipping.
Psidium guajava Linn.	Myrtaceae	Fruits offered to the various deities.
Punica granatum Linn.	Punicaceae	Fruits offered to the various deities.

Contd...

Table 26–*Contd...*

Species	*Family*	*Use*
Pyrus communis Linn.	Rosaceae	Fruits offered to the various deities.
Pyrus malus Linn.	Rosaceae	Fruits offered to the various deities. Wine of its fruits used as 'bhog' to the various deities (especially Goddess Chandika Devi) in Kinnaur. Newly harvested fruits offered to the local deities after every harvesting season.
Pyrus pashia Buch.-Ham. *ex* D.Don.	Rosaceae	Wood used for religious worshipping.
Rosa alba Linn.	Rosaceae	Flowers offered to the various deities and also used in worshipping.
Rosa banksiae R. Br.	Rosaceae	Flowers used as offering to the various deities.
Rosa foetida Herrm.	Rosaceae	Flowers used as offering to the various deities. Also offered as floral garlands.
Rosa indica Linn.	Rosaceae	Flowers or floral garlands used as offerings to the various deities.
Rosa jacquemontii Crep. *ex* Hook. f.	Rosaceae	Flowers used as offering to the various deities. Floral garlands used as offering to Goddess Kamakhya Devi.
Rosa moschata Herrm.	Rosaceae	Flowers offered to the various deities.
Rosa multiflora Thunb.	Rosaceae	Flowers considered sacred and offered to the various deities.
Rosa webbiana Wall. *ex* Royle.	Rosaceae	Flowers or garlands used as offering to the various deities.
Saccharum officinarum Linn.	Poaceae	Plant considered sacred and used in various religious ceremonies. Leaves used for making 'havan-vedika'. Unrefined sugar ('gur'), loaf sugar ('shakkar'), 'lachidana' and 'patasha' prepared from sugar or sugar used as 'bhog' to the various deities, and also given to the visitors as 'prasadam'.
Sesamum orientale Linn.	Pedaliaceae	Seeds used as special offering to Lord Shani Dev. Seeds also used in 'havan'.
Solanum tuberosum Linn.	Solanaceae	Newly harvested tubers offered to the various deities after every harvesting season in Lahul-Spiti and other districts. Vegetable prepared from its tubers used as 'bhog' to the various deities.
Tabernaemontana coronaria (Jacq.) Willd.	Apocynaceae	Flowers considered sacred for offering to the various deities.

Contd...

Table 26–*Contd...*

Species	*Family*	*Use*
Tagetes erecta Linn.	Asteraceae	Flowers specially offered to Goddess Durga. Flowers also used for making a cap ornament, 'jholnu' by the Lahuli people. 'Gur' (spokesman) of the deity has to bear the cap studded with this ornament.
Tecoma radicans Juss.	Bignoniaceae	Flowers offered to the various deities.
Tecoma stans (Linn.) Juss. *ex* H.B.K.	Bignoniaceae	Flowers used as special offering to Goddess Bagula Mukhi. Flowers also considered sacred for offering to the other deities.
Thevetia neriifolia Juss *ex* Steud.	Apocynaceae	Flowers used for offering to appease Lord Shiva.
Thuja orientalis Linn.	Cupressaceae	Leaves used as incense in various monasteries.
Triticum aestivum Linn.+	Poaceae	'Halva' made of wheat flour ('suji') used as special 'bhog' to various deities. 'Chapati', 'manni', 'kasaar', 'rot', 'puri', 'babru' and 'prantha' made of wheat flour also used as 'bhog' to the various deities. Thick fried 'chapati' dipped in sugary syrup ('rot') offered to Baba Balak Nath, Nag Devta, Bhairav, Shiva and Hanuman. 'Luchi' prepared from wheat flour and ghee used as 'bhog' to the various deities in Kullu District. Porridge made of wheat grains used as 'bhog' to Markanda Rishi. Biscuits and bread made of wheat flour offered in various Buddhist monasteries. Newly harvested grains offered by the locals to the various deities after every harvesting season.
Tropaeolum majus Linn.	Tropaeolaceae	Flowers offered to the various deities.
Vigna mungo (Linn.) Hepper.+	Fabaceae	Seeds ('mash') used as special offering to Lord Shani Dev. 'Bhalla' made of crushed seeds used as 'special bhog' to the various deities (particularly Lord Raghunath) in Kullu District. Curry prepared from seeds used as 'bhog' to the various deities.
Vigna radiata (Linn.) Wilczek.+	Fabaceae	Seeds used as special offering to Lord Shani Dev. Curry prepared from its seeds used as 'bhog' to the various deities.
**Viola canescens* Wall.	Violaceae	Roots used as incense.
Vitis vinifera Linn.+	Vitaceae	Raisins used as 'bhog' to the various deities, and also given as 'prasadam' to the visitors. Wine of its fruits offered to appease Goddess Chandika devi in Kinnaur.

Contd...

Table 26–*Contd...*

Species	*Family*	*Use*
Zea mays Linn.	Poaceae	'Rot' made of maize flour used as 'bhog' to the various deities. Newly harvested grains offered to the various deities by locals after every harvesting season.
**Zephyranthes candida* (Lindl.) Herb.	Amaryllidaceae	Dried stigmas used as vermilion ('tikka') in various religious ceremonies. Flowers offered to the various deities.
Zephyranthes grandiflora Lindl.	Amaryllidaceae	Flowers used as offering to the various deities.
Zinnia elegans Jacq.	Asteraceae	Flowers offered to the various deities.

*: New usages.

+: Plants used for offerings and religious rituals though not occurring in temple-yards.

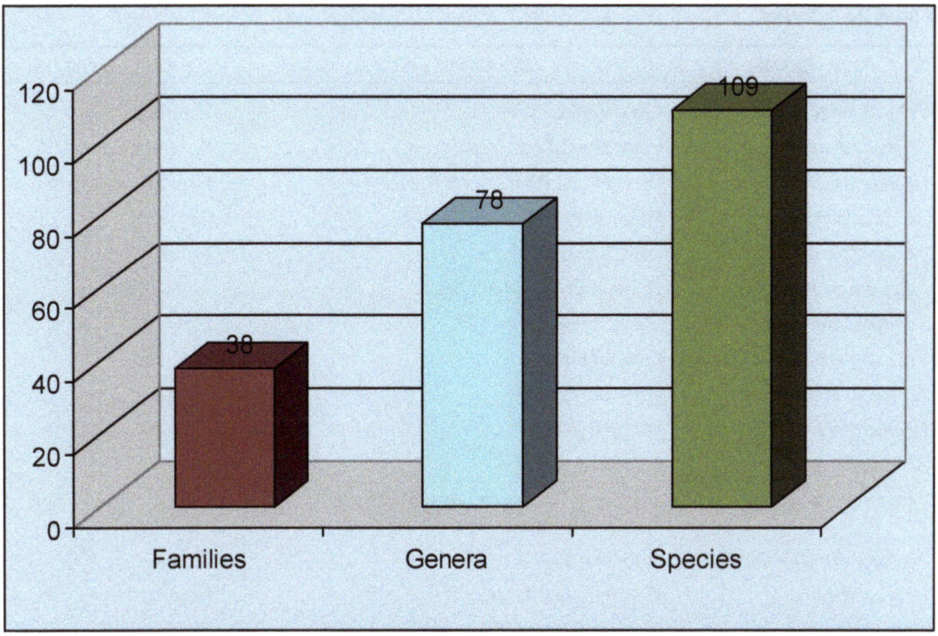

Figure 15: Families, Genera and Species of the Temple Plants Used for Offerings and Religious Rituals

To protect their families from the evil effect of Lord Shani Dev (Saturn) whom the local communities feared the most, they buy peace and well being of their families by offering seeds and seed oil of *Brassica campestris* (Brassicaceae), leaves and flowers of *Calotropis procera* (Asclepiadaceae), seeds of *Cicer arietinum* (Fabaceae), fruits of *Cucurbita hispida* (Cucurbitaceae), seeds of *Sesamum orientale* (Pedaliaceae), seeds of *Vigna mungo* and *Vigna radiata* (Fabaceae). Nevertheless, the details of offerings to appease other major Gods, Goddesses and Local Deities are listed in Table 27. It is believed that by presenting regular offerings the reverend God will be pleased and will protect their health and peace. It is a fact little realized that much of what remains of our rich biological heritage has been preserved through the ages because of a wealth of conservation-oriented cultural and religious traditions (Gadgil and Vartak, 1981).

Table 27: Offerings for Various Local Deities, Gods and Goddesses

Plant Species	*Concerned Deity*	*Part/s Used*
Aegle marmelos	Lord Shiva, Gayatri Mata	Leaves
Althaea rosea	Trilokinath (Lord Shiva), Gautam Buddha	Flowers
Areca catechu	Goddess Kali	Nuts ('Supari')
Brassica campestris	Lord Shani Dev	Seeds
Calotropis procera	Lord Shiva, Lord Shani Dev	Leaves, Flowers
Camellia sinensis	Guru Sunder Muni Nirwan	Leaves
Cannabis sativa	Lord Shiva	Leaves

Contd...

Table 27–*Contd...*

Plant Species	*Concerned Deity*	*Part/s Used*
Chrysanthemum indicum	Lord Vishnu	Flowers
Cicer arietinum	Lord Shani Dev	Split Grains
Clerodendrum philippinum	Markanday Rishi, Lord Shiva	Flowers
Coix lachryma-jobi	Raghunath	Grains
Cucurbita hispida	Lord Shani Dev	Fruits
Cynodon dactylon	Raghunath (Lord Vishnu)	Whole Plant
Datura fastuosa	Lord Shiva	Flowers
Datura innoxia	Lord Shiva	Flowers
Datura stramonium	Lord Shiva	Flowers
Helianthus annuus	Jabru Devta	Flowers
Hibiscus-rosa sinensis	Goddess Kali	Flowers
Hordeum vulgare	Goddess Chandika Devi, Raghunath, Gautam Buddha	Grains, Grain Flour
Jasminum primulinum	Goddess Kali	Flowers
Juglans regia	Goddess Usha Devi	Kernels
Mirabilis jalapa	Goddess Durga, Suhi Mata	Flowers
Musa paradisiac	Lord Vishnu	Fruits
Nerium odorum	Lord Shiva, Lomash Rishi	Flowers
Nicandra physaloides	Mahasu Devta	Flowers
Ocimum sanctum	Raghunath (Lord Vishnu)	Leaves
Oryza sativa	Most of the deities	Puffed Grains
Piper betle	Goddess Kali	Leaves
Prunus persica	Goddess Chandika Devi	Fruits (Wine)
Pyrus malus	Goddess Chandika Devi	Fruits (Wine)
Rosa foetida	Gautam Buddha	Flowers
Rosa indica	Goddess Shiv-Shakti, Goddess Durga	Flowers
Rosa jacquemontii	Kamakhya Devi	Flowers
Rosa moschata	Kamakhya Devi	Flowers
Rosa webbiana	Padam Sambhav, Deva Gyatso	Flowers
Sesamum orientale	Lord Shani Dev	Seeds
Tagetes erecta	Goddess Bagula Mukhi, Chamunda Devi, Durga, Gautam Buddha, Goddess Shiv-Shakti, Jathiya Devi, Lord Vishnu, Mrikula Devi, Raghunath, Raja Ghepan,Lord Shani Dev, Suhi Mata, Trilokinath (Lord Shiva)	Flowers
Tecoma stans	Goddess Bagula Mukhi	Flowers
Thevetia neriifolia	Goddess Bagula Mukhi, Lord Shiva	Flowers
Triticum aestivum	Baba Balak Nath	Grain Flour ('Rot')
Vigna mungo	Lord Shani Dev	Seeds
Vigna radiata	Lord Shani Dev	Seeds
Vitis vinifera	Goddess Chandika Devi	Fruits (Wine)

Table 28: Holy Trees, Shrubs, Herbs and Climbers of the Study Area

Plant Species	*Habit*	*Associated Deity*	*Reasons for Worship*
Aegle marmelos	T	Lord Shiva	Tree considered symbolic to Lord Shiva and believed to possess thousands of powers. It is regarded as one of the trees of 'Panch Pallav'.
Berberis lycium	S	Maheshwar Devta	It is believed that the deity resides on this plant.
Bischofia javanica	T	Nagni Mata	Tree considered sacred and believed to be associated with Nagni Mata.
Bougainvillea spectabilis	C	Goddess Balasundari	Regarded as a wish granting plant for which people tie sacred threads to it.
Calotropis procera	S	Lord Shiva, Lord Shani Dev	Leaves and flowers offered to appease Lord Shiva and Lord Shani Dev.
Carissa spinarum	S	Baba Balak Nath	It is believed that Baba Balak Nath meditated under this bush. People tie sacred threads ('Dori') to the sacred bush for fulfilment of their wishes.
Cedrus deodara	T	Ghatotkach	Regarded as symbolic to the deity.
Cestrum nocturnum	S	Goddess Durga	Plant considered symbolic to God.
Citrus limon	T	Goddess Durga	Tree considered symbolic to God and worshipped by locals.
Cupressus sempervirens	T	Lord Nar Singh	Considered symbolic to God
Dendrophthoe falcata	S	Sehdev	Regarded as symbolic to God.
Elaeocarpus ganitrus	T	Lord Shiva	Considered symbolic to Lord Shiva.
Ficus benghalensis	T	Baba Balak Nath, Lord Brahma	It is said that Baba Balak Nath meditated under this tree. This tree also considered symbolic to Lord Brahma, and regarded as one of the trees of 'Panch Pallav'.
Ficus glomerata	T	Renuka Mata, Rishi Markanda	Regarded as one of the trees of 'Panch Pallav'. It is believed that the river Saraswati originated from the roots of tree present in the temple-courtyard under which Rishi Markanda meditated long ago.
Ficus palmata	T	Goddess Durga	Considered symbolic to god.
Ficus religiosa	T	Lord Brahma, Mahakali	Tree worshipped as incarnation of Lord Brahma. It is believed that 33 crore deities reside on the Peepal tree. People tie scarves and sacred thread to its branches for fulfilment of their wishes (Naina Devi Temple). It is also regarded as one of the trees of 'Panch Pallav'. A group of seven trees of *Ficus religiosa* in the courtyard of Tarna Devi Temple considered sacred as the Goddess Mahakali is believed to appear under it.

Contd...

Table 28–*Contd...*

Plant Species	*Habit*	*Associated Deity*	*Reasons for Worship*
Ficus rumphii	T	Baba Sidh Chaino	Considered symbolic to Baba Sidh Chaino.
Ficus virens	T	God	Regarded as symbolic to God.
Holoptelea integrifolia	T	God	Regarded as symbolic to God.
Jasminum grandiflorum	C	Goddess Durga	Considered symbolic to God.
Jasminum primulinum	S	Goddess Durga	Considered symbolic to the deity.
Juglans regia	T	Goddess Durga	People tie sacred threads and scarves to the tree of for fulfilment of their wishes.
Juniperus macropoda	T	Gautam Buddha	Considered symbolic to the deity.
Ligustrum lucidum	T	Pataal Devta	People believe that the deity resides on this tree and regarded as gatekeeper of Manu Devta.
Machilus gamblei	T	Jathiya Devi	Regarded as symbolic to the deity.
Magnolia grandiflora	T	Padamsambhav	Trees worshipped as sacred at Buddhist Monastery, Rewalsar; one of the trees here was planted by Dalai Lama and the other one by Pancham Lama on the same day in 1957.
Mangifera indica	T	God	Newly married couples tie bangles to the tree for prosperous married life.
Mimusops elengi	T	Lord Krishana	Considered symbolic to God.
Morus alba	T	Nag Devta	Worshipped as the tree of Nag Devta.
Morus serrata	T	God	Regarded as symbolic to God.
Musa paradisiaca	H	Lord Brihaspati	Regarded as symbol of Lord Brihaspati
Ocimum basilicum	H	Lord Vishnu	Plant considered symbolic to Lord Vishnu.
Ocimum kilimandscharicum	S	Lord Vishnu	Plant considered symbolic to Lord Vishnu.
Ocimum sanctum	H	Lord Vishnu	Worshipped as well as offered to appease Lord Vishnu. Traditionally, the plant is married to Saligram (Lord Vishnu)
Phoenix sylvestris	T	Lakhdata Pir	Locals consider this tree as symbolic to the deity.

Contd...

Table 28–*Contd...*

Plant Species	*Habit*	*Associated Deity*	*Reasons for Worship*
Pistacia integerrima	T	Goddess Durga	Tree considered symbolic to God.
Populus ciliata	T	Padam Sambhav, Deva Gyatso	Tree supposed to be the abode of God worshipped by the Buddhist people.
Prinsepia utilis	S	Mahasu Devta	Regarded as symbolic to the deity.
Pyrus pashia	T	Lord Shiva	Regarded as 'Kalpa Vriksha' and resting place of Lord Shiva.
Robinia pseudoacacia	T	Goddess Chandika, Shikhdu Devta	Believed to be the abode of God.
Salix alba	T	Goddess Chandika	Believed to be the abode of God.
Salix babylonica	T	Bhairav	Considered to be the abode of God.
Tamarindus indica	T	Swami Achutanand	Believed to be associated with the deity.
Thevetia neriifolia	T	Lord Shiva	Regarded as symbolic to Lord Shiva and one of the trees of 'Panch Pallav'.
Thuja orientalis	T	Goddess Durga	Newly married couples tie their 'bengals' (bride) and 'sehra' (bridegroom) to the tree growing in the temple-courtyard for their prosperous life
Ulmus villosa	T	Maheshwar Devta	Believed to grow at sacred places only and regarded as the abode of Maheshwar Devta.
Zizyphus mauritiana	T	Nag Devta, Baba Shibbo Nath, Baba Barbhag Singh	Believed to be associated with Nag Devta (Snake God), Baba Shibbo Nath and Baba Barbhag Singh.

Like in almost all the civilizations of the world, tree worshipping by the local communities is quite prevalent in the temple-courtyards of the State. It is believed that the trees are the abode of local deities, Gods and Goddesses, and are protected and worshipped for their ecological, economical, medicinal and social importance. Trees which are considered sacred and worshipped on religious grounds are: *Mangifera indica, Pistacia integerrima* (Anacardiaceae), *Thevetia neriifolia* (Apocynaceae), *Phoenix sylvestris* (Arecaceae), *Cupressus sempervirens, Juniperus macropoda, Thuja orientalis* (Cupressaceae), *Elaeocarpus ganitrus* (Elaeocarpaceae), *Bischofia javanica* (Euphorbiaceae), *Robinia pseudoacacia, Tamarindus indica* (Fabaceae), *Juglans regia* (Juglandaceae), *Machilus gamblei* (Lauraceae), *Magnolia grandiflora* (Magnoliaceae), *Ficus benghalensis, Ficus glomerata, Ficus palmata, Ficus religiosa, Ficus rumphii, Ficus virens, Morus alba, Morus serrata* (Moraceae), *Ligustrum lucidum* (Oleaceae), *Cedrus deodara* (Pinaceae), *Zizyphus mauritiana* (Rhamnaceae), *Pyrus pashia* (Rosaceae), *Aegle marmelos, Citrus limon* (Rutaceae), *Populus ciliata, Salix alba, Salix babylonica* (Salicaceae), *Mimusops elengi* (Sapotaceae), *Holoptelea integrifolia* and *Ulmus villosa* (Ulmaceae). The associated deity with each of the holy tree and reasons for worshipping are presented in Table 28. Besides religious importance, some of the trees are valued for their fruits, as ornamental species or for their medicinal and aromatic importance. Therefore, temple-courtyards form ideal habitats for the preservation and protection of plant diversity. Moreover, the use of wood of *Bauhinia vahlii* and *Mangifera indica* for preparing holy ash ('vibhuti') to be distributed as 'prasadam' for people visiting the temple, dried stigmas of *Zephyranthes candida* (Amaryllidaceae) as vermilion ('tikka'), and wood of *Cedrus deodara* and *Pinus wallichiana* for fabricating palanquin of the local deities are some of the practices unique only to the populace of this State, and reported for the first time. Due to their religious importance, these are being conserved in the region as cultural heritage of the local communities.

Table 29: Families, Genera and Species of Sacred Courtyard Plants

Family	*Plant Species*	*Genus/Species*
Amaryllidaceae	*Narcissus tazetta, Zephyranthes candida, Zephyranthes grandiflora*	2/3
Anacardiaceae	*Mangifera indica, Pistacia integerrima*	2/2
Apiaceae	*Daucus carota*	1/1
Apocynaceae	*Carissa spinarum, Catharanthus roseus, Nerium odorum, Plumeria acutifolia, Tabernaemontana coronaria, Thevetia neriifolia*	6/6
Arecaceae	*Phoenix sylvestris*	1/1
Asclepiadaceae	*Calotropis procera*	1/1
Asteraceae	*Artemisia moorcroftiana, Chrysanthemum indicum, Cosmos bipinnatus, Cosmos sulphureus, Dahlia rosea, Helianthus annuus, Tagetes erecta, Zinnia elegans*	7/8
Berberidaceae	*Berberis lycium*	1/1
Bignoniaceae	*Tecoma radicans, Tecoma stans*	½
Brassicaceae	*Brassica oleracea* var. *Capitata*	1/1

Contd...

Table 29–*Contd...*

Family	*Plant Species*	*Genus/Species*
Cannabinaceae	*Cannabis sativa*	1/1
Cannaceae	*Canna indica*	1/1
Cupressaceae	*Cupressus sempervirens, Cupressus torulosa, Juniperus macropoda, Thuja orientalis*	3/4
Elaeocarpaceae	*Elaeocarpus ganitrus*	1/1
Euphorbiaceae	*Bischofia javanica*	1/1
Fabaceae	*Acacia catechu, Bauhinia vahlii, Pisum sativum, Robinia pseudoacacia, Tamarindus indica*	5/5
Juglandaceae	*Juglans regia*	1/1
Lamiaceae	*Ocimum basilicum, Ocimum kilimandscharicum, Ocimum sanctum*	1/3
Lauraceae	*Machilus gamblei*	1/1
Loranthaceae	*Dendrophthoe falcata*	1/1
Magnoliaceae	*Magnolia grandiflora*	1/1
Malvaceae	*Althaea rosea, Hibiscus rosa-sinensis, Hibiscus syriacus, Malva sylvestris, Malvaviscus conzatii*	4/5
Moraceae	*Ficus benghalensis, Ficus glomerata, Ficus palmata, Ficus religiosa, Ficus rumphii, Ficus virens, Morus alba, Morus serrata*	2/8
Musaceae	*Musa paradisiaca*	1/1
Myrtaceae	*Psidium guajava*	1/1
Nyctaginaceae	*Bougainvillea glabra, Bougainvillea spectabilis, Mirabilis jalapa*	2/3
Oleaceae	*Jasminum grandiflorum, Jasminum humile, Jasminum primulinum, Jasminum sambac, Ligustrum lucidum*	2/5
Pedaliaceae	*Sesamum orientale*	1/1
Pinaceae	*Cedrus deodara, Pinus roxburghii, Pinus wallichiana*	2/3
Poaceae	*Coix lachrymal-jobi, Cynodon dactylon, Saccharum officinarum, Zea mays*	4/4
Punicaceae	*Punica granatum*	1/1
Rhamnaceae	*Zizyphus mauritiana*	1/1
Rosaceae	*Prinsepia utilis, Prunus amygdalus, Prunus armeniaca, Prunus persica, Prunus puddum, Pyrus communis, Pyrus malus, Pyrus pashia, Rosa alba, Rosa banksiae, Rosa foetida, Rosa indica, Rosa jacquemontii, Rosa moschata, Rosa multiflora, Rosa webbiana*	4/16
Rutaceae	*Aegle marmelos, Citrus limon, Citrus reticulata*	2/3
Salicaceae	*Populus ciliata, Salix alba, Salix babylonica*	2/3
Sapotaceae	*Achras zapota, Mimusops elengi*	2/2
Solanaceae	*Cestrum nocturnum, Datura fastuosa, Datura innoxia, Datura stramonium, Nicandra physaloides, Solanum tuberosum*	4/6
Ulmaceae	*Holoptelea integrifolia, Ulmus villosa*	2/2
Tropaeolaceae	*Tropaeolum majus*	1/1
Verbenaceae	*Clerodendrum philippinum*	1/1
Violaceae	*Viola canescens*	1/1

Table 30: Status of Temple Plants in the Surrounding Areas

Plant Species	*Status in Surrounding Areas*	*Temple*
Acer oblongum	Not Common	Shiva Temple (Solan)
Aegle marmelos	Absent	Bhuteshwar Mahadev Temple (Shimla), Dhankeshwar Mahadev Temple (Kullu)
Aegle marmelos	Not Common	Saptdevi Sidh Peeth Ashram (Una), Vaishno Devi Temple (Kullu)
Aesculus indica	Not Common	Sai Baba Temple (Solan)
Aristolochia elegans	Not Common	Shri Raghunath Teda Temple (Kangra)
Artocarpus lakoocha	Not Common	Shibbo Than Temple, (Kangra), Shiva Temple (Solan)
Betula utilis	Not Common	Mahabodi Temple (Kinnaur)
Bischofia javanica	Not Common	Nagni Mata Temple (Kangra)
Buxus sempervirens	Absent	Bhekhali Mata Temple (Kullu)
Calotropis procera	Not Common	Saptdevi Sidh Peeth Ashram (Una)
Carissa carandas	Absent	Dera Baba Rudranand (Una)
Casuarina equisetifolia	Absent	Shakani Ganga Temple (Solan)
Cedrus deodara	Absent	Shiva Temple (Kangra)
Celtis australis	Not Common	Baba Lakhdata Pir Temple (Chamba)
Coix lachryma-jobi	Not Common	Shri Raghunath Teda Temple (Kangra)
Cupressus sempervirens	Absent	Narsingh Temple (Sirmaur)
Elaeocarpus ganitrus	Absent	Shiva Temple (Kangra), Balasundari Temple (Sirmaur), Nirwan Ashram (Sirmour)
Eriobotrya japonica	Not Common	Gurudwara Shri Guru Gobind Singh (Mandi)
Ficus benghalensis	Absent	Dhankeshwar Mahadev Temple (Kullu), Bhuteshwar Mahadev Temple (Shimla), Mahish Mardini Mahalakshmi Temple (Shimla)
Ficus benghalensis	Not Common	Shri Sidh Baba Balak Nath Temple (Solan), Saptdevi Sidh Peeth Ashram (Una)
Ficus glomerata	Not Common	Machhial Mahadev Temple (Kangra)
Ficus glomerata	Absent	Dhankeshwar Mahadev Temple (Kullu), Rishi Markanda Dham (Sirmaur)
Ficus religiosa	Absent	Dhankeshwar Mahadev Temple (Kullu), Bhuteshwar Mahadev Temple (Shimla), Hatkeshwari Durga Mata Temple (Shimla), Kali Mata Temple (Shimla), Mahish Mardini Mahalakshmi Temple (Shimla), Shri Sankatmochan Hanuman Temple (Shimla)
Ficus religiosa	Not Common	Vaishno Devi Temple (Kullu), Shri Sidh Baba Balak Nath Temple (Solan), Sidheshwari Temple (Solan)
Ficus roxburghii	Absent	Rishi Markanda Dham (Sirmaur)
Ficus rumphii	Not Common	Sidh Chaino Temple (Kangra)
Ficus virens	Not Common	Shiva Temple (Solan)
Jacaranda acutifolia	Absent	Shiva Temple (Kangra)

Contd...

Table 30–*Contd...*

Plant Species	*Status in Surrounding Areas*	*Temple*
Juglans regia	Absent	Lakshmi Narayan Temple (Chamba), Kamna Devi Temple (Shimla)
Ligustrum lucidum	Absent	Manu Temple (Kullu)
Litsea sebifera	Not Common	Shri Brijraj Swami Temple (Kangra)
Magnolia grandiflora	Absent	Ningmapa Budh Monastery (Mandi)
Mangifera indica	Absent	Mahish Mardini Mahalakshmi Temple (Shimla)
Mimusops elengi	Absent	Jwalamukhi Temple (Kangra), Shibbo Than Temple (Kangra), Shri Brijraj Swami Temple (Kangra), Balasundari Temple (Sirmaur), Jagan Nath Temple (Sirmaur), Kalisthan Temple (Sirmour), Narsingh Temple (Sirmour), Shiva Temple (Sirmaur), Dera Baba Rudranand (Una)
Narcissus tazetta	Not Common	Shri Raghunath Teda Temple (Kangra)
Platanus orientalis	Absent	Vaishno Devi Temple (Kullu)
Populus ciliata	Not Common	Shashur Monastery (Lahul-Spiti)
Pterospermum acerifolium	Absent	Shibbo Than Temple (Kangra)
Putranjiva roxburghii	Absent	Shiva Temple (Kangra)
Pyrus pashia	Not Common	Baba Lakhdata Pir Temple (Chamba), Bijli Mahadev Temple (Kullu)
Ribes orientale	Not Common	Kee Gompa (Lahul-Spiti)
Rosa webbiana	Not Common	Kee Gompa (Lahul-Spiti)
Santalum album	Absent	Saptdevi Sidh Peeth Ashram (Una)
Sapindus mukorossi	Not Common	Gurudwara Shri Guru Gobind Singh (Mandi)
Syzygium jambos	Not Common	Chamunda Nandikeshwar Temple (Kangra)
Tamarindus indica	Not Common	Narsingh Temple (Sirmaur)
Tectona grandis	Not Common	Sadashiv Temple (Una)
Terminalia chebula	Not Common	Shri Brijraj Swami Temple (Kangra)
Thevetia neriifolia	Not Common	Mamaleshwar Mahadeh Temple (Mandi)
Ulmus villosa	Absent	Chandika Devi Temple (Kinnaur), Usha Devi Temple (Kinnaur)
Ulmus villosa	Not Common	Maheshwar Devta Temple (Kinnaur), Bhekhali Mata Temple (Kullu)
Wrightia tomentosa	Not Common	Garuna Jhadi Temple (Bilaspur)

Of the plants recorded in the temple-courtyards (presently), 185 species are the commonly cultivated plants (Tables 31) and 117 species occur as wild (Tables 32). Taxa like *Duranta repens, Mangifera indica, Sapindus mukorossi* and *Tamarindus indica* grow as both cultivated and wild plants. Species, for instance *Acer oblongum, Aegle marmelos, Aesculus indica, Aristolochia elegans, Artocarpus lakoocha, Betula utilis, Bischofia javanica, Calotropis procera, Celtis australis, Coix lachryma-jobi, Eriobotrya japonica, Ficus benghalensis, Ficus glomerata, Ficus religiosa, Ficus rumphii, Ficus virens, Litsea sebifera, Narcissus tazetta, Populus ciliata, Pyrus pashia, Ribes orientale, Rosa webbiana, Sapindus mukorossi, Syzygium jambos, Tamarindus indica, Tectona grandis, Terminalia chebula, Thevetia neriifolia, Ulmus villosa* and *Wrightia tomentosa* are uncommon whereas *Aegle marmelos, Buxus sempervirens, Carissa carandas, Casuarina equisetifolia, Cedrus deodara, Cupressus sempervirens, Elaeocarpus ganitrus, Ficus benghalensis, Ficus glomerata, Ficus religiosa, Ficus roxburghii, Jacaranda acutifolia, Juglans regia, Ligustrum lucidum, Magnolia grandiflora, Mangifera indica, Mimusops elengi, Platanus orientalis, Pterospermum acerifolium, Putranjiva roxburghii, Santalum album* and *Ulmus villosa* are absent in the surrounding areas (Table 30). Species like *Betula utilis* ('bhojpatra'), *Pinus gerardiana* ('chilgoza'), and *Zanthoxylum armatum* ('tirmir') though protected in the temple-yards are facing danger to their survival in their natural abode in the absence of management practices for their regulated and scientific exploitation, protection, conservation and multiplication. These plant species have been declared threatened and endangered species for H.P. as per the Red Data Books of Indian Plants (Jain and Sastry, 1979). The data provide sufficient clues regarding the genetic diversity in species related to our crop plants and status of the temple plants of H.P. in the surrounding areas.

Table 31: Cultivated Temple-Courtyard Plants

Plant Species	*Family*
Acacia dealbata	Fabaceae
Acalypha wilkesiana	Euphorbiaceae
Achras zapota	Sapotaceae
Aloe vera	Liliaceae
Alpinia nutans	Zingiberaceae
Alstonia scholaris	Apocynaceae
Althaea rosea	Malvaceae
Amaryllis reginae	Amaryllidaceae
Antirrhinum majus	Scrophulariaceae
Araucaria columnaris	Araucariaceae
Aristolochia elegans	Aristolochiaceae
Asclepias curassavica	Asclepiadaceae
Asclepias physocarpa	Asclepiadaceae
Asparagus densiflorus	Liliaceae
Beloperone guttata	Acanthaceae
Betula utilis	Betulaceae
Bougainvillea glabra	Nyctaginaceae
Bougainvillea spectabilis	Nyctaginaceae
Brassica oleracea var. *capitata*	Brassicaceae

Contd...

Table 31–*Contd...*

Plant Species	*Family*
Bryophyllum calycinum	Crassulaceae
Buxus sempervirens	Buxaceae
Calendula officinalis	Asteraceae
Callistemon lanceolatus	Myrtaceae
Callistephus chinensis	Asteraceae
Canna indica	Cannaceae
Capsicum annuum	Solanaceae
Carica papaya	Caricaceae
Carissa carandas	Apocynaceae
Caryota urens	Arecaceae
Cassia glauca	Fabaceae
Casuarina equisetifolia	Casuarinaceae
Catharanthus roseus	Apocynaceae
Celosia argentea var. *cristata*	Amaranthaceae
Centaurea cyanus	Asteraceae
Cestrum diurnum	Solanaceae
Cestrum nocturnum	Solanaceae
Chamaerops humilis	Arecaceae
Chlorophytum capense	Liliaceae
Chrysanthemum indicum	Asteraceae
Chrysanthemum leucanthemum	Asteraceae
Cinnamomum camphora	Lauraceae
Citrus aurantifolia	Rutaceae
Citrus maxima	Rutaceae
Citrus reticulata	Rutaceae
Clerodendrum inerme	Verbenaceae
Clerodendrum splendens	Verbenaceae
Clerodendrum thomsonae	Verbenaceae
Coleus blumei	Lamiaceae
Colocacia esculenta	Araceae
Cosmos bipinnatus	Asteraceae
Cosmos sulphureus	Asteraceae
Cupressus sempervirens	Cupressaceae
Cupressus torulosa	Cupressaceae
Cycas revoluta	Cycadaceae
Cyperus flabelliformis	Cyperaceae

Contd...

Table 31–*Contd...*

Plant Species	*Family*
Dahlia rosea	Asteraceae
Daucus carota	Apiaceae
Delphinium ajacis	Ranunculaceae
Dianthus barbatus	Caryophyllaceae
Diospyros lotus	Ebenaceae
Dolichos lablab	Fabaceae
Duranta repens	Verbenaceae
Elaeocarpus ganitrus	Elaeocarpaceae
Eriobotrya japonica	Rosaceae
Eucalyptus tereticornis	Myrtaceae
Euphorbia milii	Euphorbiaceae
Euphorbia pulcherrima	Euphorbiaceae
Ficus carica	Moraceae
Ficus elastica	Moraceae
Gaillardia pulchella	Asteraceae
Gomphrena globosa	Amaranthaceae
Grevillea robusta	Proteaceae
Hamelia patens	Rubiaceae
Hedychium coronarium	Zingiberaceae
Helianthus annuus	Asteraceae
Hemerocallis fulva	Liliaceae
Hibiscus mutabilis	Malvaceae
Hibiscus rosa-sinensis	Malvaceae
Hibiscus syriacus	Malvaceae
Humulus lupulus	Cannabinaceae
Hydrangea macrophyla	Hydrangeaceae
Hymenocallis littoralis	Amaryllidaceae
Impatiens balsamina	Balsaminaceae
Ipomoea quamoclit	Convolvulaceae
Iris germanica	Iridaceae
Ixia chinensis	Iridaceae
Jacaranda acutifolia	Bignoniaceae
Jasminum grandiflorum	Oleaceae
Jasminum humile	Oleaceae
Jasminum primulinum	Oleaceae
Jatropha panduraefolia	Euphorbiaceae

Contd...

Table 31–*Contd...*

Plant Species	*Family*
Juglans regia	Juglandaceae
Juniperus chinensis	Cupressaceae
Justicia gendarussa	Acanthaceae
Kochia trichophylla	Chenopodiaceae
Lagerstroemia indica	Lythraceae
Lawsonia alba	Lythraceae
Lespedeza sericea	Fabaceae
Litchi chinensis	Sapindaceae
Lonicera japonica	Caprifoliaceae
Lycopersicum esculentum	Solanaceae
Magnolia grandiflora	Magnoliaceae
Malva sylvestris	Malvaceae
Malva verticillata	Malvaceae
Malvaviscus conzatii	Malvaceae
Mangifera indica	Anacardiaceae
Melia azadirachta	Meliaceae
Mentha piperita	Lamiaceae
Michelia champaca	Magnoliaceae
Mirabilis jalapa	Nyctaginaceae
Mimosa pudica	Fabaceae
Mimusops elengi	Sapotaceae
Momordica charantia	Cucurbitaceae
Murraya paniculata	Rutaceae
Musa paradisiaca	Musaceae
Narcissus tazetta	Amaryllidaceae
Nerium odorum	Apocynaceae
Nyctanthes arbortristis	Oleaceae
Ocimum basilicum	Lamiaceae
Ocimum kilimandscharicum	Lamiaceae
Ocimum sanctum	Lamiaceae
Papaver rhoeas var. *latifolia*	Papaveraceae
Passiflora caerulea	Passifloraceae
Pedilanthus tithymaloides	Euphorbiaceae
Pelargonium hortorum	Geraniaceae
Petunia hybrida	Solanaceae
Phaseolus vulgaris	Fabaceae

Contd...

Table 31–*Contd...*

Plant Species	*Family*
Pisum sativum	Fabaceae
Platanus orientalis	Platanaceae
Plumeria acutifolia	Apocynaceae
Polianthes tuberosa	Amaryllidaceae
Polyalthia longifolia	Annonaceae
Pongamia pinnata	Fabaceae
Populus ciliata	Salicaceae
Populus nigra	Salicaceae
Prunus amygdalus	Rosaceae
Prunus armeniaca	Rosaceae
Prunus domestica	Rosaceae
Prunus persica	Rosaceae
Psidium guajava	Myrtaceae
Pterospermum acerifolium	Sterculiaceae
Punica granatum	Punicaceae
Pyrostegia venusta	Bignoniaceae
Pyrus communis	Rosaceae
Pyrus malus	Rosaceae
Quisqualis indica	Combretaceae
Rosa alba	Rosaceae
Rosa banksiae	Rosaceae
Rosa foetida	Rosaceae
Rosa indica	Rosaceae
Rosa jacquemontii	Rosaceae
Rosa moschata	Rosaceae
Rosa multiflora	Rosaceae
Ruellia tuberosa	Acanthaceae
Russelia equisetiformis	Scrophulariaceae
Saccharum officinarum	Poaceae
Salix babylonica	Salicaceae
Salvia coccinia	Lamiaceae
Salvia splendens	Lamiaceae
Sambucus nigra	Sambucaceae
Sansevieria trifasciata	Liliaceae
Santalum album	Santalaceae
Sapindus mukorossi	Sapindaceae

Contd...

Table 31–*Contd...*

Plant Species	*Family*
Schefflera arboricola	Araliaceae
Sempervivum arboreum	Crassulaceae
Sesamum orientale	Pedaliaceae
Solanum tuberosum	Solanaceae
Syzygium jambos	Myrtaceae
Tabernaemontana coronaria	Apocynaceae
Tagetes erecta	Asteraceae
Tamarindus indica	Fabaceae
Tecoma radicans	Bignoniaceae
Tecoma stans	Bignoniaceae
Tecomaria capensis	Bignoniaceae
Tectona grandis	Verbenaceae
Thevetia neriifolia	Apocynaceae
Thuja orientalis	Cupressaceae
Thuja plicata	Cupressaceae
Thunbergia erecta	Acanthaceae
Tithonia tagetiflora	Asteraceae
Tropaeolum majus	Tropaeolaceae
Viola tricolour	Violaceae
Wedelia calendulacea	Asteraceae
Wisteria sinensis	Fabaceae
Zephyranthes candida	Amaryllidaceae
Zephyranthes grandiflora	Amaryllidaceae
Zinnia elegans	Asteraceae
Zea mays	Poaceae

Oyadomari (1990) indicated that shrine and temple groves in Japan not only provide an important reverential landscape for the shrines or temples but are also used as places of festivals and community gathering. The present study also shows the same. Surprisingly, none of the temple courtyards in H.P. excepting Boirus Temple, Kothi (Kinnaur), Machhial Mahadev Temple, Andreta (Kangra), Shiv Temple, Bachhretu (Bilaspur) harbour sacred water tanks to support and conserve any aquatic vegetation.

Based upon the work on 22 temple-yards of the Buddhist temples in Yunnan, China, Shengji (1993) reported 58 different species of commonly cultivated plants and grouped these into three types, *viz.* Group I–Ritual Plants; Group II–Fruit Trees; Group III–Ornamental Plants. However, the plants in the temple-courtyards of H.P. (present study) can be clearly delineated into six groups. These categories are: Sacred plants, edible plants, ornamental plants, medicinal and aromatic plants, fodder plants and material culture plants. It is worthy of mention that the diversity held in the temple-courtyards of H.P. is surprisingly diverse, rich, and unique. Basically, these temple-yards serve as unrecognized centres for local biodiversity management and conservation.

Table 32: Wild Temple-Courtyard Plants

Plant Species	*Family*
Acacia catechu	Fabaceae
Acacia nilotica	Fabaceae
Acer oblongum	Aceraceae
Aegle marmelos	Rutaceae
Aesculus indica	Hippocastanaceae
Albizia lebbeck	Fabaceae
Albizia stipulata	Fabaceae
Alnus nitida	Betulaceae
Artemisia moorcroftiana	Asteraceae
Artocarpus heterophyllus	Moraceae
Artocarpus lakoocha	Moraceae
Asparagus adscendens	Liliaceae
Asparagus racemosus	Liliaceae
Bauhinia vahlii	Fabaceae
Bauhinia variegata	Fabaceae
Berberis lycium	Berberidaceae
Bergenia ciliata	Saxifragaceae
Bischofia javanica	Euphorbiaceae
Boerhaavia diffusa	Nyctaginaceae
Bombax ceiba	Bombacaceae
Broussonetia papyrifera	Moraceae
Calotropis procera	Asclepiadaceae
Cannabis sativa	Cannabinaceae
Carissa opaca	Apocynaceae
Carissa spinarum	Apocynaceae
Cassia fistula	Fabaceae
Cedrela toona	Meliaceae
Cedrus deodara	Pinaceae
Celtis australis	Ulmaceae
Centella asiatica	Apiaceae
Citrus jambhiri	Rutaceae
Citrus limon	Rutaceae

Contd...

Table 32–*Contd...*

Plant Species	*Family*
Clematis grata	Ranunculaceae
Clerodendrum philippinum	Verbenaceae
Coix lachryma-jobi	Poaceae
Convolvulus arvensis	Convolvulaceae
Cordia dichotoma	Cordiaceae
Cynodon dactylon	Poaceae
Dalbergia sissoo	Fabaceae
Datura fastuosa	Solanaceae
Datura innoxia	Solanaceae
Datura stramonium	Solanaceae
Debregeasia hypoleuca	Urticaceae
Deeringia amaranthoides	Amaranthaceae
Dendrophthoe falcata	Loranthaceae
Diospyros montana var. *cordifolia*	Ebenaceae
Dodonaea viscosa	Sapindaceae
Duranta repens	Verbenaceae
Ehretia acuminata	Boraginaceae
Emblica officinalis	Euphorbiaceae
Erigeron alpinus	Asteraceae
Euphorbia royleana	Euphorbiaceae
Ficus benghalensis	Moraceae
Ficus glomerata	Moraceae
Ficus palmata	Moraceae
Ficus religiosa	Moraceae
Ficus roxburghii	Moraceae
Ficus rumphii	Moraceae
Ficus virens	Moraceae
Fraxinus xanthoxyloides	Oleaceae
Grewia oppositifolia	Tiliaceae
Holoptelea integrifolia	Ulmaceae
Hyoscyamus niger	Solanaceae
Hypericum cernuum	Hypericaceae
Ipomoea cairica	Convolvulaceae
Ipomoea purpurea	Convolvulaceae
Jasminum sambac	Oleaceae
Juniperus macropoda	Cupressaceae

Contd...

Table 32–*Contd...*

Plant Species	*Family*
Lannea coromandelica	Anacardiaceae
Leucaena leucocephala	Fabaceae
Ligustrum lucidum	Oleaceae
Litsea polyantha	Lauraceae
Litsea sebifera	Lauraceae
Machilus gamblei	Lauraceae
Mallotus philippinensis	Euphorbiaceae
Mangifera indica	Anacardiaceae
Melia azedarach	Meliaceae
Mentha longifolia	Lamiaceae
Morus alba	Moraceae
Morus serrata	Moraceae
Murraya koenigii	Rutaceae
Nicandra physaloides	Solanaceae
Ocimum gratissimum	Lamiaceae
Oenothera drummondii	Onagraceae
Olea glandulifera	Oleaceae
Phoenix sylvestris	Arecaceae
Phytolacca acinosa	Phytolaccaceae
Pinus gerardiana	Pinaceae
Pinus roxburghii	Pinaceae
Pinus wallichiana	Pinaceae
Pistacia integerrima	Anacardiaceae
Polygonum alpinum	Polygonaceae
Prinsepia utilis	Rosaceae
Prunus puddum	Rosaceae
Putranjiva roxburghii	Euphorbiaceae
Pyrus pashia	Rosaceae
Quercus dilatata	Fagaceae
Quercus incana	Fagaceae
Ribes orientale	Saxifragaceae
Ricinus communis	Euphorbiaceae
Robinia pseudoacacia	Fabaceae
Rosa webbiana	Rosaceae
Roylea cinerea	Lamiaceae
Salix alba	Salicaceae

Contd...

Table 32–*Contd...*

Plant Species	*Family*
Salix fragilis	Salicaceae
Salix oxycarpa	Salicaceae
Salix tetrasperma	Salicaceae
Sapindus mukorossi	Sapindaceae
Sapium insigne	Euphorbiaceae
Sapium sebiferum	Euphorbiaceae
Shorea robusta	Dipterocarpaceae
Syzygium cumini	Myrtaceae
Tamarindus indica	Fabaceae
Terminalia bellirica	Combretaceae
Terminalia chebula	Combretaceae
Ulmus villosa	Ulmaceae
Viola canescens	Violaceae
Withania somnifera	Solanaceae
Wrightia tomentosa	Apocynaceae
Zanthoxylum armatum	Rutaceae
Zizyphus mauritiana	Rhamnaceae

It is seen that the plants worshipped are linked up with the local livelihood, and play an important role in the local environment. By attaching some taboos and religious faith, the local communities are protecting and conserving the useful genetic diversity in the temple-courtyards. There is a strict prohibition of carrying any iron implements or cutting of any of the courtyard trees at Baba Achutanand Thakurdwara, Bhekhali Mata Temple, Chandika Devi Temple, Chaurasi Temple, Dera Baba Rudranand, Dhankeshwar Mahadev Temple, Dron Shiv Temple, Duttatrey Temple, Hadimba Devi Temple, Kalinath Kaleshwar Mahadev, Kalisthan Temple, Katasan Devi Temple, Lakshmi Narayan Temple (Chanor), Lankadvir Temple, Lohara Temple, Maheshwar Devta Temple, Mrikula Devi Temple, Mutru Mahadev Temple, Narsingh Temple, Renuka Mata Temple, Rishi Markanda Dham, Sadashiv Temple, Shakani Ganga Temple, Usha Devi Temple and Vindhyavasini Mata Temple for fear of inviting wrath of the ruling deity resulting in bringing serious illness, physical abnormalities, ill luck, and/or death in the family. All these beliefs and practices add to the conservation of these species.

'Patra-pooja' is an important ritual of offering various leaves to various Gods and Goddesses in many traditional forms of worship. Presently, species such as *Aegle marmelos* (Rutaceae), *Calotropis procera* (Asclepiadaceae), *Cannabis sativa* (Cannabinaceae), *Cynodon dactylon* (Poaceae), *Ficus glomerata* (Moraceae) and *Mangifera indica* (Anacardiaceae) have been identified as requisite resources for performing the 'pooja' ceremonies (Table 26).

Out of the 205 temple-courtyards identified presently in H.P., 10 of these, *viz.*, Lakshmi Narayan temple, Chanor (Kangra), Kalinath Kaleshwar Mahadev temple, Kaleshwar (Kangra), Mahakal Temple, Mahakal (Kangra), Shiva Temple, Amrella (Kangra), Shri Raghunath Teda Temple, Chakvan Kalidhar (Kangra), Kamakhaya Devi Temple, Kamru (Kinnaur), Hadimba Devi Temple, Manali (Kullu), Kamru Nag Temple, Kamrah (Mandi), Narsingh Temple, Nahan (Sirmaur), Dron Shiv Temple, Ambota (Una) have patches of forest attached to them in addition to the temple-yards (Table 33). And, these serve as centres of both *in-situ* and *ex situ* conservation. The rich genetic diversity of sacred and economic plants held in temple groves is being protected and conserved on religious grounds as the removal of any produce or cutting of any tree is considered a taboo. Both *in-situ* and *ex situ* methods of conservation are necessarily the major components of any comprehensive conservation strategy. As pointed out by Gadgil and Vartak (1981) that it is important for us to preserve this diversity for posterity, not only for compelling scientific and aesthetic reasons, but also for sound economic reasons as potential sources of significant chemicals as yet undiscovered, or as carriers of potentially valuable genetic traits for the future tree-breeding programmes and so on. It is needless to say that the local communities are quite aware of the traditional culture and beliefs related to biodiversity conservation and management, particularly phytodiversity and at the same time ensure perpetual propagation of the available natural resources.

A total of 60 plant species (Figure 17) belonging to 54 genera of 35 families (present study) are used for 28 material uses by the local inhabitants of the study area (Table 34). Their quantitative reflection is provided in Figure 16. Five predominantly employed families of plants used for this category belong to Fabaceae (8 spp), Salicaceae (4 spp), Euphorbiaceae (3 spp), Moraceae (3 spp) and Myrtaceae (3 spp)

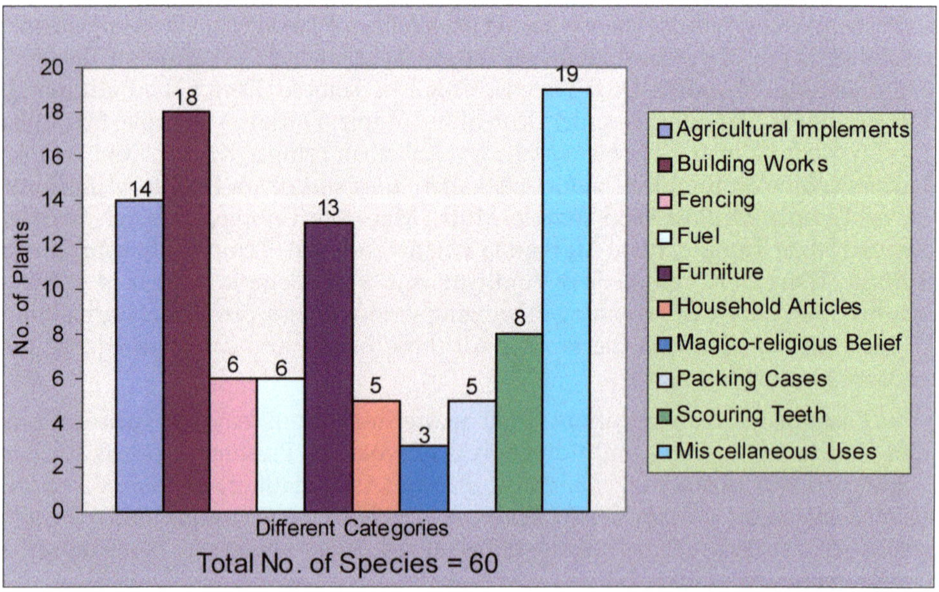

Figure 16: Quantitative Reflection of Plants Used Under Material Cultures

Table 33: Temples Having Courtyards, Orchards and Patches of Forests

Sl.No.	*Temple*	*Cultivation Land/Orchard/Extent of Forest*
1.	Lakshmi Narayan Temple, Chanor (Kangra)	Thick forests (spreading to an area of 9 hectare) attached to the temple protected on religious grounds.
2.	Kalinath Kaleshwar Mahadev Temple, Kaleshwar (Kangra)	Temple land includes cultivation land (about 1 hectare), an orchard of mango trees (1 hectare) and the temple-courtyard of about 3000 sq m.
3.	Mahakal Temple, Mahakal (Kangra)	Of the total 4 hectare of temple land, a large portion includes the forest area surrounding the temple-courtyard and the rest is cultivated. Forest trees include 'Jamun' (*Syzigium cumini*), 'Biyunsh' (*Salix tetrasperma*), 'Poplar' (*Populus nigra*), 'Khajur' (*Phoenix sylvestris*) and Bamboo (*Bambusa* sp.) plants, and given protection on religious grounds.
4.	Shiva Temple, Amrella (Kangra)	A small orchard attached to the temple contains trees of mango (*Mangifera indica*), guava (*Psidium guajava*), litchi (*Litchi chinensis*), 'loucat' (*Eriobotrya japonica*), lemon (*Citrus aurantifolia*), banana (*Musa paradisiaca*), and also trees of eucalyptus (*Eucalyptus* sp.), 'tuni' (*Cedrela toona*) and 'shisham' (*Dalbergia sissoo*).
5.	Shri Raghunath Teda Temple, Chakvan Kalidhar (Kangra)	Total land of 7 hectare comprise the forest land surrounding the temple and an orchard of mango (*Mangifera indica*), guava (*Psidium guajava*), banana (*Musa paradisiaca*), pear (*Pyrus communis*), lemon (*Citrus aurantifolia*), papaya (*Carica papaya*) and jack fruit (*Artocarpus heterophyllus*) trees and about 1500 sq m of temple-courtyard.
6.	Kamakhaya Devi Temple, Kamru (Kinnaur)	Thick sacred forest attached to the temple covers an area of about 4 hectare.
7.	Hadimba Devi Temple, Manali (Kullu)	Temple surrounded by forest land of about 5 hectare and harbours trees of *Cedrus deodara* only.
8.	Kamru Nag Temple, Kamrah (Mandi)	Land surrounding the temple courtyard has a thick forest harbouring 'Tosh' (*Abies pindrow*), 'Rye' (*Picea smithiana*), 'Kharshu' (*Quercus semecarpifolia*) and 'Rakhaal' (*Taxus baccata*). Herbs and shrubs in the forest are represented by characteristic temperate flora.
9.	Narsingh Temple, Nahan (Sirmour)	Thick sacred forest around the temple covers an area of about 10 hectare.
10.	Dron Shiv Temple, Ambota (Una)	Thick sacred forest surrounding the temple covers an area of 25 hectare.

(Table 36). Different categories of material culture with plant species used under each category depict predominant use of wood for various household chores, agricultural and building works (Table 34).

With regard to the diversity of timber yielding species and its uses by the rural communities living in the vicinity of temples in H.P., it is seen that wood is being used for agricultural implements (14 spp; *Albizia lebbeck, Albizia stipulata, Cassia fistula, Celtis australis, Eucalyptus tereticornis, Fraxinus xanthoxyloides, Mallotus philippinensis, Melia azedarach, Quercus incana, Robinia pseudoacacia, Salix fragilis, Salix tetrasperma, Syzygium cumini, Terminalia bellirica*), building works (18 spp; *Acer oblongum, Aesculus indica, Bombax ceiba, Cedrela toona, Cedrus deodara, Dalbergia sissoo, Eucalyptus tereticornis, Holoptelea integrifolia, Juglans regia, Mangifera indica, Melia azedarach, Morus alba, Pinus roxburghii, Pinus wallichiana, Populus nigra, Shorea robusta, Syzygium cumini, Tectona grandis*), furniture (13 spp; *Acer oblongum, Broussonetia papyrifera, Cedrela toona, Cedrus deodara, Dalbergia sissoo, Ehretia acuminata, Eucalyptus tereticornis, Melia azedarach, Morus alba, Morus serrata, Pinus roxburghii, Populus nigra, Tectona grandis*), household articles (5 spp; *Eucalyptus tereticornis, Fraxinus xanthoxyloides, Salix fragilis, Salix tetrasperma, Tamarindus indica*), packing cases (5 spp; *Broussonetia papyrifera, Eucalyptus tereticornis, Populus nigra, Salix babylonica, Syzygium cumini*) and plywood (1 sp; *Broussonetia papyrifera*) (Table 34). However, for fuelwood the requirements are largely met from *Debregeasia hypoleuca* (Urticaceae), *Eucalyptus tereticornis* (Myrtaceae), *Morus alba* (Moraceae), *Populus nigra* (Salicaceae), *Quercus incana* (Fagaceae), *Robinia pseudoacacia* (Fabaceae) (Table 34). Living close to nature, the locals never harvest fuelwood and timber in unsustainable manner thereby indicating their awareness and concern for management and conservation of the local plant diversity.

Besides, other categories of the material culture and the plants used are *viz.*, basketry (*Salix babylonica*), betel chewing (*Acacia catechu*), colouring hands, hair and beard (*Lawsonia alba*), decorative purpose (*Pinus roxburghii*), detergent (*Sapindus mukorossi*), facial cosmetic (*Citrus aurantifolia*), insect repellent (*Mentha longifolia*), magico-religious belief (*Capsicum annuum, Cedrus deodara, Hyoscyamus niger*), mats and brooms (*Phoenix sylvestris*), match-sticks (*Bombax ceiba*), meal plates (*Bauhinia vahlii, Pterospermum acerifolium*), preparation of bear (*Humulus lupulus*), rearing silkworms (*Morus alba*), reddening lips (*Juglans regia*), ropes (*Grewia oppositifolia*), rosaries (*Elaeocarpus ganitrus*), snake repellent (*Cyperus flabelliformis*), stuffing pillows and mattresses (*Bombax ceiba*), killing termites (*Althaea rosea*) and tooth brush (scouring teeth) (*Acacia catechu, Juglans regia, Mangifera indica, Melia azadirachta, Murraya koenigii, Psidium guajava, Ricinus communis, Zanthoxylum armatum*) (Table 34). Interestingly, the use of *Citrus aurantifolia* as facial cosmetics, *Mentha longifolia* as insect repellent, *Hyoscyamus niger* for magico-religious belief, *Pterospermum acerifolium* for meal plates, *Broussonetia papyrifera* for plywood manufacturing, *Cyperus flabelliformis* as snake repellent and *Althaea rosea* for killing termites are new uses to these plants and have not been recorded earlier. As a matter of fact the region is rich in plants of great economic utilities and the resultant information is of great significance for achieving the goal of economic benefits for the local communities by setting several cottage and rural industries in the region.

Table 34: Temple-Courtyard Plants for Material Culture

Use	*Species*	*Genera/Species*
Agricultural implements	*Albizia lebbeck, Albizia stipulata, Cassia fistula, Celtis australis, Eucalyptus tereticornis, Fraxinus xanthoxyloides, Mallotus philippinensis, Melia azedarach, Quercus incana, Robinia pseudoacacia, Salix fragilis, Salix tetrasperma, Syzygium cumini, Terminalia bellirica*	12/14
Basketry	*Salix babylonica*	1/1
Betel chewing	*Acacia catechu*	1/1
Building works	*Acer oblongum, Aesculus indica, Bombax ceiba, Cedrela toona, Cedrus deodara, Dalbergia sissoo, Eucalyptus tereticornis, Holoptelea integrifolia, Juglans regia, Mangifera indica, Melia azedarach Morus alba, Pinus roxburghii, Pinus wallichiana, Populus nigra, Shorea robusta, Syzygium cumini, Tectona grandis*	17/18
Colouring hands, hair and beard	*Lawsonia alba*	1/1
Decorative purpose	*Pinus roxburghii*	1/1
Detergent	*Sapindus mukorossi*	1/1
Facial cosmetic	*Citrus aurantifolia**	1/1
Fencing	*Acacia catechu, Bougainvillea spectabilis, Carissa opaca, Duranta repens, Euphorbia royleana, Zizyphus mauritiana*	6/6
Fuel	*Debregeasia hypoleuca, Eucalyptus tereticornis, Morus alba, Populus nigra, Quercus incana, Robinia pseudoacacia*	6/6
Furniture	*Acer oblongum, Broussonetia papyrifera, Cedrela toona, Cedrus deodara, Dalbergia sissoo, Ehretia acuminata, Eucalyptus tereticornis, Melia azedarach, Morus alba, Morus serrata, Pinus roxburghii, Populus nigra, Tectona grandis*	12/13
Household articles	*Eucalyptus tereticornis, Fraxinus xanthoxyloides, Salix fragilis, Salix tetrasperma, Tamarindus indica*	4/5
Insect repellent	*Mentha longifolia**	1/1
Magico-religious belief	*Capsicum annuum, Cedrus deodara, Hyoscyamus niger**	3/3
Mats and brooms	*Phoenix sylvestris*	1/1

Contd...

Table 34–*Contd...*

Use	*Species*	*Genera/Species*
Match-sticks	*Bombax ceiba*	1/1
Meal plates	*Bauhinia vahlii, Pterospermum acerifolium**	2/2
Packing cases	*Broussonetia papyrifera, Eucalyptus tereticornis, Populus nigra, Salix babylonica, Syzygium cumini*	5/5
Plywood	*Broussonetia papyrifera**	1/1
Prayer beads (rosaries)	*Elaeocarpus ganitrus*	1/1
Preparation of Bear	*Humulus lupulus*	1/1
Rearing silkworms	*Morus alba*	1/1
Reddening lips	*Juglans regia*	1/1
Ropes	*Grewia oppositifolia*	1/1
Snake repellent	*Cyperus flabelliformis**	1/1
Stuffing pillows and mattresses	*Bombax ceiba*	1/1
Killing Termites	*Althaea rosea**	1/1
Tooth brush (scouring teeth)	*Acacia catechu, Juglans regia, Mangifera indica, Melia azadirachta, Murraya koenigii, Psidium guajava, Ricinus communis, Zanthoxylum armatum*	8/8

*: New Uses.

From data based on classification of plants according to use (Table 35), it is seen that several multipurpose species like *Eucalyptus tereticornis* (Myrtaceae, 8 uses), *Morus alba* (Moraceae, 8 uses), *Populus nigra* (Salicaceae, 7 uses), *Acacia catechu* (Fabaceae, 6 uses), *Juglans regia* (Juglandaceae, 6 uses), *Pinus roxburghii* (Pinaceae, 6 uses), *Zizyphus mauritiana* (Rhamnaceae, 6 uses), *Aesculus indica* (Hippocastanaceae, 5 uses), *Bombax ceiba* (Bombacaceae, 5 uses), *Broussonetia papyrifera* (Moraceae, 5 uses), *Cedrus deodara* (Pinaceae, 5 uses), *Mangifera indica* (Anacardiaceae, 5 uses), *Melia azedarach* (Meliaceae,

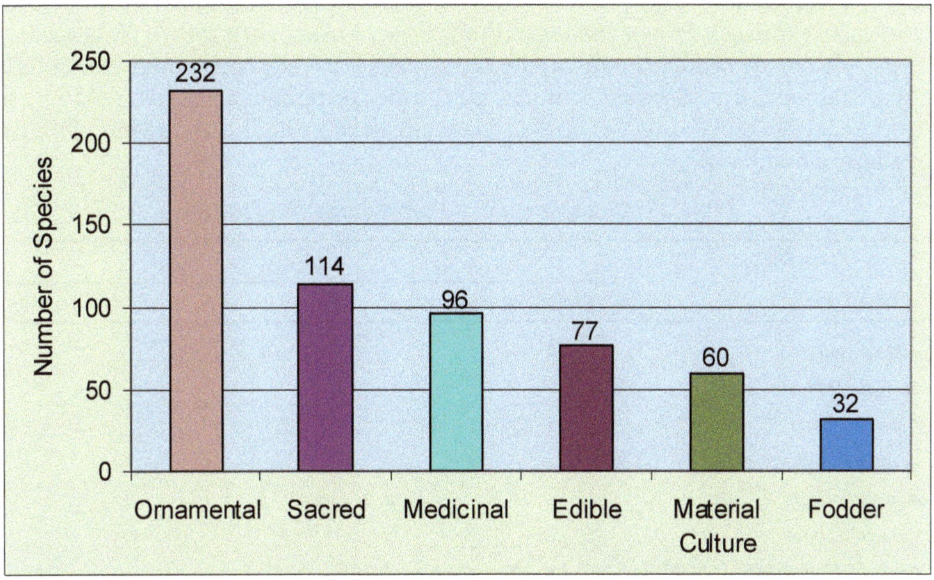

Figure 17: Quantitative Uses of the Temple Plants in H.P.

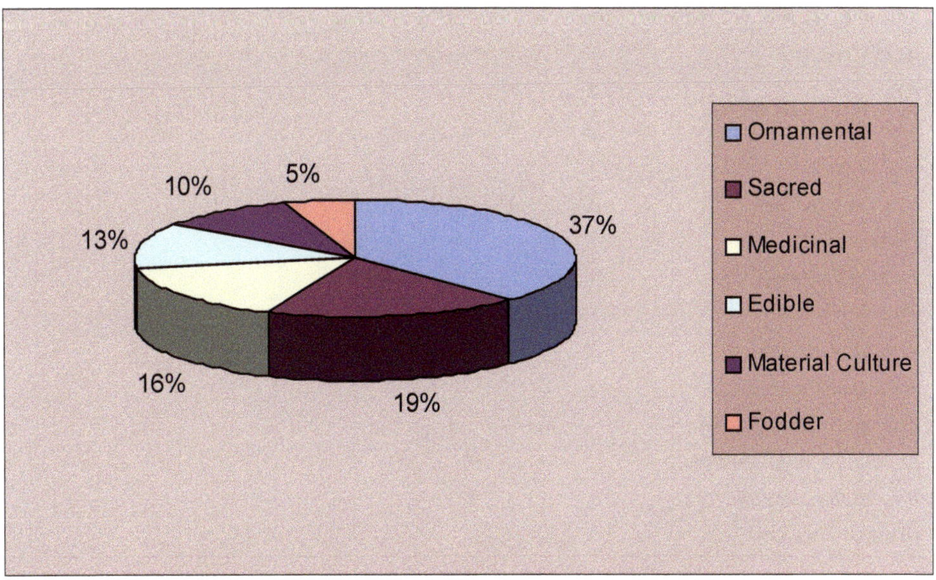

Figure 18: Per cent Uses of the Temple Plants in H.P.

5 uses), *Robinia pseudoacacia* (Fabaceae, 5 uses), *Salix babylonica* (Salicaceae, 5 uses), *Syzygium cumini* (Myrtaceae, 5 uses), *Tamarindus indica* (Fabaceae, 5 uses), *Bauhinia variegata* (Fabaceae, 4 uses), *Bougainvillea spectabilis* (Nyctaginaceae, 4 uses), *Cassia fistula* (Fabaceae, 4 uses), *Cedrela toona* (Meliaceae, 4 uses), *Celtis australis* (Ulmaceae, 4 uses), *Cynodon dactylon* (Poaceae, 4 uses), *Dalbergia sissoo* (Fabaceae, 4 uses), *Debregeasia hypoleuca* (Urticaceae, 4 uses), *Ehretia acuminata* (Boraginaceae, 4 uses), *Ficus palmata* (Moraceae, 4 uses), *Holoptelea integrifolia* (Ulmaceae, 4 uses), *Murraya koenigii* (Rutaceae, 4 uses), *Ocimum sanctum* (Lamiaceae, 4 uses), *Phoenix sylvestris* (Arecaceae, 4 uses), *Punica granatum* (Punicaceae, 4 uses), *Salix fragilis* (Salicaceae, 4 uses), *Sesamum orientale* (Pedaliaceae, 4 uses), etc. are well represented in the study area. Undoubtedly, the resulted information on the germplasm materials/wild genetic resources of useful plants varieties of the area can be profitably made use of for the welfare of mankind.

Table 35: Classification of Plants According to Use

Species	*Use*														
	1	*2*	*3*	*4*	*5*	*6*	*7*	*8*	*9*	*10*	*11*	*12*	*13*	*14*	*15*
Acacia catechu		X	X		X				X	X					X
Acacia dealbata	X														
Acacia nilotica	X		X		X										
Acalypha wilkesiana	X														
Acer oblongum	X					X		X							
Achras zapota	X			X											
Aegle marmelos		X	X	X											
Aesculus indica	X		X	X	X	X									
Albizia lebbeck					X		X								
Albizia stipulata			X		X		X								
Alnus nitida	X				X										
Aloe vera	X		X	X											
Alpinia nutans	X														
Alstonia scholaris	X		X												
Althaea rosea	X	X													X
Amaryllis reginae	X														
Antirrhinum majus	X														
Araucaria columnaris	X														
Aristolochia elegans	X														
Artemisia moorcroftiana		X													
Artocarpus heterophyllus				X											
Artocarpus lakoocha			X	X											
Asclepias curassavica	X														

Contd...

Table 35–*Contd...*

Species	*Use*														
	1	*2*	*3*	*4*	*5*	*6*	*7*	*8*	*9*	*10*	*11*	*12*	*13*	*14*	*15*
Asclepias physocarpa	X														
Asparagus adscendens			X												
Asparagus densiflorus	X														
Asparagus racemosus	X														
Bauhinia vahlii	X	X													X
Bauhinia variegata	X		X	X	X										
Beloperone guttata	X														
Berberis lycium		X	X	X											
Bergenia ciliata	X		X												
Betula utilis	X														
Bischofia javanica		X	X												
Boerhaavia diffusa			X	X											
Bombax ceiba	X		X	X		X									X
Bougainvillea glabra	X														
Bougainvillea spectabilis	X	X	X							X					
Brassica oleracea var. *capitata*				X											
Broussonetia papyrifera				X	X			X					X		X
Bryophyllum calycinum	X		X												
Buxus sempervirens	X														
Calendula officinalis	X	X													
Callistemon lanceolatus	X														
Callistephus chinensis	X														
Calotropis procera		X	X												
Cannabis sativa		X	X												
Canna indica	X	X													
Capsicum annuum				X										X	
Carica papaya	X		X	X											
Carissa carandas	X			X											
Carissa opaca				X	X					X					
Carissa spinarum		X	X	X											
Caryota urens	X														
Cassia fistula	X		X	X			X								
Cassia glauca	X														
Casuarina equisetifolia	X														
Catharanthus roseus	X	X	X												

Contd...

Table 35–*Contd...*

Species	*Use*														
	1	*2*	*3*	*4*	*5*	*6*	*7*	*8*	*9*	*10*	*11*	*12*	*13*	*14*	*15*
Cedrela toona	X		X			X		X							
Cedrus deodara	X	X				X		X						X	
Celosia argentea var. *cristata*	X														
Celtis australis	X			X	X		X								
Centaurea cyanus	X														
Centella asiatica			X												
Cestrum diurnum	X														
Cestrum nocturnum	X	X													
Chamaerops humilis	X														
Chlorophytum capense	X														
Chrysanthemum indicum	X	X													
Chrysanthemum leucanthemum	X														
Cinnamomum camphora	X	X	X												
Citrus aurantifolia			X	X											X
Citrus jambhiri				X											
Citrus limon		X	X	X											
Citrus maxima				X											
Citrus reticulata			X	X											
Clematis grata	X														
Clerodendrum inerme	X														
Clerodendrum philippinum	X	X	X												
Clerodendrum splendens	X														
Clerodendrum thomsonae	X														
Coix lachryma-jobi	X	X			X										
Coleus blumei	X														
Colocasia esculenta	X			X											
Convolvulus arvensis	X				X										
Cordia dichotoma			X	X											
Cosmos bipinnatus	X	X													
Cosmos sulphureus	X	X													
Cupressus sempervirens	X	X													
Cupressus torulosa	X	X													
Cycas revoluta	X														
Cynodon dactylon		X	X		X										X
Cyperus flabelliformis	X														X

Contd...

Table 35–*Contd...*

Species	*Use*														
	1	*2*	*3*	*4*	*5*	*6*	*7*	*8*	*9*	*10*	*11*	*12*	*13*	*14*	*15*
Dahlia rosea	X	X													
Dalbergia sissoo	X		X			X		X							
Datura fastuosa	X	X	X												
Datura innoxia	X	X													
Datura stramonium	X	X													
Daucus carota				X											
Debregeasia hypoleuca	X			X	X						X				
Deeringia amaranthoides	X														
Delphinium ajacis	X														
Dendrophthoe falcata		X													
Dianthus barbatus	X														
Diospyros lotus	X														
Diospyros montana var. *cordifolia*	X														
Dodonaea viscosa	X				X										
Dolichos lablab				X											
Duranta repens	X		X							X					
Ehretia acuminata	X			X	X			X							
Elaeocarpus ganitrus	X	X													X
Emblica officinalis			X	X											
Erigeron alpinus	X														
Eriobotrya japonica	X			X											
Eucalyptus tereticornis	X		X			X	X	X			X	X	X		
Euphorbia milii	X														
Euphorbia pulcherrima	X														
Euphorbia royleana	X		X							X					
Ficus benghalensis		X	X												
Ficus carica	X			X											
Ficus elastica	X														
Ficus glomerata		X	X	X											
Ficus palmata		X	X	X	X										
Ficus religiosa		X	X												
Ficus roxburghii		X		X	X										
Ficus rumphii		X													
Ficus virens		X													

Contd...

Table 35–*Contd...*

Species	*Use*														
	1	*2*	*3*	*4*	*5*	*6*	*7*	*8*	*9*	*10*	*11*	*12*	*13*	*14*	*15*
Fraxinus xanthoxyloides	X						X					X			
Gaillardia pulchella	X														
Gomphrena globosa	X														
Grevillea robusta	X														
Grewia oppositifolia					X										X
Hamelia patens	X														
Hedychium coronarium	X		X												
Helianthus annuus	X	X													
Hemerocallis fulva	X	X													
Hibiscus mutabilis	X														
Hibiscus rosa-sinensis	X	X													
Hibiscus syriacus	X	X													
Holoptelea integrifolia	X	X			X	X									
Humulus lupulus	X														X
Hydrangea macrophyla	X	X													
Hymenocallis littoralis	X														
Hyoscyamus niger	X		X											X	
Hypericum cernuum	X														
Impatiens balsamina	X														
Ipomoea cairica	X														
Ipomoea purpurea	X		X												
Ipomoea quamoclit	X														
Iris germanica	X														
Ixia chinensis	X														
Jacaranda acutifolia	X														
Jasminum grandiflorum	X	X													
Jasminum humile	X	X	X												
Jasminum primulinum	X	X													
Jasminum sambac	X	X	X												
Jatropha panduraefolia	X														
Juglans regia		X	X	X		X			X						X
Juniperus chinensis	X														
Juniperus macropoda	X	X													
Justicia gendarussa	X														
Kochia trichophylla	X														

Contd...

Table 35–*Contd...*

Species	*Use*														
	1	*2*	*3*	*4*	*5*	*6*	*7*	*8*	*9*	*10*	*11*	*12*	*13*	*14*	*15*
Lagerstroemia indica	X		X												
Lannea coromandelica			X												
Lawsonia alba	X		X												X
Lespedeza sericea	X														
Leucaena leucocephala	X				X										
Ligustrum lucidum	X	X													
Litchi chinensis				X											
Litsea polyantha	X		X												
Litsea sebifera	X														
Lonicera japonica	X		X												
Lycopersicum esculentum			X	X											
Machilus gamblei	X	X													
Magnolia grandiflora	X	X													
Mallotus philippinensis			X				X								
Malva sylvestris	X														
Malva verticillata	X														
Malvaviscus conzatii	X														
Mangifera indica		X	X	X		X			X						
Melia azadirachta	X		X						X						
Melia azedarach	X		X			X	X	X							
Mentha longifolia			X												X
Mentha piperita	X		X	X											
Michelia champaca	X		X												
Mimosa pudica	X		X												
Mimusops elengi	X	X		X											
Mirabilis jalapa	X	X	X												
Momordica charantia			X												
Morus alba	X	X		X	X	X		X			X				X
Morus serrata		X		X				X							
Murraya koenigii	X		X	X					X						
Murraya paniculata	X														
Musa paradisiaca		X	X	X											
Narcissus tazetta	X														
Nerium odorum	X	X													
Nicandra physaloides	X	X													

Contd...

Table 35–*Contd...*

Species	*Use*														
	1	*2*	*3*	*4*	*5*	*6*	*7*	*8*	*9*	*10*	*11*	*12*	*13*	*14*	*15*
Nyctanthes arbor-tristis	X														
Ocimum basilicum	X	X													
Ocimum gratissimum			X												
Ocimum kilimandscharicum	X	X	X												
Ocimum sanctum	X	X	X	X											
Oenothera drummondii	X														
Olea glandulifera	X														
Papaver rhoeas var. *latifolia*	X														
Passiflora caerulea	X														
Pedilanthus tithymaloides	X														
Pelargonium hortorum	X														
Petunia hybrida	X														
Phaseolus vulgaris				X											
Phoenix sylvestris	X	X		X											X
Phytolacca acinosa	X			X											
Pinus gerardiana	X			X											
Pinus roxburghii	X		X	X		X		X							X
Pinus wallichiana	X	X				X									
Pistacia integerrima		X	X												
Pisum sativum				X											
Platanus orientalis	X														
Plumeria acutifolia	X														
Polianthes tuberosa	X														
Polyalthia longifolia	X														
Polygonum alpinum	X			X											
Pongamia pinnata	X														
Populus ciliata	X	X													
Populus nigra	X	X			X	X		X			X		X		
Prinsepia utilis		X		X											
Prunus amygdalus			X	X											
Prunus armeniaca				X											
Prunus domestica				X											
Prunus persica			X	X											
Prunus puddum		X		X	X										
Psidium guajava			X	X					X						

Contd...

Table 35–*Contd...*

Species	*Use*														
	1	*2*	*3*	*4*	*5*	*6*	*7*	*8*	*9*	*10*	*11*	*12*	*13*	*14*	*15*
Pterospermum acerifolium	X														X
Punica granatum	X	X	X	X											
Putranjiva roxburghii	X		X												
Pyrostegia venusta	X														
Pyrus communis		X		X											
Pyrus malus		X		X											
Pyrus pashia		X	X	X											
Quercus dilatata	X														
Quercus incana					X		X				X				
Quisqualis indica	X														
Ribes orientale	X			X											
Ricinus communis			X						X						
Robinia pseudoacacia	X	X			X		X				X				
Rosa alba	X	X													
Rosa banksiae	X	X													
Rosa foetida	X	X	X												
Rosa indica	X	X	X												
Rosa jacquemontii	X	X													
Rosa moschata	X	X													
Rosa multiflora	X	X													
Rosa webbiana	X	X		X											
Roylea cinerea	X		X												
Ruellia tuberosa	X														
Russelia equisetiformis	X														
Saccharum officinarum		X	X	X											
Salix alba	X	X													
Salix babylonica	X	X			X								X		X
Salix fragilis	X				X		X					X			
Salix oxycarpa	X														
Salix tetrasperma	X						X					X			
Salvia coccinia	X	X													
Salvia splendens	X														
Sambucus nigra	X														
Sansevieria trifasciata	X														
Santalum album	X		X												

Contd...

Table 35–*Contd...*

Species	*Use*														
	1	*2*	*3*	*4*	*5*	*6*	*7*	*8*	*9*	*10*	*11*	*12*	*13*	*14*	*15*
Sapindus mukorossi	X		X												X
Sapium insigne	X														
Sapium sebiferum	X		X		X										
Schefflera arboricola	X														
Sempervivum arboreum	X														
Sesamum orientale		X	X	X	X										
Shorea robusta	X					X									
Solanum tuberosum				X											
Syzygium cumini			X	X		X	X						X		
Syzygium jambos	X			X											
Tabernaemontana coronaria	X	X	X												
Tagetes erecta	X	X	X												
Tamarindus indica	X	X	X	X								X			
Tecoma radicans	X														
Tecoma stans	X	X													
Tecomaria capensis	X														
Tectona grandis	X					X		X							
Terminalia bellirica			X	X			X								
Terminalia chebula			X	X											
Thevetia neriifolia	X	X	X												
Thuja orientalis	X	X													
Thuja plicata	X														
Thunbergia erecta	X														
Tithonia tagetiflora	X														
Tropaeolum majus	X														
Ulmus villosa	X	X													
Viola canescens		X													
Viola tricolour	X														
Wedelia calendulacea	X														
Wisteria sinensis	X														
Withania somnifera			X												
Wrightia tomentosa	X														
Zanthoxylum armatum	X		X						X						
Zea mays				X	X										
Zephyranthes candida	X														

Contd...

Table 35–*Contd...*

Species	*Use*														
	1	*2*	*3*	*4*	*5*	*6*	*7*	*8*	*9*	*10*	*11*	*12*	*13*	*14*	*15*
Zephyranthes grandiflora	X														
Zinnia elegans	X	X													
Zizyphus mauritiana	X	X	X	X	X					X					

1: Ornamental; 2: Sacred; 3: Medicinal; 4: Edible; 5: Fodder; 6: Building Works; 7: Agricultural Implements; 8: Furniture; 9: Scouring Teeth; 10: Fencing; 11: Fuel; 12: Household Articles; 13: Packing Cases; 14: Magico-religious Belief; 15: Miscellaneous (Basketry; Betel chewing; Colouring hands, hair and beard; Decorative; Detergent; Cosmetics; Insect repellent; Lawn Grass; Mats and brooms; Match: sticks; Meal plates; Plywood; Rosaries; Preparation of Bear; Rearing silkworms; Reddening lips; Ropes; Snake repellent; Stuffing pillows and mattresses; Killing Termites).

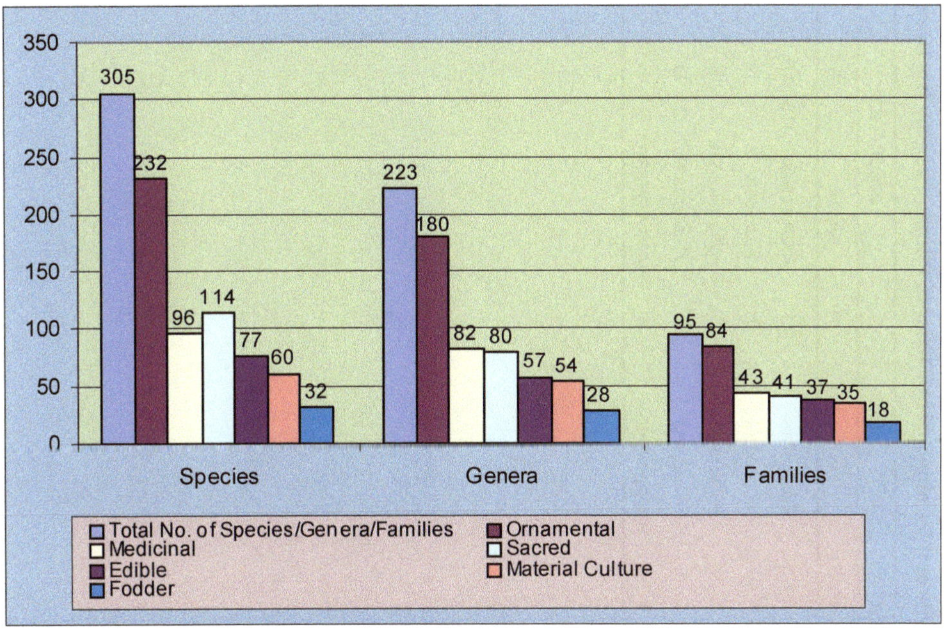

Figure 19: Total Number of Species, Genera and Families of the Temple Plants Used for Various Purposes by Rural Populace of H.P.

Tables 37 and 38 illustrate the list of exotic and native plants respectively in the temple-courtyards of Himachal Pradesh. Statistically, the ratio of exotic species has been estimated to 52.13 per cent while the representation for native species conformed to 47.87 per cent . In comparison, the exotics are more in occurrence than the natives in temple-yards of Himachal Pradesh. Habit-wise analysis of the 159 species of exotic/ aliens collected presently (Figure 20) indicates that the flora abounds in 21 annual herbs, 4 biennial herbs, 36 perennial herbs, 47 trees, 14 climbers and 37 shrubs. Among these, the herbaceous elements predominate the alien flora. Dicotyledonous

Table 36: Predominant Families of the Temple Plants Used by the Rural Populace of H. P.

Sacred			*Medicine*			*Edible*			*Ornamental*			*Fodder*			*Material Culture*		
Family	*A**	*B**	*Family*	*A**	*B**	*Family*	*A**	*B**	*Family*	*A**	*B**	*Family*	*A**	*B**	*Family*	*A**	*B**
Rosaceae	4	16	Fabaceae	7	8	Rosaceae	5	11	Asteraceae	13	15	Fabaceae	5	7	Fabaceae	7	8
Asteraceae	7	8	Euphorbiaceae	7	7	Moraceae	4	9	Fabaceae	11	14	Moraceae	3	4	Salicaceae	2	4
Moraceae	2	8	Lamiaceae	3	6	Rutaceae	3	7	Euphorbiaceae	6	9	Poaceae	3	3	Euphorbiaceae	3	3
Apocynaceae	6	6	Rutaceae	4	6	Fabaceae	6	6	Rosaceae	2	9	Salicaceae	2	3	Moraceae	2	3
Solanaceae	4	6	Apocynaceae	5	5	Apocynaceae	1	3	Apocynaceae	8	8	Ulmaceae	2	2	Myrtaceae	3	3

*A: No. of Genera; B: No. of Species.

species contribute maximum proportion (136 spp) to the exotic/alien plant species grouped under 112 genera and 53 families (Table 41), followed by Monocots (17 spp) distributed in 16 genera and 7 families (Amaryllidaceae, Arecaceae, Cannaceae, Cyperaceae, Iridaceae, Liliaceae, Poaceae; Table 42) and Gymnosperms (6 spp) represented by 5 genera and 3 families (Table 43), which include Araucariaceae (*Araucaria columnaris* (Forster) Hooker), Cupressaceae (*Cupressus sempervirens* Linn., *Juniperus chinensis* Linn., *Thuja orientalis* Linn., *Thuja plicata* D. Don) and Cycadaceae (*Cycas revoluta* Thunb.). Of the 63 alien families, Asteraceae predominate with 14 species, followed by Rosaceae (11), Fabaceae (9), Solanaceae (9), Euphorbiaceae (7), Amaryllidaceae (6), Malvaceae (6), Bignoniaceae (5), Acanthaceae (4), Convolvulaceae (4), Cupressaceae (4), Lamiaceae (4), Liliaceae (4), Myrtaceae (4), Salicaceae (4), Verbenaceae (4), Moraceae (3), Nyctaginaceae (3), Oleaceae (3), Poaceae (3), Apocynaceae (2), Asclepiadaceae (2), Crassulaceae (2), Sapindaceae (2) and Scrophulariaceae (2). Relatively, the top 20 families contribute 111 species with proportion of 63.52 per cent (Table 39). The remaining 38 families, *viz.* Amaranthaceae, Apiaceae, Araliaceae, Araucariaceae, Arecaceae, Aristolochiaceae, Brassicaceae, Buxaceae, Cannabinaceae, Cannaceae, Caprifoliaceae, Caricaceae, Caryophyllaceae, Chenopodiaceae, Cycadaceae, Cyperaceae, Ebenaceae, Geraniaceae, Hydrangeaceae, Iridaceae, Juglandaceae, Lauraceae, Lythraceae, Magnoliaceae, Onagraceae, Papaveraceae, Passifloraceae, Pedaliaceae, Platanaceae, Proteaceae, Punicaceae, Ranunculaceae, Rubiaceae, Rutaceae, Sambucaceae, Sapotaceae, Tropaeolaceae and Violaceae are represented by one species each (Table 39). Likewise, the predominant genera with highest number of aliens are *Prunus* and *Rosa* (4 each), *Clerodendrum, Hibiscus, Ipomoea, Salix* (3 each), *Asclepias, Bougainvillea, Cestrum, Chrysanthemum, Cosmos, Datura, Euphorbia, Pyrus, Salvia, Tecoma, Thuja, Zephyranthes* (2 each) (Table 40).

Table 37: Exotic Plants in the Temple-Courtyards of Himachal Pradesh

Sl.No.	*Botanical Name*	*Family*	*Origin*
1.	*Acacia dealbata* Link.	Fabaceae	Australia and Tasmania
2.	*Acalypha wilkesiana* M. Arg.	Euphorbiaceae	New Hebrides
3.	*Achras zapota* Linn.	Sapotaceae	South America
4.	*Aloe vera* (Linn.) Burm. f.	Liliaceae	Eastern and Southern Africa
5.	*Althaea rosea* (Linn.) Cav.	Malvaceae	China
6.	*Amaryllis reginae* Linn.	Amaryllidaceae	Brazil and tropical America
7.	*Antirrhinum majus* Linn.	Scrophulariaceae	Mediterranean region, from Morocco and Portugal North to Southern France, east to Turkey and Syria.
8.	*Araucaria columnaris* (Forster) Hooker	Araucariaceae	New Caledonia
9.	*Aristolochia elegans* Mast.	Aristolochiaceae	Brazil
10.	*Asclepias curassavica* Linn.	Asclepiadaceae	Tropical America
11.	*Asclepias physocarpa* (E. Mey.) Schlecht.	Asclepiadaceae	South-east Africa

Contd...

Table 37–*Contd...*

Sl.No.	*Botanical Name*	*Family*	*Origin*
12.	*Asparagus densiflorus* (Kunth) Jessop	Liliaceae	South Africa
13.	*Beloperone guttata* Brand.	Acanthaceae	Mexico
14.	*Bougainvillea glabra* Choisy	Nyctaginaceae	Brazil
15.	*Bougainvillea spectabilis* Willd.	Nyctaginaceae	Tropical and subtropical S. America
16.	*Brassica oleracea* var. *capitata* Linn.	Brassicaceae	Mediterranean region as well as southern England, Wales and northern France
17.	*Broussonetia papyrifera* L' Herit. *ex* Vent.	Moraceae	Cambodia; China; Japan; Korea; Laos; Malaysia; Myanmar; Taiwan; Thailand; Vietnam
18.	*Bryophyllum calycinum* Salisb.	Crassulaceae	Madagascar
19.	*Buxus sempervirens* Linn.	Buxaceae	AFRICA–Algeria, Libya, Morocco; ASIA-TEMPERATE–Iran, Turkey, Azerbaijan; EUROPE–United Kingdom, Austria, Belgium, Germany, Switzerland, Albania, Former Yugoslavia, Greece, Italy, France, Portugal, Spain
20.	*Calendula officinalis* Linn.	Asteraceae	Southern Europe and Australia
21.	*Callistemon lanceolatus* DC.	Myrtaceae	Australia
22.	*Callistephus chinensis* Nees	Asteraceae	China and Japan
23.	*Canna indica* Linn.	Cannaceae	Central and Southern America
24.	*Capsicum annuum* Linn.	Solanaceae	Tropical America
25.	*Carica papaya* Linn.	Caricaceae	Southern Mexico and Central America
26.	*Centaurea cyanus* Linn.	Asteraceae	ASIA-TEMPERATE–Iran, Iraq, Turkey; ASIA-TROPICAL–Pakistan; EUROPE–Albania, Bulgaria, Former Yugoslavia, Greece, Italy
27.	*Cestrum diurnum* Linn.	Solanaceae	Tropical America and the West Indies
28.	*Cestrum nocturnum* Linn.	Solanaceae	North and South America
29.	*Chamaerops humilis* Linn.	Arecaceae	Europe and N. Africa
30.	*Chlorophytum capense* Kuntze	Liliaceae	S. Africa
31.	*Chrysanthemum indicum* Linn.	Asteraceae	China and Japan
32.	*Chrysanthemum leucanthemum* Linn.	Asteraceae	Europe (including U.K.), North America
33.	*Cinnamomum camphora* (Linn.) Nees and Eberm.	Lauraceae	China and Japan
34.	*Citrus maxima* (Burm. f.) Merrill	Rutaceae	Malaysia and Polynesia
35.	*Clerodendrum philippinum* Schauer	Verbenaceae	Southern China
36.	*Clerodendrum splendens* G. Don.	Verbenaceae	Tropical Africa

Contd...

Table 37–*Contd...*

Sl.No.	*Botanical Name*	*Family*	*Origin*
37.	*Clerodendrum thomsonae* Balf.	Verbenaceae	W. Africa
38.	*Convolvulus arvensis* Linn.	Convolvulaceae	AFRICA–Algeria, Egypt, Libya, Morocco, Tunisia; ASIA-TEMPERATE–Afghanistan, Armenia, Azerbaijan, China, Cyprus, Egypt, Iran, Israel, Jordan, Kazakhstan, Kyrgyzstan, Lebanon, Mongolia, Syria, Tajikistan, Turkey, Turkmenistan, Uzbekistan; ASIA-TROPICAL–Nepal; Pakistan; EUROPE–Albania, Austria, Belarus, Belgium, Bulgaria, Czechoslovakia, Denmark, Estonia, Finland, Former Yugoslavia, France Germany, Greece, Hungary, Ireland, Italy, Latvia, Lithuania, Moldova, Netherlands, Norway, Poland, Romania, Russian Federation, Spain, Sweden, Switzerland, Ukraine, United Kingdom
39.	*Cosmos bipinnatus* Cav.	Astreaceae	Mexico
40.	*Cosmos sulphureus* Cav.	Astreaceae	Mexico
41.	*Cupressus sempervirens* Linn.	Cupressaceae	AFRICA–Libya; ASIA–TEMPERATE–Cyprus; Iran; Israel; Jordan; Lebanon; Syria; Turkey; EUROPE–Greece
42.	*Cycas revoluta* Thunb.	Cycadaceae	China and Japan
43.	*Cynodon dactylon* (Linn.) Pers.	Poaceae	Tropical Africa
44.	*Cyperus flabelliformis* Rottb.	Cyperaceae	Madagascar
45.	*Dahlia rosea* Cav.	Asteraceae	Mexico and C. America
46.	*Datura fastuosa* Linn.	Solanaceae	Central America
47.	*Datura inoxia* Mill.	Solanaceae	Central and South America
48.	*Daucus carota* Linn.	Apiaceae	AFRICA–Algeria, Morocco, Tunisia; ASIA-TEMPERATE–Afghanistan, Armenia; Azerbaijan; China; Georgia; Iran, Iraq, Israel, Jordan, Lebanon, Kazakhstan, Kyrgyzstan, Syria, Tajikistan, Turkmenistan, Uzbekistan; EUROPE
49.	*Delphinium ajacis* Linn.	Ranunculaceae	Europe and Australia
50.	*Dianthus barbatus* Linn.	Caryophyllaceae	China, Korea and Europe
51.	*Diospyros lotus* Linn.	Ebenaceae	Armenia; Azerbaijan; China; Georgia; Iran; Korea; Nepal; Pakistan; Tajikistan; Turkey; Turkmenistan; Uzbekistan
52.	*Dodonaea viscosa* (Linn.) Jacq.	Sapindaceae	Australia, North and South America
53.	*Dolichos lablab* Linn.	Fabaceae	China; Ethiopia; Ghana; Kenya; Madagascar; Peninsular Malaysia; South Africa; Sudan; Tanzania; Zambia; Zimbabwe

Contd...

Table 37–*Contd...*

Sl.No.	*Botanical Name*	*Family*	*Origin*
54.	*Duranta repens* Linn.	Verbenaceae	C. America
55.	*Erigeron alpinus* Linn.	Asteraceae	AFRICA–Ethiopia, Kenya; ASIA-TEMPERATE–Azerbaijan, Russian Federation; EUROPE–Albania; Austria; Belarus; Bulgaria; Czechoslovakia; Former Yugoslavia; France; Germany; Greece; Italy; Poland; Romania; Spain; Switzerland
56.	*Eriobotrya japonica* (Thunb.) Lindl.	Rosaceae	China and Japan
57.	*Eucalyptus tereticornis* Sm.	Myrtaceae	Australia
58.	*Euphorbia milii* Des Moul.	Euphorbiaceae	Madagascar
59.	*Euphorbia pulcherrima* Willd. *ex* Klotzsch.	Euphorbiaceae	C. America
60.	*Ficus carica* Linn.	Moraceae	AFRICA–Algeria, Morocco, Tunisia; ASIA-TEMPERATE–Afghanistan, Cyprus, Egypt, Iran, Iraq, Israel, Jordan, Lebanon, Syria, Turkey, Azerbaijan, Tajikistan, Turkmenistan; ASIA-TROPICAL–Pakistan; EUROPE–Greece, Italy, France, Spain
61.	*Gaillardia pulchella* Foug.	Asteraceae	Northern America
62.	*Gomphrena globosa* Linn.	Amaranthaceae	America
63.	*Grevillea robusta* A. Cunn. *ex* R. Br.	Proteaceae	Queensland and New South Wales
64.	*Hamelia patens* Jacq.	Rubiaceae	Tropical and subtropical America
65.	*Helianthus annuus* Linn.	Asteraceae	W. United States
66.	*Hibiscus mutabilis* Linn.	Malvaceae	China; Japan; Taiwan
67.	*Hibiscus rosa-sinensis* Linn.	Malvaceae	China
68.	*Hibiscus syriacus* Linn.	Malvaceae	Syria
69.	*Humulus lupulus* Linn.	Cannabinaceae	Europe
70.	*Hydrangea macrophylla* (Thunb.) DC.	Hydrangeaceae	Japan
71.	*Hymenocallis littoralis* (Jacq.) Salisb.	Amaryllidaceae	Southern America
72.	*Ipomoea cairica* (Linn.) Sweet	Convolvulaceae	Tropical Africa and Asia
73.	*Ipomoea purpurea* (Linn.) Roth.	Convolvulaceae	Tropical America.
74.	*Ipomoea quamoclit* Linn.	Convolvulaceae	Tropical America
75.	*Iris germanica* Linn.	Iridaceae	Southern Europe
76.	*Jacaranda acutifolia* Humb. and Bonpl.	Bignoniaceae	Western South America: Peru
77.	*Jasminum primulinum* Hemsl. *ex* Baker	Oleaceae	China

Contd...

Table 37–*Contd...*

Sl.No.	*Botanical Name*	*Family*	*Origin*
78.	*Jatropha panduraefolia* Andr.	Euphorbiaceae	W. Indies
79.	*Juglans regia* Linn.	Juglandaceae	ASIA-TEMPERATE–Afghanistan, Azerbaijan, Iran, Iraq, Turkey, Tajikistan, Turkmenistan; ASIA-TROPICAL–Pakistan; EUROPE–Bulgaria; Former Yugoslavia; Greece
80.	*Juniperus chinensis* Linn.	Cupressaceae	China, Mangolia, Manchuria, Korea and Japan
81.	*Justicia gendarussa* Linn. f.	Acanthaceae	China
82.	*Kochia trichophylla* Voss.	Chenopodiaceae	ASIA-TEMPERATE–Armenia, Azerbaijan, China, Cyprus, Iran, Japan, Kazakhstan, Korea, Kyrgyzstan, Russian Federation, Tajikistan, Turkey, Turkmenistan, Uzbekistan; ASIA-TROPICAL–Nepal, Pakistan; EUROPE–Belarus, Ukraine
83.	*Lagerstroemia indica* Linn.	Lythraceae	China
84.	*Leucaena leucocephala* (Lam.) de Wit	Fabaceae	Cyprus; Mexico; Peninsular Malaysia
85.	*Ligustrum lucidum* Aiton	Oleaceae	China
86.	*Litchi chinensis* Sonn.	Sapindaceae	China
87.	*Lonicera japonica* Wall.	Caprifoliaceae	Japan and Korea
88.	*Lycopersicum, esculentum* Mill.	Solanaceae	South America
89.	*Magnolia grandiflora* Linn.	Magnoliaceae	Southeastern United States (N. Carolina, Florida and Texas)
90.	*Malva verticillata* Linn	Malvaceae	China
91.	*Malvaviscus conzatii* Green.	Malvaceae	Mexico
92.	*Mentha piperita* Linn.	Lamiaceae	Europe
93.	*Mimosa pudica* Linn.	Fabaceae	Barbados; Bolivia; Brazil; Colombia; Cuba; Jawa; Mexico; Moluccas; Panama; Peninsular Malaysia; Peru; Sumatera; Surinam; Venezuela
94.	*Mirabilis jalapa* Linn.	Nyctaginaceae	Mexico
95.	*Morus alba* Linn.	Moraceae	China, Europe and North America
96.	*Narcissus tazetta* Linn.	Amaryllidaceae	AFRICA–Algeria, China, Egypt, Libya, Morocco; ASIA-TEMPERATE–Cyprus, Iraq, Israel, Jordan, Lebanon, Syria, Turkey Russian Federation (Ciscaucasia, Western Siberia); EUROPE–Russian Federation (European part), Albania, Former Yugoslavia, Greece, Italy, France, Portugal, Spain

Contd...

Table 37–*Contd...*

Sl.No.	Botanical Name	Family	Origin
97.	*Nicandra physaloides* (Linn.) Gaertn.	Solanaceae	Southern America
98.	*Ocimum kilimandscharicum* Gurke	Lamiaceae	East Africa
99.	*Oenothera drummondii* Hook f.	Onagraceae	Northern America
100.	*Olea glandulifera* Wall.	Oleaceae	Australia and Malaysia
101.	*Papaver rhoeas* Linn. var. *latifolia* Prain	Papaveraceae	Great Britain and adjacent areas of Europe.
102.	*Passiflora caerulea* Linn.	Passifloraceae	Native of Central and Western S. America
103.	*Pedilanthus tithymaloides* (Linn.) Poit.	Euphorbiaceae	Mexico and S. America
104	*Pelargonium hortorum* Bailey	Geraniaceae	South Africa
105.	*Petunia hybrida* Vilm.	Solanaceae	South America
106.	*Phaseolus vulgaris* Linn.	Fabaceae	Mexico
107.	*Pisum sativum* Linn.	Fabaceae	Afghanistan; Algeria; Armenia; Azerbaijan; Bulgaria; Cyprus; Egypt; Ethiopia; France; Greece; Iran; Iraq; Israel; Italy; Kenya; Libya; Morocco; Portugal; Romania; Spain; Syria; Tunisia; Ukraine; Yugoslavia
108.	*Platanus orientalis* Linn.	Platanaceae	Albania; Bulgaria; Former Yugoslavia; Greece; Italy
109.	*Plumeria acutifolia* Poiret	Apocynaceae	Northern and Southern America
110.	*Polianthes tuberosa* Linn.	Amaryllidaceae	Central America
111.	*Populus nigra* Linn.	Salicaceae	AFRICA–Algeria, Morocco, Tunisia; ASIA-TEMPERATE–Armenia, Azerbaijan, China, Georgia, Iran, Kazakhstan, Russian Federation (Ciscaucasia, Dagestan, Eastern Siberia, Western Siberia), Tajikistan, Turkey; EUROPE–Albania, Austria, Belarus, Belgium, Bulgaria, Czechoslovakia, Former Yugoslavia, France, Germany, Greece, Hungary, Italy, Netherlands, Poland, Romania, Russian Federation (European part), Spain, Switzerland, Ukraine
112.	*Prunus amygdalus* Batsch.	Rosaceae	South Europe and Algeria
113.	*Prunus armeniaca* Linn.	Rosaceae	North China, California, Washington, France, Turkey and Australia
114.	*Prunus domestica* Linn.	Rosaceae	Eurasia
115.	*Prunus persica* (Linn.) Batsch	Rosaceae	USA, Australia and Temperate Asia
116.	*Psidium guajava* Linn.	Myrtaceae	Tropical America
117.	*Punica granatum* Linn.	Punicaceae	Persia, Afganistan and Baluchistan

Contd...

Table 37–*Contd...*

Sl.No.	*Botanical Name*	*Family*	*Origin*
118.	*Pyrostegia venusta* (Ker.-Gawl.) Miers	Bignoniaceae	Brazil
119.	*Pyrus communis* Linn.	Rosaceae	Eurasia
120.	*Pyrus malus* Linn.	Rosaceae	Europe, West Asia
121.	*Ricinus communis* Linn.	Euphorbiaceae	North Africa
122.	*Robinia pseudoacacia* Linn.	Fabaceae	Canada; United States
123.	*Rosa alba* Linn.	Rosaceae	China and Japan
124.	*Rosa banksiae* R. Br.	Rosaceae	China
125.	*Rosa indica* Linn.	Rosaceae	China
126.	*Rosa multiflora* Thunb.	Rosaceae	Japan and China
127.	*Ruellia tuberosa* Linn.	Acanthaceae	Southern America
128.	*Russelia equisetiformis* Cham. and Schl.	Scrophulariaceae	Mexico
129.	*Saccharum officinarum* Linn.	Poaceae	South Pacific (probably New Guinea)
130.	*Salix alba* Linn.	Salicaceae	Western Asia and Europe
131.	*Salix babylonica* Linn.	Salicaceae	China
132.	*Salix fragilis* Linn.	Salicaceae	ASIA-TEMPERATE–Georgia, Russian Federation (Ciscaucasia), Turkey; EUROPE–Albania, Austria, Belarus, Belgium, Bulgaria, Czechoslovakia, Estonia, Former Yugoslavia, France, Germany, Greece, Hungary, Italy, Latvia, Lithuania, Moldova, Netherlands, Poland, Romania, Russian Federation (European part), Spain, Switzorland, Ukraine
133.	*Salvia coccinia* Juss.	Lamiaceae	C. and S. America
134.	*Salvia splendens* Sellow *ex* Roemer and J.A. Schultes	Lamiaceae	Brazil
135.	*Sambucus nigra* Linn.	Sambucaceae	Europe
136.	*Sansevieria trifasciata* Prain	Liliaceae	Africa
137.	*Sapium sebiferum* (Linn.) Roxb.	Euphorbiaceae	China
138.	*Schefflera arboricola* (Hayata) Merr.	Araliaceae	China and Taiwan
139.	*Sempervivum arboreum* Linn.	Crassulaceae	Portugal, Morocco and eastwards to Crete
140.	*Sesamum orientale* Linn.	Pedaliaceae	Africa
141.	*Solanum tuberosum* Linn.	Solanaceae	Chile, Bolivia and Peru
142.	*Syzygium jambos* (Linn.) Alston	Myrtaceae	Malaysia
143.	*Tagetes erecta* Linn.	Asteraceae	Mexico

Contd...

Table 37–*Contd...*

Sl.No.	*Botanical Name*	*Family*	*Origin*
144.	*Tamarindus indica* Linn.	Fabaceae	Ethiopia; Ghana; Guinea; Jawa; Kenya; Madagascar; Nigeria; Peninsular Malaysia; Sierra Leone; Somalia; Sudan; Sumatera; Tanzania; Uganda; Zaire; Zambia; Zimbabwe
145.	*Tecoma radicans* Juss.	Bignoniaceae	Southeastern United States
146.	*Tecomaria capensis* (Thunberg) Spach.	Bignoniaceae	S. Africa
147.	*Tecoma stans* (Linn.) Juss. *ex* H.B.K.	Bignoniaceae	Tropical America
148.	*Thevetia neriifolia* Juss. *ex* Steud.	Apocynaceae	Mexico and Central America
149.	*Thuja orientalis* Linn.	Cupressaceae	China and Korea
150.	*Thuja plicata* D. Don.	Cupressaceae	North America
151.	*Thunbergia erecta* (Benth.) T. Anders.	Acanthaceae	Tropical to S. Africa
152.	*Tithonia tagetiflora* Desf.	Asteraceae	Mexico
153.	*Tropaeolum majus* Linn.	Tropaeolaceae	Mexico and South America
154.	*Viola tricolor* Linn.	Violaceae	ASIA-TEMPERATE–Iran, Russian Federation (Western Siberia), Turkey; EUROPE–Albania, Austria, Belarus, Belgium, Bulgaria, Czechoslovakia, Denmark, Estonia, Finland, Former Yugoslavia, France, Germany, Greece, Hungary, Iceland, Ireland, Italy, Latvia, Lithuania, Moldova, Netherlands, Norway, Poland, Romania, Russian Federation (European part), Spain, Sweden, Switzerland, United Kingdom, Ukraine
155.	*Wisteria sinensis* (Sims) Sweet	Fabaceae	China
156.	*Zea mays* Linn.	Poaceae	Tropical America
157.	*Zephyranthes candida* (Lindl.) Herb.	Amaryllidaceae	Southern America
158.	*Zephyranthes grandiflora* Lindl.	Amaryllidaceae	Central America and Mexico
159.	*Zinnia elegans* Jacq.	Asteraceae	Mexico

With regard to the nativity status of the exotic aliens of the temple-courtyards of Himachal Pradesh (Figure 21; Table 44), present study revealed that North and South American (Brazil, Chile, Mexico, Venezuela, West Indies, United States, etc.) elements contribute the greatest percentage of species (41.51 per cent), followed by Asia (Afganistan, Bangladesh, Bhutan, China, Japan, Myanmar, Nepal, Pakistan, Sri Lanka, Thailand, etc.) 38.99 per cent, Europe (Austria, Belgium, Bulgaria, France, Germany, Greece, Italy, Romania, Spain, Switzerland, Ukraine, etc.) 23.27 per cent, Common Origin (Asia, America, Africa, Europe and Australia) 23.27 per cent, African

Table 38: Indigenous Plants in the Temple-Courtyards of Himachal Pradesh

Sl.No.	*Botanical Name*	*Family*	*Origin*
1.	*Acacia catechu* (Linn. f.) Willd.	Fabaceae	Bangladesh; Bhutan; China; India; Myanmar; Nepal; Pakistan; Sri Lanka; Thailand
2.	*Acacia nilotica* (Linn.) Willd. *ex* Delile subsp. *indica* (Benth.) Brenan	Fabaceae	Algeria; Bangladesh; Egypt; Ethiopia; India; Iran; Iraq; Israel; Kenya; Libya; Nepal; Pakistan; Sudan; Syria; Tanzania; Yemen; Zambia; Zimbabwe
3.	*Acer oblongum* Wall.	Aceraceae	Bhutan; India–Assam, Himachal Pradesh, Jammu and Kashmir, Sikkim, Uttar Pradesh; Nepal; Pakistan
4.	*Aegle marmelos* (Linn.) Correa	Rutaceae	India
5.	*Aesculus indica* Colebr.	Hippocastanaceae	India
6.	*Albizia lebbeck* (Linn.) Benth.	Fabaceae	Bangladesh; Bhutan; India; Jawa; Myanmar; Nepal; Pakistan; Peninsular Malaysia; Sri Lanka
7.	*Albizia stipulata* (Roxb.) Boiv.	Fabaceae	Bangladesh; Bhutan; Burma; Cambodia; China; India; Jawa; Myanmar; Nepal; Pakistan; New Guinea; Philippines; Sri Lanka; Thailand; Vietnam
8.	*Alnus nitida* (Spach) Endl.	Betulaceae	Native to India–Himachal Pradesh, Jammu and Kashmir, Uttar Pradesh; Nepal
9.	*Alpinia nutans* Rosc.	Zingiberaceae	North-eastern India
10.	*Alstonia scholaris* (Linn.) R. Br.	Apocynaceae	Australia; China; India; Indonesia; Malaysia; Myanmar; Nepal; Papua New Guinea; Philippines; Sri Lanka; Thailand; Vietnam
11.	*Artemisia moorcroftiana* Wall. *ex* DC.	Asteraceae	China; Bhutan; India–H.P., J. and K., U.P.; Nepal; Pakistan
12.	*Artocarpus heterophyllus* Lam.	Moraceae	Malaysia and the Western Ghats of India
13.	*Artocarpus lakoocha* Roxb.	Moraceae	India
14.	*Asparagus adscendens* Roxb.	Liliaceae	India
15.	*Asparagus racemosus* Willd.	Liliaceae	India
16.	*Bauhinia vahlii* Wight and Arn.	Fabaceae	Bhutan; India; Nepal; Pakistan
17.	*Bauhinia variegata* Linn.	Fabaceae	Burma; China; Hong Kong; India; Myanmar; Nepal; Pakistan; Thailand; Vietnam
18.	*Berberis lycium* Royle	Berberidaceae	India–Himachal Pradesh, Jammu and Kashmir, Madhya Pradesh, Tamil Nadu, Uttar Pradesh; Nepal; Pakistan

Contd...

Table 38–*Contd...*

Sl.No.	*Botanical Name*	*Family*	*Origin*
19.	*Bergenia ciliata* (Haw.) Sternb. forma *ligulata* Yeo	Saxifragaceae	Afghanistan; Bhutan; India–Assam; Nepal; Pakistan
20.	*Betula utilis* D. Don	Betulaceae	Afghanistan; China; Bhutan; India–Himachal Pradesh, Jammu and Kashmir, Sikkim, Uttar Pradesh; Nepal; Pakistan
21.	*Bischofia javanica* Blume	Euphorbiaceae	China; Japan; Taiwan; Bhutan; India–Andaman and Nicobar, Andhra Pradesh, Assam, Himachal Pradesh, Karnataka, Kerala, Orissa, Sikkim, Tamil Nadu, Uttar Pradesh, West Bengal; Nepal; Cambodia; Myanmar; Thailand; Indonesia; Malaysia; Philippines; Australia–Queensland; Fiji; New Caledonia
22.	*Boerhaavia diffusa* Linn.	Nyctaginaceae	AFRICA–Botswana, Cape Verde, Egypt, Ghana, Kenya, Liberia, Malawi, Mozambique, Namibia, Nigeria, Rwanda, Senegal, Sierra Leone, Somalia, South Africa Tanzania, Togo, Uganda, Yemen, Zambia, Zimbabwe; ASIA-TEMPERATE–Oman, Saudi Arabia, United Arab Emirates, Yemen, China, Japan, Taiwan; ASIA-TROPICAL–India, Nepal, Cambodia, Laos, Myanmar, Thailand, Vietnam, Indonesia, Malaysia, Philippines; AMERICA–United States (Florida, Georgia, South Carolina), Mexico, Fiji, New Caledonia, Belize, Guatemala, Panama, Anguilla, Cuba, Jamaica, French Guiana, Guyana, Suriname, Venezuela, Bolivia, Ecuador, Peru, Argentina, Chile, Paraguay, Uruguay
23.	*Bombax ceiba* Linn.	Bombacaceae	Australia; Bhutan; Cambodia; China; India; Indonesia; Laos; Malaysia; Myanmar; Nepal; Papua New Guinea; Philippines; Sri Lanka; Taiwan; Thailand; Vietnam
24.	*Calotropis procera* R. Br.	Asclepiadaceae	AFRICA–Cape Verde, Algeria, Egypt, Libya, Morocco, Eritrea, Ethiopia, Somalia, Sudan, Kenya, Tanzania, Uganda, Cameroon, Equatorial Guinea, Gambia, Ghana, Guinea-Bissau, Mauritania, Nigeria, Senegal, Sierra Leone; ASIA-TEMPERATE–Oman, Saudi Arabia, Yemen, Afghanistan, Egypt, Iran, Israel, Jordan; ASIA-TROPICAL–India, Nepal, Pakistan, Myanmar, Thailand, Vietnam
25.	*Cannabis sativa* Linn.	Cannabinaceae	Central Asia and Himalaya
26.	*Carissa carandas* Linn.	Apocynaceae	India
27.	*Carissa opaca* Stapf. *ex* Haines	Apocynaceae	India; Myanmar; Pakistan; Sri Lanka
28.	*Carissa spinarum* Linn.	Apocynaceae	China; India; Myanmar; Sri Lanka; Thailand

Contd...

Table 38–*Contd...*

Sl.No.	*Botanical Name*	*Family*	*Origin*
29.	*Caryota urens* Linn.	Arecaceae	India; Sri Lanka; Myanmar; Malaysia
30.	*Cassia fistula* Linn.	Fabaceae	Bangladesh; India; Jawa; Kalimantan; Moluccas; Myanmar; Nepal; Peninsular Malaysia; Sri Lanka; Sumatera
31.	*Cassia glauca* Lam.	Fabaceae	Bangladesh; Burma; India; Jawa; Laos; Myanmar; Peninsular Malaysia; Sri Lanka; Thailand; Vietnam
32.	*Casuarina equisetifolia* Linn.	Casuarinaceae	Australia; India; Indonesia; Malaysia; Myanmar; Pacific islands; Philippines; Thailand; Vietnam
33.	*Catharanthus roseus* (Linn.) G. Don	Apocynaceae	Madagascar and India
34.	*Cedrela toona* Roxb. *ex* Rottl.	Meliaceae	Afghanistan; Australia; Bangladesh; Bhutan; Cambodia; China; India; Indonesia; Laos; Malaysia; Myanmar; Nepal; Pakistan; Papua New Guinea; Philippines; Sri Lanka; Thailand
35.	*Cedrus deodara* (Roxb.) Loudon	Pinaceae	Afghanistan; China; India–Himachal Pradesh, Jammu and Kashmir, Uttar Pradesh; Nepal; Pakistan
36.	*Celosia argentea* var. *cristata* Haines	Amaranthaceae	Tropical Asia
37.	*Celtis australis* Linn.	Ulmaceae	North India and Temperate Himalaya
38.	*Centella asiatica* (Linn.) Urban	Apiaceae	India
39.	*Citrus aurantifolia* (Christin) Sw.	Rutaceae	India
40.	*Citrus jambhiri* Lush.	Rutaceae	India
41.	*Citrus limon* (Linn.) Burm. f.	Rutaceae	India, Australia, Mexico, America, Portugal, Italy, Sicily, France and Spain
42.	*Citrus reticulata* Blanco	Rutaceae	South-eastern Asia
43.	*Clematis grata* Wall.	Ranunculaceae	Afghanistan; India–Himachal Pradesh, Jammu and Kashmir, Uttar Pradesh; Nepal; Pakistan

Contd...

Table 38–*Contd...*

Sl.No.	*Botanical Name*	*Family*	*Origin*
44.	*Clerodendrum inerme* (Linn.) Gaertn.	Verbenaceae	Australia; Bangladesh; China; India–Andaman and Nicobar; Andhra Pradesh, Gujarat, Karnataka, Kerala, Madhya Pradesh, Maharashtra, Orissa, Rajasthan, Tamil Nadu, West Bengal; Indonesia; Japan; Malaysia; Myanmar; New Caledonia; Papua New Guinea; Philippines; Sri Lanka; Taiwan; Thailand
45.	*Coix lachryma-jobi* Linn.	Poaceae	China; India; Malaysia; Myanmar Nepal; Pakistan; Philippines; Sri Lanka
46.	*Coleus blumei* Benth.	Lamiaceae	Australia; Cambodia; China; India; Indonesia; Laos; Malaysia; Myanmar; Papua New Guinea; Pacific Islands Philippines; Taiwan; Thailand; Vietnam
47.	*Colocasia esculenta* (Linn.) Schott.	Araceae	India, Southeast Asia
48.	*Cordia dichotoma* Forst.	Cordiaceae	Australia; Cambodia; China; India; Indonesia; Japan; Laos; Malaysia; Myanmar; New Caledonia; Pakistan; Papua New Guinea; Sri Lanka; Taiwan; Thailand; Vietnam
49.	*Cupressus torulosa* D. Don	Cupressaceae	China; India–Himachal Pradesh, Uttar Pradesh; Nepal
50.	*Dalbergia sissoo* Roxb.	Fabaceae	Afghanistan; Bangladesh; Bhutan; India; Iran; Iraq; Myanmar; Nepal; Oman; Pakistan
51.	*Datura stramonium* Linn.	Solanaceae	Caspian region of Europe, North America and Indian subcontinent (The origin of *D. stramonium* is disputed (Curtain, 1947). The Sanskrit *dhattura* and the Hindustani *dhatur* formed the basis of the general name, the origin of Jimson weed could be Asiatic. Some sources report a probable Central American origin, due to *Datura's* habitation of most temperate and subtropical parts of the world.)
52.	*Debregeasia hypoleuca* Wedd.	Urticaceae	W. Asia to the W. Himalayas
53.	*Deeringia amaranthoides* (Lamk.) Merr.	Amaranthaceae	Australia; China; Bhutan; India; Indonesia; Laos; Malaysia; Myanmar; Nepal; Taiwan; Thailand; Vietnam
54.	*Dendrophthoe falcata* (Linn. f.) Ettingsh.	Loranthaceae	Bhutan; India; Nepal; Sri Lanka
55.	*Diospyros montana* Roxb. var. *cordifolia* (Roxb.) Hiern.	Ebenaceae	Australia; Cambodia; India; Indonesia; Laos; Malaysia; Myanmar; Nepal; Philippines; Sri Lanka; Taiwan; Thailand; Vietnam
56.	*Ehretia acuminata* R. Br.	Boraginaceae	Australia, Bhutan; China; India; Indonesia; Taiwan; Vietnam
57.	*Elaeocarpus ganitrus* Roxb.	Elaeocarpaceae	Australia; Bangladesh; Bhutan; Cambodia; China; India–Arunachal Pradesh, Assam, Bihar, Maharashtra, Manipur, Nagaland, Sikkim, West Bengal; Indonesia; Malaysia; Myanmar; Nepal; Papua New Guinea; Thailand

Contd...

Table 38–*Contd...*

Sl.No.	*Botanical Name*	*Family*	*Origin*
58.	*Emblica officinalis* Gaerth.	Euphorbiaceae	Cambodia; China; Taiwan; Bangladesh; Bhutan; India–Andhra Pradesh, Assam, Bihar, Gujarat, Haryana, Himachal Pradesh, Jammu and Kashmir, Karnataka, Kerala, Madhya Pradesh, Maharashtra, Manipur, Orissa, Punjab, Rajasthan, Tamil Nadu, Tripura, Uttar Pradesh, West Bengal; Indonesia; Laos; Malaysia; Myanmar; Nepal; Pakistan; Sri Lanka; Thailand; Vietnam
59.	*Euphorbia royleana* Boiss.	Euphorbiaceae	Bhutan; India–Assam, Himachal Pradesh, Meghalaya, Nagaland, Punjab, Rajasthan, Tripura, Uttar Pradesh; Myanmar; Nepal; Pakistan
60.	*Ficus benghalensis* Linn.	Moraceae	India; Pakistan
61.	*Ficus elastica* Roxb.	Moraceae	Bhutan; India; Indonesia; Malaysia; Myanmar; Nepal
62.	*Ficus glomerata* Roxb.	Moraceae	Australia; China; India; Indonesia; Malaysia; Myanmar; Nepal; Pakistan; Papua New Guinea; Sri Lanka; Thailand; Vietnam
63.	*Ficus palmata* Forsk.	Moraceae	Afghanistan; Egypt; Ethiopia; India; Iran; Israel; Nepal; Oman; Pakistan; Saudi Arabia; Somalia; Sudan
64.	*Ficus religiosa* Linn.	Moraceae	Bangladesh; India; Myanmar; Nepal; Pakistan; Thailand; Vietnam
65.	*Ficus roxburghii* Wall. *ex* Miq.	Moraceae	China; Bhutan; India; Nepal; Malaysia; Myanmar; Pakistan; Thailand; Vietnam
66.	*Ficus rumphii* Blume	Moraceae	Cocos Isl., Nicobar, Andaman, India (north and central), Burma, Indochina, Thailand, Malaya, Java
67.	*Ficus virens* Ait.	Moraceae	Australia; Bhutan; Cambodia; China; India; Indonesia; Japan; Laos; Malaysia; Myanmar; Nepal; Pacific Islands; Taiwan; Sri Lanka; Thailand; Vietnam; Papua New Guinea; Philippines
68.	*Fraxinus xanthoxyloides* (Wall. *ex* G. Don) DC.	Oleaceae	Afghanistan; China; India–Himachal Pradesh, Jammu and Kashmir, Uttar Pradesh; Pakistan
69.	*Grewia oppositifolia* Roxb.	Tiliaceae	Bhutan; India–Himachal Pradesh, Jammu and Kashmir, Punjab, Sikkim, Uttar Pradesh; Nepal; Pakistan
70.	*Hedychium coronarium* Koenig.	Zingiberaceae	China; India; Myanmar; Nepal; Taiwan
71.	*Hemerocallis fulva* (Linn.) Linn.	Liliaceae	Armenia; Azerbaijan; Iran; Georgia; China; India

Contd...

Table 38–*Contd...*

Sl.No.	*Botanical Name*	*Family*	*Origin*
72.	*Holoptelea integrifolia* (Roxb.) Planch.	Ulmaceae	Cambodia; India; Laos; Myanmar; Nepal; Sri Lanka;Vietnam
73.	*Hyoscyamus niger* Linn.	Solanaceae	AFRICA–Algeria, Morocco, Tunisia; ASIA-TEMPERATE–Afghanistan, Armenia, Azerbaijan, China, Georgia, Iran, Iraq, Kazakhstan, Kyrgyzstan, Lebanon, Russian Federation, Tajikistan, Turkey; ASIA-TROPICAL–India, Nepal, Pakistan; EUROPE–Albania, Austria, Belarus, Belgium, Bulgaria, Czechoslovakia, Denmark, Estonia, Former Yugoslavia, France, Germany, Greece, Hungary, Ireland, Italy, Latvia, Netherlands, Poland, Portugal, Romania, Spain, Sweden, Switzerland, Ukraine, United Kingdom
74.	*Hypericum cernuum* Roxb.	Hypericaceae	India–Himachal Pradesh, Jammu and Kashmir, Uttar Pradesh; Nepal; Pakistan
75.	*Impatiens balsamina* Linn.	Balsaminaceae	India; Myanmar
76.	*Ixia chinensis* Linn.	Iridaceae	Bhutan; China; India; Japan; Korea; Myanmar; Nepal; Philippines;Vietnam
77.	*Jasminum grandiflorum* Linn.	Oleaceae	Northern India, Pakistan and the North-Western Himalayas
78.	*Jasminum humile* Linn.	Oleaceae	Afghanistan; Bhutan; China; India; Nepal; Pakistan; Tajikistan
79.	*Jasminum sambac* (Linn.) Ait.	Oleaceae	India
80.	*Juniperus macropoda* Boiss.	Cupressaceae	Afghanistan; Armenia; Azerbaijan; Georgia; India–Himachal Pradesh, Jammu and Kashmir; Iran; Iraq; Kyrgyzstan; Oman; Pakistan; Russian Federation; Tajikistan; Turkmenistan; Uzbekistan
81.	*Lannea coromandelica* (Houtt.) Merrill	Anacardiaceae	Bangladesh; Bhutan; Cambodia; China; India–Andaman and Nicobar, Arunachal Pradesh, Assam, Bihar, Gujarat, Himachal Pradesh, Jammu and Kashmir, Karnataka, Kerala, Madhya Pradesh, Maharashtra, Meghalaya, Orissa, Rajasthan, Sikkim, Tamil Nadu, Tripura, Uttar Pradesh, West Bengal; Laos; Myanmar; Nepal; Pakistan; Sri Lanka; Thailand; Vietnam
82.	*Lawsonia alba* Lam.	Lythraceae	Middle East, North Africa and the Indian sub-continent
83.	*Lespedeza sericea* (Thunb.) Miq.	Fabaceae	Australia; Bhutan; China; Hong Kong; India; Indonesia; Japan; Korea; Myanmar; Nepal; Pakistan; Papua New Guinea; Philippines; Taiwan
84.	*Litsea polyantha* Juss.	Lauraceae	Bhutan; Cambodia; China; India; Indonesia; Laos; Myanmar; Nepal; Pakistan; Philippines; Thailand; Vietnam

Contd...

Table 38–*Contd...*

Sl.No.	*Botanical Name*	*Family*	*Origin*
85.	*Litsea sebifera* Pers.	Lauraceae	India and South China to Malaysia, Australia and the western Pacific islands.
86.	*Machilus gamblei* King *ex* Hook. f.	Lauraceae	India–Sikkim, Uttar Pradesh; Nepal
87.	*Mallotus philippinensis* (Lam.) Muell.-Arg.	Euphorbiaceae	Australia; Bangladesh; Bhutan; China; India; Japan; Malaysia; Myanmar; Nepal; Philippines; Sri Lanka; Taiwan; Vietnam
88.	*Malva sylvestris* Linn.	Malvaceae	Europe and Asia
89.	*Mangifera indica* Linn.	Anacardiaceae	South-East Asia, Indomalayan region
90.	*Melia azadirachta* Linn.	Meliaceae	India and Sri Lanka
91.	*Melia azedarach* Linn.	Meliaceae	Australia; Bhutan; China; India; Indonesia Japan; Laos; Nepal; Papua New Guinea; Pacific Islands; Philippines; Sri Lanka; Thailand; Vietnam
92.	*Mentha longifolia* (Linn.) Huds.	Lamiaceae	AFRICA–Algeria, Ethiopia, Kenya, Lesotho, Morocco, Namibia, South Africa, Swaziland, Tunisia, Zimbabwe; ASIA-TEMPERATE–Afghanistan, Cyprus, Egypt, Iran, Iraq, Israel, Lebanon, Syria, Turkey, Armenia, Azerbaijan, Georgia, Russian Federation (Ciscaucasia, Dagestan, Western Siberia); ASIA-TROPICAL–India, Nepal, Pakistan; EUROPE–Sweden, Austria, Belgium, Czechoslovakia, Germany, Hungary, Poland, Switzerland, Belarus, Estonia, Latvia, Lithuania, Moldova, Russian Federation (European part), Ukraine, Albania, Bulgaria, Former Yugoslavia, Greece, Italy, Romania, France, Portugal, Spain
93.	*Michelia champaca* Linn.	Magnoliaceae	E. Himalayas, Assam and Western Ghats
94.	*Mimusops elengi* Linn.	Sapotaceae	Peninsular India
95.	*Momordica charantia* Linn.	Cucurbitaceae	AFRICA–Angola, Cameroon, Gabon, Ghana, Kenya, Liberia, Madagascar, Nigeria, Rwanda, Sierra Leone, Tanzania, Uganda, Zaire, Zambia, Zimbabwe; ASIA-TEMPERATE–China; ASIA–TROPICAL–Cambodia, India, Indonesia, Malaysia, Nepal, Pakistan, Sri Lanka, Vietnam; AUSTRALIA–Queensland
96.	*Morus serrata* Roxb.	Moraceae	China; India–Himachal Pradesh, Jammu and Kashmir, Punjab, Uttar Pradesh; Nepal; Pakistan
97.	*Murraya koenigii* (Linn.) Spreng.	Rutaceae	India and the sub-tropical forests in much of southern Asia

Contd...

Table 38–*Contd...*

Sl.No.	*Botanical Name*	*Family*	*Origin*
98.	*Murraya paniculata* (Linn.) Jack.	Rutaceae	China, southeast Asia and Malay Penisula
99.	*Musa paradisiaca* Linn.	Musaceae	Tropical Asia
100.	*Nerium odorum* Soland.	Apocynaceae	AFRICA–Algeria, Ethiopia, Libya, Morocco, Tunisia; ASIA-TEMPERATE–Afghanistan, China, Cyprus, Iran, Iraq, Israel, Jordan, Lebanon, Oman, United Arab Emirates, Syria, Turkey; ASIA-TROPICAL–India–Jammu and Kashmir, Punjab, Nepal, Pakistan; EUROPE–Albania, Former Yugoslavia, Greece, Italy, Malta; France; Portugal, Spain
101.	*Nyctanthes arbor-tristis* Linn.	Oleaceae	India–Andhra Pradesh, Assam, Gujarat, Haryana, Karnataka, Kerala, Madhya Pradesh, Punjab, Rajasthan, Uttar Pradesh, West Bengal; Nepal; Pakistan; Thailand
102.	*Ocimum basilicum* Linn.	Lamiaceae	Tropical Asia
103.	*Ocimum gratissimum* Linn.	Lamiaceae	Africa, Angola, Ghana, Guinea, India, Kenya, Malawi, Malaysia, Namibia, Nepal, Rwanda, Somalia, South Madagascar, Sri Lanka, Tanzania, Thailand, Uganda
104	*Ocimum sanctum* Linn.	Lamiaceae	India and other tropical regions of Asia
105.	*Phoenix sylvestris* (Linn.) Roxb.	Arecaceae	India–Bihar, Punjab, Rajasthan, Tamil Nadu, Uttar Pradesh, West Bengal; Nepal
106.	*Phytolacca acinosa* Roxb.	Phytolaccaceae	Bhutan; China; India; Japan; Korea; Myanmar; Taiwan; Vietnam
107.	*Pinus gerardiana* Wall. *ex* D. Don	Pinaceae	Afghanistan; China; India–Jammu and Kashmir; Pakistan
108.	*Pinus roxburghii* Sarg.	Pinaceae	Afghanistan; China; Bhutan; India–Arunachal Pradesh, Himachal Pradesh, Jammu and Kashmir, Sikkim, Uttar Pradesh; Myanmar; Nepal
109.	*Pinus wallichiana* A.B. Jacks.	Pinaceae	Afghanistan; Bhutan; China; India; Nepal; Pakistan; Myanmar
110.	*Pistacia integerrima* J. L. Stewart *ex* Brand.	Anacardiaceae	Afghanistan; India–Arunachal Pradesh, Himachal Pradesh, Punjab, Uttar Pradesh; Nepal; Pakistan; Myanmar
111.	*Polyalthia longifolia* (Sonn.) Thwaites	Annonaceae	India and Sri Lanka
112.	*Polygonum alpinum* All.	Polygonaceae	ASIA-TEMPERATE–Afghanistan, Armenia, China, Egypt, Iran, Kazakhstan, Mongolia, Russian Federation, Turkey; ASIA-TROPICAL–India–Himachal Pradesh, Jammu and Kashmir, Pakistan; EUROPE–Albania, Austria, Bulgaria, Former Yugoslavia, France, Greece, Italy, Romania, Spain, Switzerland

Contd...

Table 38–*Contd...*

Sl.No.	*Botanical Name*	*Family*	*Origin*
113.	*Pongamia pinnata* (Linn.) Pierre	Fabaceae	Andaman Is, Australia, Bangladesh, China, Fiji, India, Japan, Myanmar, Nicobar Is, Malaysia, Philippines, Singapore, Sri Lanka, Sumatera, Taiwan, Thailand, Vietnam
114.	*Populus ciliata* Wall.	Salicaceae	China; Bhutan; India–Himachal Pradesh, Jammu and Kashmir, Sikkim, Uttar Pradesh; Nepal; Myanmar
115.	*Prinsepia utilis* Royle	Rosaceae	Temperate Himalaya
116.	*Prunus puddum* (Roxb. *ex* Ser.) Brandis	Rosaceae	Bhutan; China; India–Himachal Pradesh, Jammu and Kashmir, Meghalaya, Mizoram, Sikkim, Tamil Nadu, Uttar Pradesh, West Bengal; Laos; Myanmar; Nepal; Sri Lanka; Thailand; Vietnam
117.	*Pterospermum acerifolium* Willd.	Sterculiaceae	Bangladesh; Bhutan; India; Myanmar; Nepal; Thailand
118.	*Putranjiva roxburghii* Wall.	Euphorbiaceae	India–Andhra Pradesh, Bihar, Himachal Pradesh, Karnataka, Kerala, Madhya Pradesh, Maharashtra, Meghalaya, Orissa, Rajasthan, Tamil Nadu, Tripura, Uttar Pradesh, West Bengal; Myanmar; Sri Lanka
119.	*Pyrus pashia* Buch.-Ham. *ex* D. Don	Rosaceae	Afghanistan; Bhutan; China; India; Nepal; Pakistan; Laos; Myanmar; Thailand; Vietnam
120.	*Quercus dilatata* Lindl. *ex* A. DC.	Fagaceae	Afghanistan; India–Himachal Pradesh, Jammu and Kashmir, Uttar Pradesh; Nepal; Pakistan
121.	*Quercus incana* Roxb.	Fagaceae.	India–Assam, Himachal Pradesh, Jammu and Kashmir; Myanmar; Nepal; Pakistan; Thailand
122.	*Quisqualis indica* Linn.	Combretaceae	AFRICA–Angola, Benin, Ghana, Mali, Nigeria, Sierra Leone, Tanzania, Togo, Zaire; ASIA-TEMPERATE–China, Taiwan; ASIA-TROPICAL–Cambodia, India, Laos, Myanmar, Malaysia, Nepal, Papua New Guinea, Philippines Sri Lanka, Thailand, Vietnam
123.	*Ribes orientalis* Desf.	Saxifragaceae	India
124.	*Rosa foetida* Herrm.	Rosaceae	Afghanistan, India, Iran, Pakistan
125.	*Rosa jacquemontii* Crep. *ex* Hook. f.	Rosaceae	India
126.	*Rosa moschata* Herrm.	Rosaceae	India
127.	*Rosa webbiana* Wall. *ex* Royle	Rosaceae	Afghanistan; Mongolia; China; India; Nepal; Pakistan

Contd...

Table 38–*Contd...*

Sl.No.	*Botanical Name*	*Family*	*Origin*
128.	*Roylea cinerea* (D. Don) Baill.	Lamiaceae	India, Nepal
129.	*Salix oxycarpa* Anders.	Salicaceae	India
130.	*Salix tetrasperma* Roxb.	Salicaceae	India
131.	*Santalum album* Linn.	Santalaceae	India
132.	*Sapindus mukorossi* Gaertn.	Sapindaceae	Bhutan; Cambodia; China; India–Assam, Sikkim; Japan; Korea; Myanmar; Nepal; Taiwan; Thailand; Vietnam
133.	*Sapium insigne* (Royle) Benth.	Euphorbiaceae	India; China; Pen. Malaysia
134.	*Shorea robusta* Gaertn. f.	Dipterocarpaceae	India
135.	*Syzygium cumini* (Linn.) Skeels	Myrtaceae	Africa; Bhutan; China; India; Indonesia; Kenya; Malaysia; Nepal; Sri Lanka; Tanzania; Uganda
136.	*Tabernaemontana coronaria* (Jacq.) Willd.	Apocynaceae	Bangladesh; Bhutan; China; India; Nepal; Laos; Myanmar; Thailand
137.	*Tectona grandis* Linn. f.	Verbenaceae	India, Java, Sumatra and Indonesia
138.	*Terminalia bellirica* (Gaertn.) Roxb.	Combretaceae	India, Malaysia and the Philippines
139.	*Terminalia chebula* (Gaertn.) Retz.	Combretaceae	Tropical Asia
140.	*Ulmus villosa* Brandis	Ulmaceae	India–Himachal Pradesh, Jammu and Kashmir; Pakistan
141.	*Viola canescens* Wall.	Violaceae	Bhutan, China, India–Arunachal Pradesh, Himachal Pradesh, Jammu and Kashmir, Meghalaya, Sikkim, Tamil Nadu, Uttar Pradesh, West Bengal; Myanmar, Nepal, Pakistan
142.	*Wedelia calendulacea* Less.	Asteraceae	Cambodia; China; India; Japan; Laos; Philippines; Vietnam
143.	*Withania somnifera* (Linn.) Dunal	Solanaceae	AFRICA–Algeria, Angola, Cape Verde, Egypt, Ethiopia, Kenya, Liberia, Morocco, Nigeria, Somalia, South Africa, Sudan, Tanzania, Tunisia, Uganda, Zambia, Zimbabwe; ASIA TEMPERATE–Afghanistan, Arabia, Iran, Iraq, Israel, Jordan, Lebanon, Syria, Turkey; ASIA-TROPICAL–India, Pakistan, Sri Lanka; EUROPE–Greece, Italy, Spain
144.	*Wrightia tomentosa* (Roxb.) Roem and Schult.	Apocynaceae	Bihar, Assam, Punjab, Rajasthan, W. Peninsula

Contd...

Table 38–*Contd...*

Sl.No.	*Botanical Name*	*Family*	*Origin*
145.	*Zanthoxylum armatum* DC.	Rutaceae	Bangladesh; Bhutan; China, India–Andhra Pradesh, Himachal Pradesh, Jammu and Kashmir, Manipur, Meghalaya, Nagaland, Orissa, Uttar Pradesh; Indonesia; Japan; Korea; Laos; Myanmar; Nepal; Pakistan; Philippines; Taiwan; Thailand; Vietnam
146.	*Zizyphus mauritiana* Lamk.	Rhamnaceae	India, Myanmar

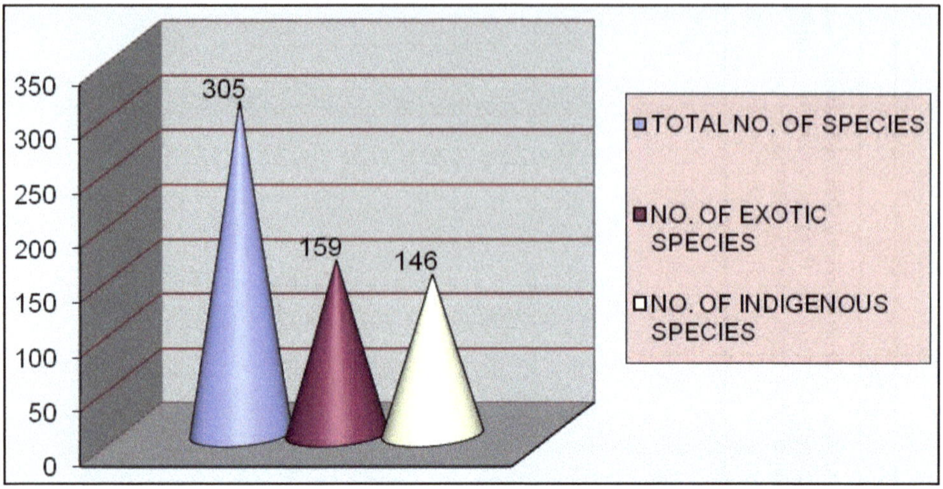

Figure 20: Total Number of Exotic and Indigenous Species in the Temple-Courtyards of H.P.

Continent (Algeria, Egypt, Ethiopia, Ghana, Kenya, Morocco, South Africa, Tunisia, Zimbabwe, etc.) 20.75 per cent, and Australia 6.92 per cent . In comparison the American elements contribute 55 per cent to the alien flora of the whole of India while as the European elements are represented by only 15 per cent (Nayar, 1977). Homogenisation of earth's biota has been reported to be due to increasing global trade and transport and consequently resulted in the translocation of species either deliberately or by accident (Drake *et al.*, 1989; Kowarik, 2003; Perrings *et al.*, 2005).

Table 39: Families of Exotic/Alien Species

Family	*Total No. of Species*	*Family*	*Total No. of Species*
Asteraceae	14	Rosaceae	11
Fabaceae	9	Solanaceae	9
Euphorbiaceae	7	Amaryllidaceae	6
Malvaceae	6	Bignoniaceae	5
Acanthaceae	4	Convolvulaceae	4
Cupressaceae	4	Lamiaceae	4
Liliaceae	4	Myrtaceae	4
Salicaceae	4	Verbenaceae	4
Moraceae	3	Nyctaginaceae	3
Oleaceae	3	Poaceae	3
Apocynaceae	2	Asclepiadaceae	2
Crassulaceae	2	Sapindaceae	2
Scrophulariaceae	2	Amaranthaceae	1

Contd...

Table 39–*Contd...*

Family	Total No. of Species	Family	Total No. of Species
Apiaceae	1	Araliaceae	1
Araucariaceae	1	Arecaceae	1
Aristolochiaceae	1	Brassicaceae	1
Buxaceae	1	Cannabinaceae	1
Cannaceae	1	Caprifoliaceae	1
Caricaceae	1	Caryophyllaceae	1
Chenopodiaceae	1	Cycadaceae	1
Cyperaceae	1	Ebenaceae	1
Geraniaceae	1	Hydrangeaceae	1
Iridaceae	1	Juglandaceae	1
Lauraceae	1	Lythraceae	1
Magnoliaceae	1	Onagraceae	1
Papaveraceae	1	Passifloraceae	1
Pedaliaceae	1	Platanaceae	1
Proteaceae	1	Punicaceae	1
Ranunculaceae	1	Rubiaceae	1
Rutaceae	1	Sambucaceae	1
Sapotaceae	1	Tropaeolaceae	1
Violaceae	1		

Table 40: Genera of Exotic/Alien Species

Genus	Total No. of Species	Genus	Total No. of Species
Prunus	4	*Rosa*	4
Clerodendrum	3	*Hibiscus*	3
Ipomoea	3	*Salix*	3
Asclepias	2	*Bougainvillea*	2
Cestrum	2	*Chrysanthemum*	2
Cosmos	2	*Datura*	2
Euphorbia	2	*Pyrus*	2
Salvia	2	*Tecoma*	2
Thuja	2	*Zephyranthes*	2
Acacia	1	*Acalypha*	1
Achras	1	*Aloe*	1
Althaea	1	*Amaryllis*	1
Antirrhinum	1	*Araucaria*	1
Aristolochia	1	*Asparagus*	1

Contd...

Table 40–*Contd...*

Genus	*Total No. of Species*	*Genus*	*Total No. of Species*
Beloperone	1	*Brassica*	1
Broussonetia	1	*Bryophyllum*	1
Buxus	1	*Calendula*	1
Callistemon	1	*Callistephus*	1
Canna	1	*Capsicum*	1
Carica	1	*Centaurea*	1
Chamaerops	1	*Chlorophytum*	1
Cinnamomum	1	*Citrus*	1
Convolvulus	1	*Cupressus*	1
Ocimum	1	*Oenothera*	1
Olea	1	*Papaver*	1
Passiflora	1	*Pedilanthus*	1
Pelargonium	1	*Petunia*	1
Phaseolus	1	*Pisum*	1
Platanus	1	*Plumeria*	1
Polianthes	1	*Populus*	1
Psidium	1	*Punica*	1
Pyrostegia	1	*Ricinus*	1
Robinia	1	*Ruellia*	1
Russelia	1	*Saccharum*	1
Cycas	1	*Cynodon*	1
Cyperus	1	*Dahlia*	1
Daucus	1	*Delphinium*	1
Dianthus	1	*Diospyros*	1
Dodonaea	1	*Dolichos*	1
Duranta	1	*Erigeron*	1
Eriobotrya	1	*Eucalyptus*	1
Ficus	1	*Gaillardia*	1
Gomphrena	1	*Grevillea*	1
Hamelia	1	*Helianthus*	1
Humulus	1	*Hydrangea*	1
Hymenocallis	1	*Iris*	1
Jacaranda	1	*Jasminum*	1
Jatropha	1	*Juglans*	1
Juniperus	1	*Justicia*	1
Kochia	1	*Lagerstroemia*	1

Contd...

Table 40–*Contd...*

Genus	Total No. of Species	Genus	Total No. of Species
Leucaena	1	*Ligustrum*	1
Litchi	1	*Lonicera*	1
Lycopersicum	1	*Magnolia*	1
Malva	1	*Malvaviscus*	1
Mentha	1	*Mimosa*	1
Mirabilis	1	*Morus*	1
Narcissus	1	*Nicandra*	1
Sambucus	1	*Sansevieria*	1
Sapium	1	*Schefflera*	1
Sempervivum	1	*Sesamum*	1
Solanum	1	*Syzygium*	1
Tagetes	1	*Tamarindus*	1
Tecomaria	1	*Thevetia*	1
Thunbergia	1	*Tithonia*	1
Tropaeolum	1	*Viola*	1
Wisteria	1	*Zea*	1
Zinnia	1		

Table 41: Exotic/Alien Dicot Species

Sl.No.	Family	Botanical Name
1.	Acanthaceae	*Beloperone guttata* Brand., *Justicia gendarussa* Linn. f., *Ruellia tuberosa* Linn., *Thunbergia erecta* T. Anders
2.	Amaranthaceae	*Gomphrena globosa* Linn.
3.	Apiaceae	*Daucus carota* Linn.
4.	Apocynaceae	*Plumeria acutifolia* Poiret, *Thevetia peruviana* (Pers.) K. Schum
5.	Araliaceae	*Schefflera arboricola* (Hayata) Merr.
6.	Aristolochiaceae	*Aristolochia elegans* Mast.
7.	Asclepiadaceae	*Asclepias curassavica* Linn., *Asclepias physocarpa* Schlecht.
8.	Asteraceae	*Calendula officinalis* Linn., *Callistephus chinensis* Nees, *Centaurea cyanus* Linn., *Chrysanthemum indicum* Linn., *Chrysanthemum leucanthemum* Linn., *Cosmos bipinnatus* Cav., *Cosmos sulphureus* Cav., *Dahlia rosea* Cav., *Erigeron alpinus* Linn., *Gaillardia pulchella* Foug., *Helianthus annuus* Linn., *Tagetes erecta* L., *Tithonia tagetiflora* Desf., *Zinnia elegans* Linn.
9.	Bignoniaceae	*Jacaranda acutifolia* Humb. and Bonpl., *Pyrostegia venusta* (Ker.-Gawl.) Miers, *Tecoma radicans* Juss., *Tecoma stans* (Linn.) H.B.K., *Tecomaria capensis* (Thunberg) Spach.
10.	Brassicaceae	*Brassica oleracea* var. *capitata* Linn.

Contd...

Table 41–*Contd...*

Sl.No.	*Family*	*Botanical Name*
11.	Buxaceae	*Buxus sempervirens* Linn.
12.	Cannabinaceae	*Humulus lupulus* Linn.
13.	Caprifoliaceae	*Lonicera japonica* Wall.
14.	Caricaceae	*Carica papaya* Linn.
15.	Caryophyllaceae	*Dianthus barbatus* Linn.
16.	Chenopodiaceae	*Kochia trichophylla* Voss.
17.	Convolvulaceae	*Convolvulus arvensis* L., *Ipomoea cairica* (L.) Sweet, *Ipomoea purpurea* (L.) Roth., *Ipomoea quamoclit* L.
18.	Crassulaceae	*Bryophyllum calycinum* Salisb., *Sempervivum arboreum* Linn.
19.	Ebenaceae	*Diospyros lotus* Linn.
20.	Euphorbiaceae	*Acalypha wilkesiana* M. Arg., *Euphorbia milii* Des Moul., *Euphorbia pulcherrima* Willd. *ex* Klotzsch., *Jatropha panduraefolia* Andr., *Pedilanthus tithymaloides* (L.) Poit., *Ricinus communis* L., *Sapium sebiferum* (L.) Roxb.
21.	Fabaceae	*Acacia dealbata* Link., *Dolichos lablab* Linn., *Leucaena leucocephala* (Lam.) de Wit, *Mimosa pudica* L., *Phaseolus vulgaris* Linn., *Pisum sativum* Linn., *Robinia pseudoacacia* L., *Tamarindus indica* Linn., *Wisteria sinensis* Sweet.
22.	Geraniaceae	*Pelargonium hortorum* Bailey
23.	Hydrangeaceae	*Hydrangea macrophylla* (Thunb.) DC.
24.	Juglandaceae	*Juglans regia* Linn.
25.	Lamiaceae	*Mentha piperita* Linn., *Ocimum kilimandscharicum* Gurke, *Salvia coccinia* Juss., *Salvia splendens* Sell.
26.	Lauraceae	*Cinnamomum camphora* (Linn.) Nees and Eberm.
27.	Lythraceae	*Lagerstroemia indica* Linn.
28.	Magnoliaceae	*Magnolia grandiflora* Linn.
29.	Malvaceae	*Althaea rosea* (L.) Cav., *Hibiscus mutabilis* Linn., *Hibiscus rosa-sinensis* Linn., *Hibiscus syriacus* Linn., *Malva verticillata* Linn., *Malvaviscus conzatii* Green.
30.	Moraceae	*Broussonetia papyrifera* L'Herit. ex Vent., *Ficus carica* L., *Morus alba* L.
31.	Myrtaceae	*Callistemon lanceolatus* DC., *Eucalyptus tereticornis* Sm., *Psidium guajava* L., *Syzygium jambos* (Linn.) Alston
32.	Nyctaginaceae	*Bougainvillea glabra* Choisy, *Bougainvillea spectabilis* Willd., *Mirabilis jalapa* L.
33.	Oleaceae	*Jasminum primulinum* Hemsl. *ex* Baker, *Ligustrum lucidum* Aiton, *Olea glandulifera* Wall.
34.	Onagraceae	*Oenothera drummondii* Hook f.
35.	Papaveraceae	*Papaver rhoeas* (L.) var. *latifolia* Prain
36.	Passifloraceae	*Passiflora caerulea* Linn.
37.	Pedaliaceae	*Sesamum orientale* L.

Contd...

Table 41–*Contd...*

Sl.No.	*Family*	*Botanical Name*
38.	Platanaceae	*Platanus orientalis* Linn.
39.	Proteaceae	*Grevillea robusta* A. Cunn. ex R. Br.
40.	Punicaceae	*Punica granatum* L.
41.	Ranunculaceae	*Delphinium ajacis* Linn.
42.	Rosaceae	*Eriobotrya japonica* (Thunb.) Lindl., *Prunus amygdalus* Baill., *Prunus armeniaca* Linn., *Prunus domestica* Linn., *Prunus persica* (L.) Batsch, *Pyrus communis* Linn., *Pyrus malus* L., *Rosa alba* Linn., *Rosa banksiae* Br., *Rosa indica* Linn., *Rosa multiflora* Thunb.
43.	Rubiaceae	*Hamelia patens* Jacq.
44.	Rutaceae	*Citrus maxima* (Burm.) Merrill
45.	Salicaceae	*Populus nigra* Linn., *Salix alba* L., *Salix babylonica* L., *Salix fragilis* Linn.
46.	Sambucaceae	*Sambucus nigra* Linn.
47.	Sapindaceae	*Dodonaea viscosa* (L.) Jacq., *Litchi chinensis* Sonn.
48.	Sapotaceae	*Achras zapota* Linn.
49.	Scrophulariaceae	*Antirrhinum majus* Linn., *Russelia equisetiformis* Cham. and Schl.
50.	Solanaceae	*Capsicum annuum* Linn., *Cestrum diurnum* Linn., *Cestrum nocturnum* Linn., *Datura fastuosa* Linn., *Datura inoxia* Mill., *Lycopersicum esculentum* Mill., *Nicandra physaloides* (Linn.) Gaertn., *Petunia hybrida* Vilm., *Solanum tuberosum* Linn.
51.	Tropaeolaceae	*Tropaeolum majus* Linn.
52.	Verbenaceae	*Clerodendrum philippinum* Schauer, *Clerodendrum splendens* G. Don., *Clerodendrum thomsonae* Balf., *Duranta repens* L.
53.	Violaceae	*Viola tricolor* Linn.

Table 42: Exotic/Alien Monocot Species

Sl.No.	*Family*	*Botanical Name*
1.	Amaryllidaceae	*Amaryllis reginae* Linn., *Hymenocallis littoralis* (Jacq.) Salisb., *Narcissus tazetta* L., *Polianthes tuberosa* Linn., *Zephyranthes candida*(Lindl.) Herb., *Zephyranthes grandiflora* Lindl.
2.	Arecaceae	*Chamaerops humilis* Linn.
3.	Cannaceae	*Canna indica* Linn.
4.	Cyperaceae	*Cyperus flabelliformis* Rottb.
5.	Iridaceae	*Iris germanica* Linn.
6.	Liliaceae	*Asparagus densiflorus* (Kunth) Jessop, *Aloe vera* L., *Chlorophytum capense* Kuntze, *Sansevieria trifasciata* Hort. *ex* Prain
7.	Poaceae	*Cynodon dactylon* (L.) Pers., *Saccharum officinarum* Linn., *Zea mays* Linn.

Table 43: Exotic/Alien Gymnosperm Species

Sl.No.	*Family*	*Botanical Name*
1.	Araucariaceae	*Araucaria columnaris* (Forster) Hooker
2.	Cupressaceae	*Cupressus sempervirens* Linn., *Juniperus chinensis* Linn., *Thuja orientalis* Linn., *Thuja plicata* D. Don
3.	Cycadaceae	*Cycas revoluta* Thunb.

Table 44: Status of Exotic/Alien Plants

Sl.No.	*Continent/Country*	*Species*	*Genera*	*Families*
1.	Asia (Afganistan, Bangladesh, Bhutan, China, Japan, Myanmar, Nepal, Pakistan, Sri Lanka, Thailand, etc.)	62	52	31
2.	North and South American Continent (Brazil, Chile, Mexico, Venezuela, West Indies, United States, etc.)	66	57	31
3.	African Continent (Algeria, Egypt, Ethiopia, Ghana, Kenya, Morocco, South Africa, Tunisia, Zimbabwe, etc.)	33	32	24
4.	Europe (Austria, Belgium, Bulgaria, France, Germany, Greece, Italy, Romania, Spain, Switzerland, Ukraine, etc.)	37	33	27
5.	Australian Continent	11	10	9
6.	Common Origin (Asia, America, Africa, Europe and Australia)	37	32	21

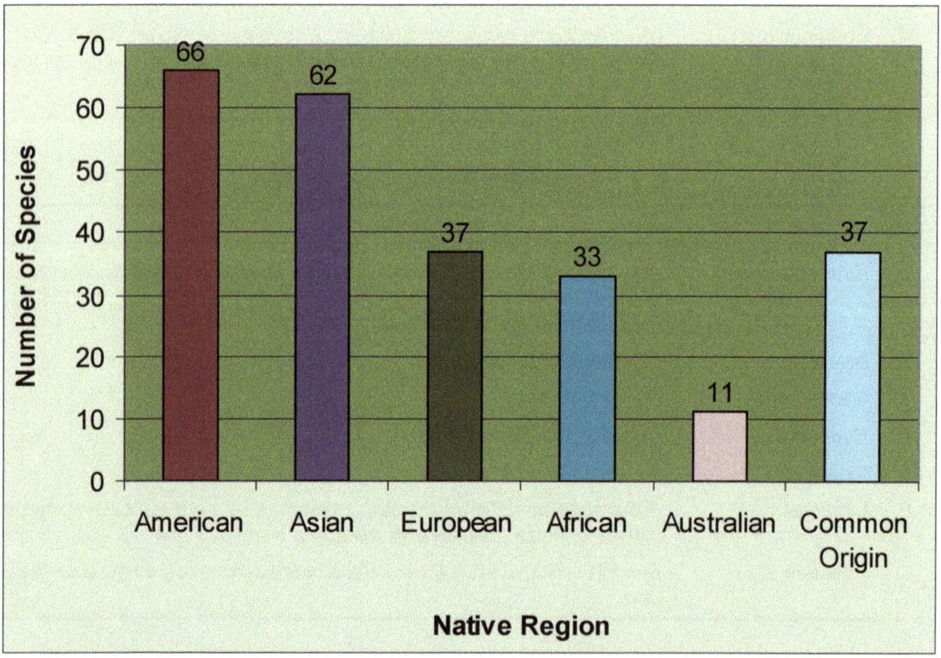

Figure 21: Native Status of Exotic/Alien Plants

The study also brought to light that a large number of plant species have been introduced intentionally in the temple-courtyards of Himachal Pradesh for various purposes, *e.g.*, 134 as ornamentals, 47 as sacred plants, 32 as food plants, 31 as medicinal plants, 18 for their material culture, 16 for horticultural purposes, 12 as fodder plants, and 11 as crop plants. Enumerative details of various species employed for the aforesaid categories have also been enlisted (*see* Tables 45–53). It is evident that majority of the exotic plant species occurring in the temple-courtyards of Himachal Pradesh were introduced intentionally for ornamental purposes.

Table 45: Ornamental Alien Plant Species

Family	*Plant Species*
Acanthaceae	*Beloperone guttata, Justicia gendarussa, Ruellia tuberosa, Thunbergia erecta*
Amaranthaceae	*Gomphrena globosa*
Amaryllidaceae	*Amaryllis reginae, Hymenocallis littoralis, Narcissus tazetta, Polianthes tuberosa, Zephyranthes candida, Zephyranthes grandiflora*
Apocynaceae	*Plumeria acutifolia, Thevetia neriifolia*
Araliaceae	*Schefflera arboricola*
Araucariaceae	*Araucaria columnaris*
Arecaceae	*Chamaerops humilis*
Aristolochiaceae	*Aristolochia elegans*
Asclepiadaceae	*Asclepias curassavica, Asclepias physocarpa*
Asteraceae	*Calendula officinalis, Callistephus chinensis, Centaurea cyanus, Chrysanthemum indicum, Chrysanthemum leucanthemum, Cosmos bipinnatus, Cosmos sulphureus, Dahlia rosea, Erigeron alpinus, Gaillardia pulchella, Helianthus annuus, Tagetes erecta, Tithonia tagetiflora, Zinnia elegans*
Bignoniaceae	*Jacaranda acutifolia, Pyrostegia venusta, Tecoma radicans, Tecoma stans, Tecomaria capensis*
Buxaceae	*Buxus sempervirens*
Cannabinaceae	*Humulus lupulus*
Cannaceae	*Canna indica*
Caprifoliaceae	*Lonicera japonica*
Caricaceae	*Carica papaya*
Caryophyllaceae	*Dianthus barbatus*
Chenopodiaceae	*Kochia trichophylla*
Convolvulaceae	*Convolvulus arvensis, Ipomoea cairica, Ipomoea purpurea, Ipomoea quamoclit*
Crassulaceae	*Bryophyllum calycinum, Sempervivum arboreum*
Cupressaceae	*Cupressus sempervirens, Juniperus chinensis, Thuja orientalis, Thuja plicata*
Cycadaceae	*Cycas revoluta*
Cyperaceae	*Cyperus flabelliformis*
Ebenaceae	*Diospyros lotus*

Contd...

Table 45–*Contd...*

Family	*Plant Species*
Euphorbiaceae	*Acalypha wilkesiana, Euphorbia milii, Euphorbia pulcherrima, Jatropha panduraefolia, Pedilanthus tithymaloides, Sapium sebiferum*
Fabaceae	*Acacia dealbata, Leucaena leucocephala, Mimosa pudica, Robinia pseudoacacia, Tamarindus indica, Wisteria sinensis*
Geraniaceae	*Pelargonium hortorum*
Hydrangeaceae	*Hydrangea macrophyla*
Iridaceae	*Iris germanica*
Lamiaceae	*Mentha piperita, Ocimum kilimandscharicum, Salvia coccinia, Salvia splendens*
Lauraceae	*Cinnamomum camphora*
Liliaceae	*Aloe vera, Asparagus densiflorus, Chlorophytum capense, Sansevieria trifasciata*
Lythraceae	*Lagerstroemia indica*
Magnoliaceae	*Magnolia grandiflora*
Malvaceae	*Althaea rosea, Hibiscus mutabilis, Hibiscus rosa-sinensis, Hibiscus syriacus, Malva verticillata, Malvaviscus conzatii*
Moraceae	*Ficus carica, Morus alba*
Myrtaceae	*Callistemon lanceolatus, Eucalyptus tereticornis, Syzygium jambos*
Nyctaginaceae	*Bougainvillea glabra, Bougainvillea spectabilis, Mirabilis jalapa*
Oleaceae	*Jasminum primulinum, Ligustrum lucidum, Olea glandulifera*
Onagraceae	*Oenothera drummondii*
Papaveraceae	*Papaver rhoeas* var. *latifolia*
Passifloraceae	*Passiflora caerulea*
Platanaceae	*Platanus orientalis*
Proteaceae	*Grevillea robusta*
Punicaceae	*Punica granatum*
Ranunculaceae	*Delphinium ajacis*
Rosaceae	*Eriobotrya japonica, Rosa alba, Rosa banksiae, Rosa indica, Rosa multiflora*
Rubiaceae	*Hamelia patens*
Salicaceae	*Populus nigra, Salix alba, Salix babylonica, Salix fragilis*
Sambucaceae	*Sambucus nigra*
Sapindaceae	*Dodonaea viscosa*
Sapotaceae	*Achras zapota*
Scrophulariaceae	*Antirrhinum majus, Russelia equisetiformis*
Solanaceae	*Cestrum diurnum, Cestrum nocturnum, Datura fastuosa, Datura innoxia, Nicandra physaloides, Petunia hybrida*
Tropaeolaceae	*Tropaeolum majus*
Verbenaceae	*Clerodendrum philippinum, Clerodendrum splendens, Clerodendrum thomsonae, Duranta repens*
Violaceae	*Viola tricolour*

Table 46: Families of Sacred Alien Plants

Family	*Species*
Apocynaceae	*Thevetia neriifolia*
Asteraceae	*Calendula officinalis, Chrysanthemum indicum, Cosmos bipinnatus, Cosmos sulphureus, Dahlia rosea, Helianthus annuus, Tagetes erecta, Zinnia elegans*
Bignoniaceae	*Tecoma stans*
Cannaceae	*Canna indica*
Cupressaceae	*Cupressus sempervirens, Thuja orientalis*
Fabaceae	*Robinia pseudoacacia, Tamarindus indica*
Hydrangeaceae	*Hydrangea macrophyla*
Juglandaceae	*Juglans regia*
Lamiaceae	*Ocimum kilimandscharicum, Salvia coccinia*
Lauraceae	*Cinnamomum camphora*
Magnoliaceae	*Magnolia grandiflora*
Malvaceae	*Althaea rosea, Hibiscus rosa–sinensis, Hibiscus syriacus*
Moraceae	*Morus alba*
Nyctaginaceae	*Bougainvillea spectabilis, Mirabilis jalapa*
Oleaceae	*Jasminum primulinum, Ligustrum lucidum*
Pedaliaceae	*Sesamum orientale*
Poaceae	*Cynodon dactylon, Saccharum officinarum*
Punicaceae	*Punica granatum*
Rosaceae	*Pyrus communis, Pyrus malus, Rosa alba, Rosa banksiae, Rosa indica, Rosa multiflora*
Salicaceae	*Populus nigra, Salix alba, Salix babylonica*
Solanaceae	*Cestrum nocturnum, Datura fastuosa, Datura innoxia, Nicandra physaloides*
Verbenaceae	*Clerodendrum philippinum*

Table 47: Edible Alien Plant Species

Species	*Edible Part/s*
Achras zapota	Fruits
Aloe vera	Leaves
Brassica oleracea var. *capitata*	Leaves, Stem
Broussonetia papyrifera	Fruits
Capsicum annuum	Fruits
Carica papaya	Fruits
Citrus maxima	Fruits
Daucus carota	Roots
Dolichos lablab	Pods

Contd...

Table 47–*Contd...*

Species	*Edible Part/s*
Eriobotrya japonica	Fruits
Ficus carica	Fruits
Juglans regia	Kernels
Litchi chinensis	Fruits
Lycopersicum esculentum	Fruits
Mentha piperita	Leaves
Morus alba	Fruits
Phaseolus vulgaris	Pods
Pisum sativum	Seeds
Prunus amygdalus	Seeds
Prunus armeniaca	Fruits
Prunus domestica	Fruits
Prunus persica	Fruits
Psidium guajava	Fruits
Punica granatum	Dried Seeds, Fruits
Pyrus communis	Fruits
Pyrus malus	Fruits
Saccharum officinarum	Cane juice
Sesamum orientale	Seed oil
Solanum tuberosum	Tubers
Syzygium jambos	Fruits
Tamarindus indica	Fruits
Zea mays	Grains

Table 48: Families of Alien Plant Species Used for Curing Various Ailments

Family	*Genera/Species*
Apocynaceae	*Thevetia neriifolia*
Asteraceae	*Tagetes erecta*
Caprifoliaceae	*Lonicera japonica*
Caricaceae	*Carica papaya*
Convolvulaceae	*Ipomoea purpurea*
Crassulaceae	*Bryophyllum calycinum*
Euphorbiaceae	*Ricinus communis, Sapium sebiferum*
Fabaceae	*Mimosa pudica, Tamarindus indica*
Juglandaceae	*Juglans regia*

Contd...

Table 48–*Contd...*

Family	*Genera/Species*
Lamiaceae	*Mentha piperita, Ocimum kilimandscharicum*
Lauraceae	*Cinnamomum camphora*
Liliaceae	*Aloe vera*
Lythraceae	*Lagerstroemia indica*
Myrtaceae	*Eucalyptus tereticornis, Psidiun guajava*
Nyctaginaceae	*Bougainvillea spectabilis, Mirabilis jalapa*
Pedaliaceae	*Sesamum orientale*
Poaceae	*Cynodon dactylon, Saccharum officinarum*
Punicaceae	*Punica granatum*
Rosaceae	*Prunus amygdalus, Prunus persica, Rosa indica*
Solanaceae	*Datura fastuosa, Lycopersicum esculentum*
Verbenaceae	*Clerodendrum philippinum, Duranta repens*

Table 49: Alien Temple-Plants of H.P. Used for Curing Various Ailments

Ailment	*Plant Species*
Allergic disorders	*Lagerstroemia indica*
Anaemia	*Lycopersicum esculentum, Punica granatum*
Anthelmintic	*Prunus persica*
Anticancerous	*Cynodon dactylon*
Bed wetting	*Aloe vera*
Body massage	*Prunus amygdalus, Sesamum orientale*
Body swelling	*Ricinus communis*
Boils	*Mirabilis jalapa*
Cholesterol level	*Juglans regia*
Cold	*Clerodendrum philippinum, Eucalyptus tereticornis*
Constipation	*Aloe vera, Ipomoea purpurea, Rosa indica, Ricinus communis*
Cough	*Aloe vera, Clerodendrum philippinum, Eucalyptus tereticornis*
Cuts	*Tagetes erecta*
Dental problems (toothache, gum-bleeding, etc.)	*Psidiun guajava, Thevetia neriifolia*
Diabetes	*Aloe vera, Bougainvillea spectabilis, Duranta repens, Psidium guajava*
Earache	*Tagetes erecta*
Face pimples	*Tagetes erecta*
Fever	*Lagerstroemia indica, Tamarindus indica*
Furuncles	*Bryophyllum calycinum*

Contd...

Table 49–*Contd...*

Ailment	*Plant Species*
Gastrointestinal disorders (dysentery, indigestion, diarrhoea)	*Lagerstroemia indica, Lonicera japonica, Mentha piperita, Psidium guajava, Punica granatum*
Hair-care (dandruff, hair–tonic)	*Cinnamomum camphora, Prunus amygdalus*
Heart ailments	*Juglans regia*
Insect bites	*Tagetes erecta*
Iron deficiency	*Psidium guajava*
Joint pains (rheumatism)	*Duranta repens, Juglans regia, Ricinus communis, Sapium sebiferum*
Kidney stone	*Bryophyllum calycinum, Mimosa pudica*
Liver problems (hepatic complaints, jaundice)	*Carica papaya, Lagerstroemia indica, Saccharum officinarum*
Memory	*Prunus amygdalus*
Mental disorders	*Datura fastuosa*
Pains and aches (body pain, backache, waist pain)	*Aloe vera*
Paralysis	*Sesamum orientale*
Piles	*Bryophyllum calycinum, Cinnamomum camphora,*
Prostate	*Aloe vera*
Skin ailments	*Carica papaya, Juglans regia*
Sprains	*Ocimum kilimandscharicum*
Stomach disorders (acidity, stomachache)	*Aloe vera, Mentha piperita, Ricinus communis, Rosa indica*
Urinary troubles	*Tagetes erecta*
Vomiting	*Mentha piperita*

Table 50: Alien Temple-Plants of H.P. Used for Material Culture

Use	*Species*
Agricultural implements	*Eucalyptus tereticornis, Robinia pseudoacacia, Salix fragilis*
Basketry	*Salix babylonica*
Building works	*Eucalyptus tereticornis, Juglans regia, Morus alba, Populus nigra*
Fencing	*Bougainvillea spectabilis, Duranta repens*
Fuel	*Eucalyptus tereticornis, Morus alba, Populus nigra, Robinia pseudoacacia,*
Furniture	*Broussonetia papyrifera, Eucalyptus tereticornis, Morus alba, Populus nigra*
Household articles	*Eucalyptus tereticornis, Salix fragilis, Tamarindus indica*
Magico-religious belief	*Capsicum annuum*

Contd...

Table 50–*Contd...*

Use	*Species*
Packing cases	*Broussonetia papyrifera, Eucalyptus tereticornis, Populus nigra, Salix babylonica*
Plywood	*Broussonetia papyrifera*
Preparation of Bear	*Humulus lupulus*
Rearing silkworms	*Morus alba*
Reddening lips	*Juglans regia*
Snake repellant	*Cyperus flabelliformis*
Killing termites	*Althaea rosea*
Tooth brush (scouring teeth)	*Juglans regia, Psidium guajava, Ricinus communis*

Table 51: Alien Horticultural Plants in the Temple-Courtyards of H.P.

Botanical Name	*Family*
Achras zapota	Sapotaceae
Carica papaya	Caricaceae
Citrus maxima	Rutaceae
Eriobotrya japonica	Rosaceae
Ficus carica	Moraceae
Juglans regia	Juglandaceae
Litchi chinensis	Sapindaceae
Prunus amygdalis	Rosaceae
Prunus armoniaoa	Rosaceae
Prunus domestica	Rosaceae
Prunus persica	Rosaceae
Psidium guajava	Myrtaceae
Punica granatum	Punicaceae
Pyrus communis	Rosaceae
Pyrus malus	Rosaceae
Syzygium jambos	Myrtaceae

Of the 173 invasive alien species documented for India (Reddy, 2008), eight species namely *Asclepias curassavica* (Asclepiadaceae), *Calotropis procera* (Asclepiadaceae), *Datura innoxia* (Solanaceae), *Ipomoea quamoclit* (Convolvulaceae), *Leucaena leucocephala* (Fabaceae), *Mimosa pudica* (Fabaceae), *Mirabilis jalapa* (Nyctaginaceae) and *Ruellia tuberosa* (Acanthaceae) do occur in the temple-courtyards of Himachal Pradesh. Contrarily, it is worthy of mention that excepting *Datura innoxia*, none of the remaining species enjoy the status of invasive aliens and posed serious threat to the native flora in the State of Himachal Pradesh.

Table 52: Alien Fodder Plants in the Temple-Courtyards of H.P.

Plant Species	*Family*	*Plant Part*
Broussonetia papyrifera	Moraceae	Leaves
Convolvulus arvensis	Convolvulaceae	Whole Plant
Cynodon dactylon	Poaceae	Whole Plant
Dodonaea viscosa	Sapindaceae	Leaves
Leucaena leucocephala	Fabaceae	Leaves
Morus alba	Moraceae	Leaves
Populus nigra	Salicaceae	Leaves
Robinia pseudoacacia	Fabaceae	Leaves
Salix babylonica	Salicaceae	Branches
Salix fragilis	Salicaceae	Young Branches
Sesamum orientale	Pedaliaceae	Seeds
Zea mays	Poaceae	Whole Plant

Table 53: Alien Crop Plants in the Temple-Courtyards of H.P.

Botanical Name	*Family*
Brassica oleracea var. *capitata*	Brassicaceae
Capsicum annuum	Solanaceae
Daucus carota	Apiaceae
Dolichos lablab	Fabaceae
Helianthus annuus	Asteraceae
Lycopersicum esculentum	Solanaceae
Phaseolus vulgaris	Fabaceae
Sacchharum officinarum	Poaceae
Sesamum orientale	Pedaliaceae
Solanum tuberosum	Solanaceae
Zea mays	Poaceae

It is obvious from the foregoing account that the temple-courtyards in Himachal Pradesh hold a wide range of wild and natural plant species used as food, fibre, fuel, fodder, medicine, ornamental and for religious purposes. The use of such a large number of plants for various purposes by the local communities reveal their great awareness, knowledge and intimate relationship with the plant kingdom. Moreover, the rich plant diversity is being protected and managed by following a wealth of conservation-oriented cultural and religious traditions. There is compelling need to apprise the people, especially the younger generation about the role of indigenous people in biodiversity conservation, the utility of temple-courtyards as the natural nurseries of useful plant species, including wild relatives of cultivated crops as well as incorporating the traditional beliefs in a new enlightened approach to nature conservation. Hopefully, for evolving a meaningful sustainable rural development plan for the region, the data generated presently would be highly beneficial to develop a mechanism for conservation of natural resources through the fusion of local people's knowledge system with that of resource personnel.

Literature Cited

Agenda 21. 1992. U.N. Conference on Environment and Development. Rio de Janeiro, Brazil.

Agharkar, S.P. 1961. *Maharashtra. State Gazetteer (General: series) Volume-A: Botany Part (III) Miscellaneous Plants.* Govt. Central Press, Bombay.

Ahluwalia, K.S. 1952. Medicinal plants of Kangra valley. *Indian Forester 78(4)*: 181-194.

Aizhong, L., Shengji, P. and Sanyang, C. 1999. Plant worship of the Yi people in Chuxiong of Yunnan, China. *Ethnobotany 11*: 1-8.

Alcorn, J.B. 1996. Is biodiversity conserved by indigenous people?: 233-238. In Jain, S.K. (ed.): *Ethnobiology in Human Welfare.* Deep Publ., New Delhi.

Amirthalingam, M. 1998. Sacred trees of Tamil Nadu. *Eco News 4*: 12-16. C.P.R. Environmental Education Centre, Chennai.

Amrita. 1999. Temples of nature. *Discover India 12*: 20-22.

Anonymous. 1994. *Ethnobiology in India–A Status Report.* All India Co-ordinated Research Project in Ethnobotany. Ministry of Environment and Forest, Govt. of India, New Delhi.

Arora, P.K. 1996. Role of ethnobotany in the conservation and use of plant genetic resources in India: 287-296. In Jain, S.K. (ed.): *Ethnobiology in Human Welfare.* Deep Publications, Delhi.

Arora, R.K. 1995. Ethnobotany and its role in domestication and conservation of native plant genetic resources: 79-86. In Jain, S.K. (ed.): *A Manual of Ethnobotany (2nd ed.).* Sci. Publ., Jodhpur.

Arora, R.K. 1997. Ethnobotany and its role in the conservation of plant genetic resources in India. *Ethnobotany 9(1 and 2)*: 6-15.

Arora, R.K. and Nayar, E.R. 1984. Wild relatives of crop plants in India. *Sci. Monogr. 7*: 90 pp.

Arora, R.K., Maheswari, M.L., Chandel, K.P.S. and Gupta, R. 1980. Mano (*Inula racemosa*): Little known aromatic plants of Lahaul valley, India. *Eco. Bot. 34(2)*: 175-180.

Aswal, B.S. and Mehrotra, B.N. 1994. *Flora of Lahaul-Spiti (A Cold Desert in North–West Himalayas).* Bishen Singh Mahendra Pal Singh, Dehra Dun.

Atkinson, E.T. 1882. *The Himalayan District of North West Provinces of India.* Cosmo Publ., New Delhi, Reprinted (1973).

Babu, C.R. 1977. *Herbaceous Flora of Dehradun.* CSIR, New Delhi.

Bachmann, V.K. 1992. Mythen. Geo. Wissen., Hamburg.

Badola, H.K. 2001. Medicinal plant diversity in Himachal Pradesh: 407-430. In Samant, S.S., Dhar, U. and Palani, L.M.S. (eds.): *Himalayan Medicinal Plants: Potential and Prospects.* Himvikas Occasional Publication, No. 14, Gyanodya Prakashan.

Bailey, L.H. 1949. *Manual of Cultivated Plants.* MacMillan, New York.

Bamber, C.J. 1916. *Plants of the Punjab.* Govt. Printing Press, Lahore. Reprinted 1976. Bishen Singh Mahendra Pal Singh, Dehra Dun and Periodical Experts, Delhi.

Banerjee, B.N. 1979. *Hindu Culture, Custom and Ceremony.* Agni Kala Prakashan, Delhi.

Banerjee, D.K. 1974. Magico-religious beliefs about plants among some *Adibasis* of India. *Jl. Myth. Soc. 65(3)*: 5-8.

Bedi, R. 2000. *Floral Spectrum of India.* Pelican Creations International, Noida, India.

Belapure, K.M., Maheshwari, J.K. and Tandon, R.K. 1987. Plants of Ramayana. *Ancient Sci. Life 7(2)*: 76-84.

Bennet, S.S.R. 1986. *Name Changes in Flowering Plants of India and Adjacent Regions.* Triseas Publishers, Dehra Dun, India.

Bennet, S.S.R., Gupta, P.C. and Rao, R.V. 1992. *Venerated plants.* Indian Council of Forestry Research and Education, New Forest, Dehra Dun, India.

Bentley, R. and Trimen, H. 2008. *Medicinal Plants,* Vols. I-IV. International Book Distributors, Dehra Dun, India.

Berry, S. and Bradley, S. 2000. *Illustrated Encyclopedia: Garden Plants.* Anness Publishing Ltd., London.

Bhatia, N., Mukherjee, T. and Singh, G. 1984. Plants: traditional worshipping. *Indian Jl. Hist. Sci. 19(1)*: 37-42.

Bhatnagar, I. and Lall, M. 1965. Tree symbol worship in Punjab. *Folklore 6*: 105-108.

Bhatt, D.C., Mitaliya, K.D. and Mehta, S.K. 2001. Some sacred plants of Gujarat and their medicinal uses. *Ethnobotany 13*: 146-149.

Bhattacharjee, S.K. 2006. *Advances in Ornamental Horticulture*, 6 Vols. Pointer Publishers, Jaipur (Rajasthan), India.

Blatter, E.J. and Hallberg, F. 1918. The flora of the Indian Desert (Jodhpur and Jaisalmer). *Proc. Nat. Inst. Sci. India 26*: 218-246.

Bole, P.V. and Vaghani, Y. 1986. *Field Guide to the Common Trees of India*. Oxford Univ. Press.

Booth, C.O. 1999. *Encyclopaedia of Garden Plants,* 2 Vols. Daya Publishing House, New Delhi.

Bor, N.L. and Raizada, M.B. 1999. *Some Beautiful Indian Climbers and Shrubs*. Oxford University Press.

Bor, N.R. 1960. *The Grasses of Burma, Ceylon, India and Pakistan (Excluding Bambuseae).* International Book Distributors, Dehra Dun.

Borthakur, S.K., Sarma, T.R., Nath, K.K. and Deka, P. 1998. The house gardens of Assam: a traditional Indian experience of management and conservation of biodiversity-1. *Ethnobotany 10(1 and 2)*: 32-37.

Bose, T.K. and Chowdhury, B. 1991. *Tropical Garden Plants in Colour.* Horticulture and Allied Publishers, Calcutta, India.

Bose, T.K., Pal, P., Chowdhury, B. and Sharma, S.P. (eds.) 2004. *Garden Plants in Colour: House Plants*. Naya Udyog, Kolkata, India.

Borrini-Feyerabend, G. 1996. *Collaborative Management of Protected Areas Tailoring the Approach to the Context. Issues in the Social Policy*. IUCN, Gland (Switzerland).

Brandis, D. 1874. *The Forest Flora of the North-West and Central India*. London. Reprinted 1972. Bishen Singh Mahendra Pal Singh, Dehra Dun.

Brandis, D. 1897. Forestry in India: Origin and Early Development. Working Oriental Institute.

Brandis, D. 1906. *Indian Trees*. Bishen Singh Mahendra Pal Singh, Dehra Dun, India.

Brickell, C., Zuk, J.D. and Zuk, J. 1997. *The American Horticultural Society A-Z Encyclopedia of Garden Plants*. American Horticultural Society, USA.

Burkill, I.H. 1965. *Chapters on the History of Botany in India*. Government of India Press, Delhi.

Burman, J. 1769. Flora Malabarica. Amsterdam.

Burman, N.L. 1768. Flora Indica. Ludg, Batavia (Indonesia).

Cambessedes, J. 1844. *Plantae Rarias quas in Indae Orientali collegit Victor Jacquemont*: 1-56. In Jacquemont, Victor (ed.): Voyage dans L' Indae, Paris.

Chakraborty M.K. and Pal, D.C. 2004. Plant conservation through tribal deities in some sacred groves of Purulia District, West Bengal. *Ethnobotany 16*: 99-102.

Champion, H.G. and Seth, S.K. 1968. *A Revised Survey of the Forest types of India*. Govt. India Publ., Delhi.

Chandervaker, P. 1965. Tree-cult in Saurashtra: 41-46. In Sen Gupta, S. (ed.): *Tree Symbols in India, A New Survey of Pattern of Folk Religion*. Indian Folklore Series No. 5, Calcutta, India.

Chandra, S. 1990. *Foundations of Ethnobotany (Pre-1900 Ethnobotany). A Review and Bibliography*. Deep Publ., New Delhi.

Chandrakanth, M.G., Gill, G. and Nagaraja. 1990. Temple forests in India's forest department. *Agrofor. Syst. 11*: 199-211.

Chauhan, N.S. 1974. Medicinal plants of Una forest division, Una district, H.P. lying in Shivalik range. *Nagarjun 27*: 31-39.

Chauhan, N.S. 1983. Some rare and interesting medicinal plants of H.P. All India Summer Inst. on Agroforestry in the Western Himalaya, July 22-August 7, 1983.

Chauhan, N.S. 1989a. Medicinal and aromatic plants of Himachal Pradesh and their role in social Forestry programme IIIrd Forestery Conference (proceeding) Vol. II: 89 (1), 89 (9): FRI, Dehra Dun.

Chauhan, N.S. 1989b. Potential of aromatic plants flora in Himachal Pradesh. *Indian Perf. 33(2)*: 118-122.

Chauhan, N.S. 1989c. Endangered ayurvedic pharmacopoeial plant resources of H.P.: 199-205. In Kaushik, P. (ed.): *Indiginous Medicinal Plants Including Microbes and Fungi*. Today and Tommorow's Printers and Publ., New Delhi.

Chauhan, N.S. 1990. Medicinal orchids of Himachal Pradesh. *J. Orchid Soc. India 4(2)*: 99-105.

Chauhan, N.S. 1995. *Plant Resources of Economic use in Himachal Pradesh*. Deptt. of Forest Products, Dr. Y.S. Parmar UHF, Solan (H.P.).

Chauhan, N.S. 1999. *Medicinal and Aromatic Plants of Himachal Pradesh*. Indus Publ. Co., New Delhi.

Chauhan, N.S. 2003. Important medicinal and aromatic Plants of Himachal Pradesh. *Indian For. 129(8)*: 979-998.

Chaurasia, O.P., Ahmed, Z. and Ballabh, B. 2007. *Ethnobotany and Plants of Trans-Himalaya*. Satish Serial Publication, New Delhi.

Chaurasia, O. P., Singh, B. and Ballabh, B. 2001. Potential aromatic floras of Himalayan cold district: Ladakh and Lahul-Spiti. *J. Econ. Taxon. Bot. 25(1)*: 91-97.

Chowdhery, H.J. 1998. *Orchid Flora of Arunachal Pradesh*. Bishen Singh Mahendra Pal Singh, Dehra Dun, India.

Chowdhery, H.J and Wadhwa, B.M. 1984. *Flora of Himachal Pradesh*, Vols. 1-3. Bot. Sur. India, Calcutta.

Collett, H. 1902. *Flora Simlensis*. Thacker Spink and Co., Calcutta and Shimla, Reprinted 1971. B.S.M.P.S. Dehra Dun.

Cooke, T. 1901-1908. *The Flora of the Presidency of Bombay*, Vols. 1-2. London. Reprinted 1958. Vols. 1-3. Botanical Survey of India, Calcutta.

Coventry, E.M. 1901. *A Catalogue of the Trees and Shrubs of Punjab*. Lahore.

Cowen, D.V. 1950. *Flowering Trees and Shrubs in India*. Thacker and Co., Bombay.

Dagar, J.C. 1995. Trees and shrubs in religion and mythology. *Tree World 3*: 3-4.

Dalzell, N.A. and Gibson, A. 1861. *The Bombay Flora*. Bombay Education Society Press, Mumbai, India.

Dalzell, N.A. 1858. *Catalogue of the Indigenous Flowering Plants of the Bombay Presidency*. Surat.

Dang, R. 1998. *Flowers of the Western Himalayas*. Wilderness Films, India.

Dasmann, R.F. 1995. The Importance of Cultural and Biological Diversity: Culture, Conservation and Eco-development. Westview Press Bonlder, San Francisco, Oxford.

Deb, D.B. (1981, 1983). *The Flora of Tripura State,* Vols. I and II. Today and Tomorrow's Printer and Publ., New Delhi.

Deorani, S.C. and Naithani, H.B. 1995. *Orchids of Nagaland*. Oriental Enterprises, Dehra Dun, India.

Dhaliwal, D.S. and Sharma, M. 1999. *Flora of Kullu District (Himachal Pradesh)*. Bishen Singh Mahendra Pal Singh, Dehra Dun.

Dhar, U. and Kachroo, P. 1983. *Alpine Flora of Kashmir Himalaya*. Scientific Publishers, Jodhpur, India.

Dhiman, A.K. 2003. *Sacred Plants and their Medicinal Uses*. Daya Publishing House, Delhi.

Dhiman, D.R. 1976. *Himachal Pradesh ki Vanoshdiya Sampada*. Imperial Printing Press, Dharmshala. H.P.

Dixit, G. 1997. Fire Sacrificial Plants. *Geobios News Reports 16(8)*: 47-48.

Dobriyal, R.M., Singh, G.S., Rao, K.S. and Saxena, K.G. 1997. Medicinal plant resources in Chakinal watershed in north-western Himalaya. *J. Herbs, Spices and Medicinal Plants 5*: 15-27.

Drake, J.A., Mooney, H.A., di Castri, F., Groves, R.H., Kruger, F.J., Rejmanek, M. and Williamson, M. (ed.) 1989. *Biological Invasions: A Global Perspective.* John Wiley and Sons, New York.

Drury, H. 1864. *Handbook of Indian Flora,* Vols. I and III. Travancore Sircar Press, Travancore.

Dube, R.K. 1998. *Nakshatra Vriksha.* For. Emp. Welf. Org., Lucknow.

Duthie, J.F. 1882. Plants found in Kumaon, Garhwal and the adjacent parts of Tibet by Capt. (now Genl.) Richard Strachey and Mr. Winterbottom: 403-670. In Atkinson, E.T.: *Gazetteer of North Western Provinces*. Vol. 10.

Duthie, J.F. 1903-1929. *Flora of the Upper Gangetic Plain and of the Adjacent Siwalik and Sub-Himalayan tracts,* Vols. 1-3. Superintendent Govt. Printing, Calcutta. Reprinted 1960. Vols. 1-2. Botanical Survey of India, Calcutta.

Dymock, W. 1889. On the use of turmeric in Hindu ceremonies. *J. Asiat. Soc. Bombay 2*: 441-448.

Farooqui, M.I.H. 1989. *Plants of the Quran.* Sidrah Publishers, Lucknow.

Fyson, P.F. 1915. *The Flora of Nilgiri and Pulney Hill Tops,* Vols. 1-3, Madras. (Reprinted 1975, Bishen Singh Mahendra Pal Singh, Dehra Dun).

Fyson, P.F. 1932. *The Flora of the South Indian Hill Stations,* Vols. I and II, Madras.

Gadgil, M. and Vartak, V.D. 1973. Groves dedicated to Gods. *Illus. Weekly in India 35*: 19-21.

Gadgil, M. and Vartak, V.D. 1975. Sacred groves of India: A plea for continuous conservation. *Journal of the Bombay Natural History Society 72(2)*: 314-320.

Gadgil, M. and Vartak, V.D. 1976. The sacred groves of Western Ghats in India. *Eco. Bot. 30 (2)*: 152-160.

Gadgil, M. and Vartak, V.D. 1981. Sacred groves in Maharashtra: 279-294. In Jain, S.K. (ed.): *Glimpses of Indian Ethnobotany.* Oxford and IBH, New Delhi.

Gamble, J.S. 1898. The Hill Route from Mussoorie to Simla. *Indian For. 24*: 105-111.

Garcia-da-Orta. 1565. *Colloquious des Simples.* Periodical Expert Book Agency, Delhi, (Reprint 1979).

Gaur, R.D. 1999. *Flora of District Garhwal-North-West Himalaya with Ethnobotanical Notes.* Transmedia Publication, Srinagar, Garhwal, India.

Ghate, V.S. 1998. Plants in patra pooja: Notes on their identity and utilization. *Ethnobotany 10*: 6-15.

Godbole, A. 1996. Role of tribals in preservation of sacred forests: 345-348. In Jain, S.K. (ed.): *Ethnobiology in Human Welfare.* Deep Publ., Delhi.

Godbole, A., Watve, A., Prabhu, S. and Sarneik, J. 1998. Role of sacred groves in biodiversity conservation with local people's participation: A case study from Ratnagiri District, Maharashtra: 233-246. In Ramakrishanan, P.S., Saxena, K.G. and Chandrashekara, U.M. (eds.): *Conserving the Sacred for Biodiversity Management.* Oxford and IBH Publ. Co. Pvt. Ltd., New Delhi.

Gogoi, P. and Borthakur, S.K. 1991. Plants in religio-cultural beliefs of Tai Khamtis of Assam (India). *Ethnobotany 3*: 89-95.

Good, R. 1964. *The Geography of the Flowering Plants.* Longman Group Limited, London.

Govaerts, R, 2001. How many species of seed plants are there? *Taxon 50*: 1085-1090.

Graham, J. 1839. *A Catalogue of the Plants Growing in Bombay and its Vicinity.* Bombay.

Griffith, W. 1854. *Notulae und Plantas Asiaticas,* Vols. 1-4. Chrles A. Serrao Publications, Calcutta.

Gupta, R. 1961. Flora of Lam Dal. *Ind. For. 87(5)*: 316-324.

Gupta, R. 1964. Survey record of medicinal and aromatic plants of Chamba Forest Division, Himachal Pradesh. *Ind. For. 90*: 454-468.

Gupta, R. 1971. Medicinal and aromatic plants of Bhandal ranges and Churah Forest Division, Chamba district, Himachal Pradesh. *J. Bomb. Nat. Hist. Soc. 68(3)*: 791-803.

Gupta, S.M. 1996a. Plant motifs sculpted on Indian Temples: 380-383. In Jain, S.K. (ed.): *Ethnobiology in Human Welfare*. Deep Publ., Delhi.

Gupta, S.M. 1996b. *Plants in Temple Arts*. B.R.A.P. Crop, Delhi.

Gupta, S.M. 1989. Ethnobotanical aspects of art and literature: 59-68. In Jain, S.K. (ed.): *Methods and Approaches in Ethnobotany*. Soc. Ethnobot., CDRI, Lucknow.

Gupta, S.M. 1991. *Plant Myths and Traditions in India*. Munshiram Manoharlal Publishers Pvt. Ltd., New Delhi.

Gupta, S.S. 1965 *Tree Symbol Worship in India*. Indian Folklore Society, Bombay.

Gupta, A.K. and Ura, K. 1990. Blending cultural values, indigenous technology and environment: experiences in Bhutan. Paper presented at the Intl. Symp. on Strategies for Sustainable Mountain Development, Kathmandu, Nepal.

Haines, H.H. 1921-1925. *The Botany of Bihar and Orissa,* 6 Vols. London. Reprinted 1961. 1-3 Vols. Botanical Survey of India, Calcutta.

Hajra, P.K. 1987. Plants in magico-religious beliefs and in Sanskrit literature, 98-103. In Jain S.K. (ed.): *A Manual of Ethobotany*. Sci. Publ., Jodhpur.

Hajra, P.K. 1997. Natural Conservation in Khasi Folk Beliefs and Taboos: 303-306. In Jain, S.K. (ed.): *Contributions to Indian Ethnobotany*. Sci. Publ., Jodhpur.

Hajra, P.K. and Das, B.K. 1982. Vegetation of Gangtok with special reference to exotic plants. *Indian For. 107*: 554-566.

Haridasan, K. and Rao, R.R. 1985. *A Forest Flora of Meghalaya,* 2 vols. Bishen Singh Mahendra Pal Singh, Dehra Dun.

Hegde, R.N. and Patil, G.A. 1999. Panchavati-the five sacred trees of India. *Kisan World 26*: 23-26.

Henry, A.N., Chitra, V. and Balakrishnan, N.P. (eds.) 1989. *Flora of Tamil Nadu*, Series 1, Vol. 3. BSI, Coimbatore.

Hohenacker, R.F. 1849. *Plants of Canara and Neighbouring areas of Maharashtra and Malabar. Flora 32*: 556-560.

Hooker, J.D. 1872-1897. *The Flora of British India*. Bishen Singh Mahendra Pal Singh. Dehradun, India.

Hooker, J.D. and Thomson, T. 1855. *Flora Indica*. W. Pamplin, Frith street. Soho. Reprinted 1976. Bishen Singh Mahendra Pal Singh, Dehra Dun and Periodical Experts, Delhi.

Hynniewta, T.M., Kataki, S.K. and Wadhwa, B.M. 2000. *Orchids of Nagaland*. BSI, Calcutta.

Induchoodan, N.C. 1996. *Ecological Study of the Sacred Groves of Kerala*. Ph.D. Thesis submitted to the Central University of Pondicherry.

Jain, A.K.1989. Tribal clans in Central India and their role in conservation. *J. Environ. Conserv. 4*: 68-70.

Jain, A.K. 1992. Clans of Sor tribals of Madhya Pradesh and their role in conservation. *Ethnobotany 4*: 67-69.

Jain, S.K. 1963. Magico-religious beliefs about plants among the tribals of Bastar, Madhya Pradesh. *R. Jour. Mythic Soc. 54*: 73-94.

Jain, S.K. 1971. Some magico-religious beliefs about plants among *adibasis* of Orissa. *Adibasi 12*: 39-44.

Jain, S. K. 2004. Objective ethnobotany-knowledge traditional, approaches modern. *Ethnobotany 16*: 1-9.

Jain, S. K. and Goel, A. K. 1995. Workshop Exercise–1. Proforma for field work: 142-147. In Jain, S.K. (ed.): *A Manual of Ethnobotany*. Scientific Publ., Jodhpur.

Jain, S.K. and Rao, R.R. (eds.) 1977. *A Handbook of Field and Herbarium Methods*. Today's and Tommorrow's Printers and Publ., New Delhi.

Jain, S.K. and Sastry, A.R.K. 1979. *Threatened Plants of India*. BSI, Howrah, India.

Janaki Ammal, E.K. 1956. Introduction to the subsistence of economy of India: 324-335. In William, L.T. (ed.): *Man's Role in Changing Face of the Earth*. Univ. of Chicago Press, Chicago.

Jha, V. and Goel, A.K. 2004. Botanical rosaries and non-flower garlands in folklore life of Mithila region (North Bihar). *Ethnobotany 16*: 27-32.

Jha, V., Saraswati, K.C., Kargupta, A.N. and Mishra, S. 1994. Plants in cultural life of Mithila. *Adv. Pl. Sci. Res. 2*: 92-119.

Joseph, J. 1986. *Flora of Nongpoh and Vicinity*. Botanical Survey of India, Calcutta.

Joseph, J. 1987. *Orchids of Nilgiris*. Botanical Survey of India, Calcutta.

Joshi, H.C., Arya, S.C. and Samant, S.S. 2001. Diversity, distribution and indigenous uses of plant species in Pindari area of Nanda Devi Biosphere Reserve II. *Indian J. For. 24 (4)*: 514-536.

Kachroo, P., Sapru, B.L. and Dhar, U. 1977. *Flora of Ladakh*. Bishen Singh Mahendra Pal Singh, Dehra Dun.

Kala, C.P. 2003. *Medicinal Plants of Indian Trans-Himalaya*. Bishen Singh Mahendra Pal Singh, Dehra Dun.

Kanjilal, P.C. 1933. *A Forest Flora of Pilibhit, Oudh, Gorakhpur and Bundelkhand*. Government Printing Press, Calcutta.

Kanjilal, U.N. 1901. *Forest Flora of the School Circle, North-West Province*. Government Printing Press, Calcutta.

Kanjilal, U.N. 1911. *Forest Flora of the Siwalik and Jaunsar Forest Divisions of United Provinces of Agra and Oudh*. Government Printing Press, Calcutta.

Kanjilal, U.N. and Gupta, B.L. 1928. *Forest Flora of Chakrata, Dehradun and Saharanpur Forest Divisions, United Provinces of Agra and Oudh*. Government Printing Press, Calcutta.

Kanjilal, U.N., Kanjilal, P.C., Das, A., De, R.N. and Bor, N.L. 1934-1940. *Flora of Assam,* 5 Vols. Govt. Press, Shillong.

Kapoor, L.D., Sharma, P.C. and Kapoor, S.L. 1971. Medicinal plant wealth around religious shrines of India. *J.Res. Indian Med. 6*: 87-94.

Kapur, S.K. 1985. Observations on the floristic composition of Kangra Valley (Himachal Pradesh). *New Botanist 12*: 157-165.

Kaul, M.K. 1986. *Weed Flora of Kashmir Valley.* Sci. Publ., Jodhpur.

Kaur, H. and Sharma M. 2004. *Flora of Sirmaur (Himachal Pradesh).* Bishen Singh Mahendra Pal Singh, Dehra Dun.

Khiewtam, R.S. and Ramakrishnan, P.S. 1989. Socio-cultural studies of the sacred groves at Cherrapunji and adjoining areas in the north-eastern India. *Man in India 69*: 64-71.

Khuroo, A.A., Dar, G.H., Khan, Z.S. and Nakshi, A.R. 2004. Floristic diversity in the phanerogams of Langate (J. and K.), India. *J. Econ. Taxon. Bot. 28*: 532-544.

Kowarik, I. 2003. Human agency in biological invasions: secondary releases foster naturalisation and population expansion of alien plant species. *Biol. Invasions 5*: 293-312.

Krisa, B. 1982. *The Illustrated Encyclopedia of Flowering Plants.* Octopus Books Limited, London.

Kulkarni, D.K. and Kumbhojkar, M.S. 2000. Plant Conservation in Maharashtra during 17th Century. *Ethnobotany 12*: 29-38.

Kulkarni, D.K., Barve, J.P., Jagdale, R.P. and Inamdar, A.C. 1993. Floristics of a sacred forest patch from Sundargad district, Orissa State. *J. Econ. Tax. Bot. 17(2)*: 415-419.

Kumar, A. and Yadav D.K. 2004. Significance of sacred plants in shraddh rituals (Pindadan) in Gaya, Bihar. *Ethnobotany 16*: 103-107.

Kumar, B., Chandra, S., Bargali, K. and Pangtey, Y.P.S. 2007. *Ethnobotany of Religious Practices in Kumaun (Havan).* Bishen Singh Mahendra Pal Singh, Dehra Dun.

Kumar, S. 2001. *Flora of Haryana (Materials).* Bishen Singh Mahendra Pal Singh, Dehra Dun.

Kumbhojkar, M.S. and Vartak, V.D. 1988. Ethnobotanical studies on wild edible grapes from sacred groves in Western Maharashtra. *J. Econ. Tax. Bot. 12(2)*: 257-263.

Kurz, S. 1870. *Report on the Vegetation of the Andaman Islands.* Government Printing Press, Calcutta.

Kurz, S. 1876. A sketch of the vegetation of the Nicobar Islands. *J. Asiat. Soc. Bengal 45(2)*: 105-164.

Llamas, K.A. 2003. *Tropical Flowering Plants: A Guide to Identification and Cultivation.* Timber Press.

Lucas, G. and Synge, H. 1978. *The IUCN Plant Red Data Book.* Surrey, England.

Maheshwari, J.K. 1963. *The Flora of Delhi*. CSIR, New Delhi.

Maheshwari, J.K. 1966. *Illustrations to the Flora of Delhi*. CSIR, New Delhi.

Maheshwari, J.K. 1987. Interdisciplinary approaches in ethnobotany: 19-34. In Jain, S.K. (ed.): *A Manual of Ethnobotany*, Sci. Publishers, Jodhpur, India.

Maheshwari, J.K. 1995. Ethnobotany in development and conservation of resources: 114-120. In Jain, S.K. (ed.): *A Manual of Ethnobotany, 2nd ed.*, Sci. Publ., Jodhpur.

Maheshwari, J.K. and Paul, S.R. 1975. The exotic flora of Ranchi. *J. Bombay Nat. Hist. Soc. 72(1)*: 158-188.

Maithani, G.P., Bahuguna, V.K., Negi, J.D.S. and Nautiyal, S. 1991. *Handbook of Some Important Himalayan Shrubs*. ICFRE-1, F.R.I., Dehra Dun, India.

Majupuria, T.C. and Joshi D.P. 1989. *Religious and Useful Plants of India and Nepal*. Gwalior.

Manilal, K.S. 1988. Flora of Silent Valley, Tropical Rain Forest of India. *J. Trop. For. 4*: 402.

Manilal, K. S. 1989. Linkages of ethnobotany with other sciences and discipline *Ethnobotany 1*: 15-24.

Manilal, K.S. and Sivarajan, V.V. 1982. *Flora of Calicut*. Bishen Singh Mahendra Pal Singh, Dehra Dun.

Mao, A.A. 1993. A preliminary report on the folklore botany of Mao Nagas of Manipur (India). *Ethnobotany 5*: 143-147.

Marglin, F.A. and Mishra, P.C. 1993. Sacred groves regenerating the body, the land, the community. In Sachs, W. (ed.): *Global Ecology: A New Arena of Political Conflict*. Zed Books, London.

Mashelkar, R.A. 2001. Intellectual property rights and the third world. CSIR, New Delhi: Cur. Sci. 81(8), 25 October, 2001.

Matthew, K.M. 1969. Exotic flora of Kodaikanal and Palni hills. *Rec. Bot. Surv. India 20(1)*: 1-241.

Maya, S., Kumari, S. P. and Menon, V. S. 2003. Ethnobotanical notes on the flora of sacred tanks of Kerala. *Ethnobotany 15*: 55-59.

Mc Neely, J.A. 1995. The interaction between biological diversity and cultural diversity: 1-15. Paper presented at the Intl. Conf. on Indigenous Peoples, Environment and Development, Zurich, May 15-18, 1995.

Mc Neely, J.A. and Pit, D. (eds.) 1985. *Culture and Conservation. The Human Dimension in Environmental Planning*. Croom Helm, London, New York, Sydney.

Menninger, E.A. 1962. *Flowering Trees of the World (for Tropics and Warm Climates)*. Hearthside Press Inc., New York.

Menninger, E.A. 1970. *Flowering Vines of the World: An Encyclopedia of Climbing Plants*. Hearthside Press, New York.

Menon, V.S. and Sasidharan, A. 1994. Studies on the photosynthetic performance of selected trees and woody vines in the sacred groves of Kerala-*An attempt to evaluate the optimum productivity of these ecosystems.* Final report submitted by Tropical Botanic Garden and Research Institute to the State Committee on Science, Technology and Environment, Kerala State, Thiruvananthapuram.

Merat, M.M. 2002. Plants of religious importance of Buldana district of Maharashtra state. *Geobios. 29*: 61-62.

Mitra, R. 1994. The sacred Asvatha-the most worshipped tree in India. *Ethnobotany 6*: 95-98.

Mitra, S.C. 1928. Notes on the tree cults in district of Patna in South Bihar. *J. Bihar and Orissa Res. Soc. 14*: 278-281.

Mohanty, R.B. 1997. Tree marriage in rural Orissa. *Ethnobotany 9*: 130.

Moldenke, H.N. 1954. The economic plants of the Bible. *Econ. Bot. 8*: 152-163.

Moorcroft, W. and Trebeck, G. 1841. Travels in the Himalayan Provinces (1819-1835), Vols 1-2. Illustrated. John Murray, London.

Murthy, E.N., Raju, V.S. and Reddy, C.S. 2007. Occurrence of exotic *Hyptis suaveolens. Curr. Sci. 93(9)*: 1203.

Murti, S.K. 2001. *Flora of Cold Deserts of Western Himalaya (Vol. 1) Monocotyledons.* BSI, Calcutta.

Nair, G., Gopikumar, K., Krishnan, P.G. and Kumar, Sunil. 1997. Sacred groves of India-vanishing greenery. *Curr. Sci. 72*: 697-698.

Nair, G.K. 2006. Project to turn vacant temple lands green. The Hindu.

Nair, P.T. 1965. Tree marriage in India. *Folklore 6*: 155-161.

Nair, R. C. 1977. *Flora of Bashahr Himalaya.* Int. Biosci. Publ., Hissar.

Nayak, H.M. 1965. Trees cultivated in Karnataka. *Folklore 6*: 236-240.

Nayar, M.P. 1977. Changing patterns of the Indian flora. *Bull. Bot. Surv. Ind. 19*: 145-154.

Nayar, M.P. 1997. Biodiversity challenges in Kerala and science of conservation biology: 7-80. In Pushpangadan, P. and Nair, K.S.S. (eds.): *Biodiversity and Tropical Forests: The Kerala Scenario.* The State Committee on Science, Technology and Environment, Kerala.

Nayar, M. P., Ramamurthy, K. and Agarwal, V. S. 1989. *Economic Plants of India,* Vols. I and II. Bot. Survey India, Calcutta.

Nipunage, D.S., Vartak, V.D. and Khumbhojkar, M.S. 1992. Cultural heritage of sacred grooves. *Proc. of South Indian History Congress, Pune.* pp 170-176.

Nonhare, B.P., Sahu, T.R. and Sahu, P. 2002. Deogudis-unrecognised centres for biodiversity conservation in Bastar (Chhattisgarh). *Ethnobotany 14*: 93-99.

Oakman, H. 1990. *Shrubs for Tropical and Sub-tropical Gardens.* Jacaranda Press, Auckland, NZ.

Oommachan, M. 1977. *The Flora of Bhopal (Angiosperm)*. J.K. Jain Brothers, Bhopal (India).

Osmaston, A.E. 1927. *A Forest Flora of Kumaon*. Govt. Press United Provinces, Allahbad. Reprinted ed. 1978. Bishen Singh Mahendra Pal Singh, Dehra Dun.

Oyadomari, M. 1990. *Cultural Perception of Nature and its Application Toward Preserving Cultural and Natural Heritage in Japan*. Chiba Nat. Univ., Tokyo, Japan.

Pal, B.P. 1972. *Beautiful Climbers of India*. Indian Council of Agricultural Research, New Delhi.

Pal, B.P. and Krishnamurthi, S. 1967. *Flowering Shrubs*. Indian Council of Agricultural Research, New Delhi.

Pal, D.C. 1970. Plants associated with *Durga Puja* Ceremony in West Bengal. *Nat. Hist. Soc. and Nat. Hist. Mus. 36*: 61-67.

Pal, D. C. and Mudgal, V. 1985. Conservational practices of tribals of Purulia and their relevance in eco-development Programmes. Proc. Develt Camp. Baghmundi (Purulia) Calcutta, School of Fundamental Research, 37-42.

Pande, T. 1964. Tree Worship in Ancient India. *Folklore 5(6)*: 213-218.

Pandey, B.P. 1989. *Sacred Plants of India*. Shree Pub. House, New Delhi.

Pandey, R.P. and Parmar, P.J. 1994. The exotic flora of Rajasthan. *J. Econ. Taxon. Bot. 18(1)*: 105-121.

Parker, R.N. 1918. *A Forest Flora for the Punjab with Hazara and Delhi*. Government Press, Lahore.

Parkinson, C.E. 1923. *A Forest Flora of the Andaman Islands*. Govt. of India, Simla.

Paroda, R.S. and Arora, R.K. (eds.) 1991. *Plant Genetic Resources Conservations and Management Concepts and Approaches*. IBPGR, Publ., New Delhi.

Pennel, F.W. 1943. The Scrophulariaceae of the Western Himalayas. *Acad. Nat. Sci. Philad. Monogr. 5*: 1-163. Reprinted 1997. Bishen Singh Mahendra Pal Singh, Dehra Dun.

Perrings, C., Dehnen-Schmutz, K., Touza, J. and Williamson, M. 2005. How to manage biological invasions under globalization. *Trends Ecol. Evol. 20*: 212-215.

Polunin, O. 1969. *A Field Guide to the Flowers of Europe*. Oxford University Press, Melbourne and London.

Polunin, O. and Stainton, A. 1984. *Flowers of the Himalaya*. Oxford University Press, Delhi.

Polunin, O and Stainton, A. 1987. *Concise Flowers of the Himalaya*. Oxford University Press, Delhi.

Prain, D. 1903. Flora of the Sunderbans. *Rec. Bot. Surv. India 2*: 231-370.

Pramod, C., Sivadasan, M. and Anilkumar N. 2003. Ethnobotany of religious and supernatural beliefs of Kurichya of Waynad district, Kerela, India. *Ethnobotany 15*: 11-19.

Pullaiah, T. (ed.) 2004. *Biodiversity in India*, Vols. 1-3. Regency Publications, New Delhi.

Pushpangadan, P. 1994. *Ethnobiology in India-A Status Report*. Ministry of Environment and Forests. pp 1-68. New Delhi.

Rai Chaudhuri, H.N. and Pal, D.C. 1981. Plants in folk religion and Mythology: 57-68. In Jain, S.K. (ed.): *Glimpses of Indian Ethnobotany*. Oxford and IBH Publishing Co., Delhi.

Raizada, M.B. and Saxena, H.O. 1978. *Flora of Mussoorie*, Vol. I. Bishen Singh Mahendra Pal Singh, Dehra Dun.

Ramakrishanan, P.S. 1998. Conserving the sacred: for biodiversity management conclusions and recommendations: 13-23. In Ramakrishanan, P.S., Saxena, K.G., Chandrashekara, U.M. (eds.): *Conserving the Sacred for Biodiversity Mangement*. Oxford and IBH Publ. Co. Pvt. Ltd., New Delhi.

Ramakrishanan, P. S., Saxena, K. G. and Chandrashekara, U.M. 1998. *Conserving the Sacred for Biodiversity Management*. Oxford and IBH Publ. Co. Pvt. Ltd., New Delhi.

Ramanujam, M.K., Kadomban, D., Kumaravelu, G. and Kumar, P. 2002. Sacred groves-an overview: 13-53. In Ethnobotany. Aavishkar Publishers, Distributors, Jaipur, Rajasthan (India).

Rama Rao, M. 1914. *Flowering Plants of Travancore*. Government Press, Trivandrum, India.

Ramaswamy, S.V. and Razi, B.A. 1973. *Flora of Banglore District*. Prasarenga. Mysore University, Mysore.

Ramsankar, B. 2001. The 'Soharia' festival of 'Mundas' in Purulia. *Ethnobotany 13*: 140-141.

Rana, T.S., Dutt, B. and Rao, R.R. 1996. Strategies of sustainable utilization of plant resources by the tribals of the Tons valley, Western Himalaya. *Ethnobotany 8*: 96-104.

Rana, T.S., Datt, B and Rao, R.R. 2003. *Flora of Tons Valley, Garhwal Himalayas (Uttranchal)*. BSMP, Dehra Dun.

Randhawa, G.S. 1973. *Ornamental Horticulture in India*. Scholarly Pubns.

Randhawa, M.S. 1983. *Flowering Trees*. National Book Trust, New Delhi, India.

Rao, A.S. and Verma, D.M. 1982. *Cyperaceae of North East India*. Botanical Survey of India, Calcutta.

Rao, R.R. 1990. Ethnobiology in the study and conservation of fragile ecosystems: Some issues from the Himalayan region. *Ethnobotany 2*: 45-55.

Rao, R.R. and Razi, B.A. 1974. *A Synoptic Flora of Mysore District*. Today and Tomorrow's Printers and Publishers, New Delhi.

Rao, R.S. 1985-1986. *Flora of Goa, Diu, Daman and Nagarhaveli*. Botanical Survey of India, Calcutta.

Rastogi, M.A. 1960. Medicines from the wild. A cure study of the Great Himalayan Park. *The Indian magazine of her people and culture*. pp. 74-75.

Rau, M.A. 1961. A recent survey of medicinal plant resources of Lahoul valley (Punjab, India). Pakistan J. Sci. and Res. (International Symposium) *4*: 202-205.

Reddy, C.S. 2008. Catalogue of invasive alien flora of India. *Life Sci. Jour. 5(2)*: 84-89.

Reddy, C.S. and Raju, V.S. 2002. Additions to the weed flora of Andhra Pradesh. *J. Econ. Taxon Bot. 26*: 195-198.

Reddy, C.S. and Reddy, K.N. 2004. *Cassia rotundifolia* Pers. (Caesalpiniaceae): a new record for India. *J. Econ. Tax. Bot. 28*: 73-74.

Reddy, C.S., Bhanja, M.R. and Raju, V.S. 2000. *Cassia uniflora* Miller: a new record for Andhra Pradesh, India. *Indian J. Forestry 23(3)*: 324-325.

Rodgers, W. A. 1985. Biogeography and Protected area planning in India. *In*: Thorselt, J. (ed.). *Conserving Asia's National Heritage*. IUCN, Glad.

Roxburgh, W. 1795-1819. *Plants of the Coast of Coromandel*, Vols. 1-3. W. Bulmer and Co., London.

Roxburgh, W. 1814. *Hortus Bengalensis*. Mission Press, Serampore.

Roxburgh, W. 1832. *Flora Indica*, 3 Vols. W. Thacker and Co., Serampore.

Roy, G.P., Shukla, B.K. and Bhaskar Datt. 1992. *Flora of Madhya Pradesh (Chhatapur and Damoh)*. Amerind Pub. Co. Pvt. Ltd. New Delhi.

Royle, J.F. 1833-1840. *Illustrations of the Botany and other Branches of the Natural History of the Himalayan Mountains and of the Flora of Cashmere*, 2 Vols. W.N. Allen and Co., London.

Sahay, K.N. 1964. Trees cultivated in tribal culture. *Folklore 5(7)*: 241-257.

Saikia, N. and Nath, S.C. 2003. Traditional use of some plants in religious–cultural purposes by 'Koch' Community in Assam, India *J. Econ. and Tax. Bot. 27(4)*: 997-1001.

Saklani, A. and Rao, R.R. 1996. Role of Brahmkamal [*Saussurea obvallata* (DC.) Edgew.] in the Life and Culture of Garhwalis. *Ethnobotany 8*: 75-78.

Saldanha, C.J. 1984. *Flora of Karnataka*, Vol. 1. Oxford and IBH Publishing Co., New Delhi, Bombay, Calcutta.

Saldanha, C.J. and Nicolson, D.H. (eds.) 1976. *Flora of Hassan District, Karnataka, India*, Amerind Publishing Co. Pvt. Ltd. New Delhi.

Sambamurty, A.V.S.S. 2002. *Dictionary of Angiosperm families and their Genera*. Reprint. Asiatic, New Delhi.

Sane, H. and Ghate, V. 2006. Sacred conservation practices at species level through tree worship. *Ethnobotany 18*: 46-52.

Santapau, H. 1953. The Flora of Khandala on the Western Ghats of India. *Rec. Botanical Survey of India 16*: 1-396.

Santapau, H. 1958. *History of Botanical Research in India, Burma and Ceylon, Part II (Systematic Botany of Angiosperms)*. Indian Botanical Society, New Delhi.

Santapau, H. 1962. The present state of taxonomic and floristic in India after independence. *Bull. Bot. Surv. India. 4*: 209-216.

Santapau, H. and Henry, A.N. 1973. *A Dictionary of the Flowering Plants in India.* Council of Scientific and Industrial Research, New Delhi.

Sarin, Y.K. 1967. A survey of vegetable raw material resources of Lahul. *Indian For. 93(7)*: 489-495.

Sarin, Y.K. and Chopra, C.L. 1985. Medicinal and aromatic plant resources of Himachal Pradesh, present status and future prospects. *Workshop on Sci. and Tech., Shimla*, Oct. 29-30, 1985.

Sastri, K.N. 1957. Tree-worship and tree-legend in the Indus Valley, new light on the Indus Civilization. Vol. 1 pp. 24-35, Delhi.

Saxena, H.O. and Brahmam, M. 1994-1996. *The Flora of Orissa*, Vols. I-IV. Regional Research Laboratory, Bhubaneswar and Orissa Forest Development Corporation Ltd., Bhubaneswar.

Saxena, K.G. 1991. Biological invasion in the Indian sub-continent: review of invasion by plants: 53-73. In Ramakrishnan, P.S. (ed.): *Ecology of Biological Invasion in the Tropics.* Intl. Sci. Pub., New Delhi.

Saxena, K.G., Rao, K.S. and Maikhuri, R.K. 1998. Religious and cultural perspective of biodiversity conservation in India: A review: 153-161. In Ramakrishanan, P.S., Saxena, K.G. and Chandrashekara, U.M. (eds.): *Conserving the Sacred for Biodiversity Management.* Oxford and IBH Publ. Co. Pvt. Ltd., New Delhi.

Sen Gupta, S. 1965. *Tree Symbol Worship in India.* Indian Folklore Society. Calcutta.

Sen Gupta, S. 1980. Idea of toleration and co-existence: a study of sacred trees in India. *Folklore 21(4)*: 85-92.

Sensarma, P. 1989. *Plants in the Indian Puranas-An Ethnobotanical Investigation.* Naya Prakash, Calcutta.

Sensarma, P. 1994. Ethnobotanical investigation into the 'Neelatantram'. *Ethnobotany 7*: 51-61.

Sensarma, P. 1995. The sacred and religious plants in the 'Tantrasarah'. *Ethnobotany 7*: 51-61.

Shah, N.C. 1994. Cultural and religious aspects of 'Jatamansi' (*Nardostachys jatamansi* DC) in the Kumaon Himalaya, India: 189. In *Ethnobiology in Human Welfare*: Abstract of the IV International Congress in Ethnobiology, 17-21 Nov., National Botanical Research Institute, Lucknow.

Sharma, B.D. 1984. *Exotic flora of Allahabad.* BSI, Dehra Dun.

Sharma, K.R. and Sood, M. 1997. *Important Medicinal Plants of Himachal Pradesh,* Vols. I and II. Dr. Y.S. Parmar University of Horticulture and Forestry, Nauni, Solan, H.P.

Sharma, N.K. 2002. Ethno-medico-religious plants of Hadoti plateau (S.E. Rajasthan)-a preliminary survey: 394-411. In Trivedi, P.C. (ed.): *Ethnobotany.* Aavishkar Publishers and Distributors.

Shengji, P. 1991. Conservation of biological diversity in temple-yards and holy hills by the Dai ethnic minorities of China. *Ethnobotany 3*: 27-35.

Shengji, P. 1993. Managing for biological diversity in temple yards and holy hills: the traditional practices of the Xishuangbanna Dai Community South-western China: 112-118. In Lawrence, S.H. and Takeuchi, H.F. (eds.): *Ethics, Religion and Biodiversity: Relations between Conservation and Cultural Values.* White Horse Press, Cambridge.

Singh, G. and Kachroo, P. 1976. *Forest Flora of Srinagar.* Bishen Singh Mahendra Pal Singh, Dehra Dun and Periodical Expert, Delhi.

Singh, G.S. 1997. Sacred groves in western Himalayas: An eco-cultural imperative. *Man in India 77*: 247-257.

Singh, G. S., Rao, K.S. and Saxena, K.G. 1998. Eco-cultural analysis of sacred species and ecosystem in Chhakinal Watershed, Himachal Pradesh: 301-313. In Ramakrishnan, P.S., Saxena, K.G. and Chandrashekara, U.M. (eds): *Conserving the Sacred for Biodiversity Management.* UNESCO Vol. Oxford and IBH Publ. Co. Pvt. Ltd., New Delhi.

Singh, H. and Sharma, M. 2006. *Flora of Chamba District (Himachal Pradesh*). Bishen Singh Mahendra Pal Singh, Dehra Dun.

Singh, J.B. and Kachroo, P. 1994. *Forest Flora of Pir Panjal Range. (Northwestern Himalaya).* Bishen Singh Mahendra Pal Singh, Dehra Dun.

Singh, N. 2002. *Plants used in Worship and Festivals of Brij Mandal. An ethnobotanical Study.* Thesis submitted to Dr. B. R. Ambedkar Univ., Agra, India.

Singh, N. and Chauhan, S.V.S. 2002. Studies on plants used in 'Govardhan Puja' and 'Annakut' in Brij Mandal of Mathura. *Ethnobotany 14*: 73-77.

Singh, N. and Chauhan, S.V.S. 2004. Studies on some leaves (*Patra*) used for worship in Brij Mandal. *Ethnobotany 16*: 69-71.

Singh, P.B. 1993. Medicinal plants of Ayurvedic importance from Mandi district of Himachal Pradesh. *Bull. of Medico-Ethnobot. Res. 14(3-4)*: 126-136.

Singh, P.B. 1999. *Illustrated Field Guide to Commercially Important Medicinal and Aromatic Plants of Himachal Pradesh (with Special Reference to Mandi District).* Society for Herbal Medicine and Himalayan Biodiversity, Mandi, H.P.

Singh, P.B. and Aswal, B.S. 1992. Medicinal plants of Himachal Pradesh used in Indian pharmaceutical industry. *B.M.E.B.R. 13*: 172-208.

Singh, P.K. and Singh, H.B.K. 1996. Superstition in botanical folklore with reference to Meitei culture: 367-372. In Maheshwari, J.K. (ed.): *Ethnobotany in South Asia.* J. Econ. Taxon. Bot. Additional Series, 12. Scientific Publishers, Jodhpur (India).

Singh, S. and Tiagi, B. 1979. *Flora of North-East Rajasthan.* Kalyani Publishers, New Delhi.

Singh, S. K. and Rawat G. S. 2000. *Flora of Great Himalayan National Park–Himachal Pradesh.* Bishen Singh Mahender Pal Singh, Dehra Dun.

Sinha, A. 2003. Sacred landscapes in India: forms and meanings. *Geograph. Rev. India 65*: 241-250.

Sinha, B.C. 1979. *Tree Worship in Ancient India.* Books Today, New Delhi.

Sinha, R.K. 1996. Conservation of cultural diversity of indigenous people essential for protection of biological diversity: 280-283. In Jain, S.K. (ed.): *Ethnobiology in Human Welfare.* Deep Publ., New Delhi.

Sithole, R.V. 1976. Plants represented in ancient Indian sculpture. Geophy. *6(1)*: 15-26.

Sood, S.K. and Thakur, S. 2004. *Ethnobotany of Rewalsar Himalaya (Dist. Mandi, Himachal Pradesh, India).* Deep Publ., New Delhi.

Sood, S.K., Nath, R., and Kalia, D.C. 2001. *Ethnobotany of Cold Desert Tribes of Lahaul-Spiti (N. W. Himalaya).* Deep Publ., New Delhi.

Sood, S.K., Sharma, R. and Mattu, V.K. 2006. *Honey Plants of the North-West Himalaya.* Indus Publishing Company, New Delhi.

Srivastava, T.N. 1976. *Flora Gorakhpurensis.* Today and Tommorrow's Publishers, New Delhi.

Stainton, A. 1988. *Flowers of Himalaya-A Supplement.* Oxford University Press, Delhi.

Temple Wright, R. 2007. *Flowers and Gardens in India-A Manual for Beginners.* Trollope Press.

Trapp, L.A. 2006. *Ornamental Plants and Flowers of Tropical Mexico.* Abbott Publications.

Uniyal, M.R. and Chauhan, N.S. 1971. Medicinal plants of Uhal valley in Kangra Forest Division, Himachal Pradesh. *J. Res. Indian Med. 6*: 287-299.

Uniyal, M.R. and Chauhan, N.S. 1972. Commercially important medicinal plants of Kullu Forest Division, H.P., *Nagarjun 4(2)*: 28-32.

Uniyal, M.R. and Chauhan, N.S. 1973 Traditionally important medicinal plants of Kangra valley in Dharamshala Forest Circle, H.P. *J. Res. Indian Med. 8*: 7-85.

Uniyal, M.R., Bhatt, A.B. and Chaturvedi, P.N. 1973. Preliminary observations on medicinal plants of Lahoul Spiti Forest Division in Himachal Pradesh. *Bull. Med. Ethnobot. Res. 3*: 1-26.

Uniyal, V.K., Pandey, S. and Sawarkas, W.B. 1996. Buffer-Zones–Key to success of protected areas: 313-327. In Ramakrishanan, P.S., Purohit, A.N., Saxena, K.G., Rao, K.S. and Maikhuri, R.K. (eds.): *Conservation and Management of Biological Resources in Himalaya.* Oxford and IBH, Delhi.

Unnikrishnan, E. 1995. *Sacred Groves of North Kerala-An Eco Folklore Study in Malayalam.* Jeevarekha, Thrissur.

Upadhye, A.S., Kumbhojkar, M.S. and Kulkarni, D.K. 1997. Ethno-medico-botany of some sacred plants of Western Maharashtra. *Ethnobotany 9*: 65-68.

Van Rheede, H. 1667. *Hortus Indicus Malabaricus,* Vols. I-XII. Amsterdam.

Vartak, V.D. 1996. Sacred groves for *in-situ* conservation: 300-302. In Jain, S.K. (ed.): *Ethnobiology in Human Welfare*. Deep Publ., New Delhi (India).

Vartak, V.D. and Gadgil, M. 1981. Studies on sacred groves along the western ghats from Maharashtra and Goa: role of beliefs and folklore: 272-278. In Jain, S.K. (ed.): *Glimpses of Indian Ethnobotany*. Oxford and IBH, New Delhi.

Vartak, V.D., Kumbhojkar, M.S. and Vandana, D. 1986. Sacred groves-a sanctuary lofty trees and lianas: 55-59. In *Proc. Seminar on Ecodevelopment of Western Ghats.* Kerela Forest Research Institute, Peechi.

Ved, D.K. and Tandon, V. (eds.) 1998. *CAMP Report for High Altitudinal Medicinal Plants of Jammu Kashmir and Himachal Pradesh.* FRLHT, Banglore.

Ved, D.K., Kinchal, G.A., Ravikumar, K., Pravharan, V., Uttakarsh, G., Vijaya Sankar, R. and Indresha, J.H. 2003. *Conservation Assessment and Management Prioritisation for the medicinal plants of Jammu and Kashmir, Himachal Pradesh and Uttranchal.* FRLHT, Banglore, India.

Warman, C.K. 1999. *Trees of India (Medicinal, Commercial, Religious and Ornamental): A Colour Atlas.* CBS Publishers and Distributors, New Delhi.

Wasson, E. (ed.) 2003. *The Complete Encyclopedia of Trees and Shrubs: Descriptions, Cultivation Requirements, Pruning, Planting*. Thunder Bay Press, San Diego, California.

Weber, E. 2003. *Invasive Plant Species of the World: A Reference Guide to Environmental Weeds.* CABI Publishing, U.K.

Wielgorskaya, T. 1995. *Dictionary of Generic Names of Seed Plants.* Bishen Singh Mahendra Pal Singh, Dehra Dun (India).

Wight, R. 1838-1853. *Icones Plantarum Indiae Orientalis,* Madras (Reprint edition 1963, Wheldon and Wesley, New York).

Wight, R. and Walker-Arnolt, G.A. 1834. *Prodromus Florae Peninsulae Indiae Orientalis.* Panbury Allen and Co. London (Reprint edition 1976, Bishen Singh Mahendra Pal Singh, Dehra Dun).

Zeven, A.C. and de Wet, J.M.J. 1982. *Dictionary of Cultivated Plants and their Regions of Diversity.* Wageningen.

Appendices

Appendix-I: Index to Families

Name of the Family	*Genus/ Species*	*Total Number of Species under a Genus*
Acanthaceae	4/4	*Beloperone* (1), *Justicia* (1), *Ruellia* (1), *Thunbergia* (1)
Aceraceae	1/1	*Acer* (1)
Amaranthaceae	3/3	*Celosia* (1), *Deeringia* (1), *Gomphrena* (1)
Amaryllidaceae	5/6	*Amaryllis* (1), *Hymenocallis* (1), *Narcissus* (1), *Polianthes* (1), *Zephyranthes* (2)
Anacardiaceae	3/3	*Lannea* (1), *Mangifera* (1), *Pistacia* (1)
Annonaceae	1/1	*Polyalthia* (1)
Apiaceae	2/2	*Centella* (1), *Daucus* (1)
Apocynaceae	8/10	*Alstonia* (1), *Carissa* (3), *Catharanthus* (1), *Nerium* (1), *Plumeria* (1), *Tabernaemontana* (1), *Thevetia* (1), *Wrightia* (1)
Araceae	1/1	*Colocasia* (1)
Araliaceae	1/1	*Schefflera* (1)
Araucariaceae	1/1	*Araucaria* (1)
Arecaceae	3/3	*Caryota* (1), *Chamaerops* (1), *Phoenix* (1).
Aristolochiaceae	1/1	*Aristolochia* (1)
Asclepiadaceae	2/3	*Asclepias* (2), *Calotropis* (1)
Asteraceae	14/16	*Artemisia* (1), *Calendula* (1), *Callistephus* (1), *Centaurea* (1), *Chrysanthemum* (2), *Cosmos* (2), *Dahlia* (1), *Erigeron* (1), *Gaillardia* (1), *Helianthus* (1), *Tagetes* (1), *Tithonia* (1), *Wedelia* (1), *Zinnia* (1)
Balsaminaceae	1/1	*Impatiens* (1)
Berberidaceae	1/1	*Berberis* (1)
Betulaceae	2/2	*Alnus* (1), *Betula* (1)
Bignoniaceae	4/5	*Jacaranda* (1), *Pyrostegia* (1), *Tecoma* (2), *Tecomaria* (1)
Bombacaceae	1/1	*Bombax* (1)
Boraginaceae	1/1	*Ehretia* (1)
Brassicaceae	1/1	*Brassica* (1)
Buxaceae	1/1	*Buxus* (1)
Cannabinaceae	2/2	*Cannabis* (1), *Humulus* (1)

Contd...

Appendix-I–*Contd...*

Name of the Family	*Genus/ Species*	*Total Number of Species under a Genus*
Cannaceae	1/1	*Canna* (1)
Caprifoliaceae	1/1	*Lonicera* (1)
Caricaceae	1/1	*Carica* (1)
Caryophyllaceae	1/1	*Dianthus* (1)
Casuarinaceae	1/1	*Casuarina* (1)
Chenopodiaceae	1/1	*Kochia* (1)
Combretaceae	2/3	*Quisqualis* (1), *Terminalia* (2)
Convolvulaceae	2/4	*Convolvulus* (1), *Ipomoea* (3)
Cordiaceae	1/1	*Cordia* (1)
Crassulaceae	2/2	*Bryophyllum* (1), *Sempervivum* (1)
Cucurbitaceae	1/1	*Momordica* (1)
Cupressaceae	3/6	*Cupressus* (2), *Juniperus* (2), *Thuja* (2)
Cycadaceae	1/1	*Cycas* (1)
Cyperaceae	1/1	*Cyperus* (1)
Dipterocarpaceae	1/1	*Shorea* (1)
Ebenaceae	1/2	*Diospyros* (2)
Elaeocarpaceae	1/1	*Elaeocarpus* (1)
Euphorbiaceae	10/13	*Acalypha* (1), *Bischofia* (1), *Emblica* (1), *Euphorbia* (3), *Jatropha* (1), *Mallotus* (1), *Pedilanthus* (1), *Putranjiva* (1), *Ricinus* (1), *Sapium* (2)
Fabaceae	15/20	*Acacia* (3), *Albizia* (2), *Bauhinia* (2), *Cassia* (2), *Dalbergia* (1), *Dolichos* (1), *Lespedeza* (1), *Leucaena* (1), *Mimosa* (1), *Phaseolus* (1), *Pisum* (1), *Pongamia* (1), *Robinia* (1), *Tamarindus* (1), *Wisteria* (1)
Fagaceae	1/2	*Quercus* (2)
Geraniaceae	1/1	*Pelargonium* (1)
Hippocastanaceae	1/1	*Aesculus* (1)
Hydrangeaceae	1/1	*Hydrangea* (1)
Hypericaceae	1/1	*Hypericum* (1)
Iridaceae	2/2	*Iris* (1), *Ixia* (1)
Juglandaceae	1/1	*Juglans* (1)

Contd...

Appendix-I–*Contd...*

Name of the Family	*Genus/ Species*	*Total Number of Species under a Genus*
Lamiaceae	5/10	*Coleus* (1), *Mentha* (2), *Ocimum* (4), *Roylea* (1), *Salvia* (2)
Lauraceae	3/4	*Cinnamomum* (1), *Litsea* (2), *Machilus* (1).
Liliaceae	5/7	*Aloe* (1), *Asparagus* (3), *Chlorophytum* (1), *Hemerocallis* (1), *Sansevieria* (1)
Loranthaceae	1/1	*Dendrophthoe* (1)
Lythraceae	2/2	*Lagerstroemia* (1), *Lawsonia* (1)
Magnoliaceae	2/2	*Magnolia* (1), *Michelia* (1)
Malvaceae	4/7	*Althaea* (1), *Hibiscus* (3), *Malva* (2), *Malvaviscus* (1)
Meliaceae	2/3	*Cedrela* (1), *Melia* (2)
Moraceae	4/14	*Artocarpus* (2), *Broussonetia* (1), *Ficus* (9), *Morus* (2)
Musaceae	1/1	*Musa* (1)
Myrtaceae	4/5	*Callistemon* (1), *Eucalyptus* (1), *Psidium* (1), *Syzigium* (2)
Nyctaginaceae	3/4	*Boerhaavia* (1), *Bougainvillea* (2), *Mirabilis* (1)
Oleaceae	5/8	*Fraxinus* (1), *Jasminum* (4), *Ligustrum* (1), *Nyctanthes* (1), *Olea* (1)
Onagraceae	1/1	*Oenothera* (1)
Papaveraceae	1/1	*Papaver* (1)
Passifloraceae	1/1	*Passiflora* (1)
Pedaliaceae	1/1	*Sesamum* (1)
Phytolaccaceae	1/1	*Phytolacca* (1)
Pinaceae	2/4	*Cedrus* (1), *Pinus* (3)
Platanaceae	1/1	*Platanus* (1)
Poaceae	4/4	*Coix* (1), *Cynodon* (1), *Saccharum* (1), *Zea* (1)
Polygonaceae	1/1	*Polygonum* (1)
Proteaceae	1/1	*Grevillea* (1)
Punicaceae	1/1	*Punica* (1)
Ranunculaceae	2/2	*Clematis* (1), *Delphinium* (1)
Rhamnaceae	1/1	*Zizyphus* (1)

Contd...

Appendix-I–*Contd...*

Name of the Family	*Genus/ Species*	*Total Number of Species under a Genus*
Rosaceae	5/18	*Eriobotrya* (1), *Prinsepia* (1), *Prunus* (5), *Pyrus* (3), *Rosa* (8)
Rubiaceae	1/1	*Hamelia* (1)
Rutaceae	4/9	*Aegle* (1), *Citrus* (5), *Murraya* (2), *Zanthoxylum* (1)
Salicaceae	2/7	*Populus* (2), *Salix* (5)
Sambucaceae	1/1	*Sambucus* (1)
Santalaceae	1/1	*Santalum* (1)
Sapindaceae	3/3	*Dodonaea* (1), *Litchi* (1), *Sapindus* (1)
Sapotaceae	2/2	*Achras* (1), *Mimusops* (1)
Saxifragaceae	2/2	*Bergenia* (1), *Ribes* (1)
Scrophulariaceae	2/2	*Antirrhinum* (1), *Russelia* (1)
Solanaceae	9/12	*Capsicum* (1), *Cestrum* (2), *Datura* (3), *Hyoscyamus* (1), *Lycopersicum* (1), *Nicandra* (1), *Petunia* (1), *Solanum* (1), *Withania* (1)
Sterculiaceae	1/1	*Pterospermum* (1)
Tiliaceae	1/1	*Grewia* (1)
Tropaeolaceae	1/1	*Tropaeolum* (1)
Ulmaceae	3/3	*Celtis* (1), *Holoptelea* (1), *Ulmus* (1)
Urticaceae	1/1	*Debregeasia* (1)
Verbenaceae	3/6	*Clerodendrum* (4), *Duranta* (1), *Tectona* (1)
Violaceae	1/2	*Viola* (2)
Zingiberaceae	2/2	*Alpinia* (1), *Hedychium* (1)

Appendix-II: Index to Local Names

Contd...

Appendix-II–*Contd...*

Local Name/s	*Botanical Name*	*Page*
Balodar	*Sapium insigne*	440
Balsam	*Impatiens balsamina*	387
Ban	*Quercus incana*	429
Band Gobi	*Brassica oleracea* var. *capitata*	338
Banda	*Dendrophthoe falcata*	368
Ban Mendi, Mehndu, Mendru	*Dodonaea viscosa*	369
Barain	*Diospyros lotus*	369
Batraj	*Clerodendrum inerme*	356
Battak Phul	*Aristolochia elegans*	329
Bekhal, Bhikhad, Bhikwa	*Prinsepia utilis*	422
Beli	*Salix fragilis*	437
Benusa	*Salix alba*	436
Besu, Biyunsh	*Salix tetrasperma*	437
Beul, Tamman	*Grewia oppositifolia*	381
Bhang	*Cannabis sativa*	341
Bhanjara	*Wedelia calendulacea*	452
Bheda	*Terminalia bellirica*	448
Bhojpattra	*Betula utilis*	335
Bhillar	*Bischofia javanica*	336
Boganvillea	*Bougainvillea glabra*	337
Boganvillea, Sada Suhagan	*Bougainvillea spectabilis*	338
Bottle Brush	*Callistemon lanceolatus*	340
Bowdu, Genda, Sartajad	*Tagetes erecta*	445
Brahmi, Minaki	*Centella asiatica*	350
Brahm Phool	*Hibiscus syriacus*	384
Cassia	*Cassia glauca*	346
Chabadu	*Clematis grata*	356
Chakotra	*Citrus maxima*	355
Chameli	*Jasminum grandiflorum*	389
Chameli	*Jasminum primulinum*	390
Chameli	*Jasminum sambac*	391

Contd...

Appendix-II–*Contd...*

Local Name/s	*Botanical Name*	*Page*
Chameli, Samsaun, Sunajui	*Jasminum humile*	390
Champa	*Michelia champaca*	404
Champa	*Plumeria acutifolia*	419
Champa, Champakali	*Magnolia grandiflora*	400
Chandan	*Santalum album*	439
Chandni	*Tabernaemontana coronaria*	445
Changma	*Salix oxycarpa*	437
Cheun, Kroon	*Morus serrata*	407
Chhoon	*Euphorbia royleana*	374
Chhui-mui	*Mimosa pudica*	405
Chiku	*Achras zapota*	323
Chil, Chir	*Pinus roxburghii*	417
Chilgoza	*Pinus gerardiana*	416
Chimbal, Timbal, Traimbal	*Ficus roxburghii*	378
Chinar	*Platanus orientalis*	419
Chini Gulab	*Hibiscus rosa-sinensis*	383
Choda	*Polygonum alpinum*	421
Cockscomb	*Celosia argentea* var. *cristata*	348
Coleus	*Coleus blumei*	358
Coral	*Russelia equisetiformis*	435
Cornflower	*Centaurea cyanus*	349
Cosmos	*Cosmos bipinnatus*	360
Cosmos	*Cosmos sulphureus*	361
Cycas, Nargig	*Cycas revoluta*	362
Daisy	*Chrysanthemum leucanthemum*	352
Dalia	*Dahlia rosea*	364
Datura, Kala-Datura	*Datura fastuosa*	364
Deaar, Deodar, Devdar	*Cedrus deodara*	348
Devidiar, Shur	*Juniperus macropoda*	393
Dhatura	*Datura innoxia*	365
Dhatura	*Datura stramonium*	365

Contd...

Appendix-II–*Contd...*

Local Name/s	*Botanical Name*	*Page*
Dheu	*Artocarpus lakoocha*	330
Dhuda, Phakura, Khasra, Phedu, Phegura	*Ficus palmata*	377
Dianthus	*Dianthus barbatus*	368
Dog Flower	*Viola tricolor*	452
Doob, Doorv, Drub	*Cynodon dactylon*	362
Drek, Kachen	*Melia azedarach*	403
Dudhal	*Jatropha panduraefolia*	391
Dumbar	*Deeringia amaranthoides*	367
Fan Palm, Khajur Palm	*Chamaerops humilis*	351
Gajar	*Daucus carota*	366
Galgal, Khatta	*Citrus limon*	354
Gamiri, Jhamirdi	*Citrus jambhiri*	354
Gandharaj, Moti-Mungra, Shiv-Darshan	*Clerodendrum philippinum*	357
Gandla, Kari patta	*Murraya koenigii*	407
Ganeera, Kaneer, Pila Kaner	*Thevetia neriifolia*	449
Ganna	*Saccharum officinarum*	436
Garna	*Carissa opaca*	344
Garuna	*Carissa spinarum*	344
Geranium	*Pelargonium hortorum*	414
Ghamera, Parvini	*Asclepias curassavica*	331
Ghari-Phool	*Passiflora caerulea*	414
Gol-gappa Phool	*Asclepias physocarpa*	331
Gonani	*Ipomoea purpurea*	387
Gudhal	*Malvaviscus conzatii*	401
Gulab	*Rosa alba*	431
Gulab	*Rosa indica*	433
Gulab	*Rosa jacquemontii*	433
Gulabri	*Rosa banksiae*	432
Gulabari	*Rosa multiflora*	434

Contd...

Appendix-II–*Contd...*

Local Name/s	*Botanical Name*	*Page*
Gulab Jamun	*Syzygium jambos*	444
Gulal	*Mirabilis jalapa*	406
Gular, Rumbal, Udumber, Umbarnae, Umro	*Ficus glomerata*	376
Guldaudi	*Chrysanthemum indicum*	352
Gulkhair	*Malva sylvestris*	401
Had kinnu	*Diospyros Montana* var. *cordifolia*	369
Hamelia	*Hamelia patens*	381
Harad	*Terminalia chebula*	448
Har-Shingar	*Lagerstroemia indica*	394
Har-Shingar	*Nyctanthes arbor-tristis*	410
Hollyhock	*Althaea rosea*	327
Hops	*Humulus lupulus*	385
Hydrangea	*Hydrangea macrophyla*	385
Iguta	*Ribes orientale*	430
Imli	*Tamarindus indica*	446
Iris	*Iris germanica*	388
Itsit, Ootshut	*Boerhaavia diffusa*	336
Jackranda	*Jacaranda acutifolia*	389
Jamun	*Syzigium cumini*	444
Jangli Pudina, Madeni	*Mentha longifolia*	403
Jangli Toot, Karyun, Vilayati Toot	*Broussonetia papyrifera*	338
Jangli Tulsi	*Ocimum gratissimum*	411
Jasmine	*Cestrum diurnum*	350
Jhadenu	*Olea glandulifera*	413
Juniperus	*Juniperus chinensis*	392
Kaamal, Kaambal	*Mallotus philippinensis*	400
Kachalu, Patedu	*Colocasia esculenta*	359
Kachnar, Karale, Karayalae	*Bauhinia variegata*	333
Kadve, Karkoi	*Roylea cinerea*	434
Kail	*Pinus wallichiana*	417

Contd...

Appendix-II–*Contd...*

Local Name/s	*Botanical Name*	*Page*
Kaimbal	*Lannea coromandelica*	395
Kainth, Shegal	*Pyrus pashia*	428
Kakad, Kakadan, Kakadeyin, Kakderan	*Pistacia integerrima*	418
Kala Rajaan	*Litsea sebifera*	398
Kali Basuti	*Justicia gendarussa*	393
Kali, Sudershan	*Canna indica*	342
Kalmi Garna	*Carissa carandas*	344
Kamini	*Murraya paniculata*	408
Kamuki	*Pyrostegia venusta*	427
Kanchak	*Alpinia nutans*	326
Kanchan	*Sempervivum arboreum*	442
Kanchan Champa, Letter Chamba	*Pterospermum acerifolium*	425
Kaner, Gandheera, Lal Kaner	*Nerium odorum*	409
Kanji	*Pongamia pinnata*	421
Kantiali	*Euphorbia milii*	374
Kapila	*Ruellia tuberosa*	435
Kapur	*Cinnamomum camphora*	353
Karela	*Momordica charantia*	406
Karkura	*Lespedeza sericea*	396
Kashmal, Sutram	*Berberis lycium*	334
Kattal	*Artocarpus heterophyllus*	330
Kela	*Musa paradisiaca*	408
Khadak, Khidak	*Celtis australis*	349
Khadamb	*Machilus gamblei*	399
Khair	*Acacia catechu*	321
Khajur	*Phoenix sylvestris*	416
Khlawa	*Wrightia tomentosa*	454
Khumani	*Prunus armeniaca*	423
Kochia	*Kochia trichophylla*	394
Kosh	*Alnus nitida*	325

Contd...

Appendix-II–*Contd...*

Local Name/s	Botanical Name	Page
Kukri, Makki	*Zea mays*	455
Kutta Phool	*Antirrhinum majus*	328
Lanjai	*Clerodendrum splendens*	357
Lasura	*Cordia dichotoma*	360
Laybala	*Rosa foetida*	432
Lily	*Amaryllis reginae*	328
Lily	*Hemerocallis fulva*	382
Lily	*Ixia chinensis*	389
Lily	*Zephyranthes grandiflora*	456
Litchi	*Litchi chinensis*	397
Local Gulab	*Hibiscus mutabilis*	383
Lohia	*Nicandra physaloides*	410
Lokat, Lugat, Lukat	*Eriobotrya japonica*	373
Lushai	*Lonicera japonica*	398
Majnu	*Salix babylonica*	436
Makhan, Pahadi-Taali	*Sapium sebiferum*	441
Makkan bean	*Phaseolus vulgaris*	415
Malgen, Tor	*Bauhinia vahlii*	333
Malti	*Jasminum grandiflorum*	389
Manjari	*Erigeron alpinus*	372
Mattar	*Pisum sativum*	418
Mendi	*Lawsonia alba*	395
Mirch	*Capsicum annuum*	343
Moldu, Mullang, Moun	*Ulmus villosa*	451
Molsari	*Mimusops elengi*	405
Morpankh	*Thuja orientalis*	449
Morpankh	*Thuja plicata*	450
Moru	*Quercus dilatata*	429
Nagar Bel	*Ipomoea quamoclit*	388
Nagan buti	*Pedilanthus tithymaloides*	414
Nag Buti	*Cyperus flabelliformis*	363

Contd...

Appendix-II–*Contd...*

Local Name/s	*Botanical Name*	*Page*
Nani	*Viola canescens*	452
Nargis	*Narcissus tazetta*	409
Nashpati	*Pyrus communis*	427
Neel Kanthi	*Duranta repens*	370
Neem, Nim	*Melia azadirachta*	402
Nimbu	*Citrus aurantifolia*	353
Oyee	*Albizia stipulata*	325
Padang, Panjole	*Acer oblongum*	323
Pahadi Babul	*Acacia dealbata*	321
Pajja	*Prunus puddum*	425
Palm	*Caryota urens*	345
Papita	*Carica papaya*	343
Papletha	*Ficus virens*	379
Papri, Pardesi	*Holoptelea integrifolia*	384
Pataal Ped	*Ligustrum lucidum*	397
Patajan, Patranjia	*Putranjiva roxburghii*	426
Pattharchat	*Bryophyllum calycinum*	339
Peela Phool	*Tecoma stans*	446
Peepal, Pipal	*Ficus religiosa*	377
Petunia	*Petunia hybrida*	415
Phulnu	*Gomphrena globosa*	380
Phultulsi	*Salvia coccinia*	438
Pinzi	*Malva verticillata*	401
Pitrika	*Hypericum cernuum*	386
Plum	*Prunus domestica*	424
Ponsetia	*Euphorbia pulcherrima*	374
Poplar, Popular	*Populus nigra*	422
Poplar, Safeda	*Populus ciliata*	422
Poppy	*Papaver rhoeas* var. *latifolia*	413
Pudina	*Mentha piperita*	404
Punarva	*Thunbergia erecta*	450

Contd...

Appendix-II–*Contd...*

Local Name/s	*Botanical Name*	*Page*
Punkesar	*Zephyranthes candida*	455
Punna	*Ehretia acuminata*	371
Rajbel	*Convolvulus arvensis*	359
Rajnigandha	*Polianthes tuberosa*	420
Ram Tulsi, Tulsi	*Ocimum sanctum*	412
Rangoon Bel	*Quisqualis indica*	430
Rat-ki-Rani	*Cestrum nocturnum*	351
Ritha	*Sapindus mukorossi*	440
Robinia, Shulu	*Robinia pseudoacacia*	431
Rocket Phool	*Delphinium ajacis*	367
Rubber plant	*Ficus elastica*	376
Rudraksh	*Elaeocarpus ganitrus*	371
Sadabahar	*Catharanthus roseus*	347
Safeda	*Eucalyptus tereticornis*	373
Safed Lily	*Hymenocallis littoralis*	385
Sagwan	*Tectona grandis*	447
Sahastrapaud, Sanspaud, Satavari	*Asparagus adscendens*	331
Sal	*Shorea robusta*	443
Salja	*Sambucus nigra*	439
Salsa	*Oenothera drummondii*	412
Salvia	*Salvia splendens*	438
Samsaj, Shamshad	*Buxus sempervirens*	339
Sangran	*Litsea polyantha*	397
Sanjal	*Fraxinus xanthoxyloides*	379
Santra	*Citrus reticulata*	355
Saplotri	*Bergenia ciliata*	335
Saru	*Cupressus torulosa*	362
Saru, Shupa	*Cupressus sempervirens*	361
Satavar	*Asparagus racemosus*	332
Satrangi	*Acalypha wilkesiana*	322

Contd...

Appendix-II–*Contd...*

Local Name/s	*Botanical Name*	*Page*
Seb, Seu	*Pyrus malus*	428
Sem	*Dolichos lablab*	370
Shaitan ka Jhad	*Alstonia scholaris*	327
Sharkhamba	*Artemisia moorcroftiana*	329
Shaybala, Socialic	*Rosa webbiana*	434
Shehnai Phool	*Tecoma radicans*	446
Shisham, Tali	*Dalbergia sissoo*	364
Shrimp	*Beloperone guttata*	334
Siaru	*Debregeasia hypoleuca*	366
Silver Fir, Silver Oak	*Grevillea robusta*	380
Simbal	*Bombax ceiba*	337
Sirin, Siris	*Albizia lebbeck*	324
Snake Plant	*Sansevieria trifasciata*	439
Spider Plant	*Chlorophytum capense*	352
Suche Moti	*Coix lachryma-jobi*	358
Surajmukhi	*Helianthus annuus*	382
Sut Saru	*Casuarina equisetifolia*	346
Tamatar	*Lycopersicum esculentum*	399
Tecomeria	*Tecomaria capensis*	447
Til	*Sesamum orientale*	442
Tindu Phool	*Tropaeolum majus*	451
Tirmir	*Zanthoxylum armatum*	454
Toot	*Morus alba*	407
Tulsi	*Ocimum basilicum*	411
Tulsi	*Ocimum kilimandscharicum*	412
Tuni	*Cedrela toona*	347
Umbrella plant	*Schefflera arboricola*	441
Vantosh	*Hedychium coronarium*	381
Vilayati Phool	*Clerodendrum thomsonae*	357
Wisteria	*Wisteria sinensis*	453

Contd...

Appendix-II–*Contd...*

Local Name/s	*Botanical Name*	*Page*
Yal	*Rosa moschata*	433
Zergul	*Calendula officinalis*	340
Zinnia	*Zinnia elegans*	456
Zorvo	*Phytolacca acinosa*	416

Appendix–III: Glossary of Words Commonly Used by the Local Populace

Akhand Jyoti	Holy fire
Amchur	Powdered dried unripe fruits of mango
Am Papad	Candy of mango fruits
Ashtmi	8th holy day of 'Navratri puja'
Ashwin	September-October
Baisakh	April-May
Bal bhog	Offering to the deity in the morning
Bhado	August-September
Bhog	Offering to the deity
Bragan	Wooden stick
Chaitra	March-April
Chana	Gram
Chapati	Bread of wheat/maize
Charnamat	Holy water possessing healing property
Chhaya patra	A concave-shaped iron vessel
Dal	Curry (An Indian dish of pulse)
Dhoop	Incense
Dhwaja	Flag carrying post
Dori	Sacred thread
Dugad Bhog	Offering of milk to the deity
Ghee	Clarified butter
Gur	Jaggery
Halwa	Sweet pudding
Havan	Sacrificial fire for performing religious rites
Havan-vedica	Platform for sacred fire
Khadawein	Wooden sleepers
Kheer	Pudding
Lachidana	Sugar balls
Langar	Community meal
Lassi	Butter milk
Magh	January-February

Makhana	Fox nut
Manni	Large wheat bread
Mishri	Sugar candy
Mohan Bhog	Offering to the deity in the morning (4 A.M.)
Murabba	Candy
Panchmeva	Five dry fruits
Patasha	Large-sized sugar balls
Pattal/Duna	Meal plate made up of leaves
Patoru	Snack
Phodi	Wooden stick
Prantha	Fried wheat chapati
Prasadam	Offerings to people
Poltu/Puri/Babru	Fried, sweet wheat chapati
Raj bhog	Offering to the deity at noon
Rot	Thick fried wheat chapati dipped in sugary syrup
Roti	Bread of wheat/maize
Sabji/Sag	Vegetable
Sakranti	Beginning of month as per Hindu calendar
Shravan	July-August
Shyam bhog	Offering to the deity in the evening
Sthapna Divas	Foundation day
Vibhuti	Sacred ash

Colour Plates

late 1: A. Baba Balak Nath temple with sacred tree of *Ficus benghalensis*. **B.** Baba Nahar Singh Dholra) temple. **C.** Garuna Jhadi temple with sacred plant of *Carissa spinarum*. **D.** Hari Devi emple. **E.** Lakshmi Narayan temple. **F.** Markanday temple. **G.** Naina Devi temple, Kharota with acred tree of *Mangifera indica* in the courtyard.

Plate 2: A. Naina Devi temple, Naina Devi. **B.** Sacred tree of *Ficus religiosa* in the courtyard Naina Devi temple. **C.** Naina Devi temple, Nihar. **D.** Nav Durga temple with sacred tree of *Fic* *religiosa*. **E.** Shiv temple.

Plate 3: A. Baba Lakhdata Pir temple. **B.** Bansi Gopal temple. **C.** Bhalai Mata temple. **D.** Bharmani Mata temple. **E.** Champawati temple. **F.** Chamunda Devi temple. **G.** Chaurasi temple. **H.** Chaurasi temple with sacred tree of *Cedrus deodara* in the courtyard. **I.** Khajji Nag temple.

Plate 4: A. Lakshmi Narayan temple, Chamba. **B.** Lakshmi Narayan temple, Dalhousi
C. Mahakali temple. **D.** Narsingh temple. **E.** Raghunath temple. **F.** Shitla Mata temple. **G.** Shiv
Shakti temple. **H.** Sita-Ram temple. **I.** Suhi Mata temple.

late 5: A. Awah Devi temple. **B.** Baba Balak Nath temple. **C.** Offerings for deity in Baba Balak ath temple. **D.** Baba Sidh Chaino temple.

Plate 6: A. Gauri Shanker temple. **B.** Murli Manohar temple. **C.** Narbadeshwar templ **D.** Floristic sculptures on the wall of Narbadeshwar temple. **E.** Santoshi Mata temple. **F.** Sha Dev temple. **G.** Shirdi Sai Baba temple. **H.** Trilokinath temple. **I.** Vaishno Mata temple.

late 7: A. Ashapuri temple. **B.** Baba Balak Rupi temple. **C.** Baba Kalinnath temple. **.** Bagulamukhi temple. **E.** Bajreshwari Devi temple. **F.** Bhagsunath temple. **G.** 'Riyol' (barley) rown in a pot for worship in Bhagsunath temple. **H.** Buddhist temple (Thek Chen Choiling). . Wooden articles used for burning lamps. **J.** Sacred tree of *Populus nigra* with a religious board. . Prayer wheels.

Plate 8: A. Buddha temple, Tilokpur. **B.** Chamunda Nandikeshwar temple. **C.** Jwalamukhi templ with sacred tree of *Mimusops elengi* in the courtyard. **D.** Kalinath Kaleshwar Mahadev templ **E.** Lakshmi Narayan temple, Alampur. **F.** Lakshmi Narayan temple, Chanor with sacred tree c *Thuja orientalis*. **G.** Machhial Mahadev temple. **H.** Mahakal temple. **I.** Mangla Mata templ **J.** Mukut Nath temple. **K.** Nag Devta temple, Chelian with sacred tree of *Zizyphus mauritiana*.

'late 9: A. Nagni Mata temple, Nagni. **B.** Nagni Mata temple, Bhati. **C.** Nag temple, Kariyada. **).** Radha Krishan temple, Baroh. **E.** Radha Krishan temple, Dehra. **F.** Rock-cut temple. **;.** Shibbo Than temple. **H.** Shiv temple, Baijnath. **I.** Shiv temple (Chhota Haridwar), Amrella. **.** Shri Brijraj Swami temple. **K.** Shri Raghunath Teda temple (Trihayu temple). **L.** Sidh Chaino :emple. **M.** Trilokinath temple. **N.** Vindhyavasini Mata temple.

Plate 10: A. Badri Narayan temple. **B.** Baring Nag temple. **C.** Boirus temple. **D.** Chandika De temple. **E.** Kamakhaya Devi temple. **F.** Mahabodi temple. **G.** Maheshwar Devta templ **H.** Shamsheer Devta temple. **I.** Usha Devi temple.

Plate 11: A. Ambika Mata temple. **B.** Bhekhali Mata temple. **C.** Bijli Mahadev temple. **D.** Chandi temple. **E.** Dakshini Mahadev temple. **F.** Dhankeshwar Mahadev Temple. **G.** Gurudwara Sahib Manikaran. **H.** Hadimba Devi temple. **I.** Sacred tree of *Cedrus deodara* near Hadimba Devi temple. **J.** Himalayan Buddhist monastery.

Plate 12: A. Place meant for burning incense at Himalayan Buddhist monastery. **B.** Maharisl Vashisht temple. **C.** Manu temple. **D.** Murlidhar temple. **E.** Naina Devi temple. **F.** Parshura temple. **G.** Raghunath temple, Kullu. **H.** Raghunath temple, Manikaran. **I.** Rama templ **J.** Palanquin of the deity at Rama temple. **K.** Shitla Mata temple. **L.** Shiva temple, Manikara **M.** Shiva temple, Nirmand.

late 13: A. Siyali Mahadev temple. **B.** Vaishno Devi temple. **C.** Vishveshwar Mahadev temple.

Plate 14: A. Kee Gompa. **B.** Labrang monastery. **C.** Mrikula Mata temple. **D.** Othang Gomp **E.** Ornamental plants cultivated in the courtyard of Othang Gompa. **F.** Raja Ghepan templ **G.** Shashur monastery. **H.** Trilokinath temple. **I.** Women collecting flowers for offerin **J.** A totem ('Jholnu' or 'Par') made from flowers of *Tagetes erecta*. **K.** Yordang Gompa.

Plate 15: A. Bhootnath temple. **B.** Ekadash Rudra temple. **C.** Gurudwara Shri Guru Gobind Singh. **D.** Hanogi Mata temple. **E.** Kamaksha Devi temple. **F.** Kamru Nag temple. **G.** Mahamaya Durga temple. **H.** Mahunag temple. **I.** Mamleshwar Mahadev temple. **J.** Naina Devi temple. **K.** Nawahi Devi temple. **L.** Ningmapa Buddh monastery. **M.** Panchvaktra Mahadev temple.

Plate 16: A. Shikari Devi temple. **B.** Shri Lomash Rishi temple. **C.** Sidh Baba Kamlahiya temple. **D.** Tarna Devi temple. **E.** Trilokinath temple.

late 17: A. Bhimakali temple. **B.** Bhuteshwar Mahadev temple. **C.** Dingu Durga temple. **).** Durga temple. **E.** Duttatrey temple. **F.** Hat Keshwari Durga Mata temple. **G.** Jakhoo Hanuman mple.

Plate 18: A. Kalibari temple. **B.** Kali Mata temple. **C.** Kamna Devi temple. **D.** Lankadvir templ **E.** Sacred tree of *Cedrus deodara* in the courtyard of Lankadvir temple. **F.** Mahish Mardi Mahalakshmi temple. **G.** Narsingh temple. **H.** Shri Ayodhya Nath temple. **I.** Shri Sankatmocha Hanuman temple. **J.** Shri Sat Narayan temple. **K.** Surya Narayan temple. **L.** Tara Devi templ **M.** Sacred tree of *Pyrus communis* in the courtyard of Tara Devi temple.

late 19: A. Balasundari temple. **B.** Sacred plant of *Bougainvillea spectabilis* in the courtyard of alasundari temple. **C.** Brahmchari Ashram. **D.** Gurudwara Shri Paonta Sahib. **E.** Jagan Nath mple. **F.** Kalisthan temple. **G.** Katasan Devi temple. **H.** Narsingh temple. **I.** Nirwan Ashram. Pujari Ghat Rama temple. **K.** Radha Krishan temple. **L.** Renuka Mata temple. **M.** Rishi Iarkanda Dham.

Plate 20: A. Sahibji Maharaj Gurudwara. **B.** Sanyas Ashram. **C.** Shiva temple. **D.** Vishwakarm temple. **E.** Yamuna temple.

Plate 21: A. Baduwada temple. **B.** Bhadrakali temple. **C.** Dev Dhar temple. **D.** Lutru Mahadev temple. **E.** Mutru Mahadev temple. **F.** Sai Baba temple. **G.** Shakani Ganga temple.

Plate 22: A. Shiva temple, Dadlaghat. **B.** Shiva temple, Jakholi. **C.** Shiva temple, Jatol **D.** Shri Sidh Baba Balak Nath temple. **E.** Shulini Mata temple. **F.** Sidheshwari Mata temple.

late 23: A. Baba Achutanand Thakurdwara. **B.** Baba Barbhag Singh Gurudwara. **C.** Bagula Iukhi temple. **D.** Bhadrakali temple. **E.** Dera Baba Rudranand. **F.** Dron Shiva temple. **.** Gugga Jahar Pir temple. **H.** Lakshmi Narayan temple. **I.** Lohara temple. **J.** Mata Chintpurni mple with sacred tree of *Ficus benghalensis* in its courtyard.

Plate 24: A. Rama temple. **B.** Sadashiv temple. **C.** Santoshi Mata temple. **D.** 'Kalash', *i.e.* cocon and leaves of mango tree placed over a pitcher filled with water and kept over barley seedlin grown specially for 'havan' ceremony. **E.** Saptdevi temple. **F.** Shitla Mata temple. **G.** Shri Hanuma temple.

late 25: A. *Acacia catechu* (L.f.) Willd. **B.** *A. dealbata* Link. **C.** *A. nilotica* (L.) Willd. *ex* Delile ubsp. *indica* (Benth.) Brenan. **D.** A flowering branch of *A. nilotica* (L.) Willd. *ex* Delile subsp. *dica* (Benth.) Brenan.

Plate 26: A. *Acalypha wilkesiana* M. Arg. **B.** *Acer oblongum* Wall. **C.** *Achras zapota* Linn.

Plate 27: A. *Aegle marmelos* (L.) Correa. **B.** *Aesculus indica* Colebr. **C.** *Albizia lebbeck* (L.) Benth.; a flowering branch magnified (inset). **D.** *A. stipulata* (Roxb.) Boiv.

Plate 28: A. *Alnus nitida* (Spach) Endl. **B.** *Aloe vera* (L.) Burm.f. **C.** *Alpinia nutans* Rosc. **D** *Alstonia scholaris* (L.) R. Br.

late 29: A. *Althaea rosea* (Linn.) Cav. **B.** *Amaryllis reginae* Linn. **C.** *Antirrhinum majus* Linn. **D.** *raucaria columnaris* (Forster) Hooker.

Plate 30: A. *Aristolochia elegans* Mast. **B.** *Artemisia moorcroftiana* Wall. *ex* DC. **C.** *Artocarp*
heterophyllus Lam. **D.** *A. lakoocha* Roxb.

late 31: A. *Asclepias curassavica* Linn. **B.** *A. Physocarpa* (E. Mey.) Schlecht. **C.** *Asparagus adscendens* oxb. **D.** *A. densiflorus* (Kunth) Jessop. **E.** *A. racemosus* Willd.

Plate 32: A. *Bauhinia vahlii* Wight and Arn. **B.** A flowering branch of *B. vahlii* Wight. and Ar
C. *B. variegata* Linn. **D.** Enlarged flower of *B. variegata* Linn. **E.** *Beloperone guttata* Brand.

Plate 33: A. *Berberis lycium* Royle. **B.** *Bergenia ciliata* (Haw.) Sternb. forma *ligulata* Yeo. **C.** *Betula utilis* D. Don. **D.** *Bischofia javanica* Blume.

Plate 34: A. *Boerhaavia diffusa* Linn. **B.** *Bombax ceiba* Linn. **C.** *Bougainvillea glabra* Choisy.

late 35: A. *Bougainvillea spectabilis* Willd. **B.** *Brassica oleracea* var. *capitata* Linn. **C.** *Broussonetia pyrifera* (Linn.) L'Herit. *ex* Vent.

Plate 36: A. *Bryophyllum calycinum* Salisb. **B.** *Buxus sempervirens* Linn. **C.** *Calendula officina*
Linn. **D.** *Callistemon lanceolatus* DC.

Plate 37: A. *Callistephus chinensis* (Linn.) Nees. **B.** *Calotropis procera* R. Br. **C.** *Cannabis sativa* Linn.

Plate 38: A. *Canna indica* Linn. **B.** *Capsicum annuum* Linn. **C.** *Carica papaya* Linn. **D.** *Cariss carandas* Linn.

late 39: A. *Carissa opaca* Stapf. *ex* Haines. **B.** *C. spinarum* Linn.; a flowering branchlet magnified
nset). **C.** *Caryota urens* Linn.

Plate 40: A. *Cassia fistula* Linn. **B.** *C. glauca* Lam. **C.** *Casuarina equisetifolia* Linn. **D.** A branch
C. equisetifolia Linn.

late 41: A. *Catharanthus roseus* (Linn.) G. Don **B.** *Cedrela toona* Roxb. *ex* Rottl. **C.** A flowering
anch of *C. toona* Roxb. *ex* Rottl. **D.** *Cedrus deodara* (Roxb.) Loudon **E.** *Celosia argentea* var. *cristata*
aines.

Plate 42: A. *Celtis australis* Linn. **B.** *Centaurea cyanus* Linn. **C.** *Centella asiatica* (Linn.) Urban.

late 43: A. *Cestrum diurnum* Linn. **B.** *C. nocturnum* Linn. **C.** *Chamaerops humilis* Linn. **).** *Chlorophytum capense* Kuntze.

Plate 44: A. *Chrysanthemum indicum* Linn. **B.** *C. leucanthemum* Linn. **C.** *Cinnamomum camphor* (Linn.) Nees and Eberm. **D.** *Citrus aurantifolia* (Christin) Sw.

Plate 45: A. *Citrus jambhiri* Lush. **B.** *C. limon* (Linn.) Burm. f. **C.** *C. maxima* (Burm. f.) Merrill **D.** *C. reticulata* Blanco.

Plate 46: A. *Clematis grata* Wall. **B.** *Clerodendrum inerme* (Linn.) Gaertn. **C.** *C. philippinum* Schaue

late 47: A. *Clerodendrum splendens* G. Don. **B.** *C. thomsonae* Balf. **C.** *Coix lachryma-jobi* Linn. **.** *Coleus blumei* Benth.

Plate 48: A. *Colocasia esculenta* (L.) Schott. **B.** *Convolvulus arvensis* Linn. **C.** *Cordia dichotoma* For

Plate 49: A. *Cosmos bipinnatus* Cav. **B.** *C. sulphureus* Cav. **C.** *Cupressus sempervirens* Linn. **D.** *C. torulosa* D. Don.

Plate 50: A. *Cycas revoluta* Thunb. **B.** *Cynodon dactylon* (Linn.) Pers. **C.** *Cyperus flabelliformis* Rott

Plate 51: A. *Dahlia rosea* Cav. **B.** *Dalbergia sissoo* Roxb. **C.** *Datura fastuosa* Linn.

Plate 52: A. *Datura innoxia* Mill. **B.** *D. stramonium* Linn. **C.** *Daucus carota* Linn. **D.** *Debregeas hypoleuca* Wedd.

late 53: A. *Deeringia amaranthoides* (Lamk.) Merr. **B.** *Delphinium ajacis* Linn. **C.** *Dendrophthoe lcata* (Linn. f.) Ettingsh.

Plate 54: A. *Dianthus barbatus* Linn. **B.** *Diospyros lotus* Linn. **C.** *D. montana* Roxb. var. *cordifo*
(Roxb.) Hiern. **D.** *Dodonaea viscosa* (Linn.) Jacq.

Plate 55: A. *Dolichos lablab* Linn. **B.** *Duranta repens* Linn. **C.** *Ehretia acuminata* R. Br.

Plate 56: A. *Elaeocarpus ganitrus* Roxb. **B.** *Emblica officinalis* Gaertn. **C.** *Erigeron alpinus* Linn **D.** *Eriobotrya japonica* (Thunb.) Lindl.

Plate 57: A. *Eucalyptus tereticornis* Sm. **B.** *Euphorbia milii* Des Moul. **C.** *E. pulcherrima* Willd. *ex* Klotzsch. **D.** *E. royleana* Boiss.

Plate 58: A. *Ficus benghalensis* Linn. **B.** *F. Carica* Linn. **C.** *F. elastica* Roxb. **D.** *F. glomerata* Roxb

late 59: A. *Ficus palmata* Forsk. **B.** *F. religiosa* Linn. **C.** *F. roxburghii* Wall. *ex* Miq. **D.** *F. rumphii* lume.

Plate 60: A. *Ficus virens* Ait. **B.** *Fraxinus xanthoxyloides* (Wall. *ex* G. Don) DC. **C.** *Gaillard pulchella* Foug.

Plate 61: A. *Gomphrena globosa* Linn. **B.** *Grevillea robusta* A. Cunn. *ex* R. Br. **C.** *Grewia oppositifolia* Roxb.

Plate 62: A. *Hamelia patens* Jacq. B. *Hedychium coronarium* Koenig. C. *Helianthus annuus* Linn.

Plate 63: A. *Hemerocallis fulva* (Linn.) Linn. **B.** *Hibiscus mutabilis* Linn. **C.** *H. rosa-sinensis* Linn.; flower (inset). **D.** *H. syriacus* Linn.; a flowering branch magnified (inset).

Plate 64: A. *Holoptelea integrifolia* (Roxb.) Planch. **B.** *Humulus lupulus* Linn. **C.** *Hydrange macrophylla* (Thunb.) DC. **D.** *Hymenocallis littoralis* (Jacq.) Salisb. **E.** *Hyoscyamus niger* Linn.

late 65: A. *Hypericum cernuum* Roxb. **B.** *Impatiens balsamina* Linn. **C.** *Ipomoea cairica* (Linn.) veet. **D.** *I. purpurea* (Linn.) Roth.

Plate 66: A. *Ipomoea quamoclit* Linn. **B.** *Iris germanica* Linn. **C.** *Ixia chinensis* Linn.

Plate 67: A. *Jacaranda acutifolia* Humb. and Bon. **B.** *Jasminum grandiflorum* Linn. **C.** *Jasminum humile* Linn. **D.** *J. Primulinum* Hemsl. *ex* Baker.

Plate 68: A. *J. sambac* (Linn.) Ait. **B.** *Jatropha panduraefolia* Andr. **C.** *Juglans regia* Linn.

Plate 69: A. *Juniperus chinensis* Linn. **B.** *Juniperus macropoda* Boiss. **C.** *Justicia gendarussa* Linn. f.

Plate 70: A. *Kochia trichophylla* Voss. **B.** *Lagerstroemia indica* Linn. **C.** *Lannea coromandelica* (Houtt. Merrill. **D.** *Lawsonia alba* Lam. **E.** *Lespedeza sericea* (Thunb.) Miq. **F.** *Leucaena leucocephala* (Lam de Wit.

ate 71: A. *Ligustrum lucidum* Aiton. **B.** *Litchi chinensis* Sonn. **C.** *Litsea polyantha* Juss.
. *L. sebifera* Pers.

Plate 72: A. *Lonicera japonica* Thunb. **B.** *Lycopersicum esculentum* Mill. **C.** *Machilus gamblei* King Hook. f.

Plate 73: A. *Magnolia grandiflora* Linn. **B.** *Mallotus philippinensis* (Lam.) Muell.-Arg. **C.** *Malva sylvestris* Linn. **D.** *M. verticillata* Linn.

Plate 74: A. *Malvaviscus conzatii* Green. **B.** *Mangifera indica* Linn. **C.** *Melia azadirachta* Linn.
D. *M. azedarach* Linn. **E.** *Mentha longifolia* (Linn.) Huds.

Plate 75: A. *Mentha piperita* Linn. **B.** *Michelia champaca* Linn. **C.** *Mimosa pudica* Linn. **D.** *Mimusops elengi* Linn. **E.** A portion of flowering branch of *M. elengi* Linn.

Plate 76: A. *Mirabilis jalapa* Linn. **B.** *Momordica charantia* Linn. **C.** *Morus alba* Roxb. **D.** *M. serrat* Roxb. **E.** *Murraya koenigii* (Linn.) Spreng.; a flowering branch (inset).

ate 77: **A.** *Murraya paniculata* (Linn.) Jack. **B.** *Musa paradisiaca* Linn. **C.** *Narcissus tazetta* Linn.

Plate 78: A. *Nerium odorum* Soland. **B.** *Nicandra physaloides* (Linn.) Gaertn. **C.** *Nyctanthes arbo tristis* Linn. **D.** *Ocimum basilicum* Linn. **E.** *O. gratissimum* Linn.

Plate 79: A. *Ocimum kilimandscharicum* Gurke. **B.** *O. sanctum* Linn. **C.** *Oenothera drummondi* Hook f. **D.** *Olea glandulifera* Wall. **E.** *Papaver rhoeas* var. *latifolia* Prain.

Plate 80: A. *Passiflora caerulea* Linn. **B.** *Pedilanthus tithymaloides* (L.) Poit. **C.** *Pelargonium hortorum* Bailey. **D.** *Petunia hybrida* Vilm.

Plate 81: A. *Phaseolus vulgaris* Linn. **B.** *Phoenix sylvestris* (L.) Roxb. **C.** *Phytolacca acinosa* Roxb. **D.** *Pinus gerardiana* Wall. *ex* D. Don.

Plate 82: A. *Pinus roxburghii* Sarg. **B.** *P. wallichiana* A. B. Jackson. **C.** *Pistacia integerrima* J. L Stewart *ex* Brand. **D.** *Pisum sativum* Linn. **E.** *Platanus orientalis* Linn. **F.** *Plumeria acutifolia* Poiret

ate 83: **A.** *Polianthes tuberosa* Linn. **B.** *Polyalthia longifolia* (Sonn.) Thwaites. **C.** *Polygonum alpinum* ll. **D.** *Pongamia pinnata* (Linn.) Pierre.; a magnified view of fruits (inset). **E.** *Populus ciliata* Wall. *P. nigra* Linn.

Plate 84: A. *Prinsepia utilis* Royle. **B.** *Prunus amygdalus* Batsch. **C.** *P. armeniaca* Lin
D. *P. domestica* Linn. **E.** *P. persica* (Linn.) Batsch.

Plate 85: A. *Prunus puddum* (Roxb. *ex* Ser.) Brandis. **B.** *Psidium guajava* Linn. **C.** *Pterospermum acerifolium* Willd. **D.** *Punica granatum* Linn.

Plate 86: A. *Putranjiva roxburghii* Wall. **B.** *Pyrostegia venusta* (Ker.-Gawl.) Miers. **C.** *Pyrus Communis* Linn. **D.** *P. malus* Linn.

Plate 87: A. *Pyrus pashia* Buch.-Ham. *ex* D. Don. **B.** *Quercus dilatata* Lindl. *ex* A. DC. **C.** *Q. incana* Roxb. **D.** *Quisqualis indica* Linn.

Plate 88: A. *Ribes orientale* Desf. **B.** *Ricinus communis* Linn. **C.** *Robinia pseudoacacia* Linn. **D.** *Ros*
alba Linn.

ate 89: A. *Rosa banksiae* R. Br. **B.** *R. foetida* Herrm. **C.** *R. indica* Linn. **D.** *R. jacquemontii* Crep.
Hook. f. **E.** *R. moschata* Herrm.

Plate 90: A. *Rosa multiflora* Thunb. **B.** *R. webbiana* Wall. *ex* Royle. **C.** *Roylea cinerea* (D. Do
Baill. **D.** *Ruellia tuberosa* Linn. **E.** *Russelia equisetiformis* Cham. and Schl. **F.** *Saccharum officinar*
Linn.

Plate 91: A. *Salix alba* Linn. **B.** *S. babylonica* Linn. **C.** *S. fragilis* Linn. **D.** *S. oxycarpa* Anders. **E.** *S. tetrasperma* Roxb.; a flowering branch (inset). **F.** *Salvia coccinia* Juss.

Plate 92: A. *Salvia splendens* Sellow *ex* Roemer and Schultes **B.** *Sambucus nigra* Linn. **C.** *Sansevieri* *trifasciata* Prain. **D.** *Santalum album* Linn.; a flowering branch (inset).

Plate 93: A. *Sapindus mukorossi* Gaertn . **B.** *Sapium insigne* (Royle) Benth. **C.** *S. sebiferum* (Linn.) Roxb. **D.** *Schefflera aeboricola* **E.** *Sempervivum arboreum* (Hayata) Merr.

Plate 94: A. *Sesamum orientale* Linn. **B.** *Shorea robusta* Gaertn. f. **C.** *Solanum tuberosum* Linn.
D. *Syzygium cumini* (Linn.) Skeels **E.** *S. jambos* (Linn.) Alston; a flowering branch (inset).

ate 95: A. *Tabernaemontana coronaria* (Jacq.) Willd. **B.** *Tagetes erecta* Linn. **C.** *Tamarindus indica* nn.; flower and fruit (inset).

Plate 96: A. *Tecoma radicans* Juss. **B.** *Tecoma stans* (Linn.) Juss. *ex* H.B.K. **C.** *Tecomaria capen* (Thunberg) Spach. **D.** *Tectona grandis* Linn. f. **E.** *Terminalia bellirica* (Gaertn.) Roxb.

Plate 97: A. *Terminalia chebula* (Gaertn.) Retz. **B.** *Thevetia neriifolia* Juss *ex* Steud. **C.** *Thuja orientalis* Linn. **D.** *T. plicata* D. Don.

Plate 98: A. *Thunbergia erecta* (Benth.) T. Anders. **B.** *Tithonia tagetiflora* Desf. **C.** *Tropaeolum majus* Linn. **D.** *Ulmus villosa* Brandis **E.** *Viola canescens* Wall.

Plate 99: A. *Viola tricolor* Linn. **B.** *Wedelia calendulacea* Less. **C.** *Wisteria sinensis* (Sims) Sweet.

Plate 100: A. *Withania somnifera* (Linn.) Dunal **B.** *Wrightia tomentosa* (Roxb.) Roem and Schult **C.** *Zanthoxylum armatum* DC. **D.** *Zea mays* Linn.

late 101: A. *Zephyranthes candida* (Lindl.) Herb. **B.** *Z. grandiflora* Lindl. **C.** *Zinnia elegans* Jacq. **.** *Zizyphus mauritiana* Lamk.; fruits (inset).

www.ingramcontent.com/pod-product-compliance
Ingram Content Group UK Ltd.
Pitfield, Milton Keynes, MK11 3LW, UK
UKHW022218051026
15624UKWH00001B/218

9 789351 241706